LIZARDS IN AN EVOLUTIONARY TREE

ORGANISMS AND ENVIRONMENTS

Harry W. Greene, Consulting Editor

LIZARDS IN AN EVOLUTIONARY TREE

Ecology and Adaptive Radiation of Anoles

Jonathan B. Losos

UNIVERSITY OF CALIFORNIA PRESS

Berkeley   Los Angeles   London

THE PUBLISHER AND AUTHOR GRATEFULLY ACKNOWLEDGE THE
GENEROUS CONTRIBUTION TO THIS BOOK PROVIDED BY

MUSEUM OF COMPARATIVE ZOOLOGY, HARVARD UNIVERSITY

University of California Press, one of the most distinguished university
presses in the United States, enriches lives around the world by advancing
scholarship in the humanities, social sciences, and natural sciences. Its activities are supported by the UC Press Foundation and by philanthropic contributions from individuals and institutions. For more information, visit
www.ucpress.edu.

Organisms and Environments, No. 10

University of California Press
Oakland, California

First paperback printing 2011

Library of Congress Cataloging-in-Publication Data

Losos, Jonathan B.
    Lizards in an evolutionary tree : ecology and adaptive
        radiation of anoles / Jonathan B. Losos.
            p. cm.
    Includes bibliographical references and index.
    ISBN 978-0-520-26984-2 (pbk : alk. paper)
        1. Anoles—Evolution. 2. Anoles—Ecology.
        3. Anoles—Adaptation.   I. Title.

QL666.L268L67 2009
597.95'48138—dc22   2008027779

16   15   14   13   12   11   10   09
10   9   8   7   6   5   4   3   2   1

Cover illustration: *Anolis monticola*. Photo copyright Eladeo Fernández.

To my wife, Melissa Losos, my parents, Carolyn and Joseph Losos, and my sisters, Carol, Elizabeth, and Louise Losos, for their love and support throughout my life

CONTENTS

FOREWORD

Lizards in an Evolutionary Tree: Ecology and Adaptive Radiation of Anoles is the tenth volume in the University of California Press's series on organisms and environments, whose unifying themes are the diversity of plants and animals, the ways they interact with each other and with their surroundings, and the implications of those relationships for science and society. We seek books that promote unusual, even unexpected connections among seemingly disparate topics, distinguished by the talents and perspectives of their authors. Previous volumes have spanned topics as diverse as grassland ecology and bison behavior, but none has encompassed the breadth and depth of scholarly coverage achieved here.

Jonathan Losos chronicles the details and historical underpinnings of an extraordinary natural legacy, the adaptive radiation of almost four hundred species of very special lizards. Thanks to their unusual diversity, abundance, and tractability, anoles have played central roles in several scientific disciplines, including physiological and community ecology, functional morphology, biogeography and molecular evolution. Losos has synthesized anole biology in lively prose, based on his own extensive studies and thousands of publications by an army of researchers over the past century. From a conceptual perspective, this book explores the cutting edges of evolutionary biology and ecology, our search for patterns and causal explanations for biodiversity. Why are there more species in some places than others, and what drives diversification? How do individuals interact with others of their species? Is competition among species important? And what will be the fate of anoles on our rapidly changing planet?

Lizards in an Evolutionary Tree also explores these fascinating, often beautiful reptiles for their own sake. Only one species is widespread in the southeastern United States, but some tropical sites perhaps boast fifteen species and Cuba has sixty-three. Anoles typically have large heads and limbs, long slender tails, small granular scales, and feet specialized for gripping. Their color patterns are generally cryptic and a few can change hues dramatically within seconds. The males of most species and the females of some have a distensible, often brightly colored dewlap used in social signaling. Any suspicion that "you've seen one, seen them all" is squashed by rare Amazonian anoles with leaf-like proboscises and by a kaleidoscopically orange-splotched Andean species with large flat scales among its granules. Certain Cuban anoles with their large eyes, prehensile tails, and slow-motion lifestyles are reminiscent of Old World chameleons.

Anoles are ecologically diverse as well. A few species are as big as a good-sized rodent and scamper among the trunks and canopy foliage of rainforest trees; some no larger than a ballpoint pen creep along twigs, while a few still smaller ones live on fallen branches and in leaf litter. Several species are semiaquatic and another hangs around cave entrances. Most anoles feed on arthropods, but at least one takes snails, and larger species sometimes add fruit and vertebrates to their diets. On the other side of the predator-prey coin, anoles are often common and they must be tasty, because spiders, frogs, other lizards, snakes, birds, and mammals eat them. Despite all this diversity, local assemblages are often predictably structured. Visit Caribbean islands, pay attention to anoles, and each place you will see so-called "crown giant," "trunk," "twig," and "grass" species, with independently evolved similarities to those same "ecomorphs" elsewhere.

A third theme lurks herein, beyond concepts and organisms, of academic lineages. Who does all this work and why would anyone devote decades to studying an adaptive radiation? Harvard's late professor Ernest Williams initiated anole work and by the end of the 1970s had supervised a string of unusually creative doctoral students working on these lizards. Meanwhile ten-year-old Jonathan, already well known to curators at his hometown zoo as an "animal nerd," had been mesmerized by anoles on a family trip to Florida. He later devoted two secondary school science projects to them, completed an undergraduate thesis on their social behavior with Williams, and his Berkeley Ph.D. was going to be on something else! My new student tried monitors, geckos, and chameleons, then succumbed to destiny. It's been anoles for more than thirty years now, and although he's briefly escaped to other lizards and even opossums, a steady stream of exciting new mysteries keeps reeling him back to those childhood favorites.

And what of the future? Anoles now stand among the most diverse and thoroughly studied of all adaptive radiations, rivaling Darwin's finches, African rift lake fishes, and other classics. Nonetheless, mysteries abound and Jonathan is candid about what we don't know. Why is one species on Isla Gorgona brilliant blue? Why are anoles so much more common in Central American rainforests than in Amazonia, and why are they even more abundant on Caribbean islands? Which species are at risk of extinction due to habitat fragmentation and climate change? Sequencing of the anole genome is at hand, ensuring that we can use the common currency of DNA to understand this radiation in ever more detail. *Lizards in an Evolutionary Tree* sets the stage for new discoveries with these wonderful animals.

Harry W. Greene

ACKNOWLEDGMENTS

My first recollection of an *Anolis* lizard is from a trip to Miami to visit a great-aunt when I was about ten. She lived near a park, where I happily chased green anoles, only to be scared out of my wits by the unexpected appearance of an enormous Cuban Knight anole (*A. equestris*). Since then, anoles have been a recurrent theme in my life. In both junior and senior high school, I conducted classroom projects on the Florida green anole. I thank my science teachers (among them Mr. Becker, Mr. Kardis, Mrs. Rosenthal and Mrs. Wolrab) and the Ladue School District more generally for awakening and nurturing my interest in science.

In college, I soon came to hang out in the laboratory of Ernest Williams and particularly with Ernest's last graduate student, Greg Mayer. Greg took me on my first field trip, gave me the idea for my honor's thesis on *Anolis*, and taught me a great deal about lizards and science. Ernest also was a great influence in developing my scientific thinking. He planned to write a book on *Anolis*, but never got around to doing it, which is a real shame, because never again will so much knowledge about *Anolis* reside in a single human head. No doubt, his book would have been different from mine; I would like to think he would have approved of what I have written. My development as a scientist was further guided by my doctoral and postdoctoral mentors, Harry Greene, Tom Schoener, and Marc Mangel, to whom I owe a great debt of gratitude.

My work over the past twenty-plus years has been highly collaborative. Much of what I know about anoles, as well as the ideas I have formed, are based on these collaborations. I am grateful to the many people with whom I've worked, including among many others, Kevin de Queiroz, Allan Larson, Tom Schoener and Dave Spiller, and the fabulous graduate students and postdocs who have passed through the lab, many of whom have worked on anoles: Duncan Irschick, Marguerite Butler, Manuel Leal, John Parks, Dave Pepin, Doug Creer, Jim Schulte, Rich Glor, Luke Harmon, Jason Kolbe, Michele Johnson, Brian Langerhans, Liam Revell, Luke Mahler, Alexis Harrison, Yoel Stuart, Todd Jackman, Delbert Hutchison, Jane Melville, Jason Knouft, Kirsten Nicholson, Tonia Hsieh, Renée Duckworth, Dave Collar, Terry Ord, and Anthony Herrel.

In addition, I thank Peter Raven for prompting me to write this book; Harry Greene and Chuck Crumly for guidance in its development; Scott Norton, Francisco Reinking, Aline Magee and the team at Michael Bass Associates for assistance in its production; Janet Browne for suggesting the title; the Organization for Tropical Studies for providing space at the La Selva Biological Station as I began writing; Roy Curtiss, Barbara Schaal, Oscar Chilson, Ralph Quatrano, Ed Macias, and the Biology Department at Washington University for fostering my career development in my 14 years at Washington University, during which time most of my work on *Anolis* was conducted; and the staffs of the biology libraries at Washington University and the Museum of Comparative Zoology which provided invaluable assistance in tracking down hard-to-find references.

I am reminded of the time I reviewed a book manuscript for a colleague and gave him 13 pages of single-spaced comments. Subsequently, the book came out and another colleague—finding it full of errors—admonished me for my lax oversight. My response was that I had my hands full, and with so many points to comment upon, couldn't be expected to have caught everything. In this spirit, I thank the many friends who helped so much by reading one or more chapters; they, indeed, had their hands full and saved me from much embarrassment. I thank these reviewers not only for catching my mistakes, but also for the many useful and insightful suggestions that had not previously occurred to me. Most or all of the book was read by Annie Chen, David Collar, Chuck Crumly, Renée Duckworth, Rich Glor, Alexis Harrison, Tonia Hsieh, Brian Langerhans, Manuel Leal, Joseph Losos, Luke Mahler, Terry Ord, Liam Revell, Yoel Stuart, Adam Wilkins and two anonymous reviewers. In addition, one to several chapters were read by Robin Andrews, Rayna Bell, Jon Chase, Kevin de Queiroz, Leo Fleishman, Harry Greene, Craig Guyer, Luke Harmon, Dror Hawlena, Anthony Herrel, Paul Hertz, Ray Huey, Duncan Irschick, Bruce Jayne, Tom Jenssen, Michele Johnson, Matt Lovern, Steve Poe, Robert Powell, Dolph Schluter, Chris Schneider, Tom Schoener, Kurt Schwenk, Ole Seehausen, Judy Stamps, Roger Thorpe, Richard Tokarz, Bieke Vanhooydonck, and Laurie Vitt. Kellar Autumn, Butch Brodie, Jim Cheverud, Hopi Hoekstra, Jason Knouft and Susan Perkins looked at particular passages. Advice, information and data of various sorts came from Craig Albertson, Gar Allen, Robin Andrews, Aaron Bauer, Jeff Boundy, Kevin Enge, Hannah Frank, Michele Johnson, Richard Lewontin, Elizabeth Losos, Joseph Losos, Don Lyman, William Magnusson, Gordon Orians, Steve Poe, Alan Pounds, Liam Revell, Gilson Rivas, Thom Sanger, Martin Schlaepfer, Kate Smith, David Spiller and Bieke Vanhooydonck. I thank Rafe Brown, Ryan Calsbeek, Jim McGuire, Steve Poe,

Peter Wainwright and Rebecca Young for providing access to their manuscripts prior to publication.

Rich Glor, Luke Harmon, Alexis Harrison, Luke Mahler and Bob Ricklefs provided extensive help with analyses presented in this book, for which I am extremely grateful. In addition, this book could not have been completed without the incredible assistance of Emily Becker, who provided great help in assembling and organizing the manuscript. Thanks, too, to Laszlo Meszoly for several drawings and to the many people who have immeasurably improved the book by providing photographs: Eldridge Adams, Chris Austin, Louis Bernatchez, Beth Brainerd, Rafe Brown, John Cancalosi, Kevin de Queiroz, Roman Dial, Eladio Fernández, Arthur Georges, Rich Glor, Harry Greene, Luke Harmon, Alexis Harrison, Anthony Herrel, Veronika Holanova, Duncan Irschick, Fred Janzen, Bruce Jayne, Tom Jenssen, Michele Johnson, Manuel Leal, Luke Mahler, Anita Malhotra, Thomas Marent, Marcio Martins, Aurélien Miralles, Luis Nieves, Wanda Parrott, Steve Poe, Margarita Ramos, Liam Revell, John Rummel, Howard Rundle, Alejandro Sanchez, Niranjan Sant, Tom Schoener, Judy Stamps, Rick Stanley, Roger Thorpe, Greg Vigle, Laurie Vitt and J.D. Willson. Particular thanks in this regard go to Kristen Crandell and Luke Mahler, who took photographs at my request.

PROLOGUE
The Case for Anolis

A green lizard sits on the bank at the edge of a tiny stream near the town of Soroa in western Cuba. It may appear unassuming, but this is not your ordinary lizard. Its head is cocked sideways as it peers into the water. Suddenly, it dives into the water, emerging with a small crayfish in its mouth.

A little later, one of these lizards basks near the water. Another lizard of the same species approaches. They nod their heads at each other, and then the larger one gives chase. As the smaller lizard flees, it comes to the water's edge. Rather than change direction, it continues straight ahead, raising its forequarters and running across the water. The larger lizard pursues and they crisscross the stream several times, sprinting upright on their hind legs, like miniature dinosaurs.

In the nearby forest, what appears to be a chameleon—sides compressed, massive head, short limbs and tail—perches 10 meters off the ground on a narrow tree trunk. But looks can be deceiving: chameleons don't occur in Cuba (or anywhere else in the New World, for that matter). Nonetheless, it thinks it's a chameleon—or at least it walks like one: moving down the tree trunk, it adopts their characteristic jerky, rocking motion: forward, backward; forward, backward—if you didn't look closely, you might think it was a leaf blowing in the wind, rather than a lizard slowly creeping its way downward. When it gets to the ground, there's another surprise: it feeds on large snails and beetle pupae in the leaf litter, crushing them with its massive molar-like teeth at the back of its mouth.

Cross the Gulf of Mexico to Central America. Up in the canopy of the rainforest, another lizard creeps along a thin twig. Its body is elongate, with short legs and a short tail. Thanks to its stubby legs, it couldn't run quickly if wanted to, so it relies on not being seen; its mottled light gray body blends in well with the tree bark. When discovered, however, it has another ploy: it launches itself into the void and sails away, limbs outstretched, to find safety elsewhere.

These are just three of the nearly 400 species in the lizard genus *Anolis*.[1] Anoles, as they are called, are one of the great evolutionary success stories of our time. The species exhibit remarkable variety in color, size, shape, habitat use, behavior, and many other attributes. As many as 15 species can be found in one place, and possibly as many as 20 at different elevations on some mountains. Some species are extraordinarily abundant, as many as one per square meter.

Anoles are notable in two other respects: their evolutionary diversity and the scientific knowledge of that diversity. With regard to the first, take a trip to the rainforest in the Luquillo Mountains of Puerto Rico. Enter the forest and sit a spell. After a moment or two, the lizards ignore you and become active, and you realize that they are all over the place. But there is order in this abundance: the different species occur in different parts of the habitat, some on tree trunks near the ground, others creeping on twigs, still others on grass stems or high in the canopy. Moreover, these species differ in their morphology and behavior, and the differences seem to make adaptive sense: the species on the trunk jumps to the ground, where it runs quickly to capture prey or confront conspecifics; its long legs seem well suited for rapid transit. The twig dweller, by contrast, has short legs, which provide stability as it creeps slowly through the thicket, and the species high in the canopy has large toepads, allowing a sure grip on the slippery leaves it uses. Each species appears well suited to the place in the environment in which it lives.

This observation in itself is not extraordinary; the phenomenon of adaptive radiation is well documented in many organisms. The situation becomes more interesting when the other islands of the Greater Antilles are visited. Go back to Soroa, in Cuba, and you'll find a twig-dwelling species that looks almost identical to the one in Puerto Rico and behaves similarly, too. Onto La Palma in the Cordillera Central of Hispaniola, and you'll observe the same phenomenon again, another twig-living doppelgänger; ditto for Southfield in the southwest of Jamaica.

But it's not just the twig species. A long-legged lizard that lives on tree trunks near the ground is present at each site, as well as a grass species (absent from Jamaica), and a large toepadded arboreal specialist. All told, there are six of these habitat specialist types that occur across islands.

DNA and other data make clear that species on different islands are not closely related. The repeated evolution of similar morphology by species using a similar

---

1. Specifically, the three species are *Anolis vermiculatus*, the Cuban "aquatic" anole; *Anolis* (Chamaeleolis) *barbatus*, the false chameleon; and *Anolis pentaprion*, a twig-inhabiting anole found from Mexico to Colombia.

environment is an old and well documented phenomenon: convergent evolution. But convergence of entire assemblages is much rarer. Although this hypothesis has been suggested for many situations, rarely is it quantitatively tested with morphological and ecological data. The anoles of the Greater Antilles may well be the best documented case of convergence of communities, and they've done it in quadruplicate across the four islands of the Greater Antilles!

Anoles are noteworthy in a second respect. Since the 1960s, they've been studied intensively by almost all manner of organismal biologist. The result is a depth and breadth of knowledge of basic biology and natural history that is shared by few other taxonomic groups. Ecology, behavior, phylogenetics, reproductive physiology, functional morphology: these subjects and many more have been studied in great detail. The result is that our understanding of the evolutionary diversification of this diverse and species-rich group is richer and more synthetic than for just about any other comparably diverse group.

So, anoles are an interesting group of animals. But are they interesting enough to write a whole book about, and for that book to be of interest to a broad audience beyond herpetologists? I hope so for two reasons. First, anoles historically have been an important group in the development of new ideas in a variety of disciplines, most notably ecology and evolutionary biology. Moreover, because they possess a fortuitous constellation of traits, anoles are excellent subjects to test current theories and develop new ones in fields as diverse as behavior, ecology, functional morphology, and evolutionary biology.

Second, no single type of research can tell us how and why a particular group of organisms evolved as it did, but by synthesizing disparate lines of evidence, we can come to the richest, most complete understanding of the patterns and processes underlying evolutionary diversification. The breadth of our knowledge of anoles makes them ideally suited for just this sort of interdisciplinary, integrative approach.

For these two reasons, anoles are an excellent group to study some of the most important questions concerning the origin and maintenance of biological diversity, such as:

- What ecological processes structure communities?
- How do these processes drive evolutionary change and adaptive radiation?
- How and why do new species arise?
- What are the relative importance of ecological opportunity and evolutionary constraint in shaping the diversification of a clade?
- Why do some clades exhibit very similar patterns of evolutionary diversification, whereas others, in seemingly similar circumstances, head down very different evolutionary paths?

Many types of organisms are suitable to answer some of these questions, but very few are appropriate for addressing all of them in a synthetic framework. Darwin's finches, East African lake cichlids, Hawaiian silverswords—these are some of the groups for which such an approach can be taken. Each has its own advantages and disadvantages, and each provides its own unique insights. In this book, I aim to show that anoles are another such group, that they have contributed much to our understanding of the factors driving evolutionary diversification, and that they will likely contribute even more in the future. Further, this background knowledge will make anoles an excellent group for the application of new methods in the fields of evolutionary developmental biology and comparative genomics. These approaches will greatly enhance our understanding of how anoles have evolved and, in particular, how and why convergent evolution has occurred repeatedly; in turn, the anole story may serve as a model of how the study of development and genomics can be integrated with the study of macroevolution.

1

EVOLUTIONARY BIOLOGY AS
A HISTORICAL SCIENCE

When we regard every production of nature as one which has had a long
history; when we contemplate every complex structure and instinct as the
summing up of many contrivances, each useful to the possessor, in the same way as
any great mechanical invention is the summing up of the labor, the experience, the
reason, and even the blunders of numerous workmen; when we thus view each
organic being, how far more interesting . . . does the study of natural history become!

CHARLES DARWIN, *ON THE ORIGIN OF SPECIES*, 1859, PP. 485–486

One of the great goals of modern science is to understand biological diversity: where it
comes from, how it evolves, and what maintains it. It has fallen to the field of evolutionary biology to try to answer these questions. In attempting to do so, evolutionary biology
does not fit the everyday view of science in which hypotheses are put forward and subjected to experimental test.

The reason is obvious. The scale in space and time is simply too large. It would be
wonderful to be able to do an experiment on, for example, the role of interspecific competition as a driving force in evolutionary diversification. Just get an island archipelago,
seed it with an ancestral finch population, and let nature take its course. Then get
another archipelago, seed with the same type of finches, but add an overabundance of
resources so that resources are not limiting and competition does not occur. Replicate
the treatments a few times (say, four archipelagoes flush with resources, four without),
come back in several million years, and, voilá, the hypothesis has been tested.

Too bad we can't do this. In trying to understand how and why evolutionary diversification has occurred, we're stuck with studying a phenomenon that has occurred over
large spatial scales through the course of thousands to millions of years. For this reason,
evolutionary biology is more akin to a social science—history—than it is to laboratory
based sciences like chemistry (Cleland, 2002; Mayr, 2004) Lacking time machines, both

historians and evolutionary biologists must draw inferences from a variety of different sources and approaches in their attempts to understand the past[2,3].

I like to compare studying evolutionary diversification to a detective story:[4] something happened in the past, and it is our job to build the best case to explain whodunit (or, at least, whathappenedtoit). In doing so, there usually is no smoking gun, no decisive experiment or single piece of evidence (Turner, 2005). Rather, we must gather as much data, from as many different sources, as possible. Then we must weave together these data to present the best explanation of what happened.[5] As in a court case, the more consistent and corroborative the data, the more compelling the case (for a generally congruent, but slightly different, view, see Cleland [2002]; also see Pigliucci [2006]).

Such explanations, of course, are more than mere stories; they are the hypotheses that guide further work. Each time we learn something new, each time we bolster our case a bit more, new hypotheses are suggested that await subsequent testing. The better supported an explanation is, the less likely it will be that a single new piece of data will discredit it. Nonetheless, given that we are trying to explain what happened in the past, we can never know for sure what happened,[6] and it is always possible that additional data will change our thinking.

SYNTHESIZING DATA FROM THE PRESENT AND FROM THE PAST

Ideally, we would like to know what processes occurred in the past and how these processes shaped the diversity we see today. This is where building the best detective case comes in. We can't directly study the processes operating in the past (Cracraft, 1981). But we can study processes in the present, and we can even observe their outcome over short evolutionary timescales.

What we can study in the past is the pattern: the history of change through time. Depending on the quality of the historical record, we can infer, with a greater or lesser degree of confidence, what happened. The key, then, is to extrapolate from our understanding of the relatively short term outcomes of ongoing processes to explain the patterns of change in the past.

---

2. Evolutionary biology is not alone among the natural sciences in taking this approach. Astronomy and some branches of geology are two others that attempt to unravel the mysteries of the past.

3. A footnote on footnotes: in an effort to make this book more readable, I will remove many detailed points and parenthetical statements from the main text and place them as footnotes. For those in a hurry or who can't be bothered, the footnotes are not crucial, although readers bypassing these notes risk missing the occasional hilarious joke or witticism.

4. An analogy made independently by Grant (1986, p.11), Cleland (2002, p.17), me (Losos, 2001, 2007), and probably many others.

5. Mayr's (2004) "historical narrative."

6. In fact, most scientists today would hold that science cannot "prove" anything. Rather, hypotheses are repeatedly tested; those that withstand every test, and for which it is inconceivable that future data will overturn current understanding, are elevated to the status of theories, such as the theory of evolution and the theory of gravity. Some would consider these overwhelmingly supported theories to be what we commonly refer to as "facts."

The historical record of evolution comes primarily in two forms: fossils and phylogenies. Each has its advantages and disadvantages. The strength of the fossil record is that we have concrete evidence of what extinct species were like and when they occurred. Were the fossil record complete, we would need nothing else to reconstruct the evolutionary history of a group.

But, of course, the fossil record is not complete. In some groups, such as horses (MacFadden, 2005) and trilobites (Fortey, 2000), it's still good enough to tell us a great deal about evolutionary history. In other groups, however, the situation is much less rosy. In *Anolis* lizards, for example, only four fossils[7]—all specimens entombed in amber—have been scientifically described (Figure 1.1; Lazell, 1965; Rieppel, 1980; de Queiroz et al., 1998; Polcyn et al., 2002), with perhaps another dozen or two undescribed (most in the hands of private collectors);[8] all but one are from the same deposit from the Dominican Republic. Clearly, what we can learn from anole fossils is limited (although valuable).

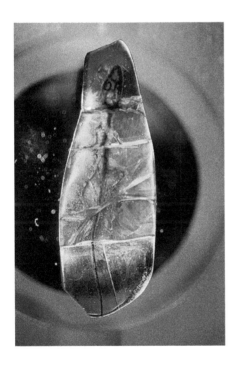

FIGURE 1.1

One quarter of the fossil record for *Anolis*. This 15–20-million-year-old juvenile lizard from the amber mines of the Dominican Republic is indistinguishable in skeletal anatomy from the green anoles found on Hispaniola today (de Queiroz et al., 1998).

7. This does not include Pleistocene subfossils (e.g., Etheridge, 1964; Pregill et al., 1988; Roughgarden and Pacala, 1989).

8. Interest in amber fossils has greatly increased in recent years, perhaps in part as a result of the movie *Jurassic Park*, in which dinosaur DNA was extracted from mosquitoes entombed in Dominican amber (no matter that the dinosaurs had been extinct for over 40–50 million years by the time Dominican amber was formed!). This interest has driven the price of amber specimens through the ceiling—one anole in amber was initially marketed with an asking price of $1,000,000 and still hasn't sold despite a 90% price reduction. Unfortunately, such specimens are usually unavailable for scientific study.

The other source of historical information comes from phylogenetic inference. Phylogenies have the advantage that they can provide evolutionary insights even in the absence of fossils (although whatever fossil data are available should be incorporated into such analyses). The disadvantages are twofold: first, many types of data (e.g., DNA sequences) generally cannot be obtained from fossils, limiting analyses only to extant taxa; and, second, the results of phylogenetic analyses are only as good as their underlying assumptions, which must be made both in constructing the phylogeny and in drawing evolutionary interpretations from it.[9]

Potential shortcomings notwithstanding, phylogenies are enormously useful and can provide information on a wide variety of evolutionary patterns, including rates of evolution, extent of convergence and stasis, the order in which particular traits evolve, and the timing of evolutionary events; they can inform biogeographic scenarios; and they can suggest hypotheses such as character displacement, taxon cycles, and cospeciation (those looking for an entrée to this literature might start with Felsenstein [2004] or by picking up a recent issue of a journal such as *Evolution, American Naturalist, Journal of Evolutionary Biology, Systematic Biology* or many others).

PRESENT-DAY PROCESSES

Historical patterns are fascinating and suggestive, but processes can only be studied directly among extant taxa. The processes that are evolutionarily important are those that affect how species interact with each other and with the environment, and how these interactions lead to evolutionary change.

Ecological interactions can be studied in many ways (Diamond, 1986). Recent years have seen an emphasis on manipulative experiments. Such experiments are extremely useful, but by necessity are limited in both length and size. Other useful information can be obtained from detailed observational studies and by following the results of species introductions or other changes to the environment brought on by human activities. By some combination of these approaches, scientists can investigate which processes operate in a given system, and how the operation of such processes may vary in different circumstances.

The evolutionary outcome of ecological processes also can be studied in a variety of ways (reviewed in Fuller et al., 2005; Reznick and Ghalambor, 2005). Experimental laboratory studies of evolution are common, and a few controlled and replicated experimental

---

9. To summarize: you can hold a fossil in your hand and inspect it directly, but usually there aren't enough of them; you can't hold phylogenetic inferences in your hand and are potentially misled by the assumptions you make in deriving them, but they're a lot easier to obtain. Of course, in practice these two approaches are not alternatives because in the absence of a complete fossil record (i.e., always), a phylogenetic framework is required to interpret fossil data.

studies have been conducted in the field. Studies of the microevolutionary change following from human manipulations and introductions also can document the evolutionary response to ecological processes operating in some circumstances (e.g., Carroll et al., 1998; Huey et al., 2000; Hendry, 2001).

MUTUAL ILLUMINATION

Given that we have data on both pattern and process, how should these data be integrated? Over the past 2–3 decades, ecologists have pioneered this approach. Starting with an observation from the natural world that suggests a mechanistic hypothesis, ecologists often design a manipulative experiment to test whether the process produces the predicted result. Less frequently, the reverse occurs: an experiment suggests that a process may work in a certain way; field work is then conducted to see if variation among study samples conforms to predictions.

A similar mutual illumination occurs between historical and present-day studies. On one hand, historical studies can identify a pattern. Studies in the present can then examine whether a hypothesized process can produce such a pattern. If, for example, historical analysis (either examination of fossils or phylogenetic inference) reveals that sister taxa have evolved differences in body size whenever they have become sympatric, then the hypothesis that the species have diverged in size to minimize resource competition (a phenomenon termed "character displacement" [Chapter 7]) can be tested in several ways: An experiment can be established to investigate whether species similar in body size compete for resources; if competition does occur, natural selection can be measured to determine whether selection favors divergence in body size, and cross-generational studies can be conducted to see if the species begin to diverge.

This procedure can work equally well in the opposite direction. Studies can reveal that a particular process plays an important role in extant populations. Then historical analyses can investigate whether evolutionary diversification has occurred in the manner predicted if the process in question has played an important role. Reversing the example from the previous paragraph, if data from extant populations suggested that character displacement occurred, a historical test might involve examining whether evolutionary changes in size occurred when similar-sized ancestral taxa came into contact due to colonization or earth history events.

A CAVEAT

A uniformitarian assumption underlies this reasoning. That is, if a given process—say, interspecific competition—leads to microevolutionary change in a predictable direction today, then I assume that interspecific competition would have affected ancestral species in the same way. I take this uniformitarianism one step further by extrapolating from the relatively small changes that occur over limited evolutionary time scales in the present to the much greater changes that have occurred over longer periods through the past.

But is the present really the key to the past? Not necessarily. Species and environments in the past may be different from those that exist today, and as a result, the ways ancient species interacted with each other and with their environment may have been fundamentally different from the interactions that occur today. For example, a common scenario of adaptive radiation begins with an ancestral species colonizing an island poor in competitors and rich in resources. The processes that affect such a species and the evolutionary outcome of such processes may be very different from what happens today among specialized descendant species in an environment in which resources are much less abundant. Thus, studies of the microevolutionary effects of interspecific competition among species in the Greater Antilles today may not be informative about how the initial stages of anole adaptive radiation proceeded.[10]

This is an important problem for any study trying to infer what happened in the past. Certainly, to the extent that what happened in the past is a series of historically unique events operating under rules that do not apply today, scientific study of the past will be difficult, if not impossible. To get around this difficulty, we must distinguish between the basic rules—e.g., those underlying the basic tenets of genetics and population biology— that we assume applied to extinct lizards in the same way that they apply to modern ones from those rules that may be contingent upon the situations in which they occur. Similarly, we must be sensitive to the many ways that past situations may have differed from present situations, and how these differences may have affected the evolutionary outcomes of ecological processes. This approach will not be easy, but is not impossible, either; it is an issue to which I will return periodically throughout the book.

More generally, I must acknowledge that not all scientists subscribe to the "detective story" approach to studying historical phenomena. Some scientists hold that the only way to study a process is to measure it directly, which, of course, is impossible for past events (Cracraft, 1981; Leroi et al., 1994). Others even more strictly restrict their view of science to those questions that can be investigated experimentally.[11]

These views are understandable, but excessively restrictive. No doubt one would have greater confidence that natural selection had operated if one measured selection directly on an extant population, rather than inferring its action from other sorts of data. But to

---

10. A related example of the inability to extrapolate from the present to the past is the hypothesis for the amazing diversity of morphological forms in the famous Burgess Shale fauna of the Cambrian Period. Gould (1989) argued that the abundance of resources led to the evolution of a variety of morphological forms and further suggested that the explanation for this great morphological disparity, which has never since been rivaled in magnitude, is that genetic regulatory systems were much more flexible at that time and have since become more canalized, so much so that the morphological variety that can be produced by mutation today is constrained. As a result, an extant population experiencing comparable ecological opportunities to those that occurred in the Cambrian would be evolutionarily incapable of producing comparable morphological variety (for an alternative view, see Conway Morris [1998]).

11. In fact, some have referred derisively to field-based studies of ecology and evolution as "Boy Scout science."

suggest that in the absence of direct measurement or experimentation, investigation of the processes that occurred in the past is impossible amounts to throwing the baby out with the bathwater. Taking this view means that some of the most fascinating questions facing humanity—How has life evolved? Did a Big Bang occur, and what happened afterward?—could not be investigated.

Moreover, neither direct measurement nor experimentation is a panacea. For many reasons, from technical to interpretational, the results of contemporary studies may be no more definitive than those obtained from historical studies (Cleland, 2002; Turner, 2005). These problems may tend to be greater for historical studies than for non-historical ones (compare Cleland [2002] to Turner [2005]; also see Pigliucci [2006]); regardless, the fact that the same issues bedevil both indicates that historical and non-historical science differ in degree, not in kind.

The bottom line is that historical hypotheses are investigated in the same way as non-historical ones. Hypotheses are developed and then tested with further data. Some tests are stronger than others; we must not overstate the confidence we have in particular inferences, and we must acknowledge the limitations and assumptions of any test, as well as alternative interpretations of the data. These caveats apply to all scientific studies, not just historical ones. By taking this approach, we can best understand what happened in the past, and why. That is not to say that we will be able to study all past events, just as we can't study all non-historical phenomena. But we will be able to learn much about how life has evolved through time.

OUTLINE OF THE BOOK

The thesis of this book is simple: interspecific interactions—primarily, but not exclusively, competition—among extant *Anolis* species play a dominant role in shaping their ecology and microevolution, and the historical record is consistent with the hypothesis that interspecific interactions have been the force driving evolution throughout anole history.

While presenting this thesis, I hope to integrate the enormous body of research conducted on anoles over the course of the past four decades. My goal will not be to exhaustively review all of this work. Rather, by synthesizing it, I hope to substantiate my claim in the prologue that the breadth and depth of our knowledge of all aspects of the organismal biology of such an ecologically and phenotypically diverse group makes *Anolis* one the best subjects for the study of evolutionary diversification.

The book is organized in the following order. First, in Chapters 2–4, I will introduce the ecological and evolutionary diversity of anoles. Although of central significance to the discussion of evolutionary diversification, phylogenetic issues will be postponed so that anole diversity can be fully presented before its evolutionary pathways are analyzed. The subsequent phylogenetically-oriented Chapters (5–7) will serve a dual purpose, both discussing the advantages and limitations of a phylogenetic perspective, as well as examining patterns of anole diversity in a historical context. Chapters 8–13 will examine anole biology, focusing on behavior, ecology, and life history (Chapters 8–10), community ecology (Chapter 11), microevolutionary change (Chapter 12), and functional capabilities (Chapter 13). Chapters 14–16 will examine speciation and adaptive radiation, and Chapter 17 will conclude by placing discussion of anole evolution in a broader context in comparison to patterns of evolutionary radiation in other taxa.

One theme that permeates this book is the importance of natural history. Only by having a rich and deep understanding of the organisms we study can we have insights into how and why they vary and how they have evolved (Greene and Losos, 1988; Greene, 2005; Dayton and Sala, 2001). For this reason, I have not shied away from providing a wealth of detail about particular species when I feel that information is important. I also hope such descriptions will bring to life the fascinating and diverse nature of these charming creatures and, in so doing, will enliven the book as a whole.

I have two audiences in mind for this book: those deeply interested in anoles and those interested in general questions of biodiversity, evolutionary biology and ecology. I have tried to walk a fine line in keeping the book general enough to be of broad interest, yet specific enough to be useful to those working, or thinking of working, on anoles. Of course, compromises are necessary, and to both audiences, I apologize in advance. To readers with more general interests, I am sorry for what at times might seem excessive details. Some times I just can't help myself! I've tried to move as much of the anole trivia as possible to footnotes, so those not so enamored with all things *Anolis* can zip right by. To anole aficionados, I beg forgiveness for not discussing every paper and every species. Where possible, I have generalized, or used the best example of many, and I have often cited the most recent or most comprehensive paper, rather than every paper on a topic.

Given the extraordinary breadth of work on *Anolis*, my review covers many fields, from phylogenetic analysis to behavior, ecology, functional morphology and beyond. In reviewing this literature, I have tried to explain methods and approaches at a basic level, as well as to provide an entrée to the literature, where possible by way of work done on anoles. Of course, this means that for readers knowledgeable about a particular area, the discussion may seem overly simplistic. In this regard, too, I have tried to walk the fine line between making the work accessible to a broad audience, while providing at least some measure of the detail of interest to the specialist.

Finally, a disclaimer about my review of the literature. Although a goal is to make this book the first place people look when they have a question about anoles, I am not trying to be encyclopedic. The reason is simple: the literature is too vast. For example, a Web of

Science search conducted in December 2007 with the keyword "*Anolis*" retrieved 1,901 papers and a Google Scholar search yielded 13,300 results. For this reason, in many cases I cite only papers which can serve as an entrée into the literature with the hope that readers can follow from those references to other relevant works.

FUTURE DIRECTIONS

I conducted my college honors' research project on display behavior and species-recognition of a sympatric pair of closely related anoles (Losos, 1985a,b). When I left and went off to graduate school, I vowed to work on anything but *Anolis* because I perceived that we already knew the important stuff, and I wanted to blaze my own trail. It was only after two years of coursework and a dozen failed projects that I realized while on an Organization for Tropical Studies summer course what should have been obvious before: the extensive previous work on anoles, rather than leaving few interesting questions, provided the groundwork for synthesis and made anoles suitable for addressing conceptual questions of broad and general significance.

That should have been enough to teach me a lesson, but I'm not a quick study. After working on anoles for more than 10 years, I decided it was time to work on other groups, where more interesting questions remained. And so I started to do so, conducting studies on other types of lizards (e.g., Losos et al., 2002; Schulte et al., 2004), and even opossums (Harmon et al., 2005)! But again anoles drew me back. The more we learned, the more new interesting and unforeseen questions arose. And, as before, the wealth of knowledge of all things *Anolis* continues to make them an ideal group for testing new ideas.

I mention these anecdotes to highlight that there is much we do *not* know about anoles. The more we know, the more we discover we don't know. And surprisingly, even basic aspects of anole biology (e.g., diet, social structure) are not known nearly as well as one might expect. Consequently, many research areas are wide open, begging for more research.

With nearly 400 species in the genus and the possibility of studying almost any aspect of its natural history, *Anolis* welcomes new researchers. For this reason, I will end chapters with a brief discussion of questions that I think would be worth pursuing. Of course, these represent just those questions that occur to me; no doubt many others exist as well. My emphasis will be on what we don't know about anoles, rather than the broader conceptual framework in which such studies could be conducted. This is not to say that the broader context is not important; quite the contrary, many eminent biologists have made their names by studying *Anolis* to address important questions of the day. Indeed, in many areas, I would argue that *Anolis* is an ideal subject to use in testing outstanding hypotheses and in developing new ones. My hope is to entice workers—particularly students developing their dissertation ideas—to consider studying anoles both to increase our knowledge of these interesting animals and to investigate topics of wider interest. There's plenty of room in the *Anolis* world for more researchers, and I will consider this book a success if it helps to produce a new generation of anole biologists.[12]

12. Anolologists? Incidentally, this is probably a good place to point out that honorable, right-thinking people can disagree over whether the correct pronunciation is uh-nole or an-ole. I am less charitably inclined to my ninth grade biology teacher's uh-no-lee, but, although I have never heard "anole" articulated in that way by anyone else, I am told that it is common in the South, from whence she came (I was surprised to find that this is the preferred pronunciation of the Random House Unabridged Dictionary, according to www.dictionary.com).

As for the origin of the name, Daudin [1802], who named the genus, said that "anolis" was the name the indigenous Caribs used for these lizards. However, there is some possibility that in fact "anolis" may have been their name for lizards in the genus *Ameiva* and that the correct Carib word was "oulléouma" (see discussion in Breuil, 2002). Right or wrong, Daudin clearly chose the more mellifluous name to bestow upon these lizards! A more interesting, though doubtless less accurate, explanation is the Saba Tourist Bureau's statement (www.sabatourism.com) that "The scientific name of 'Anolis' comes from the popular name of "anole" for these lizards. Anole is an ancient African name, meaning "little devil", that is given to small lizards in western Africa."

2

MEET THE ANOLES!

The goal of this chapter is twofold. First, to introduce anoles: what they are, what makes them unique, and where they occur. Second, to focus on what it is to be an anole species. How do species differ from one another? How do we tell one from another? How do they tell one from another? Of course, understanding what constitutes a species is a prerequisite for studying how new species arise from old ones, so this discussion will set the stage for understanding anole evolutionary diversification.

WHAT MAKES AN ANOLE AN ANOLE?

Anoles are typical lizards in many ways. Consider the following:

Species range in adult length from 33–191 mm snout-vent length (SVL),[13] with a tail—capable of detachment and regeneration in most species[14]—that is 1–4 times the

---

13. Snout-vent length is the distance from the tip of the snout to the anterior end of the vent, or cloaca, which is the orifice through which excretion and reproduction occur.

According to Schwartz and Henderson (1991), from whom I took body size measurements listed here and elsewhere in the book, the largest West Indian anole is the Cuban crown-giant *A. luteogularis* and the smallest is a Cuban grass-bush anole, *A. cupeyalensis*. Estimates of the maximum size for species are imprecise: as a result, determining which species are truly the smallest and largest is difficult, but the extreme species almost surely belong to species in the two Cuban clades that contain *A. luteogularis* and *A. cupeyalensis*. Mainland anoles do not achieve the extremes exhibited by island species, although some species are close.

14. Many lizards drop their tails when they are grabbed by a predator. Such "autotomy" is facilitated by specializations in the vertebrae and attendant muscles and blood vessels that facilitate detachment with minimal trauma (Etheridge, 1967; Arnold, 1984; see figure 8.10).

length of the body. Young hatch out of eggs after 25–130 days at a length of 15–40 mm SVL (Andrews and Rand, 1974; Schlaepfer, 2003; Köhler, 2005; Sanger et al., 2008a). Body color is usually gray, brown, or green; color can lighten and darken, and green species can turn dark brown.

Anoles are visually-oriented lizards with excellent eyesight and color vision which extends into the ultraviolet (Fleishman and Persons, 2001). Their sense of smell is poor and their tongues are used primarily for prey capture rather than for chemoreception (Schwenk, 2000). Anoles can hear (e.g., Werner, 1972) and some species vocalize, most often upon being captured (Milton and Jenssen, 1979), but the extent to which anoles respond to sound in nature is unknown (Rothblum et al., 1979).

Very few anoles are dietary specialists and most species eat a wide range of insects. The incidence of myrmecophagy varies greatly among species; many mainland species seem to avoid ants entirely. Carnivory, frugivory, and molluscivory all occur, primarily in larger species and larger individuals of medium-sized species.

Moving up the food chain, in many respects anoles are the lunchbox of the neotropics. They are eaten by all manner of birds, mammals, snakes, other lizards (including conspecifics, as well as other anoles), frogs, even spiders and other invertebrates. Predation differs among regions, and even among habitats within regions.[15]

Overall, anoles are little different from many other types of lizards. However, some of the most interesting features that have evolved repeatedly in lizards[16]—e.g., extreme limb reduction and loss, exclusively herbivorous diet, viviparity, and parthenogenesis—do not occur among anoles.

In what ways, then, do anoles differ from other lizards? *Anolis* is characterized most obviously by two characteristics: the possession of a dewlap and of subdigital toepads. In addition, a number of other features are notable, especially aspects of the anole visual system and reproductive cycle.

THE DEWLAP

The dewlap is an extensible structure located on the throat and extending far down the belly in some species. Most of the time retracted and barely visible, it can be deployed by movements of the hyoid apparatus. The hyoid, evolutionarily derived from the gill arches of fish, is composed of a series of thin, rod-like elements, mostly cartilaginous, that are located in the floor of the mouth and throat of tetrapod vertebrates.[17] Anoles have extremely long second ceratobranchial elements (as much as 67% of snout-vent length and reaching the pelvic girdle [Font and Rome, 1990]) which lie within the ventral margin of the dewlap (Fig. 2.1). When muscles attached to the front end of the

---

15. Statements in the last two paragraphs are discussed in detail in Chapter 8.
16. Vitt and Pianka (2003) provide a nice overview of lizard ecology and evolution.
17. The Tetrapoda is the clade containing all limbed vertebrates, including their descendants that have secondarily lost their limbs, such as snakes.

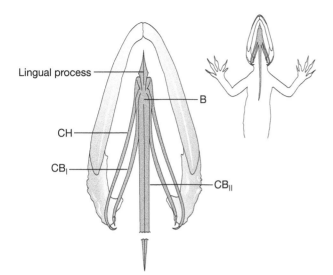

FIGURE 2.1
The hyoid of *A. equestris*.
B = basihyal; CB =
ceratobranchial; CH =
ceratohyal. Modified with
permission from Font and
Rome (1990).

hyoid contract, a lever is set in action with the result that the second ceratobranchials are rotated downward; this movement serves to unfurl the dewlap in all its glory[18] (Bels, 1990; Font and Rome, 1990; Fig. 2.2).

Dewlaps vary tremendously among species in size, color, and pattern (Nicholson et al., 2007). Except for two species,[19] males of all species possess dewlaps. By contrast, possession of a dewlap by females is variable: females of many species do not have a dewlap. Among species in which the female possesses a dewlap, its size ranges from much smaller than the male's to equal in size (Fitch and Hillis, 1984). Sometimes, the dewlaps of the sexes differ in color and appearance, with that of the male usually being more colorful. Some species have the ability to lighten and darken their dewlaps (e.g., *A. woodi* [Savage, 2002]), but in most species, the amount of change that can be produced is minimal.

Dewlaps are used for communication in a variety of contexts, including courtship and intrasexual encounters. Anoles, particularly males, regularly dewlap seemingly to no specific target; such behavior probably serves to notify any anole in the vicinity of the presence of a territory holder (Fleishman, 1992). Anoles also dewlap to potential predators, presumably serving as a "pursuit deterrent" signal (Leal and Rodríguez-Robles, 1997b).[20]

Anoles share a dewlap with *Polychrus*, which has long been thought to be the sister group to *Anolis*, partly for this reason (Frost et al., 2001); if this phylogenetic hypothesis is correct, then the dewlap would not be a synapomorphy[21] for *Anolis*, but rather for the

18. A process termed "dewlapping" by anole aficionados.
19. *A. bartschi* and *A. vermiculatus* (see Chapter 4 for information on their biology).
20. That is, a signal that informs a predator that it has been seen and thus "suggests" that it would be a waste of time and energy to attempt to capture the lizard.
21. A synapomorphy is a shared, derived character that presumably arose in the most recent common ancestor of a group of species.

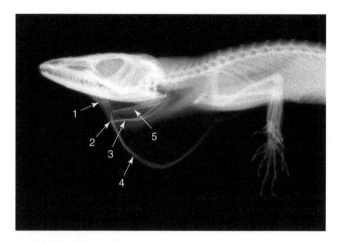

FIGURE 2.2

Dewlap extension in *A. carolinensis*. Muscular contraction pulls the ceratohyals backward. This in turn causes the anterior process of the hyoid to move up and backward, causing the second ceratobranchial to rotate forward and downward around the joint formed at the intersection of the first ceratobranchial and the basihyal, thus extending the dewlap (Font and Rome, 1990). Because the cartilaginous second ceratobranchials are so flexible, the fully-extended dewlap adopts a semicircular conformation. Photo courtesy of B. Brainerd and D. Irschick. 1 = lingual process; 2 = basihyal; 3 = ceratobranchial 1; 4 = ceratobranchial 2; 5 = ceratohyal.

clade[22] comprising *Anolis* plus *Polychrus*. However, recent molecular data do not support the existence of this clade (Frost et al., 2001; Schulte et al., 1998, 2003; Chapter 6).

Similar structures, apparently also constructed by elongation of the second cerato-branchials and other similar modifications of the hyoid apparatus, have evolved in several other types of lizards such as iguanas (members of the Iguanidae, as is *Anolis*) and a number of lizards in the family Agamidae (Fig. 2.3; Gnanamuthu, 1930; Bellairs, 1969).[23] In most cases, dewlaps of these species do not appear very similar to those of anoles, but lizards in the south Asian genera *Sitana* and *Otocryptis* have dewlaps remarkably similar in appearance, and also in function, to those of anoles (Fig. 2.3; Kästle, 1998).

22. A clade is the group that includes an ancestor and all of its descendant species. By definition, a clade is "monophyletic," whereas a group that contains the common ancestor, but not all descendant species, is termed "paraphyletic."

23. I was surprised to learn, with the help of Kurt Schwenk, that no comparative study has been conducted on the dewlap structure and function of different species of lizards. Indeed, I could not even find a list of all of the different genera that possess a dewlap, which likely has evolved numerous times.

For unknown reasons, dewlaps have only evolved within the Iguania, which is the clade of lizards that contains the Iguanidae, Agamidae, and Chamaeleonidae. Dewlaps have not evolved in any other squamate group (Squamata is the order of reptiles that includes lizards and snakes; snakes evolved from lizards and are, in evolutionary terms, limbless lizards), even though Iguania includes less than 20% of all squamate species. Why dewlaps have evolved only in the Iguania would make an interesting study of evolutionary constraints (sensu Gould, 2002; Schwenk and Wagner, 2003).

FIGURE 2.3
Dewlaps of iguanian lizards. (a) A spiny-tailed iguana, *Ctenosaura similis*, from Costa Rica. Photo courtesy of Alexis Harrison. (b) A flying dragon, *Draco jarecki*, from the Philippines. Photo courtesy of Rafe Brown. (c) The fan-throated lizard, *Sitana ponticeriana*, from Sri Lanka. Photo courtesy of Niranjan Sant.

SUBDIGITAL TOEPADS

Almost all anoles have expanded toepads underlying the digits on their fore- and hindlimbs. These pads are composed of a number of laterally expanded scales, termed lamellae (Fig. 2.4a,b). Each lamella is attached at its front (proximal to the body) end, but is free at the rear. Lamellae are covered with millions of microscopic hairlike structures termed setae which can approach 30 μm in length; these setae can end in a hook or a spatula-like structure usually less than 1 μm in width, or in an intermediate shape (Fig. 2.4c,d; Ruibal and Ernst, 1965; Peterson, 1983). Very similar structures, though more elaborate and often with a branching structure, have evolved in geckos, which also often use high, vertical structures such as trees and rock walls, and also in a small clade of skinks[24] (Ruibal and Ernst, 1965; Williams and Peterson, 1982; Irschick et al., 2006b).

Anole toepads differ in size and in the number of lamellae composing them, as well as in the fine structure, distribution, and density of setae (although the setae of relatively few species have been examined to date [Peterson, 1983; Peattie and Full, 2007]). One

24. *Prasinohaema.*

A B

FIGURE 2.4
Lizard toepads. The feet and toepads of (a) *A. chlorocyanus* and (b) *A. olssoni*. Photos courtesy of Luke Mahler.
(c) Toepad lamellae of *A. valencienni* magnified 310x. (d) Setae on the toepad of an undescribed anole species,
10,000x magnification. Toepad images courtesy of the Museum of Comparative Zoology, Harvard University.

species has completely lost its toepad, and reduction in both gross pad size and setal structure has occurred in a number of species (Peterson and Williams, 1981; Peterson, 1983). The functional significance of variation in setal structure is not well understood, although those species with small pads and pads lacking setae generally are terrestrial (Peterson, 1983; Nicholson et al., 2006).

The means by which setae provide adhesive ability has been debated for more than 175 years (Autumn, 2006). A variety of hypotheses had been suggested including interlocking into surface irregularities, suction, friction, and capillary adhesion. Hiller's (1975) clever studies, however, suggested that it was none of these. Rather, his work on geckoes suggested that adhesion was accomplished by the establishment of intermolecular bonds between the setae and the surface. Hiller (1975) supported this hypothesis by measuring the ability of a gecko to cling to a flat piece of plastic (polyethylene), then removing the lizard, zapping the plastic with a cardiac defribrillator (which would have the effect of changing the state of the electrons on the plastic), and then replacing the lizard. The resulting marked change in clinging ability strongly implicated electron-electron forces as being responsible for gecko adhesion. Sophisticated recent nanotechnological studies have confirmed this idea and have revealed that the intermolecular forces are the results of bonds formed between electrons on the setae and those on the surface. Bonds of this sort produce the weakest type of intermolecular force known, termed van der Waals forces (Autumn and Peattie, 2002; Autumn et al., 2002).

Although it is not clear what properties of the surface maximize the potential for van der Waals' forces (adhesion by means of van der Waals forces is an area of active research at the interface of biology and engineering [Autumn, 2007]), some types of surfaces are clearly better than others. Teflon, waxy plants and waxpaper are problematical for pad-bearing lizards; if an anole or gecko is placed on a waxy plant leaf and the leaf is tilted,

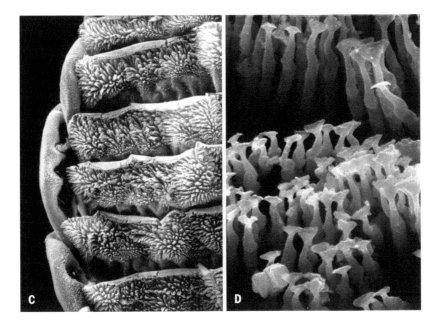

the lizard will have trouble maintaining its position and often will slide right off. By contrast, place the same lizard on a clean piece of plexiglass and the lizard will be able to hold on even at very steep angles—some geckos can even hang upside down!

The sticking ability of lizard toepads is truly phenomenal. A tokay gecko (*Gekko gecko*) clinging to a flat surface by its forefeet (the pads of which, in sum, cover less area than a dime),[25] can withstand a force greater than 20 newtons (approximately 4.5 pounds) pulling parallel to the surface (Irschick et al., 1996; Autumn, 2006). A single seta, 110 × 4 μm in length and width, can withstand 200 μnewtons of shear force (Autumn et al., 2000). As a result, geckos and anoles can hang by a single toe (Fig. 2.5)!

If the setae can generate such great adhesive ability, how is the lizard able to disengage the pad and avoid being permanently stuck to a surface? Geckos have a precise system for peeling their toepads off the surface which begins at the distal end of the pad, thus in the reverse order to the way the pad is applied to the surface. Apparently, by shifting the position of the toepad, geckos change the angle at which the setae contact the substrate: changing this angle greatly diminishes the strength of the intermolecular bond between the seta and surface, allowing it to be broken easily and allowing quick detachment of the toepad (Autumn et al., 2000, 2006).

By contrast, anoles do not remove their toepads from the surface in this fashion. Rather, pad detachment occurs as the toes lose contact with the surface during forward movement: first the sole of the foot and then the more proximal parts of the toes come

---

25. The smallest U.S. coin, with a diameter of 1.75 cm.

FIGURE 2.5

Anolis sagrei hanging from a glass slide by a single toe. Photo courtesy of Kristen Crandell/Kellar Autumn Laboratory, Lewis and Clark College.

off, with the distal end last. In other words, the sole, toes and toepads come off the surface in the same sequence in which they contacted it, which is the standard pattern in lizard locomotion (Russell and Bels, 2001). Whether anoles remove their seta by changing the angle of attachment, and why geckos and anoles have different kinematics of toepad detachment, remains to be studied.

VISUAL SYSTEM

Anoles have very good color vision. Anole eyes are placed on the sides of the head, and each eye has a monocular field of 180° with 20° of binocular overlap at the front (Fleishman, 1992). Anoles and predatory birds are the only two groups of vertebrates that have two foveae in the retina, the central fovea in the middle and the temporal fovea located in the same horizontal plane, but on the side (Fig. 2.6; Underwood, 1970; Fite and Lister, 1981; Ross, 2004). Foveae are small cuplike depressions in the retina that have high photoreceptor density and where visual acuity is highest. When an animal looks at an object, it moves its eye so that the image is positioned on its fovea. The temporal foveae of anoles and raptors appear to be used in binocular vision. For example, when an anole prepares to lunge for a food item, the lizard turns its head toward the prey so that the prey's image falls on both temporal foveae, presumably enhancing the lizard's depth perception (Fleishman, 1992; L.J. Fleishman, pers. comm.).

Anoles have high photoreceptor densities and four spectral classes of cones, but probably no rods (Fleishman et al., 1993, 1997). These observations and other aspects of retinal structure (e.g., Loew et al., 2002) suggest that anoles have good color vision in

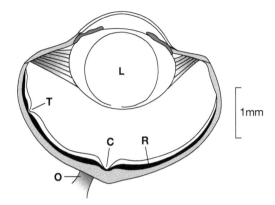

FIGURE 2.6
Horizontal slice through the eye of a
typical anole. C = central fovea; T =
temporal fovea; R = retina; L = lens;
O = optic nerve. Redrawn with permission from Fleishman (1992).

bright light at the expense of poor visual sensitivity in dim light (Fleishman, 1992), a conclusion supported by laboratory studies of the ability of anoles to detect and respond to different visual stimuli (e.g., Persons et al., 1999; Fleishman and Persons, 2001).

REPRODUCTIVE CYCLE

Anoles have a fourth diagnostic characteristic: they only lay one egg at a time, a trait which is quite unusual among lizards.[26] Most lizards lay one to several large clutches of eggs a year. Anoles, however, do not necessarily have lower reproductive output. Rather, they continuously produce eggs every 5–25 days throughout the breeding season (depending on species, size, locality, and season [Andrews and Rand, 1974; Andrews, 1985]), alternating which ovary produces the egg (Smith et al., 1972).[27]

GEOGRAPHIC DISTRIBUTION AND SPECIES DIVERSITY

Anoles range throughout the northern half of South America, through Central America and into tropical Mexico (Fig. 2.7). *Anolis carolinensis* occurs in North America as far north as North Carolina, Tennessee and Arkansas and as far west as eastern Texas. Anoles occur on just about every island in the Caribbean larger than 0.25 km², and on many that are smaller (e.g., Rand, 1969; Schoener and Schoener, 1983a,b).[28] The anoles of the Caribbean are well known; what is not so well known is that anoles also occur on

---

26. Geckos have a similar reproductive cycle, but most species lay two eggs at a time; presumably, low clutch size in anoles and geckos has evolved as an adaptation for climbing (Andrews and Rand, 1974).

27. A rare phenomenon in vertebrates that is also displayed by some primates, including humans (Jones et al., 1997). In part because of this parallel, the hormonal basis of alternating ovulation has been well studied in anoles and, as in humans, involves complex regulation via interaction between hormones from the hypothalamus, pituitary gland, and ovaries (Jones et al., 1983, 1997).

28. Some rocky islands edged with high vertical sides rather than beaches do not have anoles, presumably because these islands are difficult to colonize (Lazell, 1999).

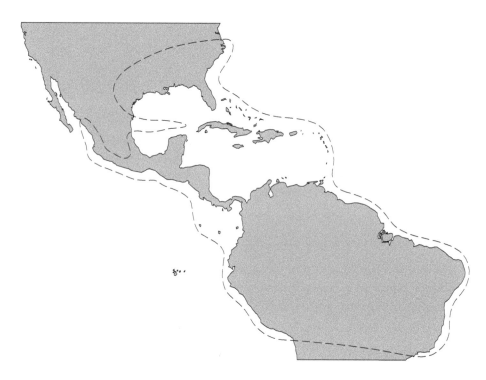

FIGURE 2.7
Geographic distribution of *Anolis*. Redrawn from Williams (1977).

several Pacific islands.[29] In addition to this natural distribution, the geographic span of *Anolis* has expanded in recent years thanks to human action to include Bermuda, the Hawaiian Islands, the Marianas, Taiwan, and the Bonin islands near Japan[30] (Wingate, 1965; Hunsaker and Breese, 1967; McCoid, 1993; Suzuki and Nagoshi, 1999; Norval et al., 2002).

At last count, approximately 361 species of *Anolis* are recognized as valid (Nicholson et al., 2005) and new species are described every year. Some new species are the result of new discoveries from less-explored areas, whereas others are the result of splitting previously described taxa into multiple species. Although some of this splitting results from morphological studies, most stems from molecular work (e.g., Glor et al., 2003, 2004), and this trend may increase greatly in the near future (a point upon which I expand in Chapter 14). Although the lion's share of research has been conducted in the West Indies, more species actually occur on the mainland (approximately 210 versus 151).

29. E.g., Cocos, Gorgona, Malpelo.
30. A. *sagrei* has also been spotted in a lawyer's office in Boston and on an airplane out of Denver, among other places. In many cases, the source of introductions may be nursery plants, which contain either lizard stowaways or their eggs (Meshaka et al., 2004).

Within the West Indies, the number of species on an island is roughly related to island area (discussed at greater length in Chapter 4). Cuba, the largest island, has 63 species, and Hispaniola, the second largest, has 41. At a more local scale, the maximum number of species that occur sympatrically also scales with island size, to a maximum of 15 in Cuba (Díaz et al., 1998), or possibly even more (Garrido and Hedges, 2001). Comparable figures for the mainland are less certain, but at least 11 species occur at Los Tuxtlas, Mexico (Vogt et al., 1997) and possibly as many as 15 at a site in Panama (S. Poe, pers. comm.).

THE *ANOLIS* SPECIES CONCEPT

Discussion of species richness and diversity naturally leads to the issue of species concepts. Few topics in evolutionary biology are more contentious than the questions of what constitutes a species and how species should be characterized and identified. Most evolutionary biologists would agree that, in theory, the term "species" should be applied to those clades which follow their own historical path, evolving independently of other clades (Simpson, 1951; Wiley, 1978; de Queiroz, 2007). In practice, however, identifying such clades and understanding the processes that maintain their evolutionary independence is not straightforward.

For many years, the Biological Species Concept (BSC)—which defines a species as "groups of populations that actually or potentially interbreed with each other" (Mayr, 1963)—was widely accepted, especially among zoologists. Species could be identified on the basis of whether they could—or would—interbreed, and their evolutionary independence from other species was understood to be a result of their lack of genetic interchange. Recently, however, the BSC has steadily lost support (for a vigorous defense, see Coyne and Orr, 2004).

Although many arguments have been made against the BSC, two are primary. First, many sympatric populations engage in substantial amounts of genetic exchange and yet still exist as distinct biological units which merit the designation "species"— i.e., they remain phenotypically and ecologically distinct through time and evolve along independent trajectories (de Queiroz, 2005, 2007; Wake, 2006). Because of the hybridization that regularly occurs among plant species, botanists have rarely been enthusiastic about the BSC (e.g., Whittemore 1993; but see Rieseberg et al., 2006). What has been surprising in recent years, however, is the extent to which animal species also engage in hybridization (e.g., Grant and Grant, 1992, 1996; Arnold, 1997). Although by no means the rule, hybridization among animal species is common enough to call into question, at least among some scientists, whether reproductive isolation is the only criterion for species definition in animals. In the place of the BSC, a wide variety of alternatives has been suggested (for recent reviews on this topic and an entrée into the literature, see Coyne and Orr [2004] and de Queiroz [2005, 2007]).[31]

---

31. My goal here is not to review the many different species concepts currently under discussion, but to focus on the BSC and its application to anoles.

The second criticism of the BSC is a combination of old and new critiques. The old critique is that the BSC is neither universal nor operational in many situations. One such situation occurs among asexual taxa for which the concept of reproductive isolation has no meaning. More pertinent here is a second situation: allopatry. Deciding whether populations that do not occur together are reproductively isolated is in most cases impossible (the words "or potentially" were inserted into the BSC for just this purpose). Captive breeding experiments, the obvious answer, are fallible: many species will mate in captivity, but exist side-by-side in nature without hybridizing. For this reason, assessing whether two allopatric populations are species under the BSC becomes an untestable judgment call.

This critique has been updated in recent years as follows: judgments about whether allopatric forms would be reproductively isolated if they ever came into contact are projections about what might happen in future situations. Such an approach, it is argued, is inherently uncertain and subjective. Wouldn't it be better to instead base a species concept on events which are more certain, namely, what has happened in the past (Frost and Hillis, 1990)? This viewpoint has led to what has become in recent years the most popular alternative to the BSC, the phylogenetic species concept (PSC). Although there are, in fact, many PSC variants (Baum and Donoghue, 1995; Coyne and Orr, 2004; de Queiroz, 2005, 2007), they generally share the underlying theme that species should be defined on the basis of historical relationships of taxa or the phylogenetic distribution of their characters, rather than by consideration of the ongoing processes affecting populations. PSCs often, though not always, require that species be monophyletic—i.e., that all populations within a species are more closely related to each other than to any population not in that species.

SPECIES RECOGNITION AND REPRODUCTIVE ISOLATION IN ANOLES

Despite these criticisms, an emphasis on reproductive isolation is an appropriate framework for consideration of the nature of anole species. Although the number of pairs of anole species that coexist in sympatry is vast, hybridization occurs extremely rarely. To my knowledge, only eight species pairs have been suggested to hybridize (Losos, 2004),[32] and the data indicate that perhaps only two of these cases result in genetic introgression (hybrids being unknown or sterile in the other six). Other undetected instances of hybridization among anoles probably exist, but given the intense fieldwork conducted on anoles over the past four decades, that number is probably small. As a nearly universal rule, sympatric anole species are reproductively isolated.

---

32. Six of these cases were discussed in Losos (2004). A seventh case involves *A. cybotes* and *A. armouri* in Haiti, but the data are inconclusive about whether hybridization, much less introgression, occurs (Schwartz, 1989). In addition, Campbell (2000) cites two reports of *A. sagrei*–*A. carolinensis* couplings, but hybrids have never been reported.

The means by which anoles distinguish conspecifics from heterospecifics is well understood. As a result, for anoles a non-arbitrary way exists for assessing whether allopatric populations should be considered conspecific, thus making the criterion of reproductive isolation applicable to such situations.

Anoles have two species-recognition mechanisms. The first is the dewlap. As previously mentioned, dewlaps vary in color, pattern, and size. The observation that sympatric species almost invariably differ in at least one of these three attributes (Fig. 2.8) has led to the idea that anoles use the dewlap as a species-recognition cue (Rand and Williams, 1970; Williams and Rand, 1977; Losos and Chu, 1998; Nicholson et al., 2007). This hypothesis is supported by an experimental study that examined interspecific aggression between males of *A. cybotes* and *A. marcanoi*, two closely related species that occur sympatrically and differ in dewlap color.[33] When unaltered males of the two species were placed together, they for the most part ignored each other. However, when the dewlap of both males was altered such that the *A. marcanoi* had a white, *cybotes*-like dewlap, and vice versa (such that each male encountered a heterospecific sporting a conspecific dewlap), levels of aggressive behavior were significantly higher (Losos, 1985a).

Anoles display not only by extending their dewlaps, but also by moving their heads up and down in a rhythmic bobbing motion.[34] The cadence of the display—e.g., how long the head is held up and how high, how quickly one bob follows another, how many total bobs—is usually species-specific (Fig. 2.9; Jenssen 1977, 1978).[35] The species-recognition significance of headbob displays was examined first in the pre-video days on *A. nebulosus*, which responded more to unaltered film footage of a male displaying than to footage edited to change the male's head-bobbing display (Jenssen, 1970a). In a similar vein, a more recent comparative study found that interspecific differences in head-bobbing patterns are related to the number of sympatric congeners with which a species occurs (Ord and Martins, 2006). Recent studies have demonstrated that anoles will not only watch TV, but respond to videos of lizards displaying (Macedonia and Stamps, 1994; Macedonia et al., 1994; Clark et al., 1997; Yang et al., 2001). Studies using video technology may be a promising way to further investigate how changes in head-bobbing patterns affect species-recognition.[36]

Because visual signals play an important role in anole species recognition, allopatric populations can be evaluated readily for conspecificity: populations differing in dewlap

33. Presumably, species recognition in male-female encounters uses the same signals as used in male-male aggressive interactions.

34. As well as using a variety of other movements and visual signals, discussed in Chapter 9.

35. The two species known to not have stereotyped, species-specific head-bobbing displays are *A. opalinus* (Jenssen, 1979a) and A. Chamaelinorops *barbouri* (Jenssen and Feely, 1991). Why their lack of stereotypy has evolved is unknown.

36. Initially, researchers thought video displays could also be used to examine the role of color in behavior. However, because other organisms have visual sensitivities different from ours, they may perceive color on a video monitor differently than we do. This realization has considerably lessened enthusiasm for using video playbacks to study the role of color in behavior and evolution (Fleishman et al., 1998). An alternative approach to studying both color and head-bobbing patterns is the use of programmable robotic lizards (Martins et al., 2005; Ord and Stamps, 2008).

FIGURE 2.8 A-D
Variation in the dewlaps of sympatric anoles. These four species of trunk-ground anoles co-occur at Soroa in western Cuba. (a) *A. allogus*; (b) *A. homolechis*; (c) *A. mestrei*; (d) *a. sagrei*. Photo of *A. sagrei* courtesy of Richard Glor.

design or head-bobbing pattern are likely to be reproductively isolated. By the same token, evolutionary divergence in these traits may be a key part of the speciation process, a point to which I will return in Chapter 14.

Although allopatric populations differing in dewlap color or head-bobbing pattern are likely to be reproductively isolated, the converse is not necessarily true: individuals from populations that are indistinguishable in dewlap and head-bobbing pattern might not mate and produce fully fertile offspring, even if they had the opportunity. Other species-recognition cues—perhaps body size or body coloration, for example—may exist (Williams and Rand, 1977). In addition, populations may be reproductively isolated due to the evolution of post-mating barriers (e.g., developmental breakdown, sterility), even in the absence of pre-mating barriers. We currently have no idea about the frequency of post-mating reproductive isolation in anoles and how common this situation may be; however, in several reported cases of hybridization in *Anolis*, hybrids appear to be sterile (Losos, 2004).[37]

Thus, to the extent that post-mating reproductive isolation evolves in allopatric populations without concomitant evolution of species-recognition signals, comparison of the species-recognition signals of allopatric populations may fail to identify situations

37. Note that in my 2004 paper, I overlooked evidence (Gorman and Yang, 1975, contra Gorman et al., 1971) that some fertile backcrosses may exist between *A. aeneus* and *A. trinitatis* on Trinidad, where neither is native.

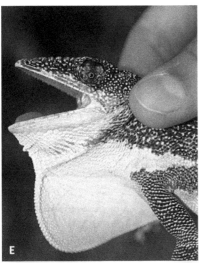

FIGURE 2.8 E-H

The other species at Soroa include (e) crown-giant
A. luteogularis. Photo courtesy of Veronika Holanova;
(f) trunk-crown *A. porcatus.* Photo courtesy Richard
Glor; (g) twig *A. angusticeps*; and (h) *A. Chamaeleolis
barbatus.* Photo courtesy of Veronika Holanova.
Three others species that occur at Soroa are the
grass-bush *A. alutaceus* (solid yellow dewlap), the
trunk *A. loysianus* (yellow and red dewlap), and the
aquatic anole *A. vermiculatus*, which does not have a
dewlap. Some of these photos were not taken at
Soroa.

in which populations have diverged to the point at which they would constitute reproductively isolated species. However, in contrast to the point made above, this situation
is amenable to experimental investigation in the laboratory: whether populations are
reproductively isolated by post-mating mechanisms can be investigated by placing
individuals together and seeing whether mating (if it occurs) leads to the production of
fertile offspring.[38]

38. One aspect of post-mating isolation—the production of hybrid offspring that are ecologically unfit—
cannot be evaluated in the lab. However, failure to produce surviving offspring or the production of sterile
offspring would be strong evidence of post-mating isolation. For these reasons, laboratory studies cannot
disprove the existence of pre-mating reproductive isolation, because individuals may mate in the lab even when
they wouldn't in nature, but they can confirm the existence of most types of post-mating isolation.

A. limifrons

A. townsendi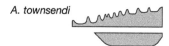

A. sericeus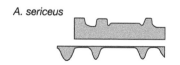

FIGURE 2.9
Headbobbing patterns of three species. The graphs—called
display-action-pattern graphs—illustrate the temporal pattern
of head bobbing (top, height indicates amplitude of head
displacement) and dewlap extension (bottom). Modified with
permission from Jenssen (1977).

PHYLOGENETIC RELATIONSHIPS AND REPRODUCTIVE ISOLATION

How useful is the application of phylogenetic information in delimiting species boundaries in *Anolis*? In some cases, the discovery of previously unsuspected phylogenetic differentiation can suggest that genetic exchange is not occurring among populations that occur in sympatry or parapatry; thus, the newly recognized clades may be reproductively isolated, a situation we will encounter in Chapter 14.

On the other hand, species defined by reproductive isolation may not always be monophyletic. It is easy to imagine situations in which one population evolves a different dewlap color for some reason, and thus becomes reproductively isolated from the remaining populations in that species. The remaining populations, however, would not form a monophyletic group and thus would not qualify as a species under many PSCs, even though they might still be fully capable, and even in the process, of exchanging genes (Harrison, 1998; Templeton, 1998; Coyne and Orr, 2004).

Such situations may occur relatively frequently in *Anolis*. Many geographically widespread species are not monophyletic because other species that are morphologically distinctive and reproductively isolated are phylogenetically nested within them. The blue-dewlapped *A. conspersus* of Grand Cayman (Jackman et al., 2002), for example, arose from within the yellow-dewlapped *A. grahami* of Jamaica. Similar examples are found in a variety of other species, including *A. porcatus* (Glor et al., 2005) and *A. cybotes* (Glor et al., 2003).

This situation highlights the difference between considerations of evolutionary process and historical pattern. When ongoing processes link together a paraphyletic group of populations, then species demarcations based on phylogenetic relationships will not be informative with regard to evolutionary processes. Given that sympatric anole species almost always are reproductively isolated and that objective criteria exist to estimate how likely allopatric populations are to be reproductively isolated, I prefer to emphasize the criterion of reproductive isolation in delimiting species of anoles, with the recognition that phylogenetic analysis provides insights concerning the history of populations and species and sometimes can identify clades that may represent previously undetected species.

FUTURE DIRECTIONS

The evolution of reproductive isolation in anoles is a topic that has received little direct study. Further research is required to establish how distinct dewlaps or head-bobbing patterns must be to cause reproductive isolation. Video and robotic techniques may be useful to address these questions, in the laboratory or even in the field (e.g., Clark et al., 1997; Fig. 2.10). More generally, relatively little work has been conducted on the head-bobbing patterns of different species, much less of conspecific populations, in the last two decades (reviewed in Ord and Martins, 2006). For this reason, we have very little idea whether many situations exist in which conspecific populations differ little in

FIGURE 2.10

A male *A. gundlachi* displays to a latex robot cast from a specimen of the same species and programmed to move its head in the *A. gundlachi* species-specific pattern (see Ord and Stamps, 2008). Photo courtesy of Terry Ord.

dewlap color, but greatly in head-bobbing pattern. By the same token, the comparative data base on dewlap variation, measured using spectrophotometric methods, is relatively small and mostly interspecific (Fleishman, 2000; but see Leal and Fleishman, 2004). Most generally, a detailed study of the extent to which dewlaps, head-bobbing patterns, and post-mating reproductive isolation coevolve, comparable to the many studies on other taxa that have investigated the evolution of pre- and post-mating isolation (Coyne and Orr, 2004), would be very interesting.

3

3

FIVE ANOLE FAUNAS, PART ONE

Greater Antillean Ecomorphs

In this and the next chapter, I break anole diversity into five groups, corresponding mostly to the anoles of different regions. "Fauna" is used loosely, as two of these faunas co-occur, and another fauna extends over the majority of the geographic distribution of these lizards. The rationale for this dissection is that these faunas exhibit different patterns of ecological and evolutionary diversity and consequently illuminate different phenomena. Moreover, the amount of study devoted to the faunas varies tremendously; as a result, much of this book will focus on the first of these faunas, the Greater Antillean ecomorphs, which are the subject of this chapter. The remaining four faunas will be discussed in Chapter 4.

GREATER ANTILLEAN ECOMORPHS

The ecomorph story was introduced in the prologue. Put simply: the same set of habitat specialists co-occur in communities throughout the Greater Antilles (Fig. 3.1). Williams (1972) coined the term "ecomorph" to refer to these habitat specialists (Fig. 3.2). A brief history of the study of *Anolis* ecological morphology is presented in Appendix 3.1 at the end of this chapter.

Williams' (1972, p. 72) definition of ecomorph: "species with the same structural habitat/niche, similar in morphology and behavior, but not necessarily close phyletically," has several components. In particular, the definition indicates that to constitute an ecomorph class, a set of species must share similarities in morphology, ecology, and

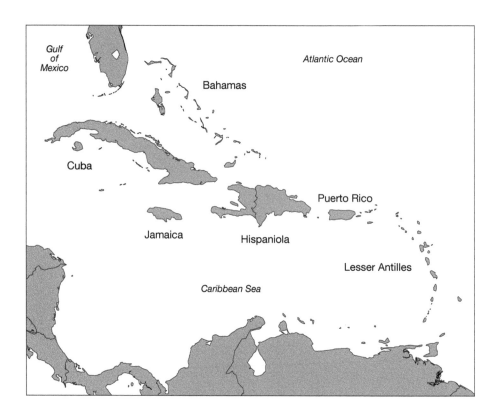

FIGURE 3.1

The West Indies. The Greater Antilles are the islands of Cuba, Hispaniola, Jamaica, and Puerto Rico and nearby smaller islands.

behavior, and these similarities must be independently derived.[39] In recent years, the term "ecomorph" has been widely applied to many types of organisms (see Appendix 3.1); however, most such designations are made only on the basis of similarity in morphology or ecology, and often without quantitative analysis. Williams' ecomorph concept is more elaborate than mere convergence; it is the idea that groups of species are recognizable as discrete and distinct entities that differ in coordinated aspects of their biology, encompassing behavior, ecology, and morphology.

Before getting into the gory statistical details concerning the existence and recognition of the anole ecomorph classes, I'll begin with a brief description of their key morphological, ecological, and behavioral attributes (summarized even more briefly in

---

39. Technically, distantly related taxa can share similarities as a result of retaining the ancestral condition, rather than from convergent evolution. However, because the ancestral anole could have been a member of only one ecomorph class (e.g., it couldn't have been both a grass-bush and a twig anole), the recognition of multiple different ecomorph classes—such as the six *Anolis* ecomorphs—implies that the similarity of species in most of the different ecomorph classes must have resulted from convergence.

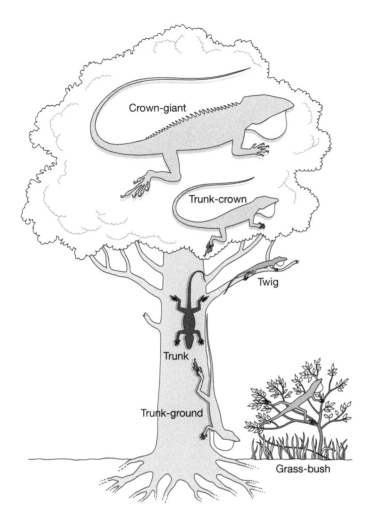

FIGURE 3.2
The ecomorphs.

Table 3.1 and illustrated for morphology in Fig. 3.3).[40] Ecomorph designations refer to the structural microhabitat[41] in which members of each ecomorph class are normally found.[42] The afterword at the end of the book provides a list of all West Indian species, including the ecomorph designations of Greater Antillean species.

40. In the table, the lower size range for crown-giants is not based on Schwartz and Hendersoni (1991) because the size they reported for *A. cuvieri:* is substantially underestimated (e.g., Losos et al., 1990).

41. "Microhabitat" refers to the attributes of the subset of the habitat used by a species. "Structural microhabitat" refers to the characteristics of the structures—e.g., trunks, branches, leaves—in the parts of the habitat a species uses (Rand, 1964a).

42. Note that these designations refer to usual structural microhabitat use of a species and do not imply that ecomorph species are exclusively found in their designated location. Ernest Williams liked to tell the story of a now well known biologist who became concerned (more accurately: freaked out) when, on a field trip, a crown-giant anole was discovered on the ground. The occasional nonconformist anole notwithstanding, field studies always clearly indicate that species in the different ecomorph classes use different parts of the structural habitat (e.g., Rand, 1964a, 1967a; Schoener and Schoener, 1971a,b; Moermond, 1979a,b; Losos, 1990c).

TABLE 3.1 Ecomorph Characteristics

Ecomorph	Body size (maximum SVL, in mm)	Limb length	Number of lamellae on toepads	Tail length	Color	Structural microhabitat	Movement rate	Type of movement
Crown-giant	Large (130–191)	Short	Intermediate	Long	Usually green	High trunks and branches	Low	Walks and runs
Grass-bush	Small (33–51)	Long hindlimbs	Intermediate	Very Long	Brown, lateral stripe	Low, narrow supports	Low	Jumps and runs
Trunk	Small (40–58)	Intermediate, relatively even ratio of fore- limbs to hindlimbs	Intermediate	Short	Gray	Trunks	High	Runs
Trunk-crown	Small to intermediate (44–84)	Short	Many	Long	Green	Trunks, branches, leaves, eye level to high	High	Walks and runs
Trunk-ground	Intermediate (55–79)	Long hindlimbs	Intermediate	Long	Brown	Broad, low surfaces	Low	Runs and jumps
Twig	Small to intermediate (41–80)	Very short	Few	Short	Gray	Narrow supports	High	Walks

NOTE: Morphological data other than SVL are relative to body size (Beuttell and Losos, 1999). Movement data from Losos (1990c), Irschick and Losos (1996) and Losos (unpubl.).

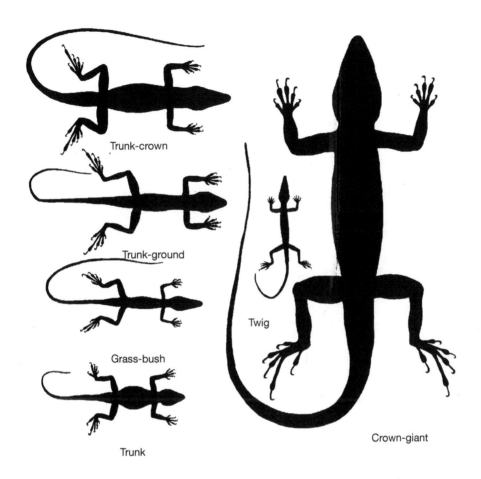

Trunk-crown

Trunk-ground

Grass-bush

Trunk

Twig

Crown-giant

FIGURE 3.3

Silhouettes of common ecomorph species of Hispaniola: the large lizard is the crown-giant, *A. ricordii*, the small one next to it is the twig anole, *A. insolitus;* the remainder, in descending order of size are trunk-crown, *A. chlorocyanus;* trunk-ground, *A. cybotes;* grass-bush, *A. bahorucoensis;* and trunk, *A. distichus.* This image was drawn from photographs of museum specimens with some slight adjustments made to correct for preservation effects.

TRUNK-GROUND ANOLES

Trunk-ground anoles are medium-sized species typically observed within a meter and a half of the ground on broad surfaces: usually tree trunks, but also walls (rock or human-made), boulders or other such objects (Fig. 3.4).[43] Often, they perch head downward,

43. A few species extensively use rocky surfaces as well as trees (e.g., *A. longitibialis, A. mestrei, A. imias, A. guafe*). These species are all closely related and morphometrically similar to more standard trunk-ground anoles. Because rock walls and tree trunks are similar in terms of the functional demands they make on lizards, I treat them all as trunk-ground anoles rather than subdividing the trunk-ground category. Some morphological differences do exist, however; for example, like rock-dwelling lizards in other genera (Revell et al., 2007b), rock-dwelling anoles have particularly long legs (Glor et al., 2003).

FIGURE 3.4

Trunk-ground anoles. (a) *A. rubribarbus,* Cuba. Photo courtesy of Richard Glor. (b) *A. cybotes,* Hispaniola; (c) *A. cristatellus,* Puerto Rico; (d) *A. lineatopus,* Jamaica.

surveying the ground. From this position, they will rapidly descend, either by foot or air, to capture prey or interact with a conspecific. Males use these prominent perches both to advertise their presence by displaying frequently, as well as to spot prey, which they often capture by a quick dash to the ground.

Trunk-ground anoles are generally a dark color, ranging from light brown to darker brown or olive. They are stocky, muscular lizards with long hindlimbs and poorly developed toepads.[44] The tail is moderately long[45] and the dewlap is usually large. Trunk-ground anoles are the most visible and seemingly the most abundant anole at most localities.

TRUNK-CROWN ANOLES

Trunk-crown anoles are wide-ranging arboreal species. They are typically found from eye level to the top of the canopy and occur regularly on the full spectrum of surface diameters, from tree trunks to narrow twigs. In addition, they regularly occur on leaves and other vegetation. Trunk-crown anoles travel over moderately large three-dimensional areas. They move relatively frequently and use both sit-and-wait and actively searching foraging modes.

Almost all trunk-crown anoles are green, some quite beautifully so, and several species are to some extent blue (Fig. 3.5). All can change color to a dark shade of brown, and one species, *A. brunneus*, has lost its verdancy entirely and can only shift in color from a light grayish brown to almost black. Trunk-crown anoles have short legs and a slender body shape, with a long snout. The toepads are extremely well developed and the tail is usually, but not always, long.

Trunk-crown anoles are often very abundant and visible, occasionally rivaling sympatric trunk-ground anoles in these regards, particularly in more open habitats. In forests, the abundance of trunk-crown species is probably underestimated because they frequently are so high up that they can be hard to see from the ground. For example, in Puerto Rico the trunk-crown *A. stratulus* was thought to be relatively uncommon until the construction of a canopy walkway at the El Verde Field Station revealed that it is extraordinarily abundant in the treetops (Reagan, 1992).

TRUNK ANOLES

Trunk anoles occur on broad tree trunks. They occur between, and overlap with, trunk-ground and trunk-crown anoles. However, unlike the former, they very rarely venture onto the ground, and unlike the latter, they do not often go out onto narrower branches or into the vegetation. Rather, they mostly stick to the trunk itself, moving up and down and round and round.

Trunk anoles only occur on the two largest Greater Antillean islands,[46] and the Hispaniolan species are more thoroughly studied, both because they are much more

44. When referring to the size of morphological attributes, descriptions are implicitly expressed relative to overall body size unless otherwise indicated.

45. Descriptions of tail length refer to unregenerated tails.

46. In addition, the Hispaniolan trunk anole *A. distichus* occurs naturally in the Bahamas and also in Florida, where some populations may be descended from natural colonists from the Bahamas, but most are the result of human introductions (Wilson and Porras, 1983; Meshaka et al., 2004).

accessible to American herpetologists and because they are much more abundant than the seemingly uncommon Cuban species (Rodríguez Schettino, 1999). Trunk anoles are fairly active, making many short movements;[47] the diet of Hispaniolan trunk anoles consists of ants to a greater extent than most anoles (little is known of diet of the Cuban trunk anole).

Trunk anoles are relatively small (Fig. 3.6). Their most obvious feature is a flattened body, with legs splayed more laterally than the legs of most anoles. Although their limbs are neither particularly long nor short compared to the limbs of other ecomorphs, trunk anoles have the longest forelimbs relative to the length of their hindlimbs, and also large forefoot toepads relative to the size of the pads on their hindfeet. These anoles have short tails and usually sport a grayish hue which blends in well on light-colored tree trunks.

47. The Hispaniolan species use an unusual form of locomotion consisting of short, spasmosdic hops in which an individual jerks forward a few centimeters, pauses briefly, and then jerks forward again, sometimes continuing for great distances (Mattingly and Jayne, 2004, 2005; Losos, unpubl.).

FIGURE 3.5
Trunk-crown anoles. (a) *A. allisoni,*
Cuba; (b) *A. chlorocyanus,* Hispaniola; (c) *A. evermanni,* Puerto Rico;
though usually found high in the
trees, in the Luquillo Mountains,
A. evermanni forages on sunny
boulders in the middle of streams;
(d) *A. grahami,* Jamaica.

CROWN GIANTS

The most obvious feature of crown giants is their size. The largest anole species—the
Cuban crown giant *A. luteogularis*—may reach a total length of well over one half a
meter, perhaps not quite dinosaurian, but menacing enough in an anole world in which
most species are 1/3 this long or shorter (and 1/20th in bulk).

Ecologically, these species do not differ much from trunk-crown anoles. They tend
to be found high in trees, usually on trunks or branches, and they use narrower
branches and leafy vegetation less than trunk-crown anoles. Probably as a result of
their size, they have the most catholic diet among ecomorphs, adding fruits and vertebrates up to the size of small birds to the standard insect fare (Dalrymple, 1980;
Meshaka et al., 2004). Similarly, their home ranges appear much larger than those of
other ecomorph species (Losos et al., 1990). Although they cover a lot of ground, moving from one tree to another by way of their interconnected canopies, they generally do
not do so rapidly.

Morphologically, crown giants are in some respects super-sized trunk-crown anoles:
they are generally green, but can change to a dark brown; their toepads are large, and

FIGURE 3.6
Trunk anoles. (a) *A. loysianus*, Cuba; (b) *A. distichus*, Hispaniola.

their limbs are moderately short (Fig. 3.7). In one respect, however, most crown giants differ substantially from trunk-crown anoles: the shape of the head. Most crown giants[48] have a massive, casqued head, a feature shared by some other large anoles. In addition, most crown-giants have a spiky crest running down their backs.

GRASS-BUSH ANOLES

Absent from Jamaica, grass-bush anoles are found on narrow vegetation near the ground, primarily grass stems and other low-lying vegetation, as well as on bushes and small tree trunks. They are agile lizards adept at moving through cluttered spaces. In some areas, particularly open grassy expanses, grass-bush anoles can occur at extremely high densities. Territories appear to be small and foraging conforms to the sit-and-wait mode. Grass-bush anoles often move by taking many short hops.

As would be expected given their structural microhabitat use, grass-bush anoles are always small. They are slender lizards with long hindlimbs, short forelimbs, and a long, narrow head. The toepads are poorly developed. Their most obvious feature, however, is their extremely long tail, which in some species can be four times the length of the body. Most grass-bush anoles are yellow and brown in color, with a light lateral stripe (Fig. 3.8).

48. *A. garmani* is the only exception.

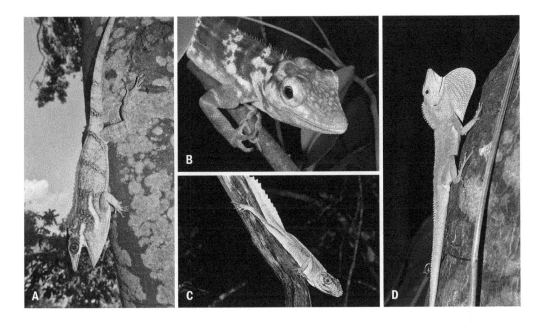

FIGURE 3.7
Crown giants. (a) *A. Smallwoodi*, Cuba. Photo courtesy of Veronika Holanova; (b) *A. baleatus*, Hispaniola. Photo courtesy of Rick Stanley; (c) *A. cuvieri*, Puerto Rico; (d) *A. garmani*, Jamaica.

TWIG ANOLES

Twig anoles are the most extreme of the ecomorphs in just about every respect. Ecologically, they use narrow surfaces more frequently than most other anoles.[49] Behaviorally, they are active searchers, often moving steadily at low speed for extended periods. They search for prey by moving slowly on narrow surfaces, investigating holes, cracks, leaves and other places in which prey may be hidden (Fig. 3.9). Light grey with a mottled pattern, twig anoles rely on crypsis for predator avoidance. Upon spotting a potential threat, they move to the opposite side of a branch and slowly creep away. Only if directly threatened will they attempt to flee by running or jumping.

Morphologically, twig anoles are also extreme. They have very slender bodies with long pointed snouts and extremely short limbs and tails (Fig. 3.10). In many species, the tail seems to be weakly prehensile,[50] and they also have a tendency to have large scales on their heads.

49. Some grass-bush and trunk-crown anoles use narrow surfaces frequently, at least in some habitats (Mattingly and Jayne, 2004). Mattingly and Jayne (2004) painstakingly measured vegetation structure and pointed out that narrow surfaces actually predominate in terms of availability in the environment.

50. Though this capability is poorly documented in the literature.

FIGURE 3.8
Grass-bush anoles. (a) *A. vanidicus*, Cuba. Photo courtesy of Kevin de Queiroz; (b) *A. olssoni*, Hispaniola; (c) *A. pulchellus*, Puerto Rico.

TESTING THE HYPOTHESIS OF THE EXISTENCE OF DISCRETE ECOMORPH CLASSES

It's one thing to assert the idea that ecomorphs exist and quite another to demonstrate it statistically. Although the idea of ecomorphs is now commonly applied in many different taxonomic groups,[51] quantitative morphological and ecological analysis supporting such a designation is still rare; such tests require investigating whether ecomorphs form

51. Prior to 1990, the term "ecomorph" was used primarily in reference to *Anolis* (see footnote 70 in the appendix at the end of this chapter for discussion of the history of the term); however, that has changed in recent years. A recent Google Scholar search found the term used in reference to ants, spiders, fish, bats, corals, and algae, and that was only on the first page of search results! Page two added crocodiles, badgers, rabbits, and pine trees.

FIGURE 3.9

Successful hunting by a twig anole. I will never forget observing this *A. valencienni* foraging in a concrete trash repository at the Discovery Bay Marine Laboratory in Jamaica. The lizard moved from one crack to the next, sticking its head in and apparently looking for concealed prey. Sure enough, it emerged from one crack with a large cockroach in its mouth.

discrete morphological clusters and whether these clusters also differ ecologically and behaviorally.

MORPHOLOGY

Two approaches have been taken to test the morphological component of the anole ecomorph hypothesis. First, a discriminant function analysis (DFA)[52] was performed on a data set comprised of morphological traits that are relevant to structural microhabitat use (see Chapter 13): radiological measurements of all limb elements plus external measurements of tail length, SVL, mass, and, for four toes, pad size and lamella number (Beuttell and Losos, 1999). Data were gathered from 32 species representing all six ecomorphs.[53] The DFA was highly significant (P less than 0.001), and all 32 species were classified a posteriori to the correct ecomorph class with probability of 1.0.

Because species are assigned to groups *a priori*, DFA is a good means of investigating whether previously established groups can be distinguished based on some combination of characters, but it is inappropriate as a means of asking whether those characters would produce those groupings if all variables were considered equally. In fact, DFA can

---

52. DFA constructs a series of linear equations that maximizes the separation of *a priori* defined groups by differential weighting of the variables.

53. Here and in the remainder of the analyses in this chapter, the data were collected only from males. The decision to focus data collection on males was made both on logistical grounds (male anoles are substantially easier to find, observe and capture than females) and because in some respects (e.g., body size [Butler et al., 2000]) differences among ecomorphs are greater for males than for females. Recent work, however, indicates that the same ecomorph categories apply to females as well as to males (Butler et al., 2007).

FIGURE 3.10

Twig anoles. (a) *A. insolitus*, Hispaniola;
(b) *A. angusticeps*, Cuba. Photo of *A. insolitus*
courtesy of Kevin de Queiroz; (c) *A. valencienni*, Jamaica. Photo courtesy of Kevin de
Queiroz; (d) *A. occultus*, Puerto Rico. Photo
courtesy of Alejandro Sanchez.

C C H J P J C C H C P C H H C C P P H P C C H H H J P H J P C H H C C C H H P H C C C J H P

Crown-giant
Grass-bush
Trunk
Trunk-crown
Trunk-ground
Twig

FIGURE 3.11

UPGMA phenogram indicating that species cluster by ecomorph in a multidimensional morphological space. Letters indicate island of origin (Cuba, Hispaniola, Jamaica, Puerto Rico). The study was based on external measurements from 46 species, including at least one representative from each ecomorph class on each island on which it occurs. Randomizing species identity across the phenogram revealed that such clustering is extremely unlikely to have occurred by chance (P ≪ 0.0001). Figure re-drawn from Losos et al. (1998) with permission.

find statistical support for particular groupings, even if most of the characters (at the extreme, all but one) do not differentiate the groups or even support very different groupings (Klecka, 1980). In other words, DFA confirms that the ecomorphs can be distinguished based on a set of morphological characters, but does not demonstrate that the characters would produce the observed ecomorph groupings in the absence of *a priori* categorization.

For this reason, the second approach for testing the ecomorph hypothesis asks whether ecomorph groupings are recovered when variables are weighted equally, thus avoiding effects of *a priori* categorization. One common means of visualizing the relative position of points in a multivariate space is to construct a similarity phenogram using the unweighted pair group method with arithmetic means (UPGMA).[54] Two different studies (Losos et al., 1998; Beuttell and Losos, 1999) have used UPGMA to summarize the position of anole species in morphometric space. Using different methods and sets of species that overlapped to a moderate extent, both studies revealed perfect clustering by ecomorph class (Fig. 3.11).

The UPGMA method has its own shortcomings, however, because it represents the position of species as a nested hierarchy of groups, which may distort the actual multidimensional position of the species (Sneath and Sokal, 1973; de Queiroz and Good,

54. UPGMA phenograms are constructed by joining the two points separated by the smallest Euclidean distance in a phenogram; these two points are replaced by their average, and again the two closest remaining points are joined, and so on, until all points have been connected into a single, bifurcating phenogram.

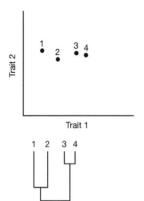

FIGURE 3.12

The problem with UPGMA phenograms. If species in a morpho-logical space were distributed as in (a), a UPGMA analysis would produce a phenogram like the one shown in (b), which would fail to indicate that species 2 and 3 occupy intermediate positions.

1997). To see why this is, consider a situation in which objects do not form discrete clusters (Fig. 3.12). UPGMA will portray the intermediate points as clustering with whichever of the other points is slightly closer. As a result, what may in reality be a continuum will be represented as a set of groups, with the implication that all members of one group are equidistant from all members of other groups; in such a hierarchical rendering, it is not possible to indicate that one object is intermediate between others.

For this reason, the results of UPGMA analysis are best thought of as a first pass, useful for heuristic purposes, but requiring subsequent corroboration. To this end, we examined the actual Euclidean distances separating each pair of species in multivariate space. In both studies, all species had as their nearest neighbor in morphological space another member of the same ecomorph class (with one exception[55]). Moreover, all species are closer to the centroid for their own ecomorph class than they are to the centroid for any other ecomorph.[56]

Taken together, these analyses provide strong corroboration of the hypothesized existence of discrete ecomorph classes that occupy distinct regions of morphological space. The nearest neighbor analysis reveals that ecomorph species are not uniformly distributed throughout morphological space; that is, all species are closest in morphological space to a species that is a member of their own ecomorph class. Combined with the perfect clustering revealed by the UPGMA analysis, this observation suggests that the boundaries of the ecomorph classes correspond with gaps in the occupation of morphological space.

55. In the Beuttell and Losos (1999) study, the Hispaniolan trunk-crown A. aliniger and the Hispaniolan twig A. darlingtoni were slightly closer to each other than each was to a member of its own ecomorph class. This result is unexpected, because both species appear morphologically to be typical members of their ecomorph classes and A. aliniger exhibits habitat use typical of trunk-crown anoles (Williams, 1965; Rand and Williams, 1969; Losos, unpubl.). Ecological data are not available for A. darlingtoni, which is from Haiti and has not been collected frequently (Thomas and Hedges, 1991). Unlike most species in these analyses, our data for A. darlingtoni came from a single specimen, which may have been a source of error.

56. The centroid is calculated as the mean position of all members of that ecomorph class in multivariate space. This analysis was only conducted in the Beuttell and Losos (1999) study, which was a considerably more in-depth morphometric analysis than Losos et al. (1998).

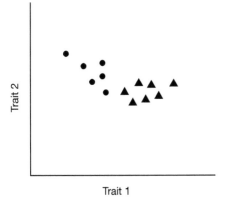

FIGURE 3.13
Cartoon representation of ecomorph clusters. Although two discrete clusters exist, some species in each cluster are closer to each other than they are to extreme members of their own cluster.

Nonetheless, the extent of ecomorph convergence should not be overstated. In particular, ecomorph classes are not so distinct that each species is more similar to all species in its own class than it is to any species from another class. Rather, in many cases a species is more similar in morphology to some members of another ecomorph class than it is to some members of its own class.[57] This occurs because extensive intra-ecomorph variation exists: e.g., trunk-ground anoles vary from being moderately to extremely long-legged; grass-bush anoles from having long tails to extraordinarily long tails; trunk-crown anoles from being rather small to moderately large. The result is that less extreme species may be more morphologically different from the more extreme members of their own class than they are from less extreme members of other ecomorph classes (Fig. 3.13).

ECOLOGY AND BEHAVIOR

Williams' (1972) formulation of the ecomorph categories referred to sets of species that are similar in ecology and behavior, as well as morphology. Because behavioral and ecological data are not as easy to collect as morphological data, many studies that discuss ecomorphs in other taxa include quantitative data for morphology, but not for ecology and behavior. This is not the case for anoles, however, as a result of the abundance of many species, and some hard work to find the scarcer ones. The discussion below is based on ecological data for 49 species and behavioral studies of 28 species (Losos, 1990c; Irschick and Losos, 1996; Johnson et al., 2008).

The two most frequently reported habitat variables are the height and diameter at which lizards are initially seen. These two variables do a nice job of separating most of

---

57. As indicated by direct examination of Euclidean distances between species; this is an example of how a UPGMA phenogram can be misleading by suggesting that all members of an ecomorph class are more similar to all other members of that class than any is to a member of another class.

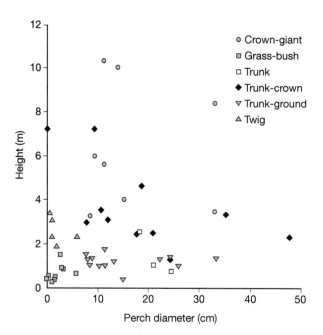

FIGURE 3.14
Perch height and diameter of
ecomorph species.

the ecomorphs, with the exception that crown-giant and trunk-crown anoles are broadly overlapping and that the trunk anoles fall within the space of both trunk-crown and trunk-ground anoles (Fig. 3.14).

Two other, less widely reported, variables serve to further distinguish the habitat use of the ecomorphs. The first is a measure of how cluttered the immediate environment is around a lizard, and correspondingly, how far away a lizard is from a support to which it could jump (Pounds, 1988; Losos, 1990c). This measure (distance to nearest perch) separates twig and grass-bush anoles, which live in highly cluttered habitats, from trunk anoles, which occur on large tree trunks with no other vegetation nearby (Fig. 3.15). The other three ecomorphs are intermediate.

The second measure is the use of leaves and other herbaceous vegetation (e.g., grass) during locomotion. This measure clearly separates the trunk-crown anoles, which often move onto leafy vegetation, from the other arboreal ecomorphs, which tend to stay on the woody branches and trunk. It also distinguishes grass-bush from trunk-ground anoles (Fig. 3.15).

The ecomorphs also differ in their locomotor and display behavior (Fig. 3.15). As a proportion of all of their movements, twig anoles walk much more often than the other ecomorphs, whereas trunk and trunk-ground anoles walk least often. Trunk-ground and grass-bush anoles are the most frequent jumpers, whereas the other ecomorphs jump relatively little. In terms of overall movement rate, the ecomorphs group into those ecomorphs that move frequently (trunk, trunk-crown, and twig), and those that move much less (the others). Display frequency also differs among the ecomorphs.

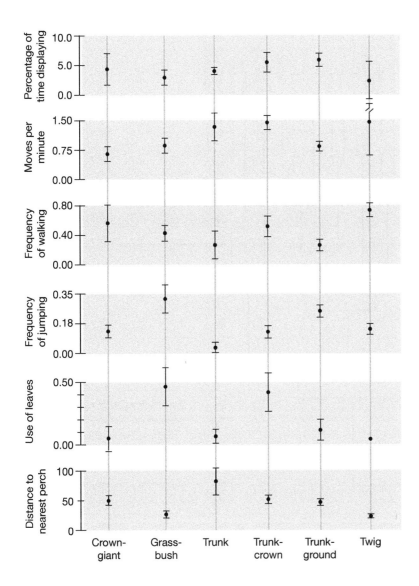

FIGURE 3.15

Ecological and behavioral differences among ecomorphs. Use of leaves is the proportion of individuals observed to use leaves or other herbaceous structures during behavioral observations. Distance to nearest support is a composite measure of the distance to the nearest object to which a lizard could jump. Frequency of walking and jumping are the proportion of all movements that were categorized as walks or jumps. Moves per minute is the number of movements per minute. Values are means plus one standard error. Data on use of leaves and distance to nearest support were not collected for all species; leaf use data were available for only one twig anole (hence, the lack of an error bar). Values are means for each ecomorph based on mean values for each species.

As would be expected given these ecomorph differences, interspecific variation in morphology, ecology, and behavior are related. For example, among species, relative hindlimb length is correlated positively with perch diameter and negatively with rate of walking. Similarly, the more often a species is observed using leaves, the greater the

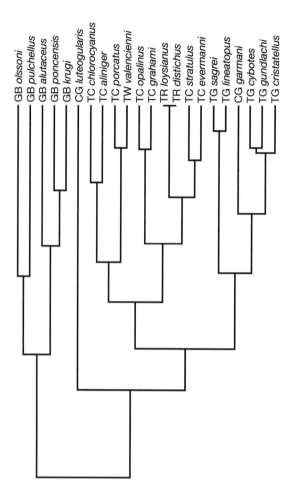

FIGURE 3.16
UPGMA phenogram of behavioral and ecological data. Similar results are obtained when ecological and behavioral data are examined separately, and when different ways of analyzing the data are used (e.g., reducing data dimensionality with principal components analysis).

number of its toepad lamellae, relative to body size. These relationships are discussed in greater detail in Chapter 13.

Overall, these data make a strong case that the ecomorphs represent a syndrome of morphologically, ecologically, and behaviorally distinctive types. Indeed, just as with the morphological data, a discriminant function analysis on behavioral and ecological data also classified all species to the correct ecomorph class.[58] Nonetheless, when a multidimensional ecological or behavioral space is examined, species do not sort out precisely along ecomorph lines. A representative UPGMA phenogram is presented in Fig. 3.16; the general groupings correspond to the ecomorph classes, but there are exceptions.

58. The analysis is based on 22 species for which the following data are available: percentage of movements that are runs and walks; movement rate; perch height and diameter; degree of clutter; and use of leaves. All data were log- or arcsine-square root transformed as appropriate. The analysis was highly significant (p less than 0.0001).

This result could be viewed in two ways. On one hand, it could be taken to indicate that the ecomorphs are defined primarily by morphology—ecology and behavior are related to morphology, but with so much variability that the ecomorphs do not represent discretely different ecological and behavioral entities. On the other hand, the field data I have analyzed were for the most part collected from few populations (often just one) over a short period of time. As I will discuss in Chapters 8 and 11, anoles alter their behavior and habitat use seasonally and as a result of many factors, such as which other species are present. Moreover, structural habitat use strongly depends on what habitat is available at a particular site (Johnson et al., 2006): in areas with many big trees, perch diameter of many species will be much greater than if the same species is studied in a scrubby area. For example, in the Bahamas, average perch diameter of *A. sagrei* varied four-fold across islands that differed in vegetation type (Losos et al., 1994). Given this variability, perhaps it is not surprising that ecomorphs are not found to cluster perfectly in ecological or behavioral space; the noise resulting from the limited extent of sampling of many species may have obscured otherwise clearer distinctions among the ecomorphs. More extensive sampling would be useful to get a more precise characterization of the habitat use and behavior of ecomorph species, both across their geographic range and over seasons and years.

ECOMORPH CLASS AND INTERSPECIFIC VARIATION

The ecomorph classes were initially defined based on a limited number of morphological characters, most of which obviously relate to an anole's position in and movement through the habitat, as well as its habitat use and foraging behavior (Williams, 1972; 1983). Recent work, however, has shown that the ecomorphs differ in a wide variety of other characteristics, including head dimensions, pelvic and pectoral girdle shape (Harmon et al., 2005), limb muscle mass (Vanhooydonck et al., 2006a), and sexual dimorphism in both size and shape (Butler et al., 2000; Butler and Losos, 2002; Losos et al., 2003a). Although variation in the girdles and limb muscle mass probably has functional significance relevant to locomotion (Peterson, 1972; Vanhooydonck et al., 2006a; Herrel et al., 2008), head shape variation probably is related more to other activities, such as eating and fighting (although species that use narrow surfaces may need narrow heads for balance and crypsis [Harmon et al., 2005]). Differences in sexual dimorphism among ecomorphs, particularly in size, also probably aren't related to sexual differences in locomotor ecology (Butler, et al., 2000; Losos et al., 2003a; see Chapter 9). These findings indicate that the morphology-ecology-behavior ecomorph syndrome likely results from more than the demands and constraints on locomotion determined by different structural microhabitats, a point that will be explored in Chapter 15.

Despite the seeming pervasiveness of ecomorph class as an explanation for interspecific differences, variation in many traits does not fall out along ecomorph lines.

Examples of morphological traits that vary tremendously among species, but for which ecomorph class does not explain a statistically significant portion of the variation, include tail crest height (Beuttell and Losos, 1999) and dewlap size, color and pattern (Losos and Chu, 1998; Nicholson et al., 2007). An important ecological trait that is independent of ecomorph class is microclimate; i.e., the temperature and humidity of the microhabitat used by a species. That microclimate does not vary by ecomorph should not be surprising, because the structural microhabitats themselves do not correspond to climatic microhabitats. Trunk-ground anoles, for example, occur in open, hot and sunny parts of the environment, but also in cool, shady, and mesic areas. Moreover, the spatial scale at which microclimate varies is great enough that lizards in a particular spot have relatively little latitude to select a preferred microclimate; for example, a trunk-ground anole in the deep forest does not have within its territory a large range of different microclimates from which to choose. The result is that substantial variation in microclimate occurs among populations and between species within all of the ecomorph classes.

An interesting, but surprisingly understudied, question concerns whether ecomorphs differ in prey type and size. Prey use could differ among ecomorphs for two reasons: either the prey available may differ among structural microhabitats or, because of behavioral or morphological adaptations, ecomorphs may utilize different portions of the prey resource spectrum, even if prey availability were the same in different structural microhabitats. Certainly, the diet of crown-giants and grass-bush anoles would differ in prey size and at least to some extent in prey type, even if prey availability were the same in their microhabitats. More generally, foraging behavior differs among the ecomorphs, predisposing them to encounter and attack different types of prey. Nonetheless, other than effects attributable to body size, little evidence of consistent differences among ecomorphs in diet has been found, although this question has not been studied in detail (see Chapter 8).

SPECIES DIVERSITY WITHIN ECOMORPHS

Most of the ecomorph classes are represented by more than one species per island,[59] although some ecomorph classes have greater species richness than others, a topic I will discuss in Chapter 15. Species within ecomorph classes on an island sometimes are ecologically distinct, even though they share the same structural microhabitat. This diversity occurs in several ways:

1. Some species are restricted to particular habitat types, such as pine forests, semi-deserts, or xeric rock outcrops.[60]

---

59. A list of all ecomorph species can be found in the Afterword at the back of the book.
60. Glor et al. (2003) refer to these as "macrohabitats."

2. Species co-occur by partitioning climatic microhabitats. In Cuba, for example, the widespread trunk-ground anoles *A. sagrei* and *A. homolechis* co-occur throughout the island, with *A. sagrei* always using hotter and more open microhabitats than *A. homolechis* (Ruibal, 1961; Hertz et al., in prep.).

Climatic microhabitat partitioning has both a spatial and an elevational component. In xeric southeastern Cuba, for example, *A. jubar* occurs in the hottest, most open microclimates, *A. sagrei* occurs in open shade, and *A. homolechis* occurs in deep shade. Conversely, at higher elevations, *A. sagrei* is absent and *A. homolechis* occurs in the open, with other species in more closed microhabitats (Hertz et al., in prep.).

3. Co-occurring species differ in body size. In a number of cases, two members of the same ecomorph class occur in sympatry without dramatic differences in microclimate, but with substantial differences in body size. Because prey size strongly correlates with body size in anoles (Chapter 8), these coexisting species probably differ in diet. Diet data are only available for one pair of sympatric ecomorphs that differ in size, the trunk-crown anoles of Puerto Rico, *A. evermanni* and *A. stratulus*. As expected, these species differ in prey type and size (Lister, 1981; Dial and Roughgarden, 2004).

Sympatry of pairs of species within the same ecomorph class occurs on all islands. The only cases in which sympatric species are not known to differ substantially in some aspect of ecology (microclimate, body size, or in a few cases, specialization to particular structural microhabitats)[61] involve either species at contact zones (e.g., Webster and Burns, 1973; Williams, 1975; Hertz, 1980a) or species for which almost nothing is known of their natural history.[62] Although in most cases the maximum number of sympatric members of the same ecomorph class is two, as many as three trunk-crown (Díaz et al., 1998; Garrido and Hedges, 2001) and grass-bush (Garrido and Hedges, 1992, 2001) and four trunk-ground species (Losos et al., 2003b) can be found in sympatry.[63]

The means by which resource partitioning occurs among sympatric members of the same ecomorph class differs among the ecomorphs. Ecomorphs that occur near the ground—trunk-ground and grass-bush—divvy up the habitat along microclimate lines and exhibit little difference in body size. By contrast, arboreal ecomorphs—primarily

61. Primarily rocks. In western Cuba, for example, the trunk-ground *A. allogus* and *A. mestrei* both occur in deep shade, but *A. mestrei* is always found either on or in close proximity to large boulders or rock walls (Rodríguez Schettino, 1999).

62. For example, almost nothing is known about the ecology of the many recently described grass-bush anoles from eastern Cuba (e.g., Garrido and Hedges, 1992).

63. Garrido and Hedges (2001) suggest that four grass-bush species may occur in sympatry on the northern slope of the Sierra Maestra in Cuba.

trunk-crown, but in a few cases twig and crown-giant anoles[64]—exhibit the opposite pattern. Although sympatric trunk-crown anoles do exhibit some differences in microclimate preferences (e.g., Reagan, 1996), they still substantially overlap in habitat use and can often be seen in close proximity, in contrast to the situation for trunk-ground and grass-bush anoles, which tend to be more segregated within a locality (Schoener and Schoener, 1971b). This phenomenon is discussed in greater detail in Chapter 11.

IS SIX THE RIGHT NUMBER OF ECOMORPH CLASSES?

In 1983, Williams expanded upon his original concept, suggesting that several of his ecomorph classes should be split in two, producing nine classes. In particular, he divided both the trunk-crown and twig ecomorphs into giants and dwarves, and divided the grass-bush ecomorph into grass and bush ecomorphs. All of this was done without explanation.

My feeling is that the data do not strongly support this proposition. Williams' (1972) definition of ecomorphs, implying discretely different groups recognizable on the basis of morphology, ecology, and behavior, accurately describes the six original ecomorph classes.[65] By contrast, the division of trunk-crown and twig anoles into large and small ecomorphs is based only on one morphological difference, and the division of grass-bush into grass and bush ecomorphs is based only on ecology. Closer examination of these three cases reveals that none of them result in groupings that are distinct in morphology, ecology, and behavior.[66]

I will go through each of these ecomorph classes in turn. With regard to trunk-crown anoles, the quantitative morphological comparisons reported earlier in the chapter indicate that small and large trunk-crown species do not form morphometrically distinct groups. Some small species cluster together in morphological space, as do some of the larger species, but, overall, small and large species do not form distinct clusters (Fig. 3.17).

64. Size differences occur among trunk-crown anoles on all four Greater Antillean islands. Similar examples are rarer among the other arboreal ecomorphs. Specifically, the inaptly named crown-giant *A. pigmaequestris* co-occurs with *A. equestris* on Santa Maria. The two species of large twig anoles, *A. valencienni* on Jamaica and *A. darlingtoni* in Haiti, are not known to coexist with smaller species; although smaller twig anoles do occur on Hispaniola, they have not been recorded in *A. darlingtoni*'s range. In Cuba, the unusual anoles in the *Chamaeleolis* clade, which might be considered twig giants (as discussed in the next chapter), coexist with more typical twig anoles (e.g., *A. angusticeps*) that are several orders of magnitude smaller in mass.

65. Even the two most similar ecomorphs, crown-giants and trunk-crown anoles, differ not only in morphology (primarily size), but also in some aspects of ecology and behavior.

66. Some readers may consider the discussion in this section particularly arcane, even by the standards of this book. However, given the central role that discussion of ecomorph evolution plays in understanding anole diversity, I feel that it is important to clearly delineate the case for recognizing the specific number of ecomorph classes that are discussed throughout this book. Moreover, some papers and websites casually refer to additional ecomorph types beyond the six recognized here, usually on the basis of structural microhabitat use. As mentioned at the outset of this chapter, the "ecomorph" concept is more than shorthand for species using the same structural microhabitat or those similar in morphology; it refers to groups of species that have independently evolve similarities in ecology, morphology, and behavior. In this regard, I consider it essential to clearly identify exactly how many such entities exist, and what the evidentiary basis is for such a claim.

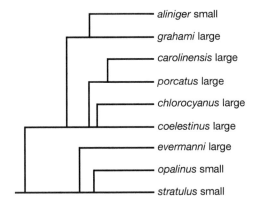

FIGURE 3.17

UPGMA phenogram of trunk crown anoles. This figure is redrawn with permission from Beuttell and Losos, 1999. A similar result, in which large and small species do not form separate clusters, was found in Losos et al. (1998).

Ecologically, few consistent differences exist between the smaller and larger trunk-crown anoles. In general, all trunk-crown anoles can be characterized as occurring at eye level and above on tree trunks, branches, and leaves. One difference in Jamaica and Puerto Rico is that the larger trunk-crown anoles use leaves much more than the smaller trunk-crown anoles (Schoener and Schoener, 1971a,b), but the smaller trunk-crown anole of Cuba, *A. isolepis*, appears to move on leaves quite often.[67] Movement patterns of large and small trunk-crown anoles also seem broadly similar (e.g., Losos, 1990c), and none of the small trunk-crown anoles exhibit the unusual movement patterns characteristic of the Hispaniolan trunk anoles.

I have heard some workers take a slightly different tack and suggest that the small Puerto Rican and Jamaican trunk-crown anoles, *A. stratulus* and *A. opalinus*, should be considered trunk anoles because they are superficially similar to members of that ecomorph category in morphology and habitat use. However, the data do not provide much support for this idea. Morphometrically, both species cluster with other trunk-crown anoles. Ecologically, both species use leaves to some extent, whereas trunk anoles rarely venture onto green matter (e.g., Schoener, 1968; Moermond, 1979a; Rodríguez Schettino, 1999).[68] *Anolis opalinus* does tend to be found at relatively low heights and in this regard is similar to trunk anoles (Rand, 1967c; Jenssen, 1973; Schoener and Schoener, 1971a; Losos, 1990c). On the other hand, detailed studies from canopy towers at the El Verde Field Station in Puerto Rico clearly demonstrate that *A. stratulus* is found on branches much more than on tree trunks (Reagan, 1992). Overall, both species appear to be good trunk-crown anoles, and the trunk-crown class as a whole does not seem to be readily divisible into small and large species.

67. I make this statement based on a comment by Williams (1969) to this effect and my unpublished observations of several *A. isolepis* from La Gran Piedra, Cuba, that repeatedly used leaves as they moved through the canopy.

68. The habitat use of the Cuban trunk anole *A. loysianus* is little known. Rodríguez Schettino (1999) does not mention use of leaves when summarizing its habitat use, but no data are provided. My unpublished data reveal that none of the 27 *A. loysianus* observed at Soroa was on a leaf (as opposed to 8 of 34 of the trunk-crown *A. porcatus* at that site); it would be interesting to know what *A. loysianus* does when it is high in the canopy, where it seems to spend a lot of its time.

Twig anoles are perhaps the least well known of the ecomorphs. It is true that most twig anoles are quite small (mean size of males less than 50 mm SVL), but *A. valencienni* of Jamaica and *A. darlingtoni* of Haiti are substantially larger (70–80 mm SVL). Unfortunately, *A. darlingtoni* is very poorly known, so it is not possible to examine whether it and *A. valencienni* share morphological, ecological and behavioral similarities relative to smaller twig anoles.

Grass-bush anoles are species which use narrow diameter vegetation near the ground, such as grass blades or the branches of bushes. One could imagine that species might adapt to one or the other, supporting Williams' (1983) decision to separate them into "bush" and "grass" ecomorph classes; alternatively, it is easy to see how the structural similarity in such supports and the fact that grasses and bushes are often found in close proximity might have led to one morphological type that is adapted to use both (the original "grass-bush" ecomorph). Although some anoles are found primarily in grassy habitats (e.g., *A. ophiolepis* of Cuba), and others usually use low-lying, narrow diameter vegetation such as bushes (e.g., *A. krugi*, Puerto Rico; *A. bahorucoensis*, Hispaniola), many grass-bush anoles use both types of habitat. Moreover, some grass-bush anoles are often found using ferns or vines (Fig. 3.18), which in some ways are structurally intermediate between grass and bushes.

The evidence upon which Williams (1983) assigned species as either "bush" or "grass" anoles is not clear to me. For example, he classified *A. pulchellus* of Puerto Rico as a bush anole, but I associate it as much with grassy as with bushy habitats (e.g., Gorman and Harwood, 1977). Morphometrically, the Puerto Rican grass-bush anoles, which Williams (1983) assigned to the "bush" category, do cluster with the bush-dwelling *A. bahorucoensis*. However, the grass-inhabiting *A. ophiolepis* clusters with this group, rather than with Williams' other putative "grass" anoles (Losos et al., 1998). For these reasons, I conclude that the case for the existence of distinct grass and bush ecomorph classes is weak.

FIGURE 3.18
The Cuban grass-bush
anole, *A. alutaceus*,
clinging to a fern.

In conclusion, I see no compelling evidence for subdividing any of the ecomorph classes. None of the divided groups are as discretely distinct and recognizable in the way the original six ecomorph classes are. Williams (1972) got it right the first time!

FUTURE DIRECTIONS

Whether the ecomorphs differ in a wide variety of other important ecological factors, including abundance, parasite load, rates of predation, social structure and foraging behavior, has yet to be studied. Making *a priori* predictions is difficult because in many cases pertinent data (e.g., whether abundance of predators or parasites varies across structural microhabitats) is unknown. Moreover, because morphology, ecology, and behavior are tightly interwoven, ecomorph differences may exist for traits that at first blush would seem unrelated to structural microhabitat. Abundance, for example, might be a function of degree of territoriality, which in turn might result from foraging mode, which is related to limb morphology, which in turn evolves adaptively in response to differences in structural microhabitat use (see Chapter 15).

By the same token, whether some aspects of morphology differ among ecomorphs also remains to be investigated. For example, no study has looked at tooth or claw morphology or aspects of the musculature (see Chapter 13). As mentioned in Chapter 2, examinations of the fine structure of the toepads suggests some differences (more terrestrial species having less developed setae), but more detailed studies are needed.

In addition, the natural history of many ecomorph species with small ranges is poorly known, particularly for species that occur in Cuba. Data on these species is needed to fully understand the ways in which the great species richness of some ecomorphs is attained on some islands.

A BRIEF HISTORY OF THE STUDY OF *ANOLIS* ECOLOGICAL MORPHOLOGY

Credit for the discovery and documentation of the anole ecomorphs goes to Ernest E. Williams, who arrived at Harvard in 1950 and served as Curator in Herpetology at the Museum of Comparative Zoology from 1957–1980. Students working under his supervision detailed the ecological, morphological, and behavioral aspects of the ecomorph phenomenon, and Williams painted the bigger picture in several synthetic papers that were in many respects well ahead of their time.[69] Indeed, one might argue that Williams' 1972 *Evolutionary Biology* paper played an important role in the development of the field of ecological morphology.[70]

Morphological and ecological differences among anoles were noted by early researchers (e.g., Oliver, 1948; Ruibal, 1961), but quantitative and comparative studies date initially to the work of a college undergraduate who made observations on anoles while visiting his parents in Cuba (Collette, 1961).[71] This work was followed by Rand's

69. For example, in the use of phylogenetic "tree thinking" in interpreting the evolution of anole ecomorphs in his 1972 paper, well before the phylogenetic revolution initiated by papers like Gittleman (1981), Lauder (1981), and Felsenstein (1985).

70. Ecological morphology has become a vibrant, multidisciplinary field that incorporates field and laboratory studies of ecology, behavior, and functional morphology, often in a phylogenetic context (e.g., Wainwright and Reilly, 1994). "Ecomorph," the term Williams coined that refers to species that share a similar set of morphological and ecological features, thus shares obvious relationships to "ecological morphology" or "ecomorphology," which is the study of the relationship between ecology and morphology.

A search of the terms "ecological morphology," "ecomorphology," and "ecomorph" in JSTOR revealed that the terms were rarely used prior to Williams' 1972 paper. In contrast to its frequent use in recent years (JSTOR reported 32 papers using the term in the period 1990–2000), "ecomorph" was only used once before Williams' definition of the term: in a 1954 paper in *Systematic Zoology*, J.G. Edwards proposed the term "ecomorph" for sympatric and synchronic interbreeding populations showing morphological and ethological differences. The term was proposed to distinguish these populations from allopatric, allochronic populations, for which the term "subspecies" would be appropriate. Why the prefix "eco" was employed was not explained.

"Ecological morphology" and "ecomorphology," too, were rarely used before Williams' work (which does not use either of those terms). JSTOR reported that "ecological morphology" was used five times prior to 1972: in a book in German by H. Fitting in 1926, *Die Ökologische Morphologie der Pflanzen*, which is about environmental forces on plants, according to a review in the *Quarterly Review of Biology* in 1927; in an obituary of the Russian plant ecologist Boris Aleksandrovich Keller published in *Science* in 1946; in a paper by Luckan in 1917 discussing the anatomical traits that allowed a plant, the velvetleaf (*Abutilon theophrasti*), to withstand a drought with little apparent ill effect; in an obscure paleontology paper that I did not look up (Kireeva, 1958); and in van der Klaauw's (1948) lengthy article, "Ecological morphology" which anticipated much of what is currently studied under the same name.

JSTOR only cites one use of the term "ecomorphology" prior to 1972, in a paper from 1902 in which I could not find "ecomorphology" (although I did find "geomorphology"). By contrast, the term was found in 153 papers from 1990–2000 (admittedly, a number of these were in the references section citing a book by that name). Of course, much of the credit for popularizing the term and the approach should go to Karr and James (1975), who were the first to use "ecomorphological," and who apparently came upon the term independently of Williams (his work was not cited). Prior to Karr and James' work, "ecomorphological" was only used, according to JSTOR, in a 1957 review of a book on Scandinavian ecology in *Ecology*, in which the term was used without explanation or definition.

71. Bruce Collette became a distinguished ichthyologist at the National Museum of Natural History, Smithsonian Institution, and is the recipient of many awards and honors. Despite his many ichthyological contributions, his 1961 paper on anoles is the second most cited of his papers. Although Collette never studied at Harvard, he was encouraged by Williams, who saw to it that the paper was published in Harvard's *Bulletin of the Museum of Comparative Zoology*.

pioneering community-wide studies on differences in microhabitat use among sympatric species in Hispaniola (Rand, 1962; Rand and Williams, 1969), Puerto Rico (Rand, 1964a) and Jamaica (Rand, 1967c), which in turn led to more detailed and sophisticated studies by Schoener and colleagues (e.g., Schoener, 1968, 1974; Schoener and Gorman, 1968; Schoener and Schoener, 1971a,b).

Rand and Williams' (1969) study of the anoles of La Palma in the Dominican Republic provided the names of the different ecomorphs, leading Williams (1972, 1983) to propose the ecomorph concept—linking morphological, ecological, and behavioral evolution—and to discuss the evolution of communities of ecomorphs.[72]

Moermond (1979a,b), working in Haiti, was the first to quantify morphological differences among sympatric anoles and to examine the relationship between morphology and behavior, an approach followed by Pounds (1988)[73] in Costa Rica. Mayer (1989) extended this quantitative approach across islands, showing that members of an ecomorph class are morphologically more similar to each other than they are to members of other ecomorph classes from their own island. I integrated these approaches by quantitatively examining the relationship between morphology, habitat use, and behavior across islands (Losos, 1990b,c,d; Irschick and Losos, 1999), and by examining the evolution of the ecomorphs in a phylogenetic context (Losos, 1992a; Losos et al., 1998).

A more detailed history of the development of anole research is provided by Rand (1999), and even greater detail can be gleaned by perusing the *Anolis* Newsletters. Starting as a 29-page grant summary report to the National Science Foundation, the idea blossomed into lengthy and informal summaries of the current work of *Anolis* researchers. Newsletters, which range in length from 29–144 pages and contain reports from 13–30 researchers, are available online.[74]

The past 15 years have seen an explosion of research investigating many aspects of ecomorph biology. The list has become too numerous to summarize here, but this work is discussed in appropriate places throughout the book.

72. No discussion of Ernest Williams would be complete without mentioning the two principles he articulated, well known to *Anolis* workers, but otherwise not widely appreciated (Williams, 1977a): "It was while walking along a hedge row in the Dominican Republic, listening to a complaint that I and some of my co-workers did not frame hypotheses every day while in the field, that I invented (or recognized) the Principle of Unsympathetic Magic. This states that, if one arrives at any firm and vivid conviction about matters of fact or theory in the field, the NEXT observation will provide a contradiction. . . . Note, however, that nature is not deceived. No opinion merely pretended to, i.e. not held with fierce conviction, will be responded to by a conclusive observation. The Malice of Nature prohibits the Principle of Unsympathetic Magic from being a source of satisfaction to the field worker."

73. J. Alan Pounds and Bruce Collette (footnote 71) are the only two researchers mentioned in this appendix who did not study under Williams, either as an undergraduate (me) or as graduate students (the rest). Pounds conducted his doctoral work on Costa Rican anole ecomorphology at the University of Florida, but he spent the summer after receiving his degree at Harvard working with Williams.

74. The URL may not remain constant, so I won't provide one. My advice is to Google them.

4

FIVE ANOLE FAUNAS, PART TWO

The Other Four

Although they've received the lion's share of research, the ecomorphs are not the whole anole story. Not even most of it. In fact, less than one anole species in three is a Greater Antillean ecomorph. In this chapter, I introduce the other elements of anole diversity, namely the unique (or non-ecomorph) anoles of the Greater Antilles, and the anoles of the smaller islands of the Greater Antilles, the Lesser Antilles, and Central America.

GREATER ANTILLEAN UNIQUE ANOLES

Ninety-five of the 120 anole species on the four large islands of the Greater Antilles belong to one of the ecomorph classes. That leaves 25 which do not, and these are a diverse and sometimes bizarre lot. They include:

- The false chameleons (clade Chamaeleolis) of Cuba mentioned in the prologue.
- The only truly terrestrial West Indian anole (Flores et al., 1994; Howard et al., 1999), a species (*Anolis* Chamaelinorops *barbouri*)[75] with an odd vertebral structure, consisting of expanded wing-like processes of the zygapophyses, that has few parallels among other vertebrates (Fig. 4.1; Forsgaard, 1983).
- Two species (the Cuban *A. vermiculatus* mentioned in the prologue and the Haitian *A. eugenegrahami*) that only are found near streams and which will enter

75. The explanation for this odd-looking scientific name and the "Clade Chamaeleolis" business will be forthcoming in the next chapter.

FIGURE 4.1
A. Chamaelinorops *barbouri.*

FIGURE 4.2
A. bartschi. Photo courtesy of Kevin de Queiroz.

water to escape predators and, at least for the Cuban species, to capture prey (Leal et al., 2002).

· A rock-wall dwelling, dewlapless species only found in a region of karstic hills (termed "mogotes") in western Cuba (*A. bartschi*; Fig. 4.2).

· A species (*A. lucius*) that lives in and near caves (found as far as 100 feet underground in one cavern) with transparent scales on its eyelids that might function like sunglasses (Williams and Hecht, 1955).

The other Greater Antillean unique anoles are less unusual in their structural microhabitats, using trunks, twigs, and rocks, but their morphology does not correspond to the ecomorphs that use those same microhabitats (Beuttell and Losos, 1999). Ecologically, the unique anoles occur in communities alongside the standard ecomorphs and exhibit no consistent differences from the ecomorphs in terms of niche breadth, behavior, abundance, or interspecific interactions.

Although some of these species are common, most are not—either being geographically restricted or uncommon where they do occur, or both—and as a result their natural history is not well known (Appendix 4.1 lists the unique anole species and briefly summarizes what is known of their natural history). All of the unique species in Hispaniola occur in the mountains, as does the one Jamaican species, but this is not true of the Cuban unique anoles.[76] Puerto Rico has no non-ecomorph species.

Other than their generally more restricted distribution, the biggest difference between the unique anoles and the ecomorphs is their lack of replication across islands. In no case are unique species on different islands particularly similar in morphology.[77]

The number of unique anoles is strongly related to island area: Hispaniola and Cuba have many; Puerto Rico and Jamaica few or none. One possible explanation is that the unique anoles on the larger islands have adapted to habitats that are not available on smaller islands. This explanation seems unlikely. For example, the montane leaf litter in which A. (Chamaelinorops) *barbouri* is found surely occurs on all islands, yet only Hispaniola has such a species. Similarly, streams meander through all four islands, yet specialized aquatic species occur only on Hispaniola and Cuba. Ditto for limestone rock walls on which several species specialize, but only in Cuba. Of course, perhaps it is not simply the occurrence of a microhabitat type, but sufficient quantity of that habitat; perhaps for example Puerto Rico does not have enough stream or leaf litter habitat to prod the evolution of stream and leaf litter specialists.[78] Advances in remote sensing and GIS capabilities should soon make this a readily testable hypothesis.

The five species in the Chamaeleolis clade deserve further discussion. These are among the largest of anoles.[79] Although like crown-giant anoles they possess a massive casqued head, they are similar to twig anoles in many respects, including limbs and tail that are extremely short for their body, a narrow head, low mass relative to body length, relatively few lamellae, and whitish-gray color (Fig. 4.3; Beuttell and Losos, 1999; Losos, unpubl.). Ecologically, they tend to be found on surfaces that are narrow relative to their body size, and the few data that exist indicate that they forage widely (Leal and Losos, 2000). In other words, a case could be made that they represent giant twig anoles. The twig ecomorph class already spans the greatest range in body size except for crown giants, from a maximum SVL of 41 mm in A. *sheplani* to 80 mm in A. *valencienni*.[80] Nonetheless, Chamaeleolis does exhibit a suite of unique features (Leal and Losos, 2000): the head casque, the chameliform body habitus,[81] molariform teeth and associated diet of hard-bodied prey such as snails and beetle pupae, and a rocking

76. Williams' (1983) postulate of an alternative montane ecomorph sequence thus lacks generality.

77. Even the aquatic anoles of Hispanolia and Cuba differ both morphologically and behaviorally (Leal et al., 2002).

78. Given that the stream and leaf litter specialists on both Cuba and Hispaniola have relatively small geographic ranges, this explanation does not seem very likely to account for the lack of such specialists on Puerto Rico and Jamaica.

79. SVL to 177 mm in A. C. *chamaeleonides*.

80. This span is dichotomous. The Hispaniolan twig anole, A. *darlingtoni*, is also relatively large (72 mm SVL), but all others are less than equal to 53 mm. In absolute terms, crown giants span a larger range (130–191 mm SVL); however, proportionally, the ratio of largest-to-smallest is far greater for twig anoles.

81. That is, body extremely compressed laterally, limbs aligned under the body during locomotion.

FIGURE 4.3
A. Chamaeleolis
guamuhaya.

back-and-forth behavior used as they walk forward, similar to that seen in chameleons and vine snakes, which presumably serves to mimic an inanimate object swaying in the breeze (Fleishman, 1985). Hass et al. (1993) suggested that Chamaeleolis should be admitted to the twig anole club; my feeling is that the application is still under review, but perhaps, just like women and the Augusta National Golf Club, it is time to dispense with tradition and let them in.[82]

ANOLES OF THE SMALLER ISLANDS OF THE GREATER ANTILLES

In addition to the four major islands, the Greater Antilles contain an enormous number of smaller islands ranging in size from nameless "rocks" a few square meters in area to Isla Juventud[83] (2419 km²) off the southwestern coast of Cuba. All but the smallest contain anoles, but great variation exists in species richness and composition. Much of this variation can be explained by the differing geological histories of these islands.

LANDBRIDGE ISLANDS

Many small islands were connected to one of the major Greater Antillean islands during periods of lower sea level, most recently within the past 10,000 years (Lighty et al., 1979; Sheridan et al., 1988). These "landbridge" or "continental"[84] islands include most of the Virgin Islands, which were part of a much larger Puerto Rican landmass; Isla Juventud

---

82. No analysis has quantitatively considered Chamaeleolis vis-à-vis ecomorph species. Because of its substantially larger size, Chamaeleolis probably falls near to, but outside, the twig ecomorph cluster. In ecology and locomotor behavior, however, Chamaeleolis exhibits data indistinguishable from other twig anoles (based on relatively little quantitative data [Leal and Losos, 2000]). Pending more extensive data and analysis, Chamaeleolis seems most usefully considered to be a giant twig anole. No other unique anole is a candidate to be a super- or sub-sized member of an existing ecomorph class.

83. Formerly knonwn as the Isle of Pines.

84. In this context, the major Greater Antillean islands are the continents.

and the many fringing islands to the north of Cuba; and Gonave, Tortuga, Saona, Beata and several other islands scattered around Hispaniola. Jamaica stands alone in having few offshore islands to which it was previously connected.[85]

Not surprisingly, given their recent connection, these landbridge islands only contain species that are also found on the major Greater Antillean island to which they were attached. In a few cases, populations have diverged to the extent that they have been described as distinct species (e.g., *A. ernestwilliamsi*, Carrot Rock, British Virgin Islands; *A. altavelensis*, Alto Velo, Dominican Republic).

The species that occur on landbridge islands are not a random subset of those that occur on their major island (Schoener, 1988; Mayer, 1989; Roughgarden, 1989). Setting aside Isla Juventud and its 11 species, only a few landbridge islands harbor a Greater Antillean non-ecomorph species or more than one species in any ecomorph class.[86] Moreover, the ecomorph species that are present on landbridge islands are almost always the most widespread member of that ecomorph class from the nearby Greater Antillean island, probably because the widespread species usually occur at sea level, in similar habitats to those found on landbridge islands.

OCEANIC ISLANDS

Oceanic islands are those that never had a connection to a larger continental landmass. By necessity, anoles found on these islands must have made their way there by overwater dispersal. That anoles are adept at overwater colonization is demonstrated by their far-flung distribution throughout the Caribbean, including many islands distant (greater than 200 km) from their ancestral home that have been colonized recently after Pleistocene submergence (Williams, 1969).

Oceanic islands in the Greater Antilles have been colonized by only six species (or species complexes): two trunk-ground (*A. cristatellus* and *A. sagrei*), two trunk-crown (*A. grahami* and the Cuban *A. porcatus–A. allisoni* complex), one twig (*A. angusticeps*) and one trunk anole (*A. distichus*). A seventh colonist may also have been a trunk-crown species, but its identity is more complicated. *Anolis acutus*, which occurs on St. Croix, forms a clade with the Puerto Rican trunk-crown anoles *A. evermanni* and *A. stratulus*, but their exact interrelationships are somewhat uncertain, making the ecomorph status of the ancestral colonist impossible to discern (Brandley and de Queiroz, 2004).

All of these ancestral species, with the exception of *A. evermanni*, are abundant lowland forms that are adapted to hot microclimates. Thus, a scenario in which one[87] or several individuals or eggs are able to survive a long voyage floating on flotsam (not to

85. Jamaica differs from the other islands also in being surrounded by a very small submerged "bank" (equivalent to a continental shelf). As a result, Jamaica's area increased to a much smaller degree than the other major Greater Antillean islands during periods of low sea level.

86. In the few instances in which more than one member of an ecomorph class occurs on a landbridge islands, the two species always differ substantially either in body size or in climatic microhabitat (Schoener, 1988).

87. By necessity an impregnated female, as parthenogenesis is not known in anoles.

mention jetsam) from one island to another is plausible.[88] Another possibility, perhaps more far-fetched, is that an anole could be transported directly by a hurricane from one island to another. Media reports of fish, frogs, and even people being transported by tornadoes are legion (Grazulis, 2001), so why not a two-gram lizard in a hurricane? *Anolis conspersus* on Grand Cayman is clearly derived from the Jamaican *A. grahami* (Jackman et al., 2002); Hurricane Gilbert traveled straight from Jamaica to Grand Cayman in 1988, so we might at least consider the possibility that a previous hurricane traveling the same route was responsible for bringing the ancestor(s) of *A. conspersus* to Grand Cayman.[89]

The species level status of some anoles on oceanic islands may require reexamination,[90] but for others there can be no doubt. *Anolis conspersus*, for example, has a dewlap radically different in color from that of *A. grahami* (Fig. 4.4) and *A. longiceps* (Navassa) and *A. maynardi* (Little Cayman), both derived from *A. porcatus* (Glor et al., 2005), differ from their ancestor not only in their remarkably pointy snouts, but also in their differently-colored dewlaps.

As with most anole faunas on landbridge islands, no oceanic island contains more than one species from the same ecomorph class. Because most of the colonizing has been done by *A. porcatus* and *A. sagrei*, communities composed of these two species, or their descendants, occur throughout the northern Caribbean.

THE BAHAMAS

The Bahamas technically are oceanic islands, because they have never been connected to a larger landmass. However, they also display the attributes of landbridge islands because many Bahamian islands that are separate today were connected into larger landmasses at times of lower sea level.

88. Mats of vegetation routinely float down large South American rivers into the Caribbean, and hurricanes throw great amounts of vegetation into the ocean (Williams, 1969; Hedges, 2001); a case of an anole clinging to floating debris has been documented (Hardy, 1982). Recent colonization of Anguilla by iguanas that drifted from Guadeloupe after two hurricanes struck there in 1996 exemplifies this scenario (Censky et al., 1998).

89. Calsbeek and Smith (2003) go so far as to suggest that hurricane-induced colonization maintains high levels of gene flow between populations of *A. sagrei* on relatively distant islands in the Great Bahama Bank. Estimated rates of gene flow in this study are remarkably high given the infrequency of hurricanes; further research with greater geographic sampling would be useful to clarify these results.

90. For example, *A. smaragdinus* in the Bahamas differs only in minor ways from its Cuban ancestor, *A. porcatus* (Glor et al., 2005); the same can be said about *A. desechensis* on Desecheo and its Puerto Rican ancestor, *A. cristatellus* (Lazell, 1983; Rodríguez-Robles et al., 2007). Similarly, *A. carolinensis*—found in the southeastern United States, rather than on an oceanic island—is for all intents and purposes a somewhat smaller version of *A. porcatus* with a very similar dewlap. I have little doubt that the two would interbreed if given the chance and, indeed, introduced *A. porcatus* in Florida appear to be hybridizing with *A. carolinensis* (Kolbe et al., 2007a). Of course, if *A. carolinensis* and *A. porcatus* were judged to be conspecific, then by the rules of the International Code for Zoological Nomenclature, the older name in the literature, *carolinensis*, would have priority, thus relegating probably the second most common Cuban lizard species to subjugation under an imperialistic American name.

FIGURE 4.4
(a) *A. conspersus* from Grand Cayman and (b) *A. grahami* from Jamaica. The dewlaps of these species are even more different than they appear because *A. conspersus*'s dewlap strongly reflects ultraviolet light, whereas *A. grahami*'s does not (Macedonia, 2001).

FIGURE 4.5
The Bahamas. Light blue areas are shallow underwater banks that were exposed during low sea levels during the last Ice Age, linking many islands together into a single large landmass. Cuba is in the lower left and Florida in the upper left. From MODIS, courtesy of NASA.

The Bahamas are composed of several island banks, which are separated by deep water and thus have never been connected (Fig. 4.5). The largest of these is the Great Bahama Bank which, when fully exposed during times of lowered sea level such as during the last Ice Age, formed a landmass rivaling Hispaniola in size. At this time, the Great Bahama Bank was also very close to Cuba, with a stretch of open water possibly as narrow as 15 km separating the two.

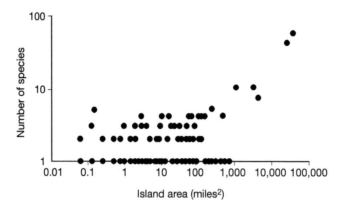

FIGURE 4.6

The species-area relationship for West Indian *Anolis*. Reprinted with permission from Losos (1996b).

Perhaps because of this proximity, or perhaps because of its size, Great Bahama (as represented by its fragmented islands today) is the only oceanic island anywhere in the Caribbean to be inhabited by more than two anole species. In particular, Great Bahama Bank islands harbor as many as four species: Cuban émigrés *A. angusticeps*, *A. smaragdinus* (descended from *A. porcatus* [Glor et al., 2005]), and *A. sagrei*, and *A. distichus* from Hispaniola.[91] All other Bahamian banks naturally contain at most two anole species.

SPECIES-AREA RELATIONSHIPS AND FAUNAL RELAXATION

The species-area relationship has been described as the closest thing to a true law in ecology (Schoener, 1976b; Lomolino, 2000): among islands and taxa of all sorts, the larger the island, the more the species. For anoles throughout the West Indies, a species-area relationship holds, although the amount of variation in species number explained by area is relatively small (29%; Fig. 4.6).

Comparison of small oceanic and landbridge islands helps explain why so little of the variation in species number is accounted for by area. Not only is no oceanic island smaller than Puerto Rico inhabited by more than two anole species, but no relationship exists among these islands between area and species number: some very large islands (e.g., Grand Bahama, 1373 km²) contain only one species naturally, whereas some much smaller islands have two species (Rand, 1969). In contrast, landbridge islands exhibit a quite strong species-area relationship, and even tiny landbridge islands can have 3–4 species (e.g., South Bimini, 8 km²).

91. The Bahamas are the only smaller islands in the West Indies colonized by twig or trunk anoles. The trunk anole *A. distichus* colonized not only the Great Bahama Bank, but several other Bahamian island banks as well.

The species-area relationship can be examined in great detail for the Virgin islands and the islands of the Great Bahama Bank due to the large number of islands and the existence of detailed records. In both cases, not only is the relationship between area and species number strong (Great Bahama Bank, r^2 = 0.67 [Losos, 1996b]; Virgin islands, r^2 = 0.38 [data from Rand, 1969]), but the pattern of species occurrence is non-random and extremely predictable (Lazell, 1983; Mayer, 1989; Roughgarden, 1989, 1995). In both cases, one-species islands are invariably occupied by a trunk-ground species (*A. sagrei* in the Bahamas, *A. cristatellus* in the Virgin Islands) and two-species islands by a trunk-ground and a trunk-crown species (*A. smaragdinus* in the Bahamas, *A. stratulus* in the Virgin Islands). The composition of three- and four-species islands differs between the banks: in the Virgin Islands, the third species is a grass-bush anole, *A. pulchellus*, whereas in the Bahamas, it is the trunk anole, *A. distichus*. No Virgin Island has four anole species,[92] whereas the four-species complement in the Bahamas is rounded out by the twig anole *A. angusticeps*. These nested patterns of species occurrence have very few exceptions (e.g., *A. cristatellus* and *A. pulchellus* on Little St. James in the Virgin Islands [Lazell, 2005]).[93]

This is a classic example of the phenomenon of faunal relaxation, in which recently fragmented areas lose species as a consequence of their smaller post-fragmentation size (Wilcox, 1978; Richman et al., 1988). On each bank, all of the species today are distributed across the geographic extent of the bank (though obviously not on every island). Consequently, we can reasonably conclude that all of these species were present across the single landmass that existed at the time of lowered sea level, which means that the observed species-area relationship must result from nonrandom extinction subsequent to island fragmentation (Rand, 1969; Mayer, 1989; Roughgarden, 1995). This pattern of extinction probably is a result of the vegetation available on small islands. For example, in the Bahamas, smaller islands tend to have fewer trees and lower vegetation; two-species islands often have nothing more than bushes and lack the habitat preferred by trunk and twig anoles; one-species islands are extremely scrubby and only support the most terrestrial of the species, *A. sagrei* (Schoener and Schoener, 1983a; Losos and Spiller, 1999; see Chapter 8).[94] Although the relationship between vegetation and

92. Although some did in the recent past, prior to the extinction of the little known crown-giant, *A. roosevelti* (Mayer, 1989).

93. Detailed examination of the fauna of landbridge islands of Hispaniola and Cuba could prove interesting. Faunal lists for a few very small islands off Hispaniola suggest that patterns of ecomorph occurrence may be different from those in the Bahamas and the Virgin Islands (Burns et al., 1992; Yeska et al., 2000).

94. Whether ecomorphs are absent from some islands because they are unable to survive in the habitat on that island or because they are competitively excluded by ecomorphs better adapted to the habitat is not known. One experimental study in the Bahamas showed that *A. smaragdinus* went extinct on many small islands that could support *A. sagrei* even when *A. smaragdinus* was the only species on the island; however, that study also showed that among populations of *A. smaragdinus* that did survive for three years, those sympatric with *A. sagrei* had lower densities than allopatric populations (Losos and Spiller, 1999). Thus, the absence of *A. smaragdinus* from small islands probably results both from the lack of appropriate habitat and from the presence of *A. sagrei*.

species occurrence has not been examined in the Virgin Islands, the same explanation is likely to hold (Mayer, 1989).[95]

Lack of suitable habitat, however, cannot explain the depauperate faunas of large oceanic islands. Presumably, the low species count on these islands is a result of limited dispersal ability of most species (recall that most successful anole dispersers are either trunk-ground or trunk-crown anoles), the inability of more than one member of an ecomorph class to coexist on these islands (presumably as a result of interspecific competition; see Chapter 11), and the failure of speciation to occur on small islands (a topic to which I will return in Chapter 14 [Rand, 1969]).[96]

ECOMORPHOLOGY

Most anoles on the small islands of the Greater Antilles are clearly descended from particular ecomorph species present on one of the four large islands of the region.[97] Given the different conditions that characterize small islands, not only differences in vegetation and topography, but also the generally depauperate anole communities, we might question whether populations on these islands have diverged from their ancestral ecology or morphology.

On the Great Bahama Bank, the answer is generally "no"; Bahamian *A. sagrei* and *A. smaragdinus* (the other two species have not been examined) appear to be typical of trunk-ground and trunk-crown species, respectively (Losos et al., 1994). The ecomorphology of populations on other land-bridge islands has not been examined, although nothing in the literature suggests notable divergence.[98]

By contrast, the situation for oceanic islands is quite different. Descendants of trunk-crown anoles seem to generally maintain their trunk-crownedness, but most trunk-ground derivatives have diverged and become more trunk-crown-like in morphology and habitat use (Losos and de Queiroz, 1997). This dichotomy and its evolutionary implications will be discussed in greater detail in Chapter 15.

ANOLES OF THE LESSER ANTILLES

The Lesser Antilles are the chain of islands that stretch from Sombrero and the Anguilla Bank east of Puerto Rico to Grenada just north of South America (Fig. 4.7). Like the

95. An alternative explanation is that species with larger territory sizes, and thus lower population densities, may require larger islands to maintain a viable population (Roughgarden, 1989). Although spacing patterns may play a role in determining species occurrence patterns (*A. sagrei* has smaller territories than other Bahamian species [Schoener and Schoener, 1982a]), habitat availability is probably a more important factor. For example, the trunk anole *A. distichus* probably could not survive on an island without broad trees regardless of the size of the island.

96. Conversely, exuberant speciation on the largest islands in the Greater Antilles also contributes to the relatively low amount of variance explained by island area in a linear regression. This, too, will be discussed in Chapter 14.

97. The one possible exception is *A. acutus* from St. Croix discussed above.

98. The one exception is *A. ernestwilliamsi* on tiny Carrot Rock in the British Virgin Islands. This species is much larger than other populations of *A. cristatellus*, from which it is derived (Lazell, 1983).

FIGURE 4.7
The Lesser Antilles.

oceanic islands in the Greater Antilles, Lesser Antillean islands contain at most two species, and no relationship exists between island area and number of species—indeed, the largest Lesser Antillean islands have only one anole species. One important difference between one- and two-species Lesser Antillean islands is the size of their species: on one-species islands, the species is almost invariably (16 of 17 islands)[99] intermediate in body size (maximum SVL of males ca. 75 mm [Roughgarden, 1995]), whereas two-species islands usually contain a small (ca. 60 mm) and a large (ca. 115 mm) species (Schoener, 1970b; Roughgarden, 1995). Null models indicate that the differences in size on two-species islands are much greater than expected by chance (Schoener, 1988; Losos, 1990a).

This pattern is consistent even though different clades occur in the northern and southern Lesser Antilles. All islands from Dominica to the north contain members of the *bimaculatus* Series,[100] which is related to the *cristatellus* Series on Puerto Rico, whereas islands from Martinique south host members of the *roquet* Series, whose affinities lie in South America. No obvious explanation exists for why these two clades should be divided by the Martinique Passage. The distance between these two islands is no greater than the distance between other Lesser Antillean Islands, nor do particularly strong currents flow between them. No other plant or animal group of which I am aware shows a similar transition at this point (e.g., bats [Jones, 1989]; birds [Ricklefs and Bermingham, 1999, 2004]; frogs [Kaiser et al., 1994; Heinicke et al., 2007]). It is as if the two clades colonized from different directions and came to a stalemate at this point, with neither clade able to invade the realm of the other.

Although the same size pattern characterizes islands in the north and south, differences do exist: species on two-species islands are larger in the south than in the north (Schoener, 1970b; Roughgarden, 1995). Moreover, the organization of ecological communities also differs. In the north, the large species perch higher than the smaller species, whereas in the south, no difference exists. Conversely, in the south, the species have different microclimate preferences, whereas in the north, these differences are less pronounced (Roughgarden et al., 1983; Buckley and Roughgarden, 2005b).

Most species on one-species islands are similar in morphology and ecology to trunk-crown anoles of the Greater Antilles (Losos and de Queiroz, 1997). By contrast, only one species from a two-species island fits well into the Greater Antillean ecomorph classification,[101] and most are not even close based on morphology (Losos and de Queiroz, 1997).

Lesser Antillean species are also notable for the great degree of intraspecific variation exhibited among populations. At the extreme, Lazell recognized 12 subspecies of

99. This figure refers to the major islands of the Lesser Antilles and does not include either the islands of the Grenadines nor small offshore islets that occur around the major islands.

100. Anole taxonomic nomenclature is discussed in Chapter 5.

101. *A. trinitatis* (which, despite its name, is native to St. Vincent and introduced to Trinidad [Lazell, 1972]) is similar in ecology and morphology to trunk-crown anoles.

A. marmoratus on Guadeloupe and nearby islands (Fig. 4.8),[102] six of *A. roquet* on Martinique and four of *A. oculatus* on Dominica (Lazell, 1972), primarily on the basis of differences in coloration, as well as on differences in scalation among populations of *A. marmoratus*.[103] Although the intergrade zones between subspecies were in some cases as large as the range of the subspecies themselves, thus calling into question the existence of subspecies as real biological units (discussed in Lazell, 1972), it certainly is true that the extent of phenotypic difference between the subspecies is as great, and often much greater, than differences among sympatric anole species elsewhere (see also Fig. 12.5).

ANOLES OF CENTRAL AND SOUTH AMERICA

Finally, that brings us to the anoles of Central and South America, which I will refer to as "mainland anoles." Lumping all of the anoles of this enormous region, with a landmass far greater than that of all West Indian islands combined, into one fauna seems preposterous, especially given that nearly 60% of all anole species are found on the mainland.

The basis for this categorization is simple: we know far less about mainland anole ecology and evolution than we do about the anoles of the West Indies. The reason for this is simple: mainland anoles are much harder to study. With some exceptions, mainland anoles are much less apparent—due either to lower density or lower visibility—than many West Indian species. The result is that data collection is vastly easier and more efficient in the West Indies than on the mainland and, consequently, much more work has been done there.[104]

This is not to say that there hasn't been any research on mainland anoles. Quite the contrary, there has been a fair amount, and some topics, such as population and reproductive biology, have been better studied on the mainland than in the West Indies. Although the natural history of many species has been reported, community level studies have been less common (but not absent; e.g., Corn, 1981; Pounds, 1988). Phylogenetic study, however, has lagged, as detailed in the next chapter.

---

102. Five of these subspecies occur on nearby islands and are considered to be different species by Breuil (2002) on the basis of the extent of their morphological differences (the reason they were described as separate subspecies in the first place). In support of this proposition, molecular work shows that one of them, *A. m. terraealtae* on Terre de Haute, Iles de Saintes, is more closely related to *A. oculatus* on Dominica than to *A. marmoratus* on Guadeloupe, and that the highly morphologically divergent *A. ferreus* of Marie Galante, initially considered a subspecies of *A. marmoratus*, but usually considered a species in recent years, also lies phylogenetically outside of the rest of *A. marmoratus* (Schneider et al. 2001; Stenson et al., 2004). The other island taxa have not yet been included in molecular studies.

103. Mention must be made of the famous "hairy" anole of northwestern Guadeloupe, *A. m. setosus*, the conical dorsal scales of which are extremely pointed and, on the neck, give "the most furred effect imaginable on an anole" (Lazell, 1964, pp. 380–381). The functional significance of this scalation is unknown.

104. This is both the blessing and the curse of West Indian anoles. They are in many respects almost ideal study organisms: remarkably abundant, often oblivious to the presence of an observer, even easy to catch. But for the evolutionary ecologist who cut his or her teeth in the West Indies, working on almost any other species anywhere else is simply too much trouble.

FIGURE 4.8
Geographic variation of A. *marmoratus* on Guadeloupe.

Mainland anole diversity certainly is no less than that seen in the West Indies. Eleven or more species can occur at a single site (Vogt et al., 1997; S. Poe, pers. comm.) and sympatric species differ in both structural and climatic microhabitats. Species occur high in the canopy, down in the leaf litter, and everywhere in between; on narrow twigs and broad tree trunks; in open sun and in deep shade (e.g., Fitch 1973a, 1975; Pounds, 1988; Vitt et al., 1995, 2001, 2003a,b; Vitt and Zani, 1996b; Birt et al., 2001). Morphologically, mainland anoles vary in the same way as West Indian species: big and small; long- and short-legged; immense and reduced toepads; long and short tails, and so on.

This variation in habitat use and morphology rivals that of the Greater Antilles (Irschick et al., 1997; Velasco and Herrel, 2007; Pinto et al., 2008), but ecomorphological diversity is organized in very different ways on the mainland and in the West Indies. Few mainland species clearly correspond to the West Indian ecomorph classes, though a few grass-bush, twig and crown-giant anoles are apparent (Fig. 4.9; Irschick et al., 1997). Many, perhaps most, mainland species appear quite different from West Indian forms, such as the seemingly longer necks of many South American species (e.g., *A. punctatus*), the sprightly and gracile morphology of *A. limifrons* or the pug-nosed and long-limbed *A. capito* (Fig. 4.10). This impression has been confirmed quantitatively with regard to toepads; for a given body size, Greater Antillean anoles generally have larger toepads than mainland species and the shape of the toepads on the forefeet also tends to differ between anoles of the two regions (Macrini et al., 2003; Velasco and Herrel, 2007).

Ecologically, too, some mainland anoles have few or no West Indian parallels, such as *A. onca*, which occurs in low-lying vegetation near beaches (Collins, 1971), or the leaf-litter inhabiting *A. humilis* and *A. nitens* (Fig. 4.11). Overall, although quantitative study is needed, it seems safe to conclude that the West Indian ecomorph syndrome does not occur within mainland communities. Whether an alternative, mainland ecomorph syndrome exists—in which the same set of microhabitat specialists (but different from those in the Greater Antilles) occurs in different mainland regions—remains to be determined.

Hopefully, in the future, when research on mainland forms begins to catch up with that for the West Indies, we can recognize and discuss multiple faunas corresponding to different geographic regions or ecological settings, such as montane and lowland faunas, or wet forest and dry forest faunas. Much work remains to be done, and time is short; some species probably have been lost already as their entire known range has been deforested (Campbell et al., 1989).

For the remainder of the book, I will include discussion of mainland anoles when the data permits.[105] Nonetheless, the vast body of comparative and synecological work has been conducted in the West Indies, and so that area will be the primary focus of this book.

FUTURE DIRECTIONS

The wealth of knowledge on *Anolis* is remarkable, yet equally remarkable is how much we don't know about anole biology and natural history. As this chapter has made clear, little is known about many interesting species. We know much about the common species, but the rarer ones may be the key to understanding what is evolutionarily possible in the *Anolis* world. For this reason, detailed autecological studies of the unique anoles could prove very insightful. Obviously, this task is orders of magnitude greater for the anole faunas of Central and South America, where many common species are little

---

105. All mainland species mentioned in the book are listed in the Afterword along with very basic information on geographic range and habitat use.

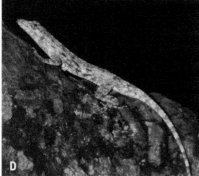

FIGURE 4.9
Examples of mainland anoles that are similar to
West Indian ecomorphs. (a) *A. auratus*, a grass-
bush anole. Photo courtesy of Eldridge Adams;
(b) *A. mittermeierorum*, a twig anole in the
Phenacosaurus clade. Photo courtesy of Steve Poe;
(c) *A. biporcatus*, a species morphologically similar
to West Indian crown-giants, though not as large;
(d) *A. pentaprion*, a twig anole. Photo courtesy
of Luke Mahler.

FIGURE 4.10
Examples of mainland anoles ecomorphologically different from West Indian species. (a) *A. punctatus.* Photo courtesy of Arthur Georges; (b) *A. gorgonae.* Photo courtesy of Thomas Marent, www.thomasmarent.com; (c) *A. capito.* Photo courtesy of Laurie Vitt.

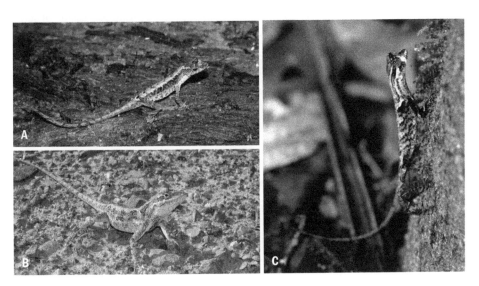

FIGURE 4.11
Ground-dwelling mainland anoles. (a) *A. humilis*; the orange spots on the lizard's side are mites. Photo courtesy of J.D. Willson; (b) *A. onca.* Photo of *A. onca* courtesy of Aurélien Miralles; (c) *A. nitens.*

known and communities have been barely investigated. Much work remains to be done, but the good news is that all that is required is patience and perseverance; the determined fieldworker is almost guaranteed of getting data that will usefully expand our understanding of anole diversity.

THE UNIQUE SPECIES OF THE GREATER ANTILLES

Except where noted, information comes from Schwartz and Henderson (1991), Rodríguez Schettino (1999) and my own observations.

CUBA

Chamaeleolis clade (*A. agueroi, A. barbatus, A. chamaeleonides, A. guamuhaya, A. porcus*) These lizards, superficially like a cross between a crown-giant anole and a chameleon (Fig. 4.3), are actually in morphometric and ecological terms super-sized twig anoles (see discussion on pp. 61–62). "Chipojo bobos" (= big, dumb lizards), as they are called in Cuba, they have a variety of unusual traits, including large molariform teeth and a diet consisting of mollusks and other hard prey (Leal and Losos, 2000). Like twig anoles, they move slowly, but cover a lot of ground.

A. argenteolus This species is similar in habitat use to its larger relative, *A. lucius*, but it uses a wider range of trees (not being restricted to very large ones) and occurs to a greater extent throughout the tree; also, it is found on rock walls less often than *A. lucius*. Morphologically, it is similar in many respects to *A. lucius*, including possession of semi-transparent scales on the eyelids and extremely long fore- and hindlimbs.

A. argillaceus, A. centralis, A. litoralis, A. pumilus, A. terueli This complex of five very similar species occurs throughout Cuba. In structural habitat, they can be found on tree trunks, but often occur on very narrow twigs in bushes, usually no higher than 2 m above the ground (but rarely on the ground). Their hindlegs are quite short, approaching those of twig anoles in their brevity, but their forelegs are average in length, much longer than those of twig anoles.

A. bartschi This beautiful, dewlapless species occurs on cliffs and other large vertical rock surfaces (Fig. 4.2). The species has long legs which are held laterally to grip vertical surfaces, like many other rock-dwelling lizards (Revell et al., 2007b). These lizards move nimbly across vertical rock surfaces.

A. lucius Ecologically similar in many respects to *A. bartschi*, *A. lucius* is also often found on the trunks of very large trees and fairly deep in caves. It is also morphologically similar to *A. bartschi* in that its limbs are very long, but it differs in having a large dewlap and semi-transparent scales on its eyelids.

A. vermiculatus Arguably the most spectacular of all anoles (Fig. 4.12a), this fairly large (123 mm SVL), dewlapless species has skin with a velvety texture (perhaps for water-proofing? [Losos, unpubl. obs.,]) and a beautiful green and black pattern on its back. Invariably found within several meters of streams, *A. vermiculatus* takes refuge in water to escape predators, either by diving into the water (occasionally by jumping out of trees from heights of several meters) or by running bipedally across it. The diet of these lizards includes fish, frogs, crayfish, and shrimp, which they some times catch underwater.

A. (Chamaelinorops) barbouri This enigmatic species is almost entirely terrestrial and is usually found in the leaf litter of montane forests (Fig. 4.1). The species is cryptic and hard to find; it seems to move mostly by making short hops (Flores et al., 1994). Its unusual vertebral structure is discussed on p. 59. The species also has highly unusual display and anti-predator behavior (Jenssen and Feely, 1991; Autumn and Losos, 1997).

A. christophei Common in montane areas of Hispaniola, *A. christophei* is morphologically most similar to a trunk-ground anole (Beuttell and Losos, 1999), but ecologically more like a trunk anole (Rand and Williams, 1969), although it does not move with the shorty, jerky movements characteristic of Hispaniolan trunk species. Its most distinctive features are the beautiful patterning of the back and the enormous purple dewlap with white stripes (Fig. 4.12b).

A. etheridgei Morphologically, this species is somewhat similar to a trunk-ground anole (Beuttell and Losos, 1999), although it is smaller (43 mm SVL) than any trunk-ground species. Ecologically, however, it is found on thin perches, usually in or near bushes (Rand and Williams, 1969). I have had difficulty finding any specimens during the day, as did Thomas and Schwartz (1967), even in areas in which they are readily found at night sleeping in bushes. I suspect that during the day they stay in the interior of heavily vegetated bushes and come out only occasionally to bask.

A. eugenegrahami This little-known species occurs on boulders along streamsides in a restricted area of Haiti (Schwartz, 1978). Morphologically, it is quite different from *A. vermiculatus* and from the Central American aquatic anoles (Leal et al., 2002); it has a morphology similar to rock-dwelling anoles and other lizards, with extremely long fore- and hindlimbs (Leal et al., 2002). This species readily enters water to escape predators and catch prey, although it does not run bipedally, perhaps due to its long hindlimbs, which shift the center of mass too far forward (Leal et al., 2002).[106]

A. fowleri This extremely poorly-known species has a very limited range in the Dominican Republic. Beautifully patterned with shades of green, brown, and grey, this species is morphologically dissimilar to all ecomorphs (Beuttell and Losos, 1999), contrary to Williams' (1983) suggestion that it is a twig anole. No published records are available concerning the ecology and behavior of this species, except for several reports of where it has been found sleeping (Schwartz, 1973; Glor, 2003).

A. monticola, A. rimarum, A. rupinae These shockingly beautiful Haitian species are little known. They appear to be found primarily on rock piles in heavily shaded habitats (Thomas and Schwartz, 1967; Moermond, 1979a,b). I have not included any of them in morphometric studies, but Moermond (1979a) showed that *A. monticola* has extremely long hindlimbs, but relatively short forelimbs, unlike other rock dwelling species (Figs. 4.12c, d).

---

106. For convenience, anoles that frequently enter the water will be referred to as "aquatic" species with the recognition that they actually spend most of their time out of water.

FIGURE 4.12
Unique anoles. (a) *A. vermicu-
latus*; (b) *A. christophei*;
(c) *A. rupinae*; (d) *A. monticola*.
Latter two photos courtesy of
Eladio Fernández.

A. reconditus Restricted to high elevations in the Blue Mountains, this species is closely related to the Jamaican trunk-ground *A. lineatopus* (Jackman et al., 2002) and is most similar morphologically to trunk-ground anoles, although it is substantially larger (100 mm SVL) and differs in a variety of other ways (Beuttell and Losos, 1999). Ecologically, it appears to be somewhat of an arboreal generalist, but relatively few data are available on its ecology and behavior (Lazell, 1966; Hicks and Jenssen, 1973).

PUERTO RICO

Puerto Rico has no unique species.

5

PHYLOGENETICS, EVOLUTIONARY INFERENCE, AND ANOLE RELATIONSHIPS

In the previous two chapters, I have described the distribution and diversity of anoles with little mention of evolution. Yet, some of the patterns of anole diversity beg, no, scream for evolutionary analysis. Are members of the same ecomorph class on different islands closely related? How many times have large and small body size evolved in the Lesser Antilles? Are the West Indian anoles descended from mainland taxa, or did it happen the other way around? In this and the next two chapters, I will discuss how information on anole phylogenetic relationships can be used to address these and other questions. This chapter will focus on conceptual issues regarding phylogenetic approaches to evolutionary questions and then will outline *Anolis* phylogeny. Based on this phylogeny, Chapter 6 will discuss anole biogeography and Chapter 7 will review ecomorphological evolution.

A revolution occurred in evolutionary biology in the 1980s and 1990s. With the development of explicit, quantitative methods for estimating phylogenetic relationships, systematics was transformed from a sleepy backwater to a rigorous and quantitative discipline at the forefront of evolutionary biology. At the same time as this reinvigoration of systematics, a new approach—termed phylogenetic "tree thinking" (O'Hara, 1988)— emerged to address evolutionary questions. This approach—which arose nearly simultaneously from a number of fields (e.g., behavior [Gittleman, 1981; Ridley, 1983]; functional morphology [Lauder, 1981]; paleontology [Cracraft, 1981])—emphasized that the history of evolutionary descent and modification is captured in a phylogeny. Thus, any study that hopes to draw insights about evolutionary patterns that extend beyond a single population must be grounded in the context of a phylogeny for the taxa under study. This new way of thinking was given a large boost by the development of new analytical methods (e.g., Ridley, 1983; Felsenstein, 1985; Huey and Bennett, 1987; Maddison, 1990) and the software programs to implement them (e.g., MacClade [Maddison and Maddison, 1992]), and entered the mainstream with the simultaneous publication of Harvey and Pagel (1991) and Brooks and McLennan (1991).

The importance of the phylogenetic perspective cannot be underestimated. In the absence of a phylogeny, we cannot tell whether the phenotypic similarity of species results from convergent evolution or from inheritance from a shared ancestor, nor can we envision whether trait evolution has occurred in a particular sequence. Moreover, because related species often share similarities inherited from their common ancestor, data from such species are not evolutionarily independent. Consequently, statistical analyses of comparative data must take account of phylogenetic non-independence (Felsenstein, 2004, and references therein). For these reasons, it is now accepted wisdom that evolutionary studies involving more than one taxon must include a phylogenetic perspective, and manuscripts that fail to do so have great trouble getting published.

But has the pendulum swung too far? Has an overbearing emphasis on phylogenetic approaches obscured their limitations and overshadowed other, complementary approaches? Phylogenetic analysis is a powerful tool, but like any tool, it has its shortcomings. In particular, phylogenetic inferences are always based on certain assumptions; when those assumptions are not met, phylogenetic perspectives may shed little insight or may even be misleading. An important area for future exploration, as I will discuss below, is determination of how to maximize what can be deduced from a phylogeny without overextending to conclusions that cannot be made confidently.

Despite these limitations, and a few resulting false steps along the way, the phylogenetic revolution has greatly advanced our understanding of anole evolution. Conversely, studies on anoles have in some ways proved to be a model system for the development of phylogenetic evolutionary analysis.

This chapter is divided into two parts. First, I will discuss the uses—and, importantly, the limitations—of phylogenetic approaches; second, I will present the current understanding of anole phylogenetic relationships. Based on this information, in the next two chapters, I will discuss what can be inferred about the evolution of the five anole faunas.

THE POWER AND PERIL OF THE PHYLOGENETIC PERSPECTIVE
PHYLOGENY: WHAT IS IT GOOD FOR?

Phylogenies can provide insights into many questions pertinent to the study of evolutionary diversification, including:

· Is phenotypic similarity of species the result of convergence or inheritance from a common ancestor (Fig. 5.1a)?

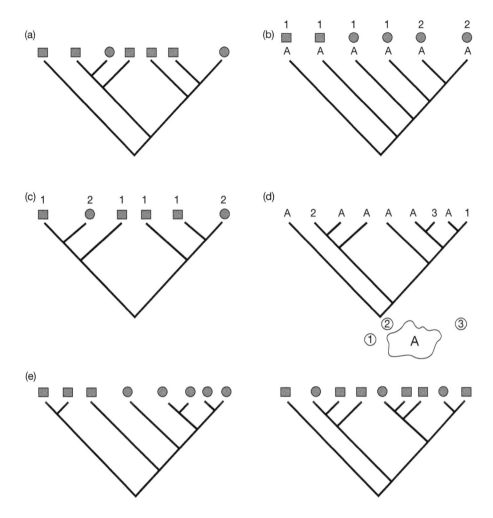

FIGURE 5.1

Cartoon illustration of the utility of a phylogenetic perspective. (a) The phylogeny suggests that the square phenotype is ancestral and that the circle phenotype has evolved convergently in two different species. (b) The A phenotype appears to have evolved prior to the circle phenotype, which in turn evolutionarily preceded the evolution of the #2 phenotype. (c) Evolution of the two traits is correlated. Transitions from #1 to #2 occurred simultaneously with evolution from squares to circles. (d) Area A is the ancestral location of the clade, with subsequent colonization of several offshore islands. (e) In the phylogeny on the left, the commonness of species with the circle phenotype is the result of higher rates of diversification of the clade that evolved that feature, whereas on the phylogeny on the right, the occurrence of this phenotype is the result of multiple evolutionary origins of the trait, none of which has sparked a high rate of diversification.

- Do traits evolve in a predictable order (Fig. 5.1b)?
- Is the evolution of two traits related (Fig. 5.1c)?
- Where did a clade originate and how has its geographic range expanded evolutionarily (Fig. 5.1d)?

- Is the commonness of a phenotype the result of many independent evolutionary events or of substantial speciation (or lack of extinction) in clades possessing that trait (Fig. 5.1e)?

Without a phylogeny, these questions are unanswerable.[107] A great deal of work over the past 25 years has been devoted to developing new methods to answer questions like these in a phylogenetic framework and the result has been a vastly enhanced understanding of evolutionary patterns and processes. In many respects, what we have learned about anole evolution is an exemplary case study of the power of a phylogenetic perspective, as I will describe shortly.

Before delving into the anole specifics, though, it's worth considering the limitations of phylogenetic studies. The following discussion is not meant to disparage phylogenetic approaches, but rather to recognize that phylogenies are useful for answering some questions, but less useful, at least sometimes, for answering others. In particular, I will suggest that in some situations, the ability to use a phylogeny to reconstruct ancestral character states will be limited. Importantly, however, this conclusion cannot be reached without evaluating patterns of character evolution on a phylogeny. Thus, phylogenetic approaches are essential, even if sometimes they will reveal their own limitations.

DIFFICULTIES WITH PHYLOGENETIC APPROACHES

PROBLEMS WITH ANCESTOR RECONSTRUCTION

Probably the biggest disappointment in the development of phylogenetic approaches has been the realization that attempts to infer ancestral character states often will be highly problematic. The reason is that when rates of change are high relative to the frequency of cladogenesis, then the confidence that can be placed in any ancestral reconstruction is bound to be low.

Consider first the simplest case, when ancestral character states are reconstructed by parsimony, which is an approach that minimizes the number of evolutionary transitions inferred to have occurred on a phylogeny. When only a few evolutionary transitions are required on a phylogeny, then the ancestral trait reconstructions may seem reasonable (Fig. 5.2a). However, when the minimum number of inferred transitions is great, then it would be unreasonable to strongly prefer one reconstruction over others that require a slightly greater number of transitions (Fig. 5.2b).

---

107. One might think that an alternative avenue for answering questions of this sort would be through examination of the fossil record. However, fossils do not come with labels on them, and so interpretation of fossils must be conducted within a phylogenetic framework as well. Moreover, for many taxa, certainly including anoles, Darwin's (1859) reservations about the imperfections of the fossil record still ring true. Finally, fossils can inform about some aspects of morphology, but insights about other aspects of the phenotype, such as ecology, behavior, and physiology, are far less reliable. Bottom line: fossils are great when you have them, but for many types of evolutionary ecological study, they usually are not a major source of information or insight.

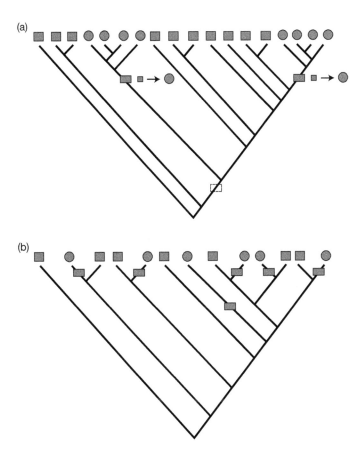

FIGURE 5.2

The reliability of ancestral reconstructions using parsimony. Parsimony reconstructs ancestral character states to minimize the number of evolutionary changes. In (a), parsimony would infer two transitions from square to circle. Of course, other reconstructions are possible. For example, the circle phenotype could have arisen independently in each of the eight species currently exhibiting that phenotype, or it could have arisen once at the point indicated by a dashed box deep in the phylogeny, followed by five instances of evolutionary reversal in each of the descendant clades that exhibit the square phenotype. Nonetheless, parsimonious inference of few evolutionary transitions, with each clade inferred to have experienced no evolutionary reversal, suggests that evolutionary change has been infrequent and that we might place high confidence in a parsimony reconstruction. By contrast, in (b), the square phenotype is again inferred to be ancestral, with six evolutionary transitions to the circle phenotype. However, a very different scenario, in which the circle phenotype is ancestral and squares are derived, requires only seven evolutionary transitions. In situations such as this, we can safely conclude that evolutionary change must have been frequent, occurring at least six times, but we probably wouldn't want to place much confidence in particular scenarios; given that evolutionary change has occurred at a high rate, a scenario requiring six transitions wouldn't seem to be much more strongly supported than another scenario requiring seven evolutionary events.

In recent years, sophisticated methods have been developed to quantify uncertainty in ancestral reconstructions (e.g., Schluter et al., 1997; Garland et al., 1999; Martins, 1999; see reviews in Ronquist, 2004; Garland et al., 2005; Hardy, 2006; Vanderpoorten and Goffinet, 2006). These methods use a model of trait evolution—often some variant of Brownian motion, which assumes that the amount of expected change is a function of time (as represented by branch lengths of the phylogeny)—to estimate the rate of change of a character based on the values of extant taxa and their phylogenetic relationships. With this rate, the methods can estimate not only the character state of ancestral taxa, but also the variance around that estimate. These methods generally produce the same conclusion arrived at for simple parsimony approaches—the more frequently character change occurs, the greater the uncertainty on estimates of ancestral character states (Fig. 5.3; Schluter et al., 1997; Oakley and Cunningham, 2000).[108]

But the news gets even worse: these models generally assume that evolutionary change has been non-directional. However, evolutionary trends, in which taxa all evolve in the same direction, are common in the fossil record. No method for reconstructing ancestral taxa can account for such trends; indeed, in the absence of fossil data, trends are undetectable. Several studies have shown that when evolutionary trends exist, ancestral reconstructions are highly inaccurate (Oakley and Cunningham, 2000; Webster and Purvis, 2002; but see Polly, 2001).

The unhappy conclusion is that we probably shouldn't have much confidence in ancestral reconstructions, except when the rate of character evolution is low relative to the frequency of cladogenesis. This exception is an important caveat, however, because many traits do, in fact, evolve slowly enough for ancestor reconstructions to be reliable. For example, the sorts of morphological characters used by systematists are often of this sort. Nonetheless, many of the characters that evolutionary ecologists work on do not evolve slowly (Frumhoff and Reeve, 1994). In particular, many studies are driven by the observation that certain traits evolve repeatedly. Although convergence is a fascinating phenomenon of great importance to evolutionary biology in general, and anole studies in particular, its widespread occurrence indicates that attempts to infer ancestral character states will often produce ambiguous outcomes.[109]

This is unfortunate, because many of the questions we would like to ask require estimation of ancestral character states: What was the ancestor like? Where did it live? How

108. Of course, there is a middle ground. Traits often evolve convergently many times in some parts of a phylogeny, and not in others; in cases such as this, ancestral reconstructions may be reliable in those parts of the tree experiencing relatively little trait evolution, but unreliable where levels of trait evolution and convergence are high.

109. A related point concerns the incorporation of phylogenetic information into statistical comparative analyses. This approach has become *de rigeur* for good reason, as many studies have shown that ignoring phylogenetic information can lead to inflated Type I error rates (Martins and Garland, 1991; Purvis et al., 1994; Díaz-Uriarte and Garland, 1996). Nonetheless, the underlying rationale for these methods is that closely related taxa are likely to be phenotypically similar because they have inherited their phenotype from a common ancestor and, consequently, possession of the same trait by two species experiencing the same environment does not constitute evidence that the trait has evolved multiple times in response to the same selective pressure.

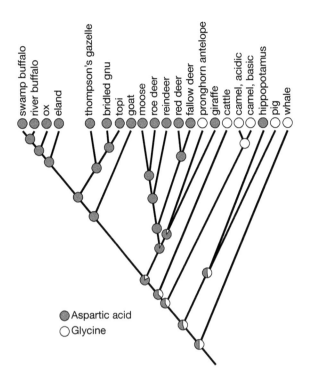

FIGURE 5.3

Maximum likelihood method for assessing support for ancestral reconstructions. Pies represent the relative strength of support for reconstructing the state of an ancestral node as one of two types of amino acid residue (note that this figure predates reconsideration of the phylogenetic position of cetaceans vis-à-vis artiodactyls). Parsimony reconstruction would infer that glycine was the ancestral state, with two transitions to aspartic acid (one on the branch leading to hippos and the other on the branch leading to the major clade in which all but pronghorns have aspartic acid) and one reversal back to glycine in the pronghorn antelope. By contrast, maximum likelihood methods reconstruct aspartic acid as the ancestral state throughout the tree with five transitions to glycine. However, these reconstructions are not strongly supported for ancestral nodes deep in the tree, as indicated by the pie charts (modified from Schluter et al. [1997] with permission).

However, if character change has been sufficiently rapid relative to the rate of speciation, then closely-related species would not necessarily be expected to be phenotypically similar. Consequently, if no relationship exists between phenotypic similarity and degree of phylogenetic relatedness, then there may be no benefit to incorporating phylogenetic information into statistical analyses. Given that using such information comes with a potential cost resulting from errors in phylogeny estimation or in misspecification of the model of evolution of the trait under study, incorporating phylogenetic information into statistical analyses might not be the best course in such situations (Gittleman and Luh, 1994; Bjørklund, 1997; Losos, 1999). This view, however, is not universally shared; some workers contend that phylogenetic information always should be used in comparative analyses (see discussion in Garland et al., 2005; Carvalho et al., 2006). Moreover, this view does not argue against the importance of a phylogenetic perspective, for only with a phylogeny can one investigate whether trait variation among species is correlated with phylogenetic relatedness (i.e., whether a "phylogenetic effect" exists [Losos, 1999]).

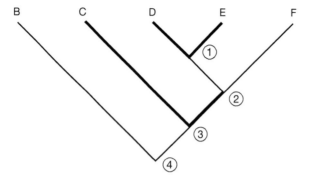

FIGURE 5.4

The difference between ancestor reconstruction and non-reconstruction approaches. In the former, ancestral character states are inferred for each node in the phylogeny and then the amount of change that occurred along each branch is calculated by subtracting the value of the ancestor from that of the descendant. In the independent contrasts approach, the difference between each pair of sister taxa—termed a "contrast"—is calculated. Pairs of sister taxa can be extant species, ancestral nodes, or one of each. The four contrasts are indicated by line shading in the figure. Note that the contrasts method includes as part of its algorithm a step in which a value is assigned to an ancestral node in the phylogeny, but this occurs solely for algorithmic purposes and should not be interpreted as an estimate of the ancestral character state (Felsenstein, 2004).

many times did the trait evolve? Evolutionary ecologists will have to accept that some questions may be unanswerable, at least with any confidence, the data erased in the fog of time.[110]

One way of getting around this problem is to ask questions in a way that does not require ancestor state reconstruction. Many (but not all) questions can be rephrased such that they only need consider a phylogeny and the character values of the taxa included in it—this is Harvey and Purvis's (1991) distinction between directional and non-directional approaches.[111] The clearest example of this is in determining whether evolution in one trait is correlated with evolution in a second trait. The ancestor reconstruction approach is to estimate ancestral traits, calculate the amount of change in both traits on each branch of the phylogeny, and then ask whether changes in one trait are correlated with changes in the second trait (e.g., Huey and Bennett, 1987; Losos, 1990b). The non-reconstruction approach is exemplified by the independent contrasts approach, which calculates the amount of difference between each pair of sister taxa—both extant species and internal nodes of the phylogeny—in a phylogeny (Fig. 5.4).

110. This highlights the major advantage of a fossil record: it provides a direct view of the past, as opposed to the inferences that must be drawn from phylogenies when one only has data on extant taxa. Of course, establishing that a fossil taxon is actually the ancestor of either another fossil taxon or a modern taxon can be problematic (see discussion in Wagner and Erwin, 1995).

111. So named because in ancestor-to-descendant comparisons, the direction of change is specified, from the ancestral state to the descendant one. By contrast, when sister taxa differ, evolution must have occurred, but such comparisons do not imply the direction in which the change occurred.

Many other questions can be framed in ways that do not require ancestral character state reconstruction. For example, instead of asking whether there are "stages" of evolutionary radiation, in which one trait evolves early in a clade's phylogenetic history and a second trait evolves more recently (see Chapter 15), one might ask whether clades that are invariant for the first trait exhibit interspecific variation in the second trait (Fig. 5.5). This would be an expected outcome if, in fact, the stages of radiation exist, but it illustrates the way a question can be turned around and investigated without requiring the reconstruction of ancestral states. Of course, the questions are not quite the same: the stages hypothesis would suggest that the second trait exhibits evolutionary change only recently, and not deep in the tree, but the non-directional approach does not shed insight on that question; rather it only investigates how the second trait diversified in clades that are fixed for the first state (see Ackerly et al., 2006).

This is the ying-and-yang of non-directional phylogenetic approaches. They avoid the need to reconstruct ancestral states, but at the cost of not being able to address, at least to some extent, hypotheses that require ancestor reconstruction. Unfortunately, some questions can only be addressed by reconstructing ancestral states. Such studies should proceed with caution.

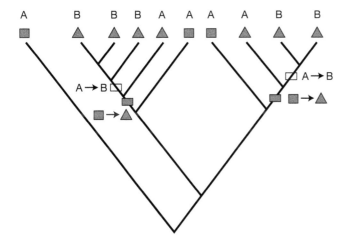

FIGURE 5.5
Approaches to investigating the temporal order of trait evolution. In the ancestor reconstruction approach, ancestral states would be reconstructed to determine whether one character state consistently evolved before another one: in this scenario, character state triangle in the first trait evolves prior to character state B in the second trait. An alternative approach that does not require ancestor reconstruction would ask whether clades that are invariant for one of the traits (in this case, clades that are invariant for state the triangle state) exhibit variation in the second trait. If evolution in one trait generally precedes evolution in the second trait, then such a pattern would be expected.

A second shortcoming of phylogenetic approaches has received detailed attention only recently. A phylogeny represents the best hypothesis for evolutionary relationships of the group under study. As such, the phylogeny likely is incorrect to some extent, and a battery of methods has been developed to assess the strength of support for different clades within a phylogenetic tree (Felsenstein, 2004).

Most comparative studies employing a phylogenetic perspective, however, take the preferred phylogenetic hypothesis as a given and base analyses and conclusions on this single phylogeny. Yet, the obvious possibility is that results would change if the analysis were performed on other, slightly less preferred, phylogenetic hypotheses.

The solution is to integrate over the universe of possible phylogenetic hypotheses, weighting the results from each phylogeny by how strongly it is supported (Felsenstein, 1988; Losos and Miles, 1994; see also Swofford, 1991; Maddison and Maddison, 1992). Although the idea has been around for more than a decade, its implementation was ad hoc and somewhat arbitrary (Richman and Price, 1992; Losos, 1994b; Martins, 1996; Donoghue and Ackerly, 1996). Now, however, the analytical and computational methods are in place to implement this approach in a sophisticated and statistically rigorous manner (e.g., Huelsenbeck et al., 2000, 2003; Pagel et al., 2004; reviewed in Ronquist, 2004).

The drawback to this approach is that it is still computationally intense, and computer programs are just now being developed. As a result, these approaches are just beginning to be used, but I predict they will become routine and expected within a few years.

ANOLIS PHYLOGENY
A BRIEF HISTORY

With these considerations in mind, I now turn to the phylogeny of *Anolis* and what we can learn from it. I first review anole phylogenetics in this chapter, and then in the next two chapters discuss the inferences we can draw about anole evolutionary history by taking a phylogenetic perspective.

Anole systematics represents in many respects a microcosm of the systematic world: as new types of data and methods have become available over the past four decades, they have been quickly put to use by anole systematists.[112] What follows in the next few paragraphs is the CliffsNotes® version; a more complete history can be had by consulting Guyer and Savage (1986), Williams (1989), Jackman et al. (1999), and Poe (2004) and working back from there.

---

112. I do not intend to review methods of phylogenetic data collection or analysis. Good entrées to the literature on these topics can be found in Hillis et al. (1996) and Felsenstein (2004).

Our understanding of anole phylogenetics traces to Etheridge's unpublished Ph.D. dissertation (1959),[113] which was based on osteological data for 12 characters, many taken from radiographs (x-rays). Etheridge divided anoles into two groups, α and β, based on the absence or presence (respectively) of transverse processes on the posterior tail vertebrae. Etheridge also identified a number of groups, termed "series," within each of the two major groups, and suggested that the other anoline genera (e.g., *Chamaeleolis*, *Chamaelinorops*, and *Phenacosaurus*) all arose from within *Anolis*. Williams (1976a,b) proposed an informal taxonomy based on Etheridge's work, as well as other data, that divided anoles into a nested hierarchy of groups ranging from sections through subspecies.[114] Immunological studies using the method of microcomplement fixation[115] agreed with much of Williams' taxonomy, but also produced some surprises: some osteologically similar species were found to be distantly related, Caribbean βs were proposed to be more closely related to Caribbean αs than to mainland βs, and *Chamaelinorops* and *Chamaeleolis* were discovered to nest well within *Anolis* (Wyles and Gorman, 1980; Shochat and Dessauer, 1981; Gorman et al., 1984; Hass et al., 1993).

Guyer and Savage (1986) were the first to investigate anole phylogenetics using quantitative phylogenetic methods. Their parsimony analysis of Etheridge's osteological data, as well as consideration of karyological and immunological data, revealed that βs were monophyletic and nested within αs, and led to the proposal that *Anolis* be divided into five genera: *Norops* (for the β anoles), *Anolis*, *Ctenonotus*, *Dactyloa*, and *Semiurus*.[116] The three other anoline genera were all found to lie outside of the clade comprising these five genera (i.e., outside *Anolis* in the former, broader sense). Guyer and Savage's work was strongly criticized for a variety of reasons, including the quality of the data employed and the methods used (Cannatella and de Queiroz, 1989; Williams, 1989; rebuttal: Guyer and Savage, 1992).[117]

113. Sometimes cited as 1960. The dissertation was completed and submitted in 1959, and the degree officially awarded the next year (R.E. Etheridge, pers. comm.).

114. The entire list of nested groups is: section, subsection, series, subseries, species group, species subgroup, superspecies, species, subspecies. Application of these names has been inconsistent through the years. The most recent and thorough list of assignments to series and lower levels was presented by Savage and Guyer (1989). I employ their classification for the remainder of the book, except in those cases in which their taxa do not represent currently recognized clades; those cases are detailed in the Afterword.

115. A method widely used in the 1970s and early 1980s to investigate phylogenetic relationships that involved measuring the extent to which proteins from one species reacted against antisera developed from the protein of a second species—the stronger the reaction, the more closely related the species were thought to be. The degree of reaction was measured using a method called quantitative microcomplement fixation; basically, the greater the concentration of the antiserum that had to be used, measured in immunological distance (ID) units, the more dissimilar were the albumins of the two species.

116. This latter genus was renamed "*Xiphosurus*" by Savage and Guyer (1991) on the basis of nomenclatural priority.

117. Two other types of data used to study anole relationships come from the study of karyology and gel electrophoresis. A great wealth of karyological data were collected from the late 1960s to early 1980s (e.g., Gorman and Atkins, 1968, 1969; Blake, 1983), but phylogenetic interpretation of these data has proven difficult (see Williams [1989] and Guyer and Savage [1992]). For the most part, electrophoretic studies conducted in the 1970s and 1980s focused on low-level clades within *Anolis* and did not address higher-level anole systematics (e.g., Yang et al., 1974; Gorman et al., 1980, 1983; Hedges and Burnell, 1990). One exception was Burnell and Hedges' (1990) study, which failed to find support for many previously described groupings (see critique in Guyer and Savage, 1992).

By the early 1990s, the state of anole phylogenetics was not a happy one. Despite 30 years of research, much remained uncertain about anole relationships. Osteological analyses, based on the few characters originally examined in the 1950s, gave a weakly supported phylogenetic signal, which was contradicted in a variety of ways by various molecular methods. Many informally described lower-level groups received support from various analyses, though results using different methods were often contradictory. Particularly frustrating was the apples-and-oranges aspect of the various studies: because these studies had used so many different techniques, with different sets of species and different analytical methods, combining the results in any sophisticated way seemed nearly impossible (see Cannatella and de Queiroz, 1989; Guyer and Savage, 1992).

THE DNA ERA

Since the early 1990s, phylogenetic work on anoles and all other organisms has been dominated by DNA sequence data. DNA-based studies of higher-level anole relationships have clarified many, but not all, issues (Hass et al., 1993; Jackman et al., 1999; Nicholson, 2002, 2005; Nicholson et al., 2005; the Afterword presents the complete 187-species phylogeny of Nicholson et al. [2005]). One clear result that has come from this work is the finding that anole phylogeny is composed of many reasonably well-supported clades that all arose within a relatively short period of time (Jackman et al., 1999; Nicholson et al., 2005; Fig. 5.6). These clades for the most part correspond to groups previously identified in the informal anole taxonomy. Whether these clades diverged sequentially, but so rapidly that it is now difficult to distinguish the pattern of branching among them, or whether the lack of phylogenetic branching structure in parts of the phylogeny are an accurate reflection of simultaneous, "star burst," origins of these clades cannot be determined (Jackman et al., 1999; see Poe and Chubb [2004] on a general approach to this question). This finding of very short branches deep in the tree

FIGURE 5.6

Anolis phylogeny. This phylogeny is based on mitochondrial DNA data (Nicholson et al., 2005), with branch lengths made proportional to time using the program r8s (Sanderson, 2003; the unabridged phylogeny is presented in the Afterword). The clades correspond to the 17 well supported clades identified by Jackman et al. (1999), except that three clades have been divided to highlight geographical differences (Dactyloa and Norops are split into three clades each and Chamaeleolis and its Hispaniolan sister taxon are indicated as separate clades). Statistical support for the existence of these clades is generally quite strong: the 17 clades have Bayesian posterior probabilities of 90–100%, and most have bootstrap support in maximum parsimony analyses greater than 80%; only some of the subclades within the three divided clades have this level of support (Nicholson et al., 2005). Some of the deeper clades within the phylogeny are well supported, but others are not. Geographic distributions do not include some smaller islands. Species counts are approximate, as new species are described every year, particularly within mainland Norops and Dactyloa. See the Afterword for more information on clade names.

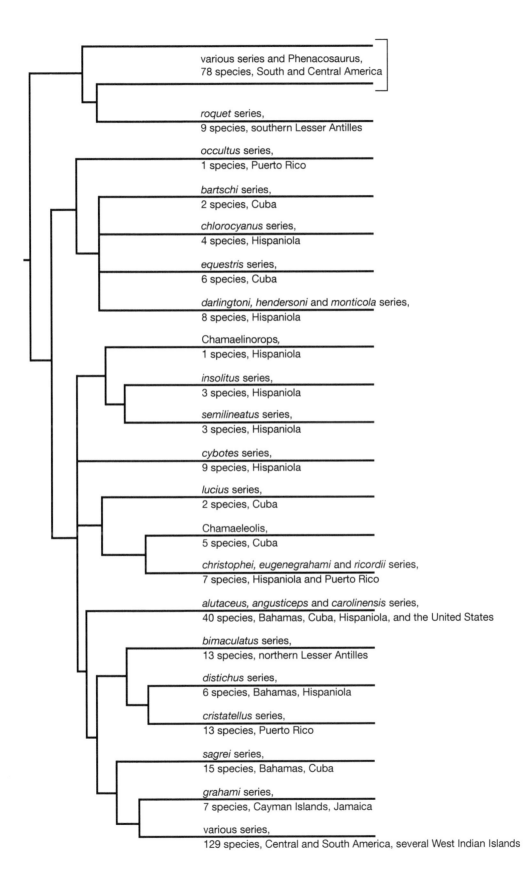

various series and Phenacosaurus,
78 species, South and Central America

roquet series,
9 species, southern Lesser Antilles

occultus series,
1 species, Puerto Rico

bartschi series,
2 species, Cuba

chlorocyanus series,
4 species, Hispaniola

equestris series,
6 species, Cuba

darlingtoni, hendersoni and *monticola* series,
8 species, Hispaniola

Chamaelinorops,
1 species, Hispaniola

insolitus series,
3 species, Hispaniola

semilineatus series,
3 species, Hispaniola

cybotes series,
9 species, Hispaniola

lucius series,
2 species, Cuba

Chamaeleolis,
5 species, Cuba

christophei, eugenegrahami and *ricordii* series,
7 species, Hispaniola and Puerto Rico

alutaceus, angusticeps and *carolinensis* series,
40 species, Bahamas, Cuba, Hispaniola, and the United States

bimaculatus series,
13 species, northern Lesser Antilles

distichus series,
6 species, Bahamas, Hispaniola

cristatellus series,
13 species, Puerto Rico

sagrei series,
15 species, Bahamas, Cuba

grahami series,
7 species, Cayman Islands, Jamaica

various series,
129 species, Central and South America, several West Indian Islands

is consistent with the findings of immunological studies (Shochat and Dessauer, 1981; Hass et al., 1993) and may explain why the results of earlier phylogenetic studies on anoles were so inconsistent with regard to the relationships of major anole clades.

The DNA data also clearly resolve the question of the position of the three other anoline genera. In contradiction to the morphological analysis of Guyer and Savage (1986), but in agreement with various earlier molecular studies, *Chamaelinorops*, *Chamaeleolis*, and *Phenacosaurus* all are nested within *Anolis*. Thus, Williams' (1969; Case and Williams, 1987) hypothesis that *Chamaelinorops* and *Chamaeleolis* are surviving relicts of a pre-*Anolis* anoline radiation in the West Indies is disproven. Rather, these genera represent highly divergent forms that have evolved from more typical anoles.

By contrast, in agreement with Guyer and Savage (1986), but contradicting some molecular studies (e.g., Shochat and Dessauer, 1981; Gorman et al., 1984; Burnell and Hedges, 1990), the DNA data indicate that β anoles (Norops) are monophyletic and nested within α anoles. All of the other genera proposed by Guyer and Savage (1986), however, are not monophyletic.

As stated above, Williams' (1976b) informal taxonomy for West Indian anoles has stood up fairly well. Many series and species groups are not monophyletic, but often this is the result of misplacement of one or a few taxa; much of Williams' lower level taxonomy is reflected in current phylogenetic hypotheses. By contrast, the series and species groups established by Williams (1976a) and others for mainland β anoles[118] (reviewed in Nicholson, 2002, 2005) find little support in recent phylogenetic work (Nicholson, 2002; Poe, 2004). Why Williams, in a pre-cladistic fashion, was so much more successful in using morphology to discover coherent groups in the West Indies than in the mainland is not altogether clear. The lack of osteological characters differentiating the mainland βs, perhaps resulting from the relative youth of the clade, probably is an important factor.[119]

Coincident with the upsurge in DNA work has been a long overdue reexamination of anole morphology. Poe (1998, 2004) has augmented Etheridge's original 12 characters,[120] producing a data set of 91 morphological characters for 174 anole species plus seven outgroups. Phylogenetic analysis of this data set finds both similarities and differences from the DNA phylogeny and from previous morphological work. As in previous studies, many informal groups are monophyletic, or nearly so, in the most parsimonious tree. In contrast to Guyer and Savage's (1986) morphological study, however, all three other anoline genera nest within *Anolis*, as with the DNA studies. In addition, in contrast to both the DNA phylogeny and Guyer and Savage (1986), the β anoles are not

118. Relationships of mainland αs have yet to be examined with molecular data.

119. Thanks to Richard Etheridge for suggesting this to me. Another possibility suggested to me is that Williams was more interested in and familiar with West Indian species. Although this may be true, Williams actually published many more papers on the taxonomy and systematics of taxa from the mainland than on those from the West Indies.

120. Treated as 15 characters by Guyer and Savage (1986).

monophyletic, but rather form separate mainland and West Indian clades. In a variety of other ways, the morphological tree differs from the DNA tree; I attribute much of this difference to ecomorphic convergence, as discussed in Chapter 7.[121]

Poe (2004) also conducted an analysis including not only the morphological data, but also DNA, electrophoretic, karyological and immunological data.[122] The preferred phylogenetic hypothesis from this analysis differs only in minor ways from the trees produced from only DNA data; many of the differences concern relationships deep in the phylogeny that are not well supported in either analysis.[123] The similarity between the combined data phylogeny and the DNA phylogeny indicates that the DNA data play a dominant role in structuring the combined analysis tree. Partly, this simply reflects the preponderance of the data; more than half of the data in the combined analysis is from mitochondrial DNA (mtDNA).[124]

ANOLE TAXONOMY

Anolis exemplifies the sort of taxonomic situation that has generated controversy in recent years. In the old days, under the rubric of evolutionary classification, systematists generally endeavored to recognize monophyletic taxa, but with the proviso that the classification system should also highlight evolutionarily distinctive taxa (Mayr, 1969; Wiley, 1981). Thus, for example, birds are so different from other reptiles that they merit being placed in their own class, even if it renders the class Reptilia paraphyletic. The advantage of this approach is that it draws attention to highly distinctive taxa. In this light, one might argue that *Chamaelinorops* and *Chamaeleolis* are so distinctive that they deserve to retain their generic level status, even if doing so renders *Anolis* paraphyletic.[125]

121. Poe (2005) shows that several ecomorph characters (e.g., hindlimb, tail length) contain phylogenetic signals (i.e., species with a similar character state are more closely related phylogenetically than would be expected by chance). The reason, as Poe (2005) notes, is that many ecomorphs are represented by clades of 2–14 species on some islands. Ecomorph characters such as hindlimb length do indicate phylogenetic relatedness at this low level. However, deeper in anole phylogeny, convergence results in ecomorph characters not reflecting phylogenetic relatedness, leading phylogenies based on morphology to produce, for example, separate clades of twig, crown-giant, and grass-bush anoles, each containing species from three or more islands (Poe, 2004).

122. This analysis, rather than the morphological one, was the centerpiece of Poe (2004). Some electrophoretic and immunological data were not used either because they overlapped with similar data in other studies or because they could not be coded as character data. More recent unpublished analyses that also include the DNA data from Nicholson et al. (2005) do not change the phylogenetic results in any important way (S. Poe, pers. comm.).

123. At least as far as shared taxa are concerned. DNA data are only available for about half of the species in the combined data phylogeny. One valuable aspect of morphological data is that it can be collected from museum specimens and thus is obtainable for species for which we cannot obtain DNA data because the species are rare or otherwise difficult to collect (obtaining DNA from museum specimens of *Anolis*, most of which were preserved in formaldehyde, is still an iffy proposition in most cases).

124. Nicholson et al. (2005), building on Jackman et al. (1999), presented data on 1408 base pairs of mitochondrial DNA for 187 species; a number of other, smaller studies have used mitochondrial DNA as well. In addition, Nicholson (2002) added 2077 base pairs of the nuclear ITS-1 region for 54 species.

125. The case for *Phenacosaurus* is not so strong (for an opposite view, see Lazell [1969]). This clade of 11 South American anoles does have some distinctive features, but in many ways, it seems not so different from other anoles, particularly twig anoles. Of course, just how distinctive a taxon needs to be to merit generic recognition is completely subjective, as is the question of which traits to consider.

The problem with evolutionary classification is that paraphyletic groups do not accurately portray phylogenetic history (Eldredge and Cracraft, 1980; Wiley, 1981; Frost and Hillis, 1990). Widespread agreement with Hennig's (1966) redefinition of the concept of monophyly has led to now nearly universal acceptance of his view that all taxa should be monophyletic. In the case of anoles, that leaves two options: either place everything in *Anolis*, or name a large number of separate genera. Indeed, Guyer and Savage's proposal was to break *Anolis* into five clades representing monophyletic subgroups. Unfortunately, although one of their genera, *Norops*, has proven monophyletic, all of the others have not. To recognize *Norops* as a genus would require establishing many other genera, probably 16 or more (Jackman et al., 1999); moreover, given the uncertainty of deeper-level anole relationships, such a classification system might need to be revised quickly, as more phylogenetic information becomes available. For this reason, most workers have favored retaining *Anolis* for all anoles.[126]

Although I understand and accept this logic, it nonetheless makes me sad. *Chamaeleolis* truly is a unique lizard, different in so many ways from all other anoles. Ditto for *Chamaelinorops*. They deserve their names! Fortunately, there is a happy solution. Currently, there is a movement to scrap the traditional nomenclatural system (at least for clades, as opposed to species), which is tied heavily to taxonomic ranks, and replace it with one in which names are more strongly tied to clade concepts than to ranks (de Queiroz and Gauthier, 1992; de Queiroz and Cantino, 2001; Donoghue and Gauthier, 2004). This system would enable systematists to name clades at any level, without placing undue emphasis on those associated with certain arbitrary ranks (e.g., genus). Thus, we might recognize:

Anolis
   Chamaeleolis
      barbatus

or

Anolis
   Norops
      grahami

The efforts to formally replace the traditional system with this new one are progressing slowly. Nonetheless, the general idea is a good one. Nicholson (2002) and Nicholson et al. (2005) adopted variants of this approach to identify taxa that belong to the Norops clade. In the following, I will insert non-italicized clade names within the scientific name where doing so provides useful information, e.g., *Anolis* Chamaelinorops *barbouri*.[127]

---

126. Although *Norops* remains popular among scientists and students working in Central America.

127. Of course, one could insert multiple, hierarchical clade names if that were useful, such as *Anolis* Norops grahami *conspersus* to indicate that *A. conspersus* is a member of the *grahami* Species Group clade within the Norops clade. This convention, though arbitrary in its use, is a way of indicating the phylogenetic relationship of a species within the context of the Linnean binomial classification system, and for this reason seems to me to be quite useful.

To my mind, this solution represents the best of all possible worlds; we can have our cake and eat it too: all taxa are monophyletic, yet evolutionarily distinctive taxa can be recognized.

FUTURE DIRECTIONS

Despite the substantial work to date, much still remains to be discovered about anole phylogenetics. A pressing question concerns higher-level relationships among anoles: Does the base of the anole tree truly represent a series of rapid and nearly simultaneous divergence events, or will the addition of data from more genes provide resolution? To date, most studies of interspecific relationships within *Anolis* have relied on mitochondrial DNA genes. However, studies that have employed both mitochondrial and nuclear genes have found a general correspondence in results from the two markers (Glor et al., 2005; Nicholson et al., 2005), and several more extensive projects currently in progress appear to indicate that phylogenies built from nuclear DNA data give much the same general result as published studies based on mtDNA. The forthcoming *A. carolinensis* genome should help identify many more gene regions suitable for addressing questions concerning higher-level anole relationships.

The phylogeny of mainland anoles also needs clarification. In the case of mainland *Norops*, greater taxon sampling as well as data from more genes may prove useful. The other mainland anoles, the basal group of α anoles sister to the rest of *Anolis*,[128] also require further work. This group is probably the most poorly sampled of any of the anoles, with molecular data for less than 20% of described species. Although Poe's (2004) combined data analysis found the basal αs to be monophyletic, the morphological analysis suggested that mainland αs occurred along the eleven most basal branches of the anole tree, forming a paraphyletic group from which Caribbean taxa arose. Addition of taxa here seems most likely to change our understanding of anole relationships by providing clarity to what occurs at the base of the anole tree.

Finally, relationships within well supported West Indian clades require further work. The *bimaculatus* (Schneider et al., 2001; Stenson et al., 2004), *carolinensis* (Glor et al., 2004, 2005), *cristatellus* (Brandley and de Queiroz, 2004), *cybotes* (Glor et al., 2003), *grahami* (Jackman et al., 2002), *roquet* (Giannasi et al., 2000; Creer et al., 2001) and *sagrei* (Knouft et al., 2006) Series have been studied recently, but even in some of these groups, questions remain.[129] In addition, many other groups have not been studied since the advent of molecular approaches (e.g., crown-giant and grass-bush clades on Cuba and Hispaniola, the *angusticeps* Series on Cuba and several montane Hispaniolan clades,

---

128. Which corresponds to Guyer and Savage's (1986) *Dactyloa*, with the inclusion of *Phenacosaurus*, and to which I henceforth will refer to as the Dactyloa clade.

129. E.g., resolution of relationships of *A. bimaculatus*, *A. gingivinus*, and *A. leachii* in the *bimaculatus* Series or of *A. gundlachi* and the grass-bush anoles in the *cristatellus* Series.

many of which were placed in the *monticola* Series, but are now known not to form a monophyletic group). Moreover, some of the larger groups indicated by the phylogenetic analyses (e.g., the Hispaniolan clade containing *A. insolitus, A. barbouri,* and others, or the clade including *A. occultus, A. darlingtoni, A. bartschi, A. vermiculatus,* and the *equestris* and *chlorocyanus* Series) need further work to investigate whether their monophyly holds up and, if so, to determine relationships within these groups.

6

PHYLOGENETIC PERSPECTIVE ON THE TIMING AND BIOGEOGRAPHY OF ANOLE EVOLUTION

Our current understanding of anole phylogeny (Chapter 5) provides substantial insight into the evolution of the anole faunas. Throughout the rest of the book, I will frequently use this knowledge to address questions concerning the origin and maintenance of anole biodiversity. In this chapter, I will focus on two seminal, if at times maddeningly inconclusive, issues: When did anoles arise? And how did they attain their current geographic distribution?

WHEN DID ANOLES ARISE?

Evolutionary biologists are accustomed to thinking of island radiations as being young in geological terms. This perception no doubt stems from the fact that many of the most famous radiations occur in relatively young localities, such as volcanic islands that emerged only within the last ten million years (e.g., the Galápagos and Hawaiian Islands [Carson and Clague, 1995; Rassman, 1997; Grant and Grant, 2008]) and many of the African Rift Lakes (Seehausen, 2006). The Greater and Lesser Antilles are considerably older than these locales, and their anoles appear to match this antiquity.

FOSSIL DATING

Ideally, we would look to the fossil record to date the evolutionary appearance of these lizards. Unfortunately, the fossil record is quite scant, consisting of four specimens in amber from Mexico and the Dominican Republic (Lazell, 1965; Rieppel, 1980; de Queiroz et al., 1998; Polcyn et al., 2002), all dating to the mid-Miocene, approximately

15–20 million years ago (mya; Iturralde-Vinent, 2001). These fossils indicate that the anole radiation is not a recent one, but provide only a minimum estimate of anole age. The Dominican fossils are indistinguishable from an extant anole clade (the green anoles of Hispaniola [*chlorocyanus* Series]) that originated some time after the earliest divergence events in anole phylogenetic history (Fig. 5.6), which suggests that the first anole must have lived at some earlier time (more on this point below; the Mexican specimen is composed only of skin and does not exhibit characters allowing it to be placed phylogenetically).

MOLECULAR DATING

The alternative approach is to try to date the origin of *Anolis* by examining the amount of molecular divergence that has occurred between species and by correlating this divergence with rates of change through time. This so-called "molecular clock" has a checkered history, but has become increasingly sophisticated in recent years (e.g., Arbogast et al., 2002; Near et al., 2005; Britton et al., 2007; Kitazoe et al., 2007; reviewed in Rutschmann, 2006).

The chronology of anole evolution has been examined using two completely independent molecular approaches. The first is the comparison of albumin proteins using the method of microcomplement fixation described in the previous chapter (footnote 115). Based on studies of a variety of species, an albumin molecular clock in which 1.7 ID [Immunological Distance Units] = 1 million years of divergence was established (Gorman et al. [1971] for reptiles; more generally, see Wilson et al. [1977]). The maximum ID difference among anoles is 67, suggesting that diversification within *Anolis* began nearly 40 million years ago (Shochat and Dessauer, 1981).[130]

More recently, molecular clock approaches have been applied to DNA sequence data. Based on analyses of another iguanian lizard clade (the Asian agamid genus *Laudakia*), Macey et al. (1998a) suggested that certain regions of mtDNA evolve at a rate of 0.65% per lineage per million years. This is consistent with rates obtained for this region in other ectotherms (Macey et al. 1998b; Weisrock et al. 2001) and with general estimates of approximately 2% pairwise divergence per million years for mtDNA in other animals (e.g., Brower, 1994). Nevertheless, these calibrations are generally based on relatively young events for which multiple changes at the same DNA site probably occur infrequently; extrapolating this rate to much older divergence events requires correcting sequence divergence estimates to account for the greater probability

130. I restrict this discussion to Shochat and Dessauer's (1981) study, which is by far the most inclusive immunological study of anole relationships with serum from seven species reacted against antigens from 40 species. Wyles and Gorman (1980) produced antiserum from one species and obtained results compatible with those of Shochat and Dessauer. By contrast, Gorman et al. (1984) developed antiserum from another species, A. Norops *gadovi* from Mexico, but their ID values are substantially higher, which suggests either an accelerated rate of evolution in the mainland Norops clade or an artifact stemming from comparisons conducted in different laboratories. The pattern of phylogenetic relationships suggested by Gorman et al. (1984) is greatly at odds with current understanding of phylogeny (Chapter 5), whereas those of Shochat and Dessauer (1981), though not completely in agreement, are much more concordant.

that multiple hits have occurred.[131] Using maximum likelihood corrected mtDNA distances in anoles and applying the Macey et al. rate yields an estimated date for the initial divergence event within *Anolis* of 66 million years (Glor, unpubl.).[132]

An alternative approach is to calibrate the rate of molecular evolution using a range of fossils, some older and others younger than the divergence being studied. A second DNA analysis has taken this approach, using the nuclear *RAG-1* gene and calibrating with a variety of lizard fossils (following Wiens et al. [2006]), including the amber anoles. This study, not as yet published, provides an estimate of the age of the initial divergence within *Anolis* that is very similar to the date estimated using mtDNA (T. Townsend, pers. comm.).

The imprecision of molecular clocks, particularly when they are calculated through simple extrapolation, as done here, has been widely noted (e.g., Crother and Guyer, 1996; Bromham and Penny, 2003; Hugall and Lee, 2004; Rutschmann, 2006); consequently, the dates suggested by these methods should not be given too much credence. Large uncertainty probably exists in the estimates I have provided, but the simple methods available for these data do not allow this uncertainty to be quantified. Certainly, as more sequence data become available, more sophisticated methods will allow much more reliable estimation of divergence events. These caveats notwithstanding, it is striking that these two independent molecular methods provide dates that are roughly congruent, at least to the extent that they indicate an ancient origin for *Anolis*.

ANOLIS BIOGEOGRAPHY

Reconstructing a group's biogeographic history is one of those endeavors that necessarily requires inferences of ancestral character states, in this case ancestral geographic distributions.[133] As a result, such questions must be addressed cautiously (see chapter 5).

Examination of anole phylogeny (Fig. 6.1) reveals several clear patterns:[134]

- The basal divergence in the anole phylogeny separates a clade of primarily mainland anoles from a clade composed of many West Indian subclades and one large and deeply nested mainland clade.

131. That is, the rate of sequence divergence between two species decreases over time because some of the changes occur at DNA sites that have already changed at least once before, and thus do not increase the net difference between the species.

132. Macey et al. (1997) calibrated their estimate using geological events that occurred within the past ten million years. Glor corrected the estimated rate of sequence evolution for substitutional saturation. As Glor notes, extrapolating this calibration to events that occurred over a much longer time period is an iffy proposition.

133. As noted in the last chapter, sometimes questions can be rephrased to avoid the need for ancestral reconstructions. For example, the minimum number of movements from one area to another can be estimated without requiring inference of the location of each ancestor.

134. Here and for the remainder of the book, I rely on the anole phylogeny based solely on DNA data (Nicholson et al., 2005) for several reasons: first, as discussed in this chapter, DNA data can be used to estimate not only the phylogeny, but also the dates at which particular events occurred; second, the DNA phylogeny and the phylogeny based on all types of data are highly congruent (see chapter 5), with few important discordances, which will be noted as relevant; and third, in some cases phylogenies based solely on the morphological data are misleading because they group together members of the same ecomorph class from different islands.

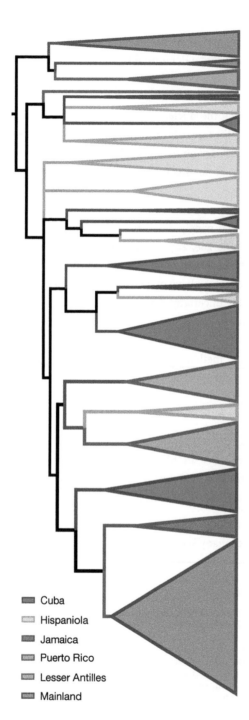

FIGURE 6.1
Anole biogeography in a phylogenetic context. The phylogeny is the same as Figure 5.6, with branch lengths set proportional to time, as in Fig. 5.6. Triangles are proportional to the number of species within each clade. Small islands and the Bahamas are not shown.

- Within the basal mainland anole clade resides the *roquet* Series anoles of the southern Lesser Antilles, indicating a separate colonization of those islands from the rest of the West Indies.
- Within the West Indian clade are a number of clades containing 1–32 species from a single island: Jamaica is occupied by one clade, Puerto Rico by three, Hispaniola by seven and Cuba by six. The northern Lesser Antilles are also occupied by a single clade.
- The mainland Norops radiation is a single clade nested well within the West Indian clade, indicating a back-colonization from the West Indies to the mainland.

These phylogenetic relationships clarify much about the biogeographic history of anoles, but leave unresolved important questions about Cuba, Hispaniola and Puerto Rico, to which I will return momentarily.

AREA OF ORIGIN OF *ANOLIS*

One common question in biogeographic studies concerns where a clade initially originated (i.e., its "center of origin"). In the case of *Anolis*, the answer appears straightforward: these lizards evolved in mainland Central or South America. This conclusion stems from two observations:

1. The closest relatives to anoles occur on the mainland. Traditionally, anoles were placed in a clade, the Polychrotinae, along with *Polychrus*, a small genus of dewlap-bearing, vaguely chameleon-like arboreal lizards from Central and South America, and a group of poorly known, primarily terrestrial, South American lizards referred to as para-anoles and anoloids (Fig. 6.2;

FIGURE 6.2

Species traditionally considered to be close to *Anolis*. (a) *Polychrus liogaster*. Photo courtesy of Greg Vigle. (b) *Diplolaemus darwinii*.

Etheridge, 1959; Etheridge and de Queiroz, 1988; Frost and Etheridge, 1989). Although reaffirmed by recent morphological analyses incorporating new fossil material (Conrad et al., 2007), the monophyly of the Polychrotinae is not supported by molecular studies, nor is the hypothesis that any of its constituent species are closely related to *Anolis* (Schulte et al., 1998, 2003; Frost et al., 2001). Nonetheless, given that *Anolis* is in the family Iguanidae, which occurs almost exclusively in North, Central, and South America (Pough et al., 2004), the sister group to anoles, whatever it is, probably occurs on the mainland.

2. A basal clade in *Anolis*, Dactyloa, occurs primarily on the mainland (Fig. 6.1). Most Dactyloa occur in South America or southern Central America. Exceptions are the *roquet* Series in the southern Lesser Antilles and *A. agassizi* on Malpelo Island in the Pacific; both of these clades represent colonizations from mainland Dactyloa ancestors, as discussed below.

Given that both the nearest relatives of *Anolis* as well as a basal clade within *Anolis* occur on the mainland, the most parsimonious interpretation is that the ancestral anole evolved somewhere on the mainland.[135] Reconstruction of ancestral character states with likelihood methods (as in Nicholson et al., 2005) confirms this impression: on the vast majority (89%) of phylogenetic trees produced in a Bayesian analysis, the ancestral anole is reconstructed as occurring on the mainland, usually with high support (Losos, Mahler, and Glor, unpubl.).

DIRECTION OF COLONIZATION

A longstanding view in biogeography is that colonization is primarily a one way street from continents to islands; island taxa are often viewed as competitively inferior to taxa continually tested by interspecific interactions in species-rich continental settings, and consequently the flow of potential colonists has been thought to be greater from continents to islands than in the reverse direction (Carlquist, 1974; Brown and Lomolino, 1998; Cox and Moore, 2000; but see Heaney [2007] for a reappraisal).

135. Two caveats to this conclusion must be mentioned. First, one recent study based primarily on mtDNA identified the curly-tailed lizards, *Leiocephalus*, which occur in the West Indies, as the sister group to Anolis (Schulte et al., 2003), although without strong statistical support. Another study using nuclear DNA did not find a close relationship between anoles and curly-tailed lizards (Townsend et al., 2004). Second, the Puerto Rican twig anole, *A. occultus*, has long perplexed anole phylogeneticists (Poe, 2004). It has many unique morphological and molecular features; Poe's (2004) combined data analysis places it as the basal clade in anole phylogeny, even though it is basal in neither morphological or DNA analyses. The possibility that the sister group to anoles and the basal clade within anoles are both West Indian suggests an alternative scenario in which the ancestor of *Anolis* + *Leiocephalus* occurred in the West Indies. This scenario would push back the mainland-to-West Indies colonization to a phylogenetically earlier point, the ancestor of the putative *Anolis* + *Leiocephalus* clade, and would indicate that the Dactyloa clade of *Anolis* recolonized the mainland, possibly several times (depending on whether the *roquet* Series was ancestrally in the West Indies or represented a re-colonization after its ancestors returned to the mainland). However, given that neither element of this scenario is strongly supported, this hypothesis is unlikely to be correct.

In *Anolis*, however, the street runs both ways. To test this hypothesis, Nicholson et al. (2005) conducted a state-of-the-art phylogenetic reconstruction exercise. This study examined the complete set of phylogenies included in the posterior distribution of a Bayesian analysis of mtDNA. On each tree, the likelihood that ancestral taxa at several nodes in the phylogeny occurred on the mainland or in the West Indies was calculated.

This analysis indicated that West Indian species are the result of two incursions from the mainland. The first is the *roquet* Series in the southern Lesser Antilles. Ancestor reconstruction supports the conclusion that the ancestor of the clade consisting of the *roquet* Series and the mainland Dactyloa occurred on the mainland (Fig. 6.3). Because *roquet* Series anoles occur on oceanic islands (Chapter 3), their presence there almost

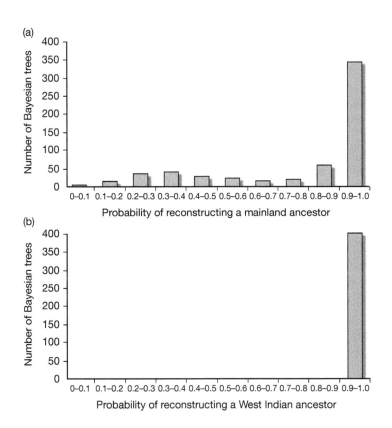

FIGURE 6.3

Ancestral location of the Dactyloa and mainland Norops clades. (a) Summing across all 539 Bayesian phylogenetic trees, Nicholson et al. (2005) found that in 80% of the phylogenies, the ancestral Dactyloa was reconstructed as occurring on the mainland; on 61% of the trees, a mainland ancestor was reconstructed with greater than 95% likelihood (modified from Nicholson et al. [2005] with permission). (b) Similarly, the ancestor of all Norops is reconstructed as occurring in the West Indies in 100% of phylogenies, and with greater than 95% likelihood in 96% of the trees.

certainly has resulted from overwater dispersal and island-hopping up the Lesser Antillean island chain. The remaining West Indian species (inhabitants of the northern Lesser Antilles, Greater Antilles and surrounding islands) belong to a second clade that entered the West Indies and diversified greatly (more on this shortly). In addition, at least three mainland-to-island overwater colonization events resulted in the species that occupy a number of Pacific and Caribbean islands (e.g., Malpelo, Cocos, San Andres [Williams, 1969]).[136]

On the other hand, anoles have reinvaded the mainland several times as well. Most notable is the Norops clade, which contains not only 22 species on Cuba, Jamaica and Grand Cayman, but also 127 species in Central and South America. Traditionally, the Norops species in the Greater Antilles had been viewed as the result of 1–2 dispersal or vicariance events from the mainland, but the phylogenetic data clearly indicate that the movement has occurred in the opposite direction (Fig. 6.3), and that, consequently, the mainland Norops clade is the result of an island-to-continent colonization event. In addition, anoles have colonized Florida and two species have established beachheads along coastal Central America (Williams, 1969).

OVERWATER COLONIZATION VERSUS VICARIANCE, AND THE GEOLOGY OF THE CARIBBEAN

When did anoles get to the West Indies, and how did they get from one island to another? It turns out that trying to answer this question is a messy business for two reasons: first, the geology of the West Indies is surprisingly uncertain, and second, current efforts to date the ages of particular divergence events produce results that are somewhat baffling.

A long-running debate in the field of biogeography concerns the extent to which species attain their current distribution by dispersing over water or by riding landmasses as they move, divide, and merge.[137] Distinguishing these possibilities requires a phylogeny for the species and an understanding of the geological history of the region. Surprisingly, at least to me, the geology of the Caribbean is not well understood due to the tectonic complexity of the region, as well as to the fact that the lack of petroleum deposits in most parts of the Caribbean has limited the amount of exploration by oil companies.

Nonetheless, the general story of Caribbean geology is understood, at least in very broad terms (see reviews in Pindell [1994], Pindell et al. [2006], Hedges [2001] and

---

136. As mentioned already, *A. agassizi* on Malpelo represents a separate colonization from mainland Dactyloa. By contrast, the anole species on the other islands originated from mainland Norops stock. These species represent island colonization from mainland ancestors that themselves are descended from ancestors that recolonized the mainland from the West Indies (as discussed in the next paragraph), whose ancestors in turn colonized the West Indies from the mainland.

137. The term "vicariance" in the title of this section refers to the isolation of populations by geographic barriers (e.g., landmasses splitting apart, mountains arising), leading to differentiation of populations on either side of the barrier.

Iturralde-Vinent [2006], but see James [2006]). The Proto-Antilles formed in the Pacific Ocean in the Cretaceous. They then moved through the passage between South and Central America and into the Caribbean by at least 70 mya (Fig. 6.4a). This block contained the landmasses that would become Cuba, Hispaniola, Puerto Rico, and the Virgin

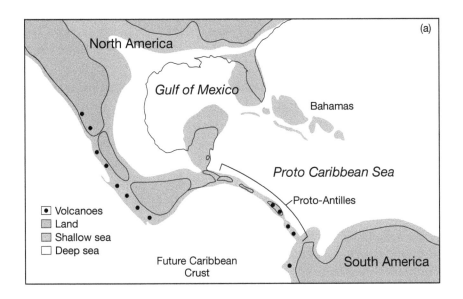

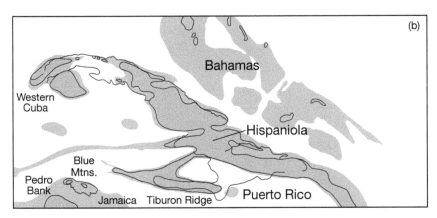

FIGURE 6.4

Geological evolution of the Caribbean. (a) Pacific Origin of the Proto-Antilles. By at least 70 mya, the landmass destined to form much of the Greater Antilles had passed through the Americas; other scenarios date this event substantially earlier in the Cretaceous (e.g., approximately 90 mya [Pindell, 1994]). (b) Proto-Antilles 33–35 mya. Landmasses now located in eastern Cuba, Hispaniola, and Puerto Rico are shown as connected. This scenario, proposed by Iturallde-Vinent (2006), also suggests that the Blue Mountains of Jamaica were emergent and connected to this landmass as well, a view that is not widely supported. Figures modified from Iturallde-Vinent (2006) with permission.

Islands. The histories of these islands are not particularly clear, and two issues are of central importance for the biogeography of terrestrial species: when were these landmasses connected and when were they above sea level?

Much uncertainty exists in this regard. Iturralde-Vinent (2006) suggested that the landmasses that today form Puerto Rico, Hispaniola, and Cuba were last connected some time around the late Oligocene (approx. 27–29 mya; Fig. 6.4b). The first separation was the formation of the Mona Passage, which separated Puerto Rico from Hispaniola. Subsequently, the Windward Passage divided what is now eastern Cuba from Hispaniola so that by the Middle Miocene, 14–16 mya, all three islands were distinct entities.

Jamaica has a different history from the rest of the Greater Antilles. It was connected to Central America until after the mid-Eocene, approximately 50 mya. It then broke off and moved eastward to its current position. However, in contrast to the other landmasses, Jamaica probably was completely submerged at various times in the past.

The first question to address is how anoles got to the Caribbean. All Greater Antillean anoles, as well as those of the northern Lesser Antilles, are the descendants of a single colonizing species. How did that species get to the Greater Antilles? One possibility is that it hopped onto the proto-Antillean block as it moved eastward from the Pacific and passed between North and South America (Fig. 6.4a). Whether the proto-Antillean block was actually connected to the continents is not clear, but at least it may have been close enough to facilitate dispersal.

The question then becomes: did *Anolis* evolve early enough to have been present when the proto-Antilles passed between the Americas? If we use the date for the first divergence in *Anolis* as being 40–66 mya according to molecular estimates, then we must conclude that the answer is "no": *Anolis* apparently had not evolved when the proto-Antilles moved through the Americas. In addition, Iturralde-Vinent (2006) suggested that the proto-Antillean islands were underwater subsequent to passing through the Americas, making the question moot. For these reasons, it seems most likely that the ancestral Antillean anole reached the islands by overwater dispersal some time after the Proto-Antilles passed from the Pacific into the Caribbean.[138]

138. Recently, Iturralde-Vinent and MacPhee (1999) have resurrected the hypothesis that a land bridge connected the Greater Antilles to the mainland some time in the past, providing a conduit for overland dispersal. Specifically, they postulate the existence of a continuous land connection along the Aves ridge extending from South America to the proto-Antillean landmass approximately 33–35 mya. This hypothesis has been criticized on a number of grounds (Hedges, 2001; rebuttal: Iturralde-Vinent, 2006). To me, the most troubling aspect of this hypothesis is the fact that, if a land bridge existed, so few taxa took advantage of it. Where are the many types of mammals that would have been expected to walk across this bridge (e.g., carnivores, lagomorphs, ungulates and marsupials), not to mention the salamanders and many types of turtles, snakes and frogs (Hedges, 2001)? With regard to anole timing, if we accept the immunologically-based age of 40 million years of *Anolis*, then anoles colonized the Greater Antilles approximately 37 mya, fairly close to the proposed date for the Aves Ridge land bridge. On the other hand, the DNA-derived date of 66 million years for the initiation of anole diversification would push the colonization so far back as to be incompatible with overland dispersal in this scenario.

Once anoles colonized the West Indies, then what? Had the proto-Antilles already fragmented such that further dispersal events were needed to populate all of the Greater Antillean islands or, alternatively, did this ancestral species radiate into many distinct species prior to separation of the landmasses? Moreover, how frequently has overwater dispersal occurred since the time of Antillean disintegration?

These questions can be addressed by examination of Figure 6.1. First, most of the species diversity on the Greater Antilles is the result of within-island diversification (i.e., diversification within the clades in Figure 6.1 whose constituent species occur only on one island). As a result, we can conclude that the 121 species of Greater Antillean anoles are the result of relatively few inter-island dispersal events. If, for example, Cuba were the home of the ancestral Antillean anole, and assuming that the anole reached the Antilles subsequent to the breakup of the proto-Antilles, then the phylogeny provides evidence of 9–11 dispersal events to other Greater Antillean islands.[139] Consequently, the great majority of Greater Antillean diversity has been produced by clades diversifying *in situ*, rather than from dispersal across islands[140].

Even this may be an overestimate of the extent of overwater dispersal, however, because it assumes that clades on different islands were established by dispersal from one island to another. An alternative possibility is that the ancestral anole arrived when (and if) the proto-Antilles were a single landmass, and that subsequent fragmentation of this landmass is responsible for the existence of related clades on different islands.

Addressing the issue of timing of diversification relative to the breakup of the proto-Antilles requires not only a phylogeny, but one in which the lengths of the branches have been adjusted to be proportional to time. The development of methods to estimate branch lengths is currently an area of active research (Arbogast et al., 2002; Near et al., 2005; Rutschmann, 2006; Marshall et al., 2006; and references therein). Thus the results I present below should be viewed cautiously and no doubt will be modified as methods improve and more molecular data are gathered.

Based on the branch lengths in Figure 6.1, we can infer that most major clades of *Anolis* originated in the first third of anole diversification. If we use the younger age for *Anolis* from immunological studies, this indicates that most clades had arisen by 27 mya, which is the time at which the Proto-Antilles began to separate and well before Hispaniola and Cuba separated (perhaps 14–16 mya). These figures indicate that the origin of anole clades on different islands cannot be explained as a result of vicariance associated

139. One to Jamaica, three to Puerto Rico, and 5–7 to Hispaniola, depending on how several polytomies in the phylogeny are resolved.

140. The logic is that if all members of a clade occur on an island, then the ancestor of that clade must have occurred on that island as well. One caveat to the statement that most speciation has occurred *in situ* is that several of the islands of the Greater Antilles were at times divided into multiple islands by high ocean levels. Some of what we recognize as *in situ* speciation may result from dispersal or vicariance producing allopatric populations on different parts of what today is a single large island (e.g., Glor et al., 2004).

with breakup of the Proto-Antillean landmass because the clades had already diverged prior to the geological separation.[141]

Using this chronology, two potential cases of relatively recent dispersal between the modern Greater Antillean islands are evident. First, a small clade of Hispaniolan twig anoles is nested within a large clade of Cuban anoles and is sister to Cuban twig anoles; using the immunological calibration, this dispersal event may have occurred 22 mya. Second, the Puerto Rican crown-giant A. *cuvieri* is sister to a clade of Hispaniolan species that includes crown-giants; this divergence also dates to about 22 mya.

Dispersal also occurred to Jamaica. This is not surprising, as most geologists believe that Jamaica represents an isolated geologic entity that was entirely submerged at some point after breaking off from Central America (reviewed in Buskirk, 1985; Robinson, 1994, but see Iturralde-Vinent, 2006). The estimated arrival date of the Jamaican anoles, about 24 mya, approximates some estimates of the emergence of the Jamaican landmass (Buskirk, 1985) and predates others (Vinent-Iturallde, 2006), but is in the same ballpark as the estimated arrival of *Eleutherodactylus* frogs (Hass and Hedges, 1991). Several other more recent dispersal events, not illustrated in Figure 6.1, are responsible for anoles of the oceanic islands near the Greater Antilles, such as Navassa, Mona, Grand Cayman, Little Cayman and St. Croix (Jackman et al., 2002; Brandley and de Queiroz, 2004; Glor et al., 2005). Divergence age estimates from the phylogeny (not shown), assuming the immunological calibration, indicate that most of these species arose 5–10 mya, which accords with more detailed molecular phylogenetic studies for some,[142] but not all,[143] of these species.

At face value, these data indicate that almost all major clades of anoles had begun radiating prior to fragmentation of the Proto-Antilles landmass, with very few dispersal events required. However, if this were the case, the obvious question arises: Why is each of these clades represented on only a single Greater Antillean island today? That is, if six Hispaniolan clades arose prior to the separation of what would become Hispaniola from the landmasses that became Cuba and Puerto Rico, why are these clades not represented on those other islands today?

The origin of most major anole clades prior to proto-Antillean fragmentation could be explained in two ways. One possibility is that even though the Proto-Antilles hadn't

141. These estimated dates of divergence are calculated by simple extrapolation. For example, if the basal divergence event within *Anolis* occurred 40 mya, and another event occurred half of the distance from the base of the phylogeny, then the second event is estimated to have occurred at 20 mya. If the older, DNA-based estimates of *Anolis* age are used, ages of divergence of the major clades would have been even older, and thus even less congruent with geological dates of island separation.

142. E.g., *A. acutus* (Brandley and de Queiroz, 2004) and *A. carolinensis* Species Group (Glor et al., 2005).

143. E.g., *A. conspersus* (Jackman et al., 2002) and *A. ernestwilliamsi* (Brandley and de Queiroz, 2004). The overestimate of divergence times for these two species from the Nicholson et al. (2005) phylogeny probably results because both of these species have arisen from within widespread species (*A. grahami* and *A. cristatellus*); these widespread species were only represented by single specimens in the Nicholson et al. phylogeny and thus the most closely related populations to *A. conspersus* and *A. ernestwilliamsi* probably were not included, leading to overestimation of their time of divergence.

separated, clades were restricted to those parts of the landmass that correspond to different islands today. Given that most of these clades are widely distributed across the islands on which they occur today, it seems hard to imagine that they would not have been equally widely distributed in the past.

The other possibility is that today's clades were widely distributed over the Proto-Antilles, but that subsequent extinction has left each clade represented only on a single Greater Antillean island. To me, this possibility seems unlikely. If the ancestors of the 16 clades represented on Cuba, Hispaniola, and Puerto were widespread across the proto-Antilles, I would expect at least some of these species to have managed to leave descendants on more than one island after vicariance.

This entire discussion, however, is based on the accuracy of molecular dating of divergence events and on the interpretation that the proto-Antilles were united into a single landmass. If we consider only the non-molecular data, the amber specimens place *Anolis* on Hispaniola minimally 15 mya. These fossils likely belong to the *chlorocyanus* species group on Hispaniola (de Queiroz et al., 1998; Polcyn et al., 2002), which diverged relatively early in the anole radiation (Fig. 5.6). If the origin of this clade is dated to 15 mya and assuming the branch lengths in Figures 5.6 and 6.1, then anole diversification may have begun as recently as about 19, rather than 40, mya.[144] One possibility, then, is that the molecular dates are too old and that most island clades originated much more recently, after the breakup of the Proto-Antilles had begun.

Even if anole diversification began 40 mya, or even earlier, another possibility is that the method of estimating branch lengths may tend to push events deep into the tree (Hugall and Lee, 2004); i.e., *Anolis* may have evolved in the distant past, but the early divergence events that produced the island clades may be more recent than estimated by molecular dating. Although possible, this explanation would require that the same problem occurs in two independent data sets, DNA and immunological, both of which suggest the existence of many divergence events early in anole history. More molecular data and more sophisticated methods for branch length estimation are needed to refine the molecular dating of anole divergence.

The other possibility is that the molecular dating is correct, but the geology is wrong. Perhaps the landmasses forming the current Greater Antilles were not actually connected and above water[145] in the proto-Antilles. Although some geologists believe this to be the case, others are not sure.

Given all of uncertainties, my own view is that the existence of clades containing species from only one Greater Antillean island is evidence that those clades arose on unconnected landmasses. This implies to me either that the age of divergence of these clades has been overestimated (though they are still not young) or that the Greater

144. These figures, however are based on assuming the youngest possible age for the fossils and the oldest phylogenetic position (i.e., at the very root of the clade); older dating of the fossils and a higher position in the phylogeny would yield substantially older estimates.

145. "sub-aerial" in geological parlance.

Antillean islands have not been connected since the time these clades emerged. Otherwise, too many ad hoc assumptions of nonrandom extinction or restricted clade distribution prior to proto-Antillean vicariance are required. This viewpoint requires more dispersal events, perhaps as many as 11, between the Greater Antilles, but still indicates that most anole diversity has arisen from within-island speciation, rather than from dispersal. It also would suggest that most dispersal occurred early in the history of anole diversification, with few recent successful dispersal events.

BIOGEOGRAPHY OF CENTRAL AMERICAN ANOLES

If you found the uncertainty about Caribbean geology and anole biogeography unsettling, then you'll be horrified by the situation in Central America. Nicholson (2005) reviewed ideas about the geology of Central America and how it may have related to anole diversification there. To date, no clear biogeographic pattern is obvious, but further research on mainland anole phylogeny will allow the history of anole diversification and geographic expansion to be better understood.

FUTURE DIRECTIONS

What clearly is needed is more data of three types. The first two—more pre-Pleistocene anole fossils and a better understanding of Caribbean geology—are probably beyond the reach of readers of this book; indeed, the probability of finding old, non-amber-encased fossils of *Anolis* is slight given the scant fossil record for neotropical reptiles.[146]

The third source of information is more readily attainable: better molecular estimates of divergence times. As phylogenetic analyses increasingly are based on sequence data from multiple genes, molecular dating will become more reliable (Dolman and Moritz, 2006). Integration of whatever fossil data are available will provide the most robust estimate of the age of anole diversification (Near et al., 2005; Donoghue and Benton, 2007; Hug and Roger, 2007).

146. Although recent finds of fragmentary Eocene fossils possibly related to anoles—one in Jamaica (Pregill, 1999), the other in North Dakota (Smith, 2006)—hold out some hope.

7

EVOLUTION OF ECOMORPHOLOGICAL DIVERSITY

Fortunately, the uncertainties about the biogeographic history of *Anolis* discussed in the previous chapter have little bearing on understanding of patterns of ecomorphological radiation and diversification, at least within the West Indies. Phylogenetic information indicates that for the most part anoles have radiated independently on each island of the Greater Antilles, regardless of how or when they got there.

EVOLUTION OF THE ECOMORPHS

Phylogenetic analysis is critical to investigating patterns of ecomorph evolution. The presence of members of the same ecomorph class on multiple islands could be explained in two ways. On one hand, each ecomorph class might have evolved a single time; in this scenario, the presence of each ecomorph class on multiple islands would be the result of overwater dispersal or vicariance. Given six ecomorph classes, this hypothesis would posit five evolutionary transitions from one ecomorph type to another (assuming one ecomorph class is ancestral). Alternatively, each ecomorph class could have arisen independently on each island upon which it occurs. This scenario would require at least 17 evolutionary transitions.[147]

The phylogeny (Fig. 7.1) clearly favors the latter hypothesis. In only one case are members of the same ecomorph class on different islands sister taxa: the twig anole

147. Recall that four ecomorph types occur on all four islands in the Greater Antilles, one occurs on three islands, and one occurs on two islands (Chapter 3).

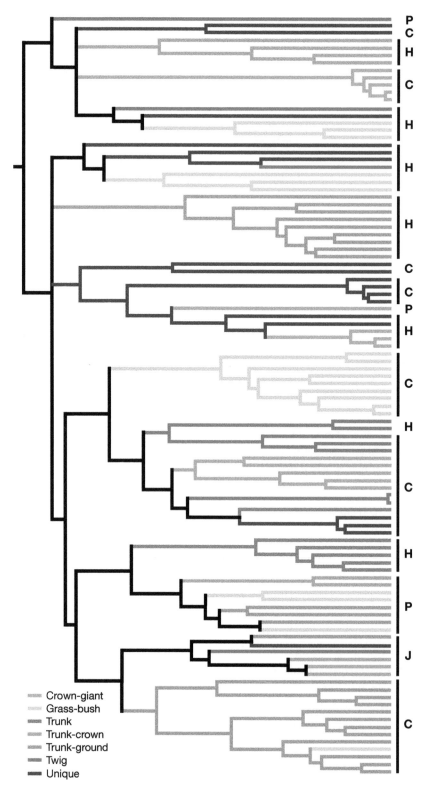

P
C
H
C

H

H

H

C
C
P
H

C

H

C

H

P

J

C

Crown-giant
Grass-bush
Trunk
Trunk-crown
Trunk-ground
Twig
Unique

FIGURE 7.1
Ecomorph evolution in the Greater Antilles. This is the same phylogeny as in Figures 5.6 and 6.1, but species other than those on Cuba (C), Hispaniola (H), Jamaica (J), and Puerto Rico (P) have been pruned out. To visualize the parsimony reconstruction of ecomorph evolution, assume that red (twig anole) is the ancestral state. Ignore unique anoles (brown) and count the number of evolutionary changes in color. Note that in Puerto Rico, two clades of trunk-ground and two clades of grass-bush anoles form a clade, but there have only been three evolutionary transitions, rather than four, because the ancestor of the larger clade containing these four clades was either a grass-bush or a trunk-ground anole. On Jamaica, the ancestor of the trunk-crown/crown-giant clade was probably a trunk-crown anole, thus requiring only two evolutionary transitions. Overall, this approach yields an estimate of 19 evolutionary transitions.

clade *A. sheplani* + *A. placidus* on Hispaniola is the sister taxon to the Cuban twig anoles; this clade likely originated in Cuba because it is nested within a larger clade of Cuban species.[148] Consequently, in the vast majority of cases, the existence of the same ecomorph on different islands is the result of convergent evolution. The phenomenon of convergence of entire communities has often been suggested (e.g., Orians and Paine, 1983; Wiens, 1989), but this is probably the best documented example in terms both of phylogenetic evidence and quantitative measures of morphology and ecology (a topic to which I return in Chapter 17).

Overall, a minimum of 19 evolutionary transitions in ecomorph are required by the phylogeny.[149] The reason that this number is greater than the minimum required for independent evolution of each ecomorph type on each island (17) is that several ecomorphs have evolved multiple times on a single island: grass-bush anoles twice on Cuba and Hispaniola[150] and either grass-bush or trunk-ground anoles twice on Puerto Rico.[151] Two clades of twig anoles occur on both Hispaniola and Cuba,[152] but in the most parsimonious reconstruction of ecomorphs, twig anoles are the ancestral state and thus are not convergent. The separation of the Cuban twig anoles into two clades is not strongly supported by the data and seems unlikely, given that the morphological differences among them are slight (e.g., Estrada and Hedges, 1995; Díaz et al., 1996).

Although the DNA data clearly indicate the convergent nature of the ecomorphs across islands, this result was not surprising; Williams' (1972) initial presentation of the ecomorphs, grounded in a phylogenetic understanding of *Anolis* based primarily on morphology, recognized their convergent nature.[153] Although Williams was misled in a few cases by morphological similarity,[154] for the most part he considered members of the same ecomorph class on different islands to be independently derived. More recently, Poe's (2004) phylogenetic analysis of morphological data—including both quantitative,

---

148. In addition, the Puerto Rican crown-giant *A. cuvieri* is closely related to, though not the sister taxon of, Hispaniolan crown-giants. One other example of closely-related members of the same ecomorph class on different islands, not indicated on Figure 7.1, is the trunk-ground *A. sagrei*, which is native to Cuba and a recent colonist, natural or otherwise, of Jamaica (Williams, 1969; Kolbe et al., 2004).

149. Again, assuming that the ancestral form was a member of one of the ecomorph classes, such that members of one ecomorph class are not convergent, but have retained the ancestral morphology. This estimate was calculated on a phylogeny on which the unique anoles were not included, and thus the Puerto Rican and Hispaniolan crown-giants were treated as sister taxa, rather than independent derivations.

150. Or possibly even three times on Hispaniola. The Haitian *A. koopmani* seems to be a grass-bush anole in both appearance and ecology (Moermond, 1979a,b), although it has not been included in recent morphometric and phylogenetic analyses due to lack of material. Williams (1976b) placed it in his *monticola* Series, which, if correct, would mean that it is unrelated to the other two clades of grass-bush anoles on Hispaniola.

151. The Puerto Rican grass-bush *A. poncensis* is nested within a clade of trunk-ground species. The sister taxon to this clade is the clade containing the other Puerto Rican grass-bush anoles. Either *A. poncensis* has evolved from a trunk-ground ancestor to a grass-bush anole, which would mean that grass-bush anoles had evolved twice on Puerto Rico, or *A. poncensis* has retained the ancestral grass-bush class and trunk-ground anoles have evolved twice (Brandley and de Queiroz, 2004; Fig. 7.1).

152. Three on Cuba if Chamaeleolis is considered a twig anole (Chapter 4).

153. Indeed, it was part of the ecomorph definition that he provided (see Chapter 3).

154. E.g., Williams thought the trunk-ground anoles of Hispaniola and Puerto Rico were closely related, and suspected a similarly close relationship between the twig anoles, *A. sheplani* and *A. occultus*, from the same two islands (Williams, 1976b).

ecomorphic characters such as limb length and standard systematic characters such as the shape, position, or presence of particular bones and scales—revealed pretty much the same result; in several cases, members of the same ecomorph class on different islands appeared as close relatives,[155] but in most cases they did not. Overall, Poe's analysis required 15 evolutionary transitions in ecomorph class, far more than the five that would have been required if each ecomorph class had only evolved a single time. In contrast, in Poe's (2004) phylogenetic analysis that combined morphological and DNA data, the number of evolutionary transitions in ecomorph class was 18.[156]

Two take home messages emerge from this consideration of morphology, ecomorphs, and phylogeny. First, even though species cluster perfectly by ecomorph class in morphometric analyses (Chapter 3), phylogenetic analysis based only on morphology, including ecomorphic characters, still leads to the conclusion that ecomorphs have for the most part evolved independently on each island, albeit with a few exceptions. Second, although most analyses of ecomorph evolution to date have been based on phylogenies derived solely from DNA data (e.g., Losos et al., 1998), phylogenetic analyses taking a "total evidence" approach and including ecomorphic characters lead to essentially the same conclusions about the independent evolution of all ecomorph types on each island of the Greater Antilles.

Chapter 3 detailed the differences in species richness across ecomorph classes and islands. As just noted, in only a few cases has an ecomorph class evolved more than once on an island and, as noted in the last chapter, the paucity of closely related members of the same ecomorph class on different islands indicates that dispersal is rare. Consequently, the existence of multiple members of an ecomorph class on an island must be primarily the result of in situ evolutionary diversification within ecomorph clades, rather than of dispersal or multiple independent evolution of the same ecomorph class. Indeed, some ecomorph clades have diversified exuberantly, such as the 14 trunk-ground anoles of Cuba, and the equally diverse 14 grass-bush anoles in the *alutaceus* Series on the same island, or the nine trunk-ground anoles of Hispaniola.

For the most part, these ecomorph clades are monophyletic—once a member of that ecomorph class, always a member of that ecomorph class. In only three cases has one ecomorph type arisen from within a clade composed of members of another ecomorph type. In two of these cases, the direction of change is clear: on Cuba, the grass-bush *A. ophiolepis* is nested within a clade of trunk-ground anoles and ancestor reconstruction strongly supports the conclusion that it evolved from a trunk-ground ancestor (Fig. 7.1).

155. Specifically, the same two examples mentioned in the previous footnote (the twig anole clade also included a second Hispaniolan twig anole, *A. insolitus,* which the DNA data indicates is not closely related to the others), as well as clades composed of Hispaniolan and Cuban grass-bush anoles plus Chamaelinorops, and of Cuban, Puerto Rican, and Hispaniolan crown-giants plus Chamaeleolis. These latter two clades are strongly contradicted by the DNA data.

156. With regard to ecomorph clades, the primary difference between the DNA and combined analyses is that some Cuban grass-bush anoles are placed in a clade with Hispaniolan grass-bush anoles in the combined analysis.

In Jamaica, the crown-giant *A. garmani* probably evolved from a trunk-crown anole, given that it occurs in the middle of a clade containing the trunk-crown anoles *A. grahami* and *A. opalinus*. In contrast, on Puerto Rico, the grass-bush *A. poncensis* occurs within a clade of trunk-ground anoles, but the direction of evolutionary change is less clear (see footnote 151).

An obvious question is: what was the ancestral ecomorph class? Going one step further: have the convergent ecomorph faunas been built up in the same way? Alas, these are just the sort of questions that cannot be answered confidently by reconstructing ancestral states. Initially, I tried to do so. Comparing the evolution of the Jamaican and Puerto Rican radiations, I reconstructed ancestral states and found that the two islands went through the same series of ancestral stages: 2 species—twig anole and generalist; 3 species—twig, crown, and trunk-ground anoles; 4 species—twig, crown-giant, trunk-ground, and trunk-crown anoles (Fig. 7.2; Losos, 1992a).[157] In their seminal paper developing a method for placing error bars on ancestral reconstructions, Schluter et al. (1997) used this as one of their examples and showed that one could have little confidence in reconstructions deep in the phylogeny (Fig. 7.3); the ancestor I had inferred to be a generalist could just as easily have been a member of almost any ecomorph class.

The inability of ancestor reconstruction to provide unequivocal reconstructions is evident in Figure 7.4. For most clades containing members of more than one ecomorph, including almost all clades deep in the tree, the ecomorph of the ancestor is ambiguous with as many as four possibilities and no ecomorph type strongly favored. For example, the ancestral Greater Antillean anole is reconstructed with approximately equal support for twig, grass-bush or trunk-ground anole as the ancestral state. Clearly, ecomorph evolution has been too labile to permit confident reconstruction of ancestral nodes whose descendants belong to several different ecomorph classes. By contrast, when the descendants of an ancestral node all belong to one ecomorph class, as occurs for the within-ecomorph radiations, then confidence in the ancestral state is much higher.

The question of parallel pathways of ecomorph evolution can be rephrased to avoid the necessity of ancestor reconstruction. We can ask, instead, whether the phylogenetic relationships among the ecomorph classes are the same across islands; for example, is a twig anole always the basal clade, sister taxon to the rest? Figure 7.1 makes clear that there are few similarities in the patterns of ecomorph relationship among the islands. Consider, for example, twig anoles. On Puerto Rico, they are the most basal clade; on Jamaica they are sister to a clade of trunk crown and crown-giant anoles; on Cuba, they

---

157. The question of whether ecomorph faunas were assembled in the same way is more complicated than it appears (or than I treated it in my paper) because only the Jamaican ecomorph radiation is monophyletic. As discussed in Chapter 6, the anole faunas of the other three Greater Antillean islands are composed of multiple clades. Consequently, only on Jamaica are all species more closely related to all of the other species on their own island than they are to a member of their same ecomorph class on another island. Thus, determining the sequence of ecomorph assembly on other islands requires information about ancestral character states for both ecomorphology and geographic location.

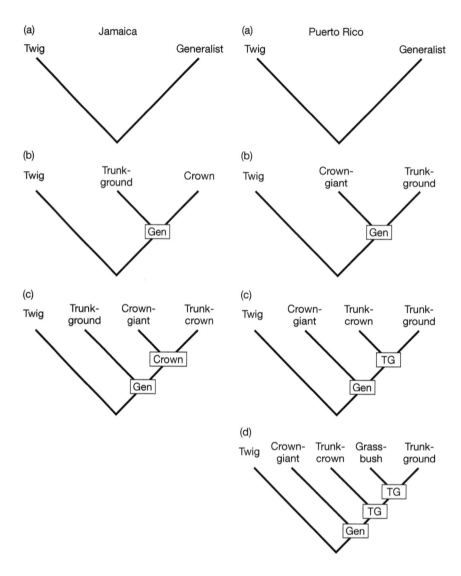

FIGURE 7.2

Reconstructed sequence of ecomorph evolution on Puerto Rico and Jamaica. Ecomorph type was reconstructed based on several quantitative characters (principal component scores from a morphometric analysis); consequently, parsimony could reconstruct a morphotype not corresponding to any of those observed in the study. The "generalist" was an ancestor reconstructed to occur in a position intermediate between the ecomorphs in morphological space. Note that the phylogeny differs from those produced in more recent analyses. Also, for purposes of ancestor reconstruction, Puerto Rican anoles were treated as monophyletic, even though the island is occupied by three anole lineages. Figure from Losos (1992a) with permission.

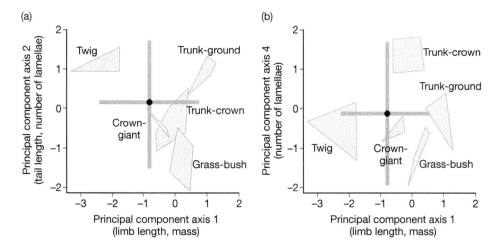

FIGURE 7.3

Ancestor reconstruction with 95% confidence limits. Maximum likelihood methods were used to estimate uncertainty around the estimate of the generalist ancestor for the Puerto Rican anole radiation. PC III, not shown, represents body size. Figure from Schluter et al. (1997) with permission.

are sister to a clade of trunk-crown, trunk and unique anoles; and the two twig anole clades on Hispaniola are related to different clades each comprised of grass-bush and unique anoles. Examination of Figure 7.1 reveals that the other ecomorphs show the same sort of dissimilarities among islands.[158]

Although this analysis does not reconstruct ancestral states, the many differences in branching pattern of the ecomorphs suggest that the history of ecomorph evolution on the four islands has probably been quite different.[159] In other words, the convergent faunas that have evolved across the Greater Antilles have achieved their similarity through different evolutionary trajectories.

We also can draw conclusions from anole phylogeny about the age of the ecomorph phenomenon. The fossil anoles in amber reveal that at least one ecomorph class, trunk-crown, was present on one island, Hispaniola, minimally 15 million years ago.[160] The phylogenetic data are perfectly consistent with this antiquity, as discussed in Chapter 6. Many of the ecomorph clades extend deep into the anole phylogeny (Fig. 7.1). For example, the Hispaniolan trunk-ground anoles in the *cybotes* series are estimated to have

158. For a statistical rejection of the hypothesis of similar phylogenetic pattern of ecomorph evolution, see Losos et al. (1998).

159. Of course, without reconstructing ancestral states, we can't say for sure. If evolution has been very non-parsimonious, then the history of ecomorph evolution could have been identical across islands, even if the phylogenetic topology of the ecomorphs differs greatly.

160. Other amber anoles from the same Dominican deposits are known, but most are in private hands. One, pictured in Poinar and Poinar (2001), does not look like a trunk-crown anole to me. All Dominican amber specimens probably are of approximately the same age (Iturralde-Vinent, 2001).

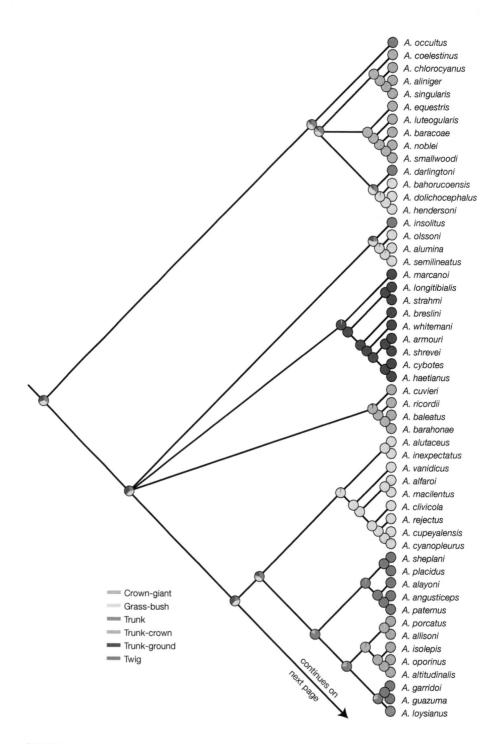

continues on next page

FIGURE 7.4

Ancestral character states were reconstructed using likelihood; the relative support for each possible ecomorph character state was calculated for each ancestral node and is indicated by the proportional coloring within each circle. Thus, the original ancestral node is approximately equally likely to have been a grass-bush, trunk-ground, or twig ecomorph, with slight support for trunk-crown and crown-giant. Analyses were conducted on the preferred phylogenetic hypothesis from Nicholson et al. (2005); only ecomorph species were included in character reconstruction.

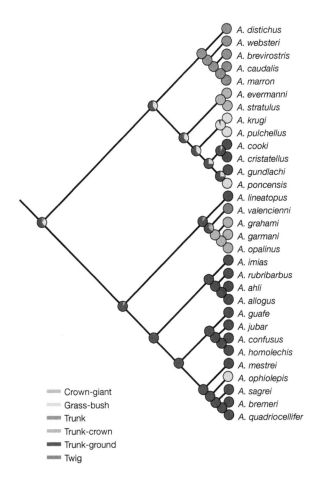

A. distichus
A. websteri
A. brevirostris
A. caudalis
A. marron
A. evermanni
A. stratulus
A. krugi
A. pulchellus
A. cooki
A. cristatellus
A. gundlachi
A. poncensis
A. lineatopus
A. valencienni
A. grahami
A. garmani
A. opalinus
A. imias
A. rubribarbus
A. ahli
A. allogus
A. guafe
A. jubar
A. confusus
A. homolechis
A. mestrei
A. ophiolepis
A. sagrei
A. bremeri
A. quadriocellifer

- Crown-giant
- Grass-bush
- Trunk
- Trunk-crown
- Trunk-ground
- Twig

arisen 34 mya, using the 40 mya age estimate for anole diversification. Using this calibration, most ecomorph clades appear to have been present on their islands for at least 25 million years (more than 40 million years if the older, DNA calibration is used). I'll return to these questions concerning the age and sequence of ecomorph evolution, as well as the identity of the ancestral ecomorph, in Chapter 15.

GREATER ANTILLEAN UNIQUE ANOLES

Unique anole species are confined almost exclusively to Cuba and Hispaniola (Chapter 4), the two islands that have all six ecomorph classes. Unique anoles do not form a clade relative to the ecomorphs; rather, unique anole clades have arisen multiple times on each island (Fig. 7.1).

One possible explanation for this distribution is that unique anoles occupy "less favorable" niches and only evolve subsequent to the evolution of the full complement of ecomorph classes. If this were the case, we might expect non-ecomorph species to be relatively young in age and perhaps even to have arisen from within ecomorph clades.

Neither prediction is borne out. Unique anoles often occur in clades that do not include ecomorph species and that are old, descending deep into the anole tree; in no case does a unique anole occur nested within a clade otherwise consisting of members of a single ecomorph.[161] In this regard, then, the evolution of the unique anoles follows the same pattern as the evolution of the ecomorphs. I'll return to examination of how and why the unique ecomorphs have evolved in Chapter 16.

ANOLES OF THE SMALLER ISLANDS OF THE GREATER ANTILLES

As discussed in Chapter 4, the phylogenetic affinities of species on the small islands of the Greater Antilles are clearcut. In many cases, species on these islands are considered to be conspecific with species on the four large islands of the Greater Antilles. In all other cases but one, species on small islands are sister taxa to, or arose from within, a species on one of the larger islands. The one exception is *A. acutus* on St. Croix, which appears to be the sister taxon to the Puerto Rican trunk-crown clade comprised of *A. evermanni* and *A. stratulus* (Brandley and de Queiroz, 2004).

LESSER ANTILLEAN ANOLES

The Lesser Antilles illustrate a pattern of adaptive evolutionary divergence only partially consistent with that seen in the Greater Antilles. As in the Greater Antilles, repeated patterns of morphological differentiation recur across islands in the Lesser Antilles, although in this case it is body size that shows regularity: on one-species islands, the species is almost invariably intermediate in size, whereas on two-species islands, the two species almost always differ greatly in size (Figs. 7.5 and 7.6; see Chapter 4).

At the largest geographic scale, these size patterns show a remarkable convergence, having been independently produced by two different evolutionary clades, the *roquet* Series in the southern Lesser Antilles and the *bimaculatus* Series in the northern Lesser Antilles. Within each clade, however, the parallel to ecomorph evolution is incomplete. The similarity to the Greater Antilles lies in the fact that convergence has occurred repeatedly, in this case in body size. However, unlike in the Greater Antilles, many cases exist in which closely related species on different oceanic islands share the same body size. For example, all of the one-species islands in the central part of the northern Lesser Antilles are occupied by intermediate-sized species that belong to a single clade. A similar pattern is exhibited by two different clades of intermediate-sized species in the southern Lesser Antilles, as well as by the clade of small species in the northern Lesser Antilles (Fig. 7.7).

161. The only possible exceptions are the Hispaniolan unique anoles *A. christophei* and *A. eugenegrahami*, themselves quite dissimilar (Chapter 4), which are sandwiched in a clade between the crown-giants of Hispaniola and Puerto Rico.

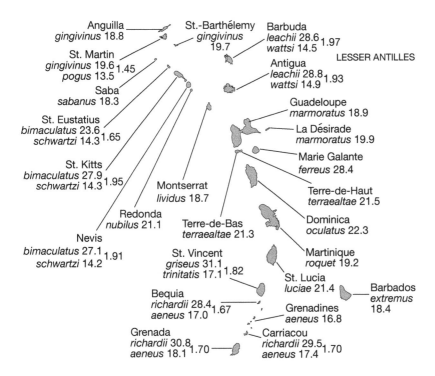

FIGURE 7.5
Size patterns in the Lesser Antilles. Sizes are jaw length (in mm) which correlates strongly with body size. Ratios of larger-to-smaller are provided for two-species islands. Modified with permission from Schoener (1970).

These patterns suggest that both evolutionary and ecological processes have been at work in shaping size distributions in the Lesser Antilles. The repeated pattern of convergence indicates the operation of evolutionary processes, whereas the occurrence of closely related species of similar size on different islands reveals a role for differential colonization success: one-species islands are occupied today by the descendants of intermediate-sized colonists, whereas species on two-species islands, particularly in the northern Lesser Antilles, are descended from large and small colonists.[162] This is a phenomenon termed "size assortment" (Grant and Abbott, 1980; Case, 1983; Case and

162. As with the clades of species belonging to the same ecomorph class, I assume here that when all of the species in a clade are phenotypically similar, then these species have inherited that phenotype from their ancestor. For example, all members of the *wattsi* Series are small, so I assume that the ancestor of this clade was small. We cannot rule out the possibility that small size evolved multiple times in parallel in members of this clade. Perhaps some ecological quirk of this clade predisposed it to evolve small size in the presence of another species; it might be something as simple as a slight ecological difference that allowed initially intermediate-sized members of this clade to coexist with another species long enough for small size to evolve. As mentioned in Chapter 5, ancestor reconstruction is particularly impotent in the face of parallel evolution among clade members. Consequently, we cannot rule out the possibility that the six populations of this clade on different islands independently evolved to small size from an intermediate-sized ancestor; the best we can do is say that there is no phylogenetic evidence supporting that hypothesis.

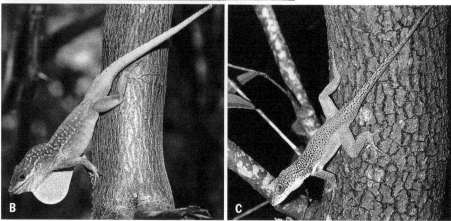

FIGURE 7.6
Anoles of the northern Lesser
Antilles. (a) Small: *A. wattsi*,
St. Kitts. (b) Intermediate:
A. lividus, Montserrat.
(c) Large: *A. leachii*, native
to Antigua. Photo from
introduced population in
Bermuda.

Sidell, 1983).[163] Here's how size assortment would work: assume that intermediate size
is optimal for a solitary species, but that large and small sizes are the optimal case for
two-species islands. If an island is initially occupied by an intermediate-sized species,
then it cannot be invaded by a large or small species. Conversely, an island occupied by
both a large and a small species could not be invaded by an intermediate-sized species.
If an island was occupied solely by a large or a small species, and then an intermediate-
sized species invaded, the initial species would become extinct. The only way that a two-
species island could become established would be if an empty island were invaded by
first a small species and then a larger one, or vice versa, before an intermediate-sized
species arrived. No island could be invaded by a species of the same size as the resident.
These rules would produce a pattern like that seen in the Lesser Antilles without any
evolutionary adjustment of coexisting species.

163. Two other lizard groups show a similar pattern of large and small species on two-species islands and
intermediate-sized species on one-species islands: *Cnemidophorus* in the Gulf of California (Radtkey et al., 1997)
and *Phelsuma* in the Praslin Island group (but not elsewhere) in the Seychelles (Radtkey, 1996). In both cases,
size divergence only occurred a single time and the occurrence of size dissimilarity on multiple two-species
islands is the result of size assortment.

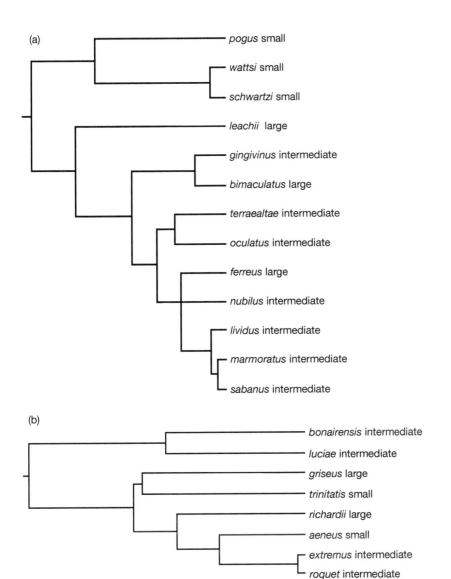

FIGURE 7.7

Phylogenetic relationships of anoles of the (a) northern and (b) southern Lesser Antilles. This phylogeny is based on sequence data from several mitochondrial and nuclear genes (C. Cunningham, pers. comm.). The results are generally consistent with previous work on these clades (Creer et al., 2001; Schneider et al., 2001; Stenson et al., 2004). Bonaire is not considered part of the Lesser Antilles, and hence *A. bonairensis,* which is intermediate in size, is not included in Fig. 7.5. The division between small, intermediate, and large size categories is somewhat arbitrary. Statistical analyses of size evolution have used quantitative measures, rather than these categories (e.g., Losos, 1990a; Butler and Losos, 1997).

The difference between the Greater and Lesser Antilles can be quantified in the following way. In the Greater Antilles, if each ecomorph type had evolved once, followed by colonization (or vicariance) to occupy each island, then five evolutionary transitions would have occurred. Conversely, if each ecomorph class evolved on each island, then at least 17 transitions would have occurred. The observed number is 19; clearly, independent evolution on each island has been responsible for the repeated patterns observed across islands.

By contrast, in the Lesser Antilles, given the different clades in the two regions, the minimum number of evolutionary transitions in body size is four (assuming three size states: small, intermediate, and large). By contrast, if evolutionary divergence in body size has occurred independently on each island bank,[164] then at least 10 transitions would be required. The actual number is 7, exactly intermediate.[165] Thus, in the Lesser Antilles evolutionary divergence has been less important than in the Greater Antilles; conversely, colonization leading to ecological similarity of closely related species on different islands has been more important in the Lesser Antilles.

CHARACTER DISPLACEMENT

But how did the size differences evolve initially? One possibility is that the size differences evolved by character displacement, the phenomenon that when two similar species come into contact, they evolve in opposite directions to minimize resource overlap, thus permitting coexistence (Brown and Wilson, 1956). Character displacement was controversial for many years because some theoretical treatments suggested that it was unlikely to occur (one species was expected to go extinct before substantial evolutionary divergence could occur) and because there were few well documented examples. However, in recent years an abundance of examples has been published and now the evolutionary significance of character displacement is well established (e.g., Schluter, 2000; Dayan and Simberloff, 2005).

Several authors have hypothesized that character displacement is responsible for size divergence in Lesser Antillean anoles (Schoener, 1970b; Williams, 1972; Lazell, 1972; Losos, 1990a). This hypothesis predicts that large and small size evolved at the same time

---

164. As in the Greater Antilles, a number of Lesser Antillean islands were united during the last period of lower sea level.

165. In the northern Lesser Antilles, the outgroup, the *cristatellus* Species Group, is primarily intermediate in size. Assuming that intermediate size is ancestral, then three transitions have occurred, to small size in the *wattsi* species group and to large size independently in *A. bimaculatus* and *A. leachii* or, alternatively, to large size in the ancestor of *bimaculatus* + *gingivinus* + *leachii* and back to intermediate size in *A. gingivinus*. The size of the outgroup for the *roquet* Series in the southern Lesser Antilles is unknown. Nonetheless, regardless of whether the ancestor of this clade was small, intermediate, or large, four evolutionary transitions are required to produce the sizes of the extant species.

These estimates are based on the best current phylogenetic hypotheses for these two clades. However, both phylogenies have weakly supported nodes in critical areas that affect the interpretation of size evolution (see discussion in Creer et al. [2001] and Schneider et al. [2001]).

and on the same island.[166] My previous test of this hypothesis supported these predictions for the anoles of the northern Lesser Antilles, but not for those of the southern Lesser Antilles (Losos, 1990a; see also Miles and Dunham, 1996; Butler and Losos, 1997). However, this analysis used a phylogeny which was pieced together based on a variety of pre-DNA studies and which differs in important ways from more recent studies (e.g., Creer et al., 2001; Schneider et al., 2001; Stenson et al., 2004; Nicholson et al., 2005).

Examination of the most recent phylogeny (Fig. 7.7) reveals three possible cases of character displacement:

1. *A. griseus* (large) and *A. trinitatis* (small) on St. Vincent. The species are sister taxa. Uncertainty about the ancestral state makes the scenario somewhat unclear. Nonetheless, at least one species has evolved away from the size of the other; if the ancestor was intermediate in size, then both species have diverged in opposite directions.

2. *A. richardii* (large) and *A. aeneus* (small) on Grenada and elsewhere in the Grenadines. *Anolis richardii* is the sister taxon to a clade composed of *A. aeneus* plus two intermediate-sized species, *A. roquet* (Martinique) and *A. extremus* (Barbados). Two biogeographic scenarios are possible: (i) the ancestor of this clade diverged into two species on Grenada or in the Grenadines, *A. richardii* and the ancestor of *aeneus + extremus + roquet*, and then the latter species gave rise to a colonist that traveled to Martinique or Barbados; or (ii) The ancestor occurred somewhere in any of these islands; colonization and allopatric speciation then gave rise to two species, *A. richardii* in Grenada and the ancestor of *aeneus + extremus + roquet* on Martinique or Barbados. Subsequently, colonization to Grenada from the Martinique/Barbados species led to another speciation event producing *A. aeneus*. Given that all of the basal clades of the *roquet* series are in the south of the SLA, this scenario would require a north-to-south dispersal event that runs counter to the general direction of movement (C. Cunningham, pers. comm.).

    These scenarios can then be combined with phylogenetic reconstruction of body size. The ancestor of the clade may have been intermediate-sized. This is supported both by ancestor reconstruction[167] and by the assumption that intermediate size is optimal on one-species islands. Given this assumption, character displacement could have occurred if biogeographic scenario (i) occurred, and, if the ancestor of *aeneus + extremus + roquet* was small, with intermediate size evolving subsequent to colonization of Martinique/

166. Character displacement doesn't have to occur equally in both species (e.g., Grant and Grant, 2006a). However, in the context of the size patterns of Lesser Antillean anoles, the character displacement hypothesis would predict that the evolution of small and large size occurred simultaneously.

167. If the evolutionary transition from small to large large or vice versa is considered to be twice as great a change as from either large or small to intermediate, then intermediate size is the most parsimonious reconstruction for the ancestor of this clade.

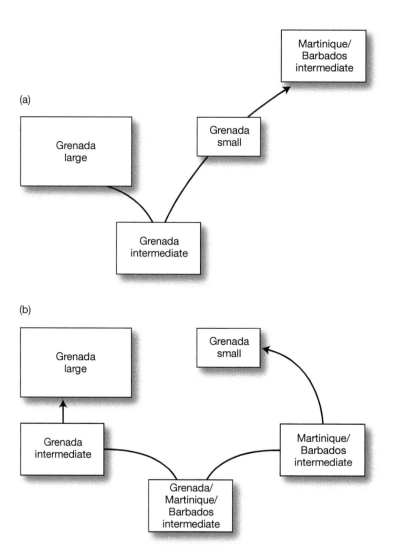

FIGURE 7.8

Two possible scenarios for size evolution in the *richardii* + *aeneus* + *extremus* + *roquet* clade of the southern Lesser Antilles. In (a), character displacement occurs between two species on Grenada. Then, the smaller species, ancestral to *A. aeneus*, gives rise to a colonist which moves to Martinique or Barbados and evolves intermediate size. In (b), allopatric speciation gives rise to two intermediate sized species, one the ancestor to *A. richardii* on Grenada, the other on Barbados or Martinique. This latter species then sends a colonist, the ancestor of *A. aeneus*, back to Grenada, where character displacement occurs between the two initially intermediate-sized species.

Barbados (Fig. 7.8a); or if biogeographic scenario (ii) occurred, the ancestor of *A. richardii* remained intermediate-sized until Grenada was colonized by the ancestor of *A. aeneus*, and the ancestor of *aeneus* + *roquet* + *extremus* was intermediate in size, with small size evolving only when the ancestor of *A. aeneus* colonized Grenada (Fig. 7.8b). Alternatively, other combinations of biogeographic and size evolution scenarios would not be consistent with character displacement.

3. The large and small anoles of the Northern Lesser Antilles. The situation here is uncertain, but in a different way than in the just discussed case of *A. richardii* and *A. aeneus*. The *wattsi* + *schwartzi* + *pogus* clade of small anoles today occurs on the St. Kitts, Antigua, and Anguilla Banks, where it is sympatric with *A. bimaculatus* (large), *A. leachii* (large) and *A. gingivinus* (intermediate), respectively. Given this, it is likely that the small anole clade has been in the presence of another anole species throughout most or all of its evolutionary history, and thus the evolution of small size is consistent with a character displacement hypothesis. But an inconsistency exists: if the ancestor of the other clade was large, then intermediate size subsequently re-evolved in *A. gingivinus*, probably in the presence of small anoles,[168] contrary to what would be expected (Fig. 7.9a). Alternatively, if the ancestral size of this clade is intermediate, then the failure of the ancestor of *A. gingivinus* to evolve larger size is contrary to the character displacement scenario. In this scenario, large size would subsequently have evolved when intermediate-sized colonists were confronted with a smaller anole on the St. Kitts and Antigua Banks (Fig. 7.9b). In either scenario, the coexistence of intermediate-sized *A. gingivinus* and small *A. pogus* is not predicted by the character displacement hypothesis.

In summary, one case, St. Vincent, unequivocally supports character displacement, and the other two cases are ambiguous—character displacement can't be ruled out, but is not definitively supported, either. Although the overall support for character displacement is not overwhelming, it has not been disproven. This is important to keep in mind, because phylogenetic examination could have produced reconstructions entirely inconsistent with character displacement. In summary, the character displacement hypothesis is alive, and has some support; this may be as much as one often can expect from phylogenetic analyses relying on ancestor reconstruction.

Two alternative hypotheses for size evolution must be briefly mentioned. The first is sympatric speciation, in some sense character displacement that occurs within a species, rather than between two species. Certainly, the situation on St. Vincent—sister taxa occurring on the same island—is consistent with sympatric speciation. However, the inferred date of the divergence predates the emergence of the island of St. Vincent, suggesting that these species evolved somewhere else—perhaps the Grenadines?— thus obscuring the geography of this divergence event (C. Cunningham, pers. comm.). Further, in both Grenada and the northern Lesser Antilles, sympatric speciation is consistent with character displacement scenarios, although alternative versions in which

---

168. *A. gingivinus* today occurs on islands on the Anguilla Bank and the small nearby island of Sombrero. *A. pogus* is present on St. Martin and occurred as recently as 1922 on Anguilla. Given that the Anguilla Bank was above seawater during the last ice age, it is probable that *A. pogus* occurred throughout the bank in the recent past.

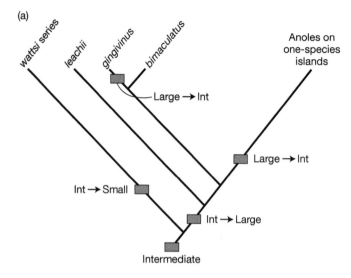

(a)

wattsi series
leachii
gingivinus
bimaculatus

Anoles on
one-species
islands

Large → Int

Large → Int

Int → Small

Int → Large

Intermediate

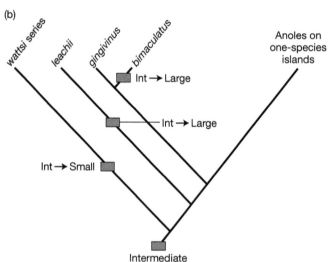

(b)

wattsi series
leachii
gingivinus
bimaculatus

Anoles on
one-species
islands

Int → Large

Int → Large

Int → Small

Intermediate

FIGURE 7.9

Two scenarios for size evolution in the northern Lesser Antilles. In (a), character displacement occurs between the two basal clades. Subsequently, intermediate size evolves from large size both in *A. gingivinus* and in the clade of species occupying one-species islands in the southern part of the northern Lesser Antilles. In (b), large size evolves independently in *A. bimaculatus* and *A. leachi* in the presence of a smaller species, consistent with the character displacement hypothesis. In both scenarios, the intermediate size of *A. gingivinus*, whether ancestral or derived, is inconsistent with predictions of the character displacement hypothesis.

speciation occurs in allopatry (e.g., in Grenada and the Grenadines) are possible as well. Sympatric speciation in anoles will be discussed in greater detail in Chapter 14.

The second possibility is the taxon cycle (Roughgarden and Pacala, 1989) or taxon loop (Roughgarden, 1992, 1995) hypotheses. The idea of these hypotheses is that intermediate size is optimal, but that sympatry of large and small species is not produced by divergence in opposite directions, but by a large species invading an island occupied by an intermediate-sized species, which then evolves to smaller size. The only phylogenetic evidence in support of this scenario is the possible evolution from large to intermediate size in *A. gingivinus*,[169] although in the scenario discussed above, this could only have

169. In fact, the taxon cycle was inspired by the *gingivinus-pogus* situation, so this doesn't constitute very strong support.

FIGURE 7.10

Mainland aquatic anoles. (a) *A. oxylophus.* Photo courtesy of J.D. Willson. (b) *A. aquaticus.* Photo courtesy of Luke Mahler. Morphological data suggest that these species are sister taxa (Poe, 2004), but molecular data imply that they are more distantly related (Nicholson et al., 2005), raising the possibility that they may have evolved their aquatic tendencies independently.

happened after small size already had evolved in the *wattsi* clade, contrary to the taxon cycle/loop hypothesis. Other criticisms and discussion of this theory can be found in Losos (1992b), Roughgarden (1992, 1995), Schneider et al. (2001) and Stenson et al. (2004).

MAINLAND ANOLES

The mainland is occupied by the Norops clade and the basal anole clade, corresponding to Savage and Guyer's (1989) Dactyloa clade with the inclusion of Phenacosaurus. Both clades exhibit a range of morphologies and ecologies, more so in Norops (Pinto et al., 2008), but that is not surprising given the approximately two-fold greater species richness of Norops. In the absence of more detailed information on the ecomorphological variation of these anoles and greater phylogenetic resolution within each of these clades, statements about patterns of ecomorphological evolution are difficult to make.

I can only come up with two cases in which we can say something definitive about mainland ecomorphological evolution. First, twig anoles seem to have evolved at least twice, once in each clade: A. Norops *pentaprion* and close relatives, and the species in the Phenacosaurus clade (Fig. 4.9). Both have twig anole morphology (Beuttell and Losos [1999] for Phenacosaurus; Losos [unpubl. obs.] for *A. pentaprion*) and the scant ecological literature on both species indicates a twig anole lifestyle (Dunn, 1944; Miyata, 1983; Losos, unpubl. for *A. pentaprion*).

Second, aquatic anoles appear to have evolved three times within mainland Norops. These species are different from the West Indian aquatic anoles, but are quite similar to each other in morphology and ecology (Leal et al., 2002; Fig. 7.10).

THE PHYLOGENETIC PERSPECTIVE: WHAT DOES IT TELL US THAT WE DIDN'T ALREADY KNOW AND WHAT HYPOTHESES DOES IT SUGGEST THAT ARE TESTABLE WITH ECOLOGICAL DATA?

Even in the absence of phylogenetic information, ecological interactions would be implicated as an important factor affecting anole communities. The regular size patterns in the Lesser Antilles and the absence of sympatric species occupying the same niche on any island (discussed in Chapter 11) implicate a deterministic factor, the most likely of which is ecological interactions among anole species. Phylogenetic information, however, much more precisely refines this notion.

First, the anole phylogeny makes clear how rarely colonization has occurred in the Greater Antilles, particularly in relatively recent times. We know that this is not for lack of trying: a few species have dispersed widely to unoccupied islands in recent geological time (Chapter 4). It seems most likely that over the course of tens of millions of years, colonists have moved from one Greater Antillean island to another, yet they have failed to become established. The most likely explanation is that they have been repelled by resident species, an example of a "priority effect" (Williams, 1969; MacArthur, 1972; reviewed in Morin, 1999; see also Chase, 2007).

The priority effect is also seen in the Lesser Antilles, where despite the high levels of colonization necessary to populate these oceanic islands, no island contains two species of the same size. Although potential lack of reproductive isolation between similarly-sized species on different islands might be partly responsible for lack of coexistence,[170] the fact that the distantly-related anoles of the southern and northern Lesser Antilles have not been able to cross into each other's domain indicates that this is not the entire explanation.[171] The two-species islands of the northern Lesser Antilles add a twist. The phylogeny indicates that the size difference on these islands has not arisen in situ, but rather results from successful co-invasion of species that evolved their size differences elsewhere. This pattern illustrates the ability of a species to invade successfully in the absence of a similar-sized species.

Second, repeated evolution of the same phenotypes in similar ecological circumstances—ecomorphs in the Greater Antilles, size classes in the Lesser Antilles—implies a deterministic cause. If each phenotype had arisen only once and subsequently had been replicated across islands via colonization, then we could not conclude that these particular phenotypes were specifically well adapted to their environment. Rather, it might be that selection favored the coexistence of any different phenotypes, and the specific ones that evolved were a result of the particular contingencies of history. But repeated evolution of the same phenotypes suggest that the environment is favoring those

170. In the absence of reproductive isolation, colonists might simply be subsumed into the native species' gene pool (Losos, 1990a).

171. Given that these two clades occur on opposite sides of the basal split in anole phylogeny and thus shared their most recent common ancestor at least 40 mya, they almost surely are reproductively isolated; the few cases of hybridization among anoles all occur among much more closely related species (Chapter 2).

specific phenotypes—that, of course, is the basis for the long-held view that convergent evolution is strong evidence of adaptation (see discussion in Chapter 13). Moreover, repeated evolution of the same phenotypes suggests the existence of niches independently of the species that fill them (see discussion in Chapter 16).

Third, the observation that the same ecomorph (Greater Antilles) or size class (Lesser Antilles) has rarely evolved more than once on an island[172] implies that interspecific interactions have a constraining, as well as a driving, role in evolutionary diversification. If island environments favor particular phenotypes, one might expect those phenotypes to evolve repeatedly. The observation that they don't suggests an evolutionary corollary of the priority effect—once an ecological niche is filled by one clade, it is evolutionarily inaccessible to other clades.

These observations lead to the following predictions:

1. Sympatric species interact ecologically.

2. The extent of interspecific competition between species is a function of how ecologically—and morphologically—similar they are.

3. The degree to which a species can colonize a new area is a function of how ecologically similar it is to resident species.

4. Ecological interactions lead to divergence in habitat use.

5. Divergence in habitat use leads to natural selection for phenotypes appropriate to the new habitat.

6. Island environments are similar and favor the evolution of the same phenotypes.

None of these hypotheses is novel, and some of them would have been suggested even in the absence of phylogenetic information. Nonetheless, phylogenetic examination clearly suggests that these hypotheses may be true. Much of the remainder of the book will be devoted to addressing them by studying the extant anole faunas.

FUTURE DIRECTIONS

Future work here echoes that of previous chapters. Pending phylogenetic issues need to be resolved to further understanding of patterns of evolutionary diversification in a number of areas, such as the mainland and the northern Lesser Antilles. Resolution of

172. Multiple instances of evolution of an ecomorph on an island may be even rarer than suggested by the 5–7 examples mentioned above (the uncertainty stems from the fact that if twig anole is the ancestral state, then it has not evolved multiple times on Cuba and Hispaniola). Three of those examples—*A. koopmani*, the grass-bush anoles of the *hendersoni* Series, and the twig anole *A. darlingtoni*—are endemic to the South Island of Hispaniola. It is unclear whether this landmass, which has a separate geological history from the North Island, was ever emergent as a separate island (for much of its history, it was underwater [Iturallde-Vinent, 2006]); after merger of the two landmasses, the two regions have been divided many times by high sea levels. Consequently, it is at least conceivable that the South Island clades may not indicate multiple evolution of ecomorph types on the same island, but rather may be the remnants of a separate South Island radiation (R. Glor, pers. comm.).

within-clade relationships will provide insight into how intra-ecomorph diversification proceeds, as well as how unique anoles evolve.

As well as the obvious need to sample more mainland species, a few West Indian taxa are also needed. Most pressing is the Haitian *A. koopmani*, a putative third clade of grass-bush anole from Hispaniola.

8

CRADLE TO GRAVE

Anole Life History and Population Biology

Before tackling the question of how anole species interact (Chapter 11), and how such interactions might drive evolutionary change (Chapter 12), I need to discuss what makes anoles tick. That is, how do anoles interact with their environment? What happens during the course of an anole lifetime and why? These questions will be the focus of this and the next two chapters.

The goal of this chapter is to review the basic aspects of anole population biology and life history, as well as to discuss the role of anoles in the ecosystem. In some sense, much of the information that I will discuss could be categorized as "natural history." In recent years, natural history has not been given a lot of respect—some contend that it does not even qualify as a science. Quite the contrary, I would argue that natural history is not only based on the important scientific foundations of careful observation and inquiry, but that it is essential if we are to formulate meaningful hypotheses about an organism's place in the environment. Moreover, to understand how species interact and evolve through time, knowledge of natural history is indispensable (Greene, 1994, 2005).

Anoles have been intensively studied for more than 40 years, and we know more about anole natural history than we do for most types of organisms. Nonetheless, the amount of information we do not know is staggering. One clear message from this chapter is that many important aspects of anole biology are still little known. As subsequent chapters will illustrate, this lack of information impedes our ability to interpret broad scale patterns of anole ecology and evolutionary diversification. This chapter is meant to be a call to arms: there's much to be discovered and no time like the present!

FIGURE 8.1

Photo of the everted hemipenes of a museum specimen of *A. magnaphallus*. As the name implies, this species' hemipenes are larger than those of most anoles. Only one hemipenis is used at a time: males alternate their use. The hemipenis used depends on whether the male swings its tail over the left or right side of the female. Why some species, such as *A. magnaphallus*, have bilobed hemipenes is not known; the lobes function as a single intromittent organ, rather than being used separately. Photo courtesy of Steve Poe.

The purpose of this chapter is to review the basic aspects of anole biology. In some cases, such as reproductive biology, anoles exhibit little variation (at least of which we are aware), and my goal is simply to report what anoles do.[173] In other cases, such as diet, considerable diversity exists within and among anole species, and my goal is to explore this diversity and to explain its ecological and evolutionary significance.

REPRODUCTION

I'll start at the beginning, with the way a young *Anolis* makes its way into the world. Anole courtship is a highly stereotyped business. Males perform a display in which they bob their heads and extend their dewlaps in a species-specific manner. The stereotypy of the head-bob cadence is important, as it appears to be a means for females to distinguish conspecifics from heterospecifics, a topic to which we will return in Chapter 14. Females respond by headbobbing or dewlapping (or both), and sometimes arch their neck to indicate receptivity. The male often bites the female on the neck and mounts on her back, swinging his tail around to the underside of the female's tail and bringing their cloacae into close proximity. The male then everts one of his two intromittent organs, termed hemipenes and stored in the base of the tail (Fig. 8.1), and inserts it into the female's

173. An important caveat is that most generalizations are based on data for relatively few species, most of which are usually West Indian. Even basic aspects of the biology of most of the 361 species of anoles are unknown. Who knows what surprises remain to be discovered?

cloaca (for a review of the reproductive behavior of *A. carolinensis*, which has been studied much more intensively than any other anole species, see Wade [2005]).[174]

Most, but not all, species have a distinct breeding season (e.g., Andrews, 1971; Fitch, 1973a; Jenssen and Nunez, 1994). In the West Indies, seasonal variation in temperature plays a role in determining its length: more northerly species and populations at higher elevations tend to have shorter breeding seasons (Licht and Gorman, 1970; Gorman and Licht, 1974). Some mainland and West Indian species also reduce reproduction in the dry season (Licht and Gorman, 1970; Sexton et al., 1971).[175]

In *A. carolinensis*, the only species whose endocrinology has been studied in detail, long days and warmer temperatures trigger growth of the testes and ovaries. Males begin reproductive behavior before females, and the sight of a displaying male serves as a cue to bring females into reproductive condition.[176] The endocrinological and neurophysiological mechanisms underlying the development and cycling of the reproductive system and governing the expression of reproductive behaviors of the green anole have been extensively studied and are a model system in laboratory research (see references in Crews [1975], Crews and Moore [2005], Lovern et al. [2004], and Wade [2005]); however, little comparative work has been conducted, except for some research on *A. sagrei* (e.g., Tokarz et al., 2002).

Unlike almost all other lizards, females produce only one egg at a time at an average interval of 5–25 days[177] (Andrews and Rand, 1974; Andrews, 1985b). In at least one species, egg size is correlated with female size (Vogel, 1984; see also Jenssen and Nunez, 1994); the only data of which I am aware suggests that inter-egg interval is not a function of female size (Jenssen and Nunez, 1994). Species vary in where they lay their eggs, in places as varied as in the leaf litter; under rocks and logs; in tree holes, rock crevices, and bromeliads; and attached to walls and ceilings of caves (reviewed in Rand, 1967a; see Andrews [1988] for a detailed study of *A. limifrons*). A number of species are known to use communal egg sites, presumably in areas in which appropriate habitat is limited (Rand, 1967a; Novo Rodríguez, 1985; Estrada and Novo Rodríguez, 1986); hatching, or at least dispersal from the communal nesting site, may be synchronized in *A. valencienni* (Hicks and Trivers, 1983).

174. Anoles (at least *A. carolinensis* and *A. sagrei*, the only species so studied) alternate the use of hemipenes. Each hemipenis is connected to its own testis. If prevented from using one hemipenis (by placing tape over one side of the cloaca), the male transfers significantly fewer sperm when it continually reuses the same hemipenis (Tokarz, 1988; Tokarz and Slowinski, 1990).

175. But see Gorman and Licht (1974), who reinterpreted patterns of reproduction in West Indian species and considered seasonal changes in temperature to be more important than changes in precipitation.

176. But see Jenssen et al. (2001), who argued that some of these conclusions are artifacts resulting from abnormally high densities in laboratory populations. They reported that in the field, males and females emerge simultaneously and become reproductively active more or less in synchrony, with male reproductive development slightly preceding that of females.

177. Although only a single egg is ovulated at a time, alternating ovaries, females retain eggs during times of drought and thus sometimes are found carrying as many as three eggs, two shelled and one in the oviduct (Stamps, 1976). Data on egg-laying intervals are scarce, primarily coming from *A. carolinensis*.

The actual egg laying process is little known. *Anolis carolinensis* begins by selecting a site and then probing it with her snout, followed by digging with the forelegs. An egg is then laid into the hole, pushed deeper with the snout, and then covered with dirt by back-to-forward movements of the forelegs. Eggs not laid in the hole are rolled in with the snout. Periodically, lizards stop digging to probe the hole with their snouts; occasionally the female abandons the hole, presumably because she has determined that conditions are unsuitable (Greenberg and Noble, 1944; Propper et al., 1991). Behavior of *A. aeneus* is very similar; in the laboratory, egg laying behavior didn't culminate in laying until a patch of ground was experimentally watered, which suggests that the lizards were examining whether the soil was sufficiently moist (Stamps, 1976).[178] *Anolis polylepis* has also been shown experimentally to prefer moister soil for egg laying (Socci et al., 2005). Andrews and Sexton (1981) compared eggs of two species that occur in hydrically different habitats and showed that eggs of the species that occupies more xeric habitats, *A. auratus*, have lower rates of water loss than do the eggs of the more mesic *A. limifrons*.

Incubation time is not well documented; in the laboratory it is 3.5–6 weeks for several Caribbean species (Greenberg and Hake, 1990; Sanger et al., 2008a), but may be as long as 130 days for montane species (Schlaepfer, 2003). Offspring are precocial and hatch at a small size (16–42 mm SVL; interspecific variation in hatchling size correlates with adult size [Andrews and Rand, 1974]).

GROWTH

Andrews (1976) measured growth rate and age of maturity of females in 13 anole species. Growth rate varied five-fold among species, with mainland species growing considerably faster than West Indian species. Schoener and Schoener (1978), however, found that some arboreal West Indian anoles had growth rates comparable to mainland species, whereas the terrestrial *A. sagrei*, which has much higher population densities, had growth rates more in line with Andrews' West Indian data. Schoener and Schoener (1978) suggested that population density, which should be related positively to degree of food limitation, may be the primary determinant of anole growth rates (see also Vogel, 1984).

Males generally grow faster than females (Schoener and Schoener, 1978; Vogel, 1984; Schlaepfer, 2006), leading to the sexual size dimorphism seen in many species (Chapter 9). Mean age of sexual maturity of females also shows a five-fold span, from 57–279 days; in line with growth rates, mainland species matured much earlier (Andrews, 1976).

DISPERSAL

Little is known about the dispersal of anoles. One study of *A. limifrons* found that most lizards dispersed very little and that the home ranges of many individuals moved little

---

178. Anole eggs, like those of many other reptiles, are very sensitive to hydric conditions in their incubating medium (Andrews and Sexton, 1981; Socci et al., 2005; Sanger et al., 2008a).

from the juvenile to adult age. The maximum dispersal distance, measured as distance from the center of the juvenile home range to the center of the adult home range, based on 148 individuals, was 45 meters. Both the mean and extremes were greater for males than for females (Andrews and Rand, 1983). *Anolis limifrons* is a small and short-lived mainland species; it is always possible that larger, longer-lived species may disperse further.

The only other data come from *Anolis aeneus*, which moves as much as 150 meters or more after hatching to occupy open clearings (Stamps, 1983b, 1990). Ultimately, the lizards move back into shadier areas when they reach subadult size, although it is not known whether they return to the vicinity of their hatching site.

A number of arboreal species are known to disperse across open ground between trees (Trivers, 1976; Hicks and Trivers, 1983; Losos and Spiller, 2005).

LIFE SPAN AND SURVIVAL RATES

Maximum life spans for most species are not known, but Meshaka and Rice (2005) estimate from growth rates that *A. equestris*, among the largest of anole species, can live for more than 10 years.

Schoener and Schoener (1982b) studied the survival of four species at 12 sites on three Bahamian islands. Some *A. sagrei* survived for at least 48 months (the length of the study), some *A. angusticeps* and *A. distichus* for 36 months, and some *A. smaragdinus*[179] for 24 months. Life expectancy of a newborn lizard varied between 0.9 and 1.9 years for most species/sex classes. In contrast, most mainland species, and some West Indian ones, have substantially lower life expectancy, with a lifespan that rarely exceeds a year (see reviews in Schoener and Schoener [1982b] and Andrews and Nichols [1990]). In the Bahamas, survival was higher on a smaller island with fewer avian predators than on larger islands with more predators; in closed forest compared to scrubby habitat; and, in *A. sagrei*, for females than for males(Schoener and Schoener, 1982b).[180] Variation in survival across sites in Panama for *A. limifrons* is greater than between-site variation in survival in Bahamian species (Andrews and Nichols, 1990).

Another, rarely studied, aspect of survival concerns mortality in the egg stage, which can vary greatly among sites (Andrews, 1988; Schlaepfer, 2003). Andrews (1988) noted that survival of eggs was 2–3 times more variable than survival of adults across the same study sites.

PREDATORS

Anoles are preyed upon by a wide variety of vertebrate and invertebrate predators (Fig. 8.2). I am unaware of any comprehensive review of predation on anoles, but many case studies and some limited reviews are available. Perhaps the most thorough is Henderson and

179. Referred to at that time as *A. carolinensis*.
180. In *A. distichus*, there was a suggestion that males survived better than females.

colleagues' analysis of the diet of West Indian snakes, which shows that most West Indian snakes eat anoles and that, collectively, anoles constitute more than 50% of the diet of West Indian snakes (Henderson and Crother, 1989; Henderson and Sajdak, 1996). By contrast, anoles constitute a much smaller fraction of snake prey on the mainland, 3% in one estimate (Henderson and Crother, 1989).[181] On Guam, the introduced brown tree snake (*Boiga irregularis*) preys heavily on *A. carolinensis* (itself introduced), and appears to have eliminated the anoles from non-urban settings (Fritts and Rodda, 1998).

Some birds, such as the American kestrel (*Falco sparverius*), the pearly-eyed thrasher (*Margarops fuscatus*), cattle egrets (*Bubulcus ibis*) and the aptly named lizard cuckoo (*Saurothera vielloti*) take anoles regularly, and many others take them at least occasionally (Wetmore, 1916; Cruz, 1976; Wright, 1981; Waide and Reagan, 1983; McLaughlin and

181. Though some mainland snakes, such as the blunt-headed vinesnakes, genus *Imantodes*, eat anoles in great quantities (e.g., Myers, 1982; Duellman, 2005).

FIGURE 8.2

Anole predators. (a) Predation by a jumping spider on *A. limifrons* at La Selva, Costa Rica. Photo courtesy of Harry Greene. (b) Predation by a basilisk (*Basiliscus basiliscus*), on *A. gorgonae* on Isla Gorgona, Colombia. Photo courtesy of Thomas Marent, www.thomasmarent.com. (c) Predation by *Siphlophis compressus* on *A. nitens* in Brazil. Photo courtesy of Marcio Martins. (d) Predation by a trogon on a green anole, possibly a new species of the *chlorocyanus* Series. Photo courtesy of Eladio Fernández.

Roughgarden, 1989; Gassett et al., 2000; Powell and Henderson 2008a).[182] Forty percent of the bird species at one study site in Grenada were observed eating anoles (Wunderle, 1981), as were many understory bird species in a Panamian rainforest (Poulin et al., 2001).

Lizards, too, can be important predators, and predation by other anole species and cannibalism have been reported in many species (Gerber, 1999). Other predators include monkeys, dogs, cats, mongooses, frogs, katydids, tarantulas, spiders, whip scorpions, and centipedes (e.g., Rand, 1967b; Waide and Reagan, 1983; Corey, 1988; Guyer, 1988a; Mitchell, 1989; Leal and Thomas, 1992; Reagan, 1996).

Probably the best documented effect of a predator on anoles involves the interaction between the curly-tailed lizard, *Leiocephalus carinatus*, and *A. sagrei* (Fig. 8.3). The curly-tailed lizard is a large, lumbering, primarily terrestrial lizard with poor climbing abilities.

182. Wetmore (1916) had the right spirit when, in his book *Birds of Porto Rico*, he stated in reference to the American kestrel (p. 32): "The only real criticism of this small hawk is its large consumption of lizards, amounting to 40.4% of its entire food." Similarly, with regard to the lizard cuckoo (p.59): "In consuming such quantities of lizards, this bird must be considered injurious, though to some extent it makes up for this by a diet of pernicious caterpillars . . ."

FIGURE 8.3

Curly-tailed lizard eating an *A. sagrei* in the Bahamas. Photo courtesy of Tom Schoener.

FIGURE 8.4

Tom Schoener on one of the small Bahamian islands near Abaco in the northern Bahamas on which predation studies were conducted. This was the smallest island used in the study.

In 1997, Tom Schoener, David Spiller and I set up an experiment in the Bahamas in which we introduced curly-tailed lizards to six small Bahamian islands, using another six islands as controls (Fig. 8.4; vegetated area of islands: 137–270 m²).[183]

We did not know what to expect from this experiment. The curly-tailed lizard has a broad diet and is known to eat anoles (Schoener et al., 1982), but we didn't know how often. Possibly such an occurrence is rare and would have minimal population level effects. We contemplated that our experiment might show no effect at all. But we couldn't have been more wrong. The response was immediate and dramatic: within two months, a two-fold difference in lizard density was established between the islands with and without predators, and this difference was maintained for the 2.5 year course of the experiment (Fig. 8.5a). In a second run of this experiment,[184] in which lizards were individually marked, survival rates on the islands in which the curly-tailed lizards were introduced were as low as 6% over the course of six months, whereas on control islands, survival was as high as 91% (Schoener et al. 2005).

183. Curly-tailed lizards occur nearby (usually within 200 m) and naturally colonize these islands (Schoener et al., 2002). Moreover, the effects of these experiments are transitory, as hurricanes periodically remove the populations (Schoener et al., 2001).

184. Initiated after the first experiment was washed away by Hurricane Floyd in 1999. This second experiment, too, was terminated abruptly, by Hurricane Frances in 2004.

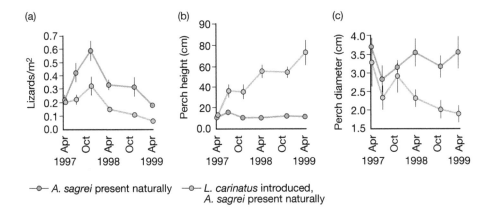

(a) (b) (c)

—○— *A. sagrei* present naturally —○— *L. carinatus* introduced,
A. sagrei present naturally

FIGURE 8.5
Effect of curly-tailed lizards on population size and habitat use of *A. sagrei*. Figure modified with permission from Schoener et al. (2002).

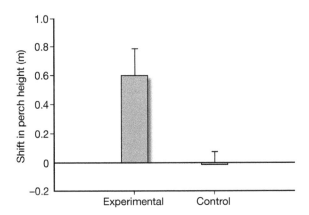

FIGURE 8.6
Immediate behavioral response of *A. sagrei* to newly-introduced curly-tailed lizards. When the predators were introduced to the island, they were placed in front of an anole, which was observed for 10 minutes. As a control, trials with the curly-tailed lizards were alternated with trials in which a piece of wood the size of a curly-tailed lizard was placed in front of an anole. Shifts in perch height were calculated as the height of the anole at the end of the experiment compared to its initial height. Figure modified with permission from Losos et al. (2004).

Behaviorally, anoles responded to the presence of the entirely terrestrial curly-tailed lizards by increasing their use of higher and thinner perches (Fig. 8.5b,c). Focal animal studies conducted when the curly-tails were first introduced revealed that anoles responded immediately to the presence of these predators, even though these islands had been curly tail-free for many generations (Fig. 8.6).

Anoles exhibit similar behavioral shifts in the presence of teid lizards in the genus *Ameiva*, which like curly-tailed lizards are relatively large, entirely terrestrial and known to eat anoles. On both Grenada and some offshore islets near Antigua, several anole species use the ground more often early and late in the day, when the more heliothermic *Ameiva* is not active (Simmons et al., 2005; Kolbe et al., 2008a). Moreover, in comparisons of *A. wattsi* on Antiguan islets with and without *Ameiva*, the anoles on the *Ameiva*-free islands use warmer microsites, are more active at midday and shift upward off the ground to a lesser extent (Kolbe et al., 2008a). These findings strongly suggest a response to the predatory lizard, but experimental manipulations would be useful to rule out other possibilities (e.g., temperature and habitat differences among islands).

Anoles behaviorally react to an approaching predator in many ways. Common responses include running around to the other side of a tree (termed "squirreling"),[185] running up a tree or other vertical object, running or jumping to the ground and running away, and jumping to another structure (Regalado, 1998; Schneider et al., 2000; Cooper, 2006; Larimer et al., 2006). Cryptic species allow a closer approach before fleeing (Heatwole, 1968; Cooper, 2006; Vanhooydonck et al., 2007), and often slowly creep to the other side of a branch or trunk and then quietly slip away (Oliver, 1948); others just hunker down against the surface and hope they will not be detected, a behavior which seems much more common in mainland species (Fitch, 1975; Andrews, 1979; Talbot, 1979). Some mainland species will freeze when a stick—supposedly mimicking a snake—is thrust toward them (Fitch, 1971, 1973a,b; Henderson and Fitch, 1975). Among Greater Antillean ecomorphs, grass-bush anoles tend to flee toward the ground, whereas trunk-crown anoles flee upward; the contradictory data for trunk-ground anoles may reflect a habitat effect: in forests, trunk-ground anoles sometimes flee upward, but in areas with few large trees, they move toward the ground (Collette, 1961; Ruibal, 1961; Rand, 1962; Schoener, 1968; Heatwole, 1968; Losos et al., 1993a; Schneider et al., 2000; Mattingly and Jayne, 2005; Cooper, 2006; Vanhooydonck et al., 2007).

Many anoles will display to an approaching human (e.g., Rand, 1962, 1967b). This behavior has probably been observed by anyone studying anoles in the West Indies (I don't know if it occurs in mainland anoles). I suspect that most workers, like me, dismissed the behavior as non-adaptive and inconsequential—probably the result of some defect in anole neural circuitry and perhaps an artifact of human disruption of the environment. What could be the use of an anole displaying at a potential predator that is so much larger than itself?

In a perceptive series of experiments, Manuel Leal showed that such behavior is adaptive in *Anolis cristatellus*, which displays not only at humans, but also to a common predator, the Puerto Rican racer, *Alsophis portoricensis* (Leal and Rodríguez-Robles, 1997b; Leal, 1999). In laboratory trials, Leal and Rodríguez-Robles (1995) showed that the snake, (which can attain a length of more than), attacked anoles much less often

185. A tactic which Wunderle (1981) showed to be markedly less successful when anoles were attacked by a flock of Carib grackles (*Quiscalus lugubris*), rather than by a single individual.

when the lizard displayed. Moreover, they demonstrated that when attacked, the lizards fought back, often biting the snake on the snout for as long as 20 minutes and managing to escape in 37% of the encounters (Leal and Rodríguez-Robles, 1995)—remarkable given the size discrepancy of the snake and the lizard. In field trials, Leal (1999) found that the extent of display behavior toward a snake model correlated with the endurance capacity of the lizard (as determined in subsequent laboratory trials); the greater the endurance capacity of the lizard, the more it displayed to an approaching snake model.

Anole displays to predators may be an example of a pursuit deterrent signal (reviewed in Caro, 2005). By signaling their endurance capability, anoles may be indicating their ability to fight back, escape, and potentially even injure a snake (Leal, 1999). Future work is needed to determine whether other anole species display toward natural predators (Leal and Rodríguez-Robles [1997a] have shown that *A. cuvieri* similarly displays toward a snake model in the field) and whether the pursuit deterrence hypothesis appears generally applicable.

PARASITES

A wide variety of internal and external parasites, including, coccidians, cestodes, nematodes, trematodes, mites and flies, has been reported in many anole species (e.g., Coy Otero and Hernández, 1982; Vogel and Bundy, 1987; Dobson et al., 1992; Cisper et al., 1995; Zippel et al., 1996; Goldberg et al., 1997; Schlaepfer, 2006; Irschick et al., 2006a). In the few cases where it has been investigated, the effect of parasitism on individual lizards appears relatively minor (e.g., Dobson et al., 1992; Schlaepfer, 2006), although infestation by larvae of the parasitic fly, *Lepidodexia blakeae*, which occurred at very high levels in one New Orleans population (Irschick et al., 2006a), is often fatal (Dial and Roughgarden, 1996).

Malaria parasites have been a particular focus of research, and several species are known to afflict anoles (Telford, 1974; Schall and Vogt, 1993; Staats and Schall, 1996; Perkins, 2001; Perkins et al., 2007). However, the individual and population level consequences of such parasitism are not well known. *Plasmodium* infection has detrimental effects on *A. gingivinus* (Schall, 1992), but little or no detectable effect on *A. gundlachi* (Schall and Pearson, 2000) or *A. sabanus* (Schall and Staats, 2002). No direct studies of the population effects of malaria have been conducted, but one correlational study related the prevalence of parasitism in *A. gingivinus* to interspecific interactions (Schall, 1992; see discussion in Chapter 11).

POPULATION DENSITY AND CONSTANCY

Anolis trinitatis has the highest density reported for any anole, more than 32,000 individuals per hectare (Hite et al., 2008), followed by *A. stratulus* (with more than 21,000 individuals per hectare [Reagan, 1992]) and *A. pulchellus* (17,000 individuals per hectare [Gorman and Harwood, 1977]).[186] Among 25 species, mean density of West Indian species (not including either *A. stratulus* or *A. pulchellus*) was five-fold greater than the mainland mean, and 2/3 of the West Indian species had a higher density than any mainland species (Stamps et al., 1997).[187]

Interpopulational variation in density can be great on both the mainland and in the West Indies. Densities exhibit six-fold variation among populations of *A. limifrons* on or near Barro Colorado Island in Panama (Andrews, 1991). Similar or greater variability is seen among populations of four species of Bahamian anoles (Schoener and Schoener, 1980a).

West Indian anole populations generally are quite stable, showing relatively little fluctuation in population size from one year to the next (Schoener, 1985, 1986a).[188] By contrast, mainland anoles show considerably greater year-to-year variability, as much as eight-fold for one population of *A. limifrons* (Fig. 8.7; Andrews, 1991). These generalities should be taken cautiously, however, as they are based on studies of only four mainland and six West Indian species.

DIET

A good place to start a review of the food anoles eat is a discussion of how they procure it. Although prey movement may be required to elicit feeding behavior in some species, others such as the aquatic anole, *A. aquaticus* recognize and rapidly eat non-moving prey (Burghardt, 1964; Goodman, 1971; reviewed in Moermond, 1981). Certainly, recognition of non-moving food occurs in the many anole species that eat fruit (see below).[189]

---

186. This figure is somewhat lower than that given by Gorman and Harwood (1977), whose numbers were inflated by several rounding errors.

187. Mainland density figures come from Central American species. Few data are available for the density of Amazonian species, primarily because they are so much scarcer even than Central American species (Vitt and Zani, 1998b).

188. Populations hit by hurricanes being an exception (Schoener et al., 2004).

189. No discussion of anole frugivory and feeding behavior is complete without mention of the famous Chuckles® experiment. On an expedition to remote Malpelo Island off the coast of Colombia, Rand et al. (1975) noted that the native anole of the island, *A. agassizi*, was attracted to the orange cap of a bottle of suntan lotion and to the orange packaging for Kodak film, and would come running from great distances and in great numbers when half of an orange was placed on the ground. The intrepid biologists wondered whether these anoles had a particular predisposition to the color orange. Fortunately, the expedition was outfitted with packages of Chuckles®—billed as "America's most popular jelly candy" in a 1949 advertisement—which conveniently contain candies in five colors: orange, yellow, red, green, and black. By placing various combinations of these sweets on the ground, the authors found that anoles are most attracted to orange and yellow candies, and least attracted to black ones.

But the story does not end there. In an effort to extend this research program to additional species, a graduate student in my laboratory tested a captive *A. grahami* with differently colored Starbursts®, a non-jellied candy that also comes in different colors (Chuckles® may not have been available in the local vending machine). Unfortunately, this experiment was stymied by other members of the lab, who removed lizard-bite sized pieces from the candies, thus briefly convincing the experimentalist that he was on to a major discovery.

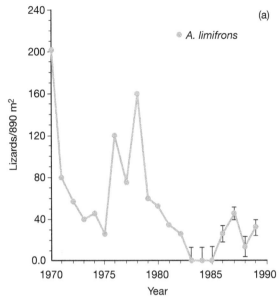

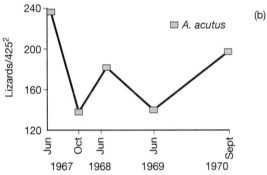

FIGURE 8.7
Population density of *A limifrons* on Barro Colorado Island, Panama over a 19-year period and *A. acutus* from St. Croix over a three-year period. Figure modified with permission from Andrews (1991); data for *A. acutus* from Ruibal and Philibosian (1974).

FORAGING MODE

How lizards forage for prey has been extensively studied (reviewed in Reilly et al., 2007). Two "modes" of foraging are recognized: "sit-and-wait," in which lizards remain in one spot, ready to pounce upon any unsuspecting prey that wanders by, and "active foraging," in which lizards seek out prey items, which oftentimes are immobile and hidden. Much ink has been spilt on whether these modes represent distinct alternatives, as opposed to being endpoints of a continuum (e.g., Perry, 1999; Butler, 2005; Cooper, 2005b, 2007).

Anoles epitomize the different foraging modes. At one extreme, trunk-ground anoles are classic sit-and-wait foragers, surveying the ground from their perches low on tree trunks and rapidly dashing or jumping to the ground to apprehend prey that move within range (e.g., Rand, 1967b). On the other hand, trunk-crown and at least some twig anoles often cruise through the arboreal matrix eating prey they come upon (see, e.g., the description of *A. valencienni* foraging in Hicks and Trivers [1983][190] and the comparison between the trunk-ground anole *A. sagrei* and the twig and trunk-crown anoles *A. angusticeps* and *A. smaragdinus*[191] in the Bahamas [Schoener, 1979]). Quantitative data confirm these impressions. The ecomorphs differ, dividing into active (trunk, trunk-crown) and sedentary (crown-giant, grass-bush, and trunk-ground) groups, with twig anoles being intermediate (Fig. 3.15; Johnson et al., 2008).[192]

Moermond (1979b) proposed that differences in foraging movement rates result from differences in the visibility in different structural microhabitats: lizards sitting in some microhabitats can keep an eye on a larger expanse than lizards in other microhabitats. For example, this hypothesis could account for differences among species that occur primarily on tree trunks; because they survey a larger area (the ground), trunk-ground anoles may need to move less than trunk anoles, which only scan a small area of tree trunk. However, Johnson et al. (2008) measured vegetation structure and found no overall relationship between visibility and movement rates.

An alternative hypothesis is that the costs and benefits of the foraging modes vary among microhabitats, potentially as a result of costs of movement, prey availability, or other factors. Comparisons across lizard families indicate that sit-and-wait foragers have a lower rate of energy acquisition than active foragers (Anderson and Karasov 1981).[193] Behavioral data suggest that the same relationship may exist for anoles; the ecomorphs that move at the lowest rates also eat less frequently (Johnson et al., 2008). Detailed data on rates of energy use and intake, and how they vary among habitats,

190. Hicks and Trivers (1983) reported on one female A. *valencienni* observed for three hours and forty minutes that moved up from the base of a tree into the vegetation at a height of 11 m, and then back down to the ground, feeding three times along the way.
Some of the danger inherent in an active foraging mode was apparent in another observation of a female moving upside down on a bromeliad, searching for prey (quoting from Trivers' field notes, p. 575): ". . . it seems to spot something on a neighboring bromeliad, also upside down. I too spot something on the second bromeliad. Starts to dart the 5 cm to the neighboring bromeliad but—as if forgetting it is upside down—it steps into thin air and falls 6 m to the ground. It appears to be uninjured."
191. Referred to as A. *carolinensis* in that paper.
192. This conclusion is based on Johnson et al.'s (2008) analysis of data for 31 species. Other studies on smaller numbers of species (Moermond, 1979b; Irschick, 2000; Cooper, 2005a) generally reach similar conclusions, with one key difference: grass-bush anoles generally were found to be among the most active species (e.g., Perry, 1999; Cooper, 2005a); the explanation for this discrepancy is not obvious (reviewed in Johnson et al., 2008).
Some studies of lizard foraging behavior use a second metric, percentage of time spent moving. When species differ in the duration of their movements, this metric is a necessary complement to measuring the number of discrete movements per unit time (Perry et al., 1990; Cooper, 2005b). However, because most movements of anoles are extremely brief, these two measures are highly congruent for anoles (Perry, 1999; Irschick, 2000; Cooper, 2005a): if anything, some of those anoles that move most frequently—trunk-crown and some twig anoles—are also the ones that have longer movement durations (Irschick et al., 2000). Hence, were data available for percentage moving for many species, they would likely reveal even greater differences among the ecomorphs than those shown in Figure 3.15.
193. In comparison to classic active foragers like teid lizards, which spend as much as 87% of their time on the move, all anoles are relatively sedentary (Perry, 1999; Butler, 2005; Cooper, 2005b). Nonetheless, even if the degree of difference is more muted, the sit-and-wait versus active searching dichotomy applies as well to anoles as it does to larger scale differences among lizard families (see discussion in Johnson et al., 2008).

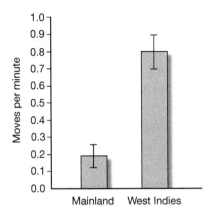

FIGURE 8.8

Differences in movement rate between mainland and West Indian anoles. Differences are statistically significant (Analysis of variance, $F_{1,44} = 41.90$, $p < 0.001$). Data from Perry (1999), Irschick et al. (2000), Cooper (2005a), and Johnson et al. (2008). When multiple values were available for a species, they were averaged. Considerable heterogeneity in methods and results exists among studies. Only Perry's (1999) study contains data for both mainland and West Indian species. Results are nearly significant ($p = 0.091$) if the analysis is confined to only the nine species studied by Perry, thus eliminating effects of inter-study differences in methodology.

could prove useful in understanding the genesis of foraging mode differences among the ecomorphs.

The data in hand also hint at a difference between West Indian and mainland anoles in movement rates. Among 46 species, mean movement rate of West Indian species is three times that of mainland anoles (Fig. 8.8).[194]

Detailed observations of foraging are surprisingly scarce in the literature. One exemplary study focused on the behavior of *A. carolinensis* in Georgia (Jenssen et al., 1995; Nunez et al., 1997). Three foraging styles were observed: sit-and-wait; eat on the run, in which a lizard captured a prey item it encountered as it was moving, usually while patrolling its territory; and active searching, in which the lizard moved very slowly

194. As discussed in Chapter 6, the Norops clade of mainland anoles is the result of a single colonization event. Almost all of the ecological and behavioral data for the mainland comes from species in this clade, which accounts for the large majority of mainland species. For this reason, as discussed in Chapter 5, data points from mainland anoles are not statistically independent; consequently, statistical analyses between mainland and island species will suffer from phylogenetic pseudoreplication and, were statistical analyses conducted in a phylogenetic context (not yet possible due to lack of a well supported phylogeny for mainland Norops), most results would be non-significant. This situation does not invalidate the finding that mainland and island anoles are different, but complicates causal explanation of such differences. That is, an analysis of variance between mainland and West Indian anoles tests the hypothesis that geographic location is related causally to foraging rates. However, because mainland Norops represent a single clade, they have inherited from their common ancestor many characteristics other than their geographic location, and thus it is statistically impossible to separate out which factors have been responsible for differences in foraging rate. In an ideal world, we would have many clades that have independently moved from one area to the other, and thus we could investigate whether a statistical association exists between change in geographic location and change in foraging rates. In the real world, however, we are stuck with the distribution of species and clades as they actually occur.

through the habitat, carefully looking for prey by, for example, inspecting the undersides of leaves. Among females, use of the three approaches was correlated with their success rate (Nunez et al., 1997): sit-and-wait (83% of feeding attempts/89% success rate), on-the-run (13%/71%) and active searching (4%/60%). Active searching behavior is also exhibited by the Jamaican twig anole, *A. valencienni*, which seeks out concealed prey items (Hicks and Trivers, 1983; see Fig. 3.9).

My guess is that the foraging behavior of trunk-ground, trunk-crown, and twig anoles conforms to the classic distinction between sit-and-wait versus active foragers: the first group sits on tree trunks and surveys its surroundings, eating what comes along, whereas the latter two move around more frequently[195] and probably search out inactive prey, as well as grabbing whatever passes by (e.g., Schoener, 1979). The other ecomorphs are more mysterious. Although crown-giants do not move at high rates, they do at times cover large distances; they have been seen stalking other anoles (P.E. Hertz, pers. comm.) and take not only fruit, but nestling birds (Dalrymple, 1980), both of which must be sought out. My impression is that crown-giants, though less active overall, are more like twig and trunk-crown anoles in their foraging patterns than they are like the other, less active ecomorphs. In turn, the highly active trunk anoles seem to have a very different strategy than twig and trunk-crown anoles, moving up and down tree trunks, but not through the arboreal matrix. The trunk anoles of the *distichus* Series are probably the most myrmecophagous of all anole species (Schoener, 1968)—an individual actively searches for ants and then "sits passively in front of a trail and gobbles the ants up as they pass by" (Schoener, 1979, p. 484). Unfortunately, the diet and foraging behavior of the Cuban trunk species, *A. loysianus*, is unknown. Finally, it is hard to speculate on the foraging behavior of grass-bush anoles given the disparate results concerning their rate of activity (see footnote 192).[196]

Seasonal shifts in foraging mode have been reported in two species. Male *A. nebulosus* in one of two wet seasons switched from a sit-and-wait to an active foraging mode in which they spent 60% of their time in "slow transit" foraging (Lister and Aguayo, 1992). Similarly, male *A. carolinensis* in the breeding season (May–July) captured prey using a sit-and-wait mode (58% of feeding events) or while they patrolled their territories (eat on the run; 42%); in the non-breeding season (August–September), 22% of feeding events occurred while actively searching for prey as described above and 74% in a sit-and-wait context. The decrease in eating on the run, to 4%, resulted from the substantial decrease in territorial patrolling that occurred in the non-breeding season (Jenssen et al., 1995).

195. Although twig anoles are highly variable. Two species, *A. valencienni* and *A. angusticeps*, are among the most active of anoles, but other twig species move considerably less (Johnson et al., 2008).
196. For completeness, I should point out that little data are available concerning the foraging behavior of West Indian unique, Lesser Antillean and mainland species.

The actual process of anole prey capture has received some study. The most common prey capture behavior is for an anole to rapidly approach a prey item, pause, turn its head toward the prey, and grab it (Moermond, 1981). Some differences in prey attack behavior correspond with morphological differences: the twig anoles *A. insolitus* and *A. angusticeps* use a similar behavior to that just discussed, but their approach is much slower than other anoles, whereas several long-legged species launch themselves toward prey, capturing it as they land (Schoener, 1979; Moermond, 1981).[197] *Anolis carolinensis* uses all of these behaviors; when feeding in the ambush mode, the anoles use the approach-pause-strike method, when feeding on the run they often jump forward to capture prey, whereas while actively searching for prey, they use the slower creeping approach typical of twig anoles (Jenssen et al., 1995; see also Monks [1881]).

One correlate of foraging mode is prey type: across all lizards, sit-and-wait foragers tend to eat active prey, whereas active foragers search out sedentary species (Huey and Pianka, 1981). Whether this trend occurs among anoles is unknown; the dietary information summarized below is insufficient to characterize the attributes of most prey items.

DIET COMPOSITION

Many studies have reported the diet of one or more species from a particular locality at a particular time. Few studies have compared how the diet of a species changes through time or across space, and no comprehensive review of anole diet exists. Most anoles appear to be generalists, eating almost anything they can get their jaws on and swallow, but some exceptions exist (Fig. 8.9).

The diet of Puerto Rican anoles has probably been studied better than the diets of lizards on other islands; these species may be representative of the general situation for anoles, at least in the West Indies. In one of the most thorough studies, Wolcott (1923) examined the stomach contents of large numbers (30–110 for most species) of Puerto Rican anoles; many of his specimens were collected near the campus of the University of Puerto Rico in Río Piedras, but others were collected elsewhere on the island. He found that they ate a wide variety of insects, as well as spiders, millipedes, centipedes, snails, seeds, and other items. Beetles, ants, flies, lepidopterans, hemipterans and homopterans were common prey items for most species, though in varying proportions.

197. Examples of this prey-catching behavior were provided for the relatively short-limbed *A. carolinensis* (under the name *A. principalis*) by Lockwood (1876, p. 7): "I have just been watching Nolie eying a fly which was walking on one of the glass panes of his house. He made a noiseless advance of about three or four inches; then followed a spring, when he was seen cleaving to the glass by his feet, and champing the captured fly. I saw him once intently watching the movements of a fly which was walking on the glass. As seemed evident to me by an ominous twitch of that little head, his mind was made up for a spring; but lo, there was a simultaneous make-up of mind on the part of the fly, which at this juncture flew towards the other side of the case. Then came—and how promptly—mental act number two of Anolis, for he sprang as the after-thought directed, and caught the insect on the fly." Dial and Roughgarden (1995) report an anole jumping from a branch one meter above a spider web, catching the spider as it passed by, before landing in the vegetation below.

FIGURE 8.9

Anoles eating. (A) *A. stratulus* staking out a
termite tunnel, Puerto Rico. Photo courtesy
of Richard Glor. (B) *A. evermanni*, Puerto
Rico. Photo courtesy of Michele Johnson.
(C) *A. chlorocyanus*, Hispaniola. Photo courtesy of Alejandro Sanchez. (D) *A. oculatus*,
Dominica. Photo courtesy of Anita Malhotra.

Some of these species have been studied at other localities or at other times; comparisons of these studies reveals great variability in diet composition. For example, for
A. evermanni, Wolcott (1923) found that beetles and homopterans numerically dominated the diet, but Lister (1981) found that beetles, ants, orthopterans, and seeds were
the most important prey by volume in the Luquillo Mountains.[198] Reagan (1996), working elsewhere in the Luquillo Mountains, found that diet varied between the sexes and
between two habitats for this species. Ants were always numerically the most important
prey, but when considered by volume, a wide variety of prey types were taken with none
dominating the diet. Finally, Dial and Roughgarden (2004), also working in the Luquillo
Mountains shortly after Hurricane Hugo devastated the forest, found that dipterans, spiders, and orthopterans were the most common prey item by biomass, but no category
constituted more than 25% of the total.

Other species show a comparable degree of variation. All studies except Dial and
Roughgarden's (2004) post-hurricane survey agree that ants are the numerically most
important prey of *A. stratulus* and also make up a substantial proportion of total biomass ingested. However, beetles, flies, hemipterans, homopterans and caterpillars are
other important prey, and their importance varies among the studies. For *A. gundlachi*,
orthoptera were a dominant prey by volume in the Luquillo Mountains (Lister, 1981;
Reagan, 1996), but were not present in the 10 animals examined by Wolcott. Ants were

198. One frustration in trying to compare studies of diets is that authors report diet composition in a myriad
of different ways. The most common are to lump all of the prey eaten by all individuals, and then calculate for
each prey type either its proportion of the total number of prey items or of total biomass.

numerically important in Wolcott's and Reagan's study, but not Lister's. Flies, caterpillars, and earthworms were also important in some of the localities.

Several messages emerge from this comparison. Most anole species tend to eat a wide variety of different prey types. Diet can differ markedly from one place to another, or even among habitats within a locality. Moreover, at least when prey are grouped at the familial or ordinal level,[199] interspecific variation in anole diet is relatively small compared to the variation that occurs within a species in different studies. One reason for the intraspecific variation in diet across studies is that anoles generally are opportunists, taking advantage of any prey type that is abundant. For example, anoles are widely known to congregate at broken termite nests, eating vast quantities of the swarming insects (e.g., Barbour, 1930; Rand et al., 1975), and in one experimental study, coccinelid beetles were released onto a tree in Bermuda, and the next day an adult *A. grahami* was caught with 26 in its stomach (Simmonds, 1958). Given that such opportunities are sporadic, variation in diet composition from samples collected at different times and areas is not surprising. Differences in prey availability probably also underlie the seasonal diet shifts reported in a number of species (e.g., Sexton et al., 1972; Fleming and Hooker, 1975; Floyd and Jenssen, 1983; Bullock et al., 1993; Fontenot et al., 2003; but see Rodríguez Schettino and Reyes [1996]). Stamps et al. (1981) suggested that *A. aeneus* eats seasonally rare types of prey disproportionately often to maintain a nutritionally balanced diet.

199. One major difficulty with most lizard diet studies (though not Wolcott's) is that prey are not identified to species or genus (sometimes this is the fault not of investigators, but of journals being unwilling to print detailed lists of prey taxa [L.J. Vitt, pers. comm.]). Great ecological variation can exist among insects within a family; lumping together such disparate species can obscure dietary differences between species (Greene and Jaksić, 1983).

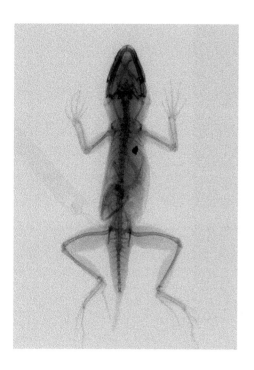

FIGURE 8.10

Carnivory in *A. cristatellus*. This 57 mm
SVL male ate a smaller *Anolis* (approximately
35 mm SVL), the head and anterior skeleton
of which is apparent in its stomach region.
Note also the lack of bones in the regenerated
portion of the tail: when a tail regenerates,
the new portion is made of a rod of cartilage
and thus lacks the intravertebral breakage
planes that enable an unregenerated tail to
autotomize. Photo courtesy of Liam Revell.

Maximum prey size and hardness, as well as the range of prey sizes taken, tends to increase with body size both within and between species (e.g., Rand, 1967b; Schoener, 1967, 1968; Schoener and Gorman, 1968; Sexton et al., 1972; Corn, 1981; Lister, 1981; Vitt et al, 2003a,b; Vitt and Zani, 2005; Whitfield and Donnelly, 2006; Herrel et al., 2006); nonetheless, even very large individuals will eat very small prey, particularly when they are abundant (e.g., Schoener and Gorman, 1968).

Dietary differences between the sexes have been noted in many species. Differences in prey size would be predicted to result from dimorphism in body size (Chapter 9), but this expectation is only sometimes fulfilled (Schoener, 1967, 1968; Schoener and Gorman, 1968; Corn, 1981; Floyd and Jenssen, 1983; Stamps et al., 1997). When the effects of body size are removed, intersexual differences in prey size exist in some species, but not in others.[200] Differences in prey type also exist between the sexes, which may reflect intersexual differences in size, microhabitat use, or energetic demands (e.g., Schoener, 1967, 1968; Reagan, 1986; Perry, 1996; Sifers et al., 2001). In terms of the overall amount of food ingested, as measured by the volume of stomach contents, females sometimes, but not always, eat more than males relative to their body size (e.g., Schoener, 1968; Schoener and Gorman, 1968; Perry, 1996; Vitt et al., 2003a; Vitt and Zani, 2005). In *A. limifrons* (the only species so studied), females ingested more

200. Males eat larger prey than females: Schoener, 1967, 1968; Schoener and Gorman, 1968; Corn, 1981; Stamps et al., 1997; females eat larger prey than males: Schoener, 1968; Schoener and Gorman, 1968; Andrews, 1971, 1979; Perry, 1996; no difference in prey size between males and females: Corn, 1981; Vitt and Zani, 1996a, 2005; Vitt et al., 1995.

than twice as many calories as males during the reproductive season, but only 18% more in the non-reproductive season (Andrews and Asato, 1977). In *A. humilis*, females responded more often and more quickly than males to supplementary food placed in front of them in the field (Parmelee and Guyer, 1995).

Carnivory (as opposed to insectivory)—including both heterospecific and conspecific anoles—is reported for many species (Fig. 8.10) and its frequency is related to size (although this has not been quantified); giant anoles will eat a wide variety of vertebrates, including birds (Dalrymple, 1980; Meshaka and Rice, 2005). However, even small anoles will occasionally eat vertebrates; the increased carnivory of larger anoles probably represents only their greater ability to capture and subdue other vertebrates, rather than indicating a size-related shift in foraging behavior or preference.[201]

One prey type of particular interest is among the smallest and least nutritious: ants. Ant consumption is high in many West Indian species: e.g., more than 80% of prey items in *A. distichus* (Schoener, 1968; Cullen and Powell, 1994) and *A. gingivinus* (Eaton et al., 2002); more than 68% in male *A. evermanni* in rainforest habitat (Reagan, 1986), *A. opalinus* (Floyd and Jenssen, 1983) and *A. homolechis* (Berovides Alvarez and Sampedro Marin, 1980); and more than 50% in *A. grahami* (Simmonds, 1958) and *A. stratulus* (Reagan, 1986). In many West Indian species, ants comprise a greater proportion of the diet of adult males than of adult females (Schoener, 1967, 1968; Schoener and Gorman, 1968; Reagan, 1986). In contrast to their prominence in the cuisine of West Indian anoles, ants form a small proportion of the diet of most, but not all, mainland anoles.[202]

Many anole species are known to eat fruits at least occasionally, and in some species at some localities, frugivory is quite common (Herrel et al., 2004). Species known to be frugivorous are larger than those not known to eat fruits, and West Indian species are frugivorous more than mainland species (30% versus zero % in Herrel et al.'s [2004] survey).[203] Among West Indian anoles, no grass-bush anoles and all crown-giants have been reported to be frugivorous; data for other ecomorphs is mixed. My hunch is that when more species are studied, almost all but the smallest species will be found to occasionally take fruit.[204] Seed eating is also reported for a number of species

201. This point was made clear to me when, while I was drafting this chapter, I observed a 38 mm SVL female *A. pulchellus* subdue and ingest a small gecko (Losos et al., unpublished).

202. Some reports of few ants in diet: e.g., Andrews, 1971; Talbot, 1979; Corn, 1981; Lieberman, 1986; Vitt and Zani, 1996a, 1998b, 2005; Duellman, 2005; many ants in diet: Fleming and Hooker, 1975; Duellman, 1978; Vitt et al., 2003a, 2008; one of these exceptions is *A. auratus* (Vitt and Morato de Carvalho, 1995; Mesquita et al., 2006), which in ecology and morphology is very similar to West Indian grass-bush anoles.

203. The lack of frugivory in mainland anoles should not be overemphasized, because dietary data are available for few large mainland species. Moreover, frugivory is known for one mainland species, *A. pentaprion* (Perez-Higareda, 1997). Nonetheless, frugivory does seem to be rare in mainland anoles; a survey of herpetological books for different mainland regions (e.g., Savage's [2002] authoritative compendium of information on the reptiles and amphibians of Costa Rica) revealed no additional reports of frugivory in mainland anoles, nor did Vitt and Zani's (1998b) study of seven Nicaraguan species.

204. For example, the fact that an *A. evermanni*, not definitively known to eat fruit, once jumped on my shoulder, ran down my arm, perched on my thumb, and bit at the red knob of the stop-watch I was holding suggests to me that this trunk-crown anole will eat red berries, just like many other anoles. Seeds (Reagan, 1996) and "seeds or fruit" (Lister, 1981) have been reported in the diet of this species, so my prediction that it is frugivorous is not very daring.

FIGURE 8.11

Anole exclusion from trees in the Luquillo
Mountains, Puerto Rico. By climbing trees
and capturing lizards, Dial and Roughgarden
(1995) were able to remove lizards from trees
isolated by Hurricane Hugo to monitor the
effect to lizards on the ecosystem. Plastic
collars on the trunks prevented lizard
recolonization. Photo © Roman Dial.

(e.g., Wolcott, 1923; Reagan, 1996). In some cases, these seeds may have been ingested incidentally, but in other instances, seeds, which are digested more slowly than pulp, may be the last remaining trace of a fruity meal in the digestive tract of an anole.[205] Nectarivory has been reported in a number of West Indian trunk-crown anoles (Liner, 1996; Perry and Lazell, 1997; Campbell and Bleazy, 2000; Echternacht and Gerber, 2000; Okochi et al., 2006; Valido, 2006), a grass-bush anole (Perry and Lazell, 2006), and two Lesser Antillean species (Timmermann et al., 2008). The greater occurrence of frugivory and nectarivory among island species compared to those on the mainland agrees with a trend seen for lizards in general (Olesen and Valido, 2003).

Andrews (1979), in a review of anole diet studies, argues that West Indian species generally eat small prey, whereas mainland species take much larger prey. Moreover, mainland anoles generally have more food in their stomach than do West Indian lizards, indicative of a higher rate of food intake. Work by Vitt and colleagues (e.g., Vitt and Zani, 1996a, 1998a,b 2005; Vitt et al., 2001, 2002, 2003a,b) indicates that extensive variation exists among mainland species in prey size and in the extent to which lizards have full stomachs; unfortunately, quantitative comparisons among published studies by various researchers are not possible.

205. Indeed, intact seeds some times make their way out the other end. When anoles are captured alive, they often defecate in the container into which they are placed, and seeds are not infrequently found in the poop.

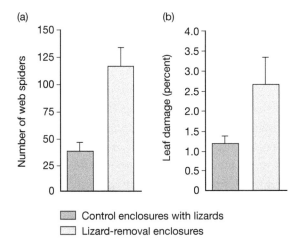

(a) (b)

Number of web spiders / Leaf damage (percent)

■ Control enclosures with lizards
□ Lizard-removal enclosures

FIGURE 8.12
Food web effects of anoles. The numbers of spiders and damage to leaves both increase when *A. sagrei* is removed from experimental enclosures. Based on data from Spiller and Schoener (1996) kindly provided by David Spiller.

PLACE IN THE ECOSYSTEM

Anoles thus eat and are eaten (as well as parasitized) by many other species. Consequently, given their great abundance in the West Indies, anoles probably play an important role in the functioning of ecosystems, both as major predators on insects and as food for predators on higher trophic levels. A number of experimental studies have confirmed this expectation. These experiments have been impressive in scope and sometimes adventurous in execution. In the Bahamas, *A. sagrei* has been introduced to small islands lacking them (Schoener and Spiller, 1999), using other lizard-less islands as controls. In Puerto Rico, Dial and Roughgarden (1995) took advantage of the aftermath of Hurricane Hugo, which fragmented the canopy into discrete tree crowns isolated from each other. Dial climbed 14 trees, removing most or all anoles from seven of them, a Herculean task given that the trees were 20–30 m tall. By placing polypropylene plastic collars around the trunks of the trees to prevent re-colonization, they were able to use each tree as an experimental unit (Fig. 8.11). Finally, on St. Eustatius, Pacala and Roughgarden (1984) created six 12 × 12 m fenced enclosures in the middle of the rainforest. Isolating these enclosures from the outside involved removing all overhanging vegetation so that lizards could not walk in and out of the enclosures through the canopy. Once the enclosures were completed, all lizards were removed from all of them, and then subsequently reintroduced into three of the enclosures to initiate the experiment. Similar experiments in approximately 9 × 9 m enclosures were conducted in the Bahamas by Spiller and Schoener (1988, 1990).

The results of these experiments have been very similar. The abundance of web spiders is generally much higher in the absence of anoles compared to in their presence (Fig. 8.12). Arthropod density is also often affected, although effects are not entirely consistent among studies and vary by type of arthropod: many types increase in the absence of anoles, presumably a result of release from predation, but others decrease in abundance, presumably due to predation from the more abundant spiders, an example of an indirect effect. Another indirect effect is an increase in plant damage in the absence of anoles, which results from an increased abundance of herbivorous insects (this effect was not observed in the St. Eustatius study [Pacala and Roughgarden, 1984]).

These three-level trophic cascades were taken a step further in the experiments involving the introduction of the predatory curly-tailed lizard discussed earlier in the chapter (Schoener et al., 2002). The presence of the predatory lizard reversed the effect of *A. sagrei* on spider density and diversity, but didn't alter any of the other effects of *A. sagrei*, perhaps because the larger predatory lizard also eats ground arthropods (Schoener et al., 2002).

The presence of anoles on Bahamian islands also decreases spider species richness, both experimentally (Spiller and Schoener, 1988, 1998; Schoener and Spiller, 1996) and among natural populations (Toft and Schoener, 1983).

Given their abundance, one might expect that anoles in the West Indies play an important role in determining the flow of energy and nutrient cycling; indeed, Reagan (1996) estimated that the three most common anoles in the Luquillo Mountains consume approximately 450,000 insects per hectare per day. However, despite the increase in ecosystem level analysis in the last 10–15 years, few studies have focused on the West Indies, and thus the role of anoles in ecosystem functioning has been little studied. The most comprehensive food web for a West Indian locality was established for St. Martin; in that web, the two anole species link to a very large number of other species, either as predators or prey (Goldwasser and Roughgarden, 1993); a food web for anoles in the Luquillo Mountains in Puerto Rico shows a similarly large number of connections (Reagan et al., 1996).

In the post-hurricane study in Puerto Rico, densities of *A. evermanni* and *A. stratulus* were strongly positively related to insect abundance, which varied among trees. Aerial insect abundance was greater in tree gaps than within the canopy: as a result, trees downwind from forest gaps supported more lizards because they had the greatest arthropod abundance, apparently as a result of insects being blown into the canopy from the tree gaps (Dial and Roughgarden, 2004). Arthropod groups that were commonly caught in sticky traps suspended across tree gaps at 12–15 m height, and thus were likely being blown into the canopy from the ground, exhibited similar abundance changes through time in trees with and without lizards. Dial and Roughgarden (2004) suggested from this observation that although anoles may reduce the abundance of these insects in the canopy, the underlying population dynamics of the insects are determined by what happens in the tree gaps where they are born. In contrast, for those arthropod groups which spend their entire life cycle in the canopy and thus are rarely caught in tree gaps, population abundances fluctuated out of synchrony between trees with and without lizards. Dial and Roughgarden (2004) suggested that the presence of anoles fundamentally changes the population dynamics of these insects.

Similar studies have not been conducted on mainland anoles. However, the lower density of mainland anoles may indicate that their ecosystems effects are generally much less than those of West Indian lizards (Rand and Humphreys, 1968).

MAINLAND—ISLAND COMPARISON

Andrews (1979), synthesizing many of the types of data reviewed in this chapter, argued that the life history and population biology of mainland and West Indian anoles are fundamentally different. She argued that West Indian anole populations are generally food limited, whereas mainland anole populations are generally limited by predation. In support of this argument, she made the following points:

- Mainland anole survivorship is lower
- Mainland anole densities are lower
- Mainland anole food intake is greater
- Mainland anole average prey size is larger, presumably reflecting the ability of mainland anoles to be more selective in what they eat
- Mainland anole growth rate is higher
- Mainland anoles spend less time foraging, which Andrews attributed both to higher average prey size and greater risk of predation of mainland anoles

From these points, Andrews (1979) concluded that, as a generality, West Indian anole life history evolution has been shaped by intraspecific competition, whereas mainland anole evolution has been determined primarily by predation.

As the literature summarized in this chapter indicates, most work in the last three decades has continued to support the dichotomies Andrews recognized. However, it is notable that work on anole population biology has diminished in the last 10–15 years; little work of this sort is currently being conducted, so to some extent Andrews' hypothesis may not have been falsified partly because few relevant data have been collected in recent years. Moreover, as Andrews (1979) noted, her conclusions were based on data from relatively few mainland species, primarily the more terrestrial species in lowland wet tropical habitats. Data from more arboreal species and from species in other habitat types are needed to test the generality of these findings. For example, the montane mainland species *A. tropidolepis* is in some respects, such as growth rates and survivorship, more like West Indian anoles (Fitch, 1972); conversely, low density arboreal species in the West Indies have high growth rates similar to those of mainland anoles (Schoener and Schoener, 1978).

Although predation may play a much larger role in regulating mainland anole populations, limited experimental data do not support the hypothesis that only West Indian species are food limited. Food supplementation experiments have been conducted on five species—four West Indian and one mainland—and all have found evidence for food limitation. West Indian species readily take supplemental food when it is provided[206] and show increased body mass and, sometimes, growth rate (Rand, 1967b; Licht, 1974; Stamps and Tanaka, 1981; Rose, 1982).[207] However, the only mainland species so investigated, *A. humilis*, also provides evidence of food limitation (Guyer, 1988b). When insect availability was increased by placing containers with rotting meat on experimental plots, male lizards grew larger and females increased egg production relative to individuals on control plots. All species studied to date have been very abundant; whether less abundant species are also food limited remains to be seen.

Of course, the predation versus competition hypothesis paints with an extremely broad brush. Substantial variation in many attributes—e.g., productivity, seasonality, predator richness—exists among habitats within both the West Indies and the mainland. For this reason, the relative importance of predation and competition is likely to vary within each region, as well as between them. Moreover, many, perhaps most, West Indian species occupy habitats that are more similar to mainland dry forest than to rainforest. Consequently, comparisons between mainland and island anoles should consider effects due to habitat, such as productivity or seasonality (e.g., Frankie et al., 1974; Opler et al., 1980), as well as those representing region-wide differences (Duellman, 1978, p. 330).

FUTURE DIRECTIONS

Despite decades of intensive work, many basic aspects of anole ecology remain to be learned, as I hope this chapter has highlighted. Dietary data are surprisingly incomplete, and considerably less is known about predation and parasitism. The role of anoles in the ecosystem has barely been examined. Moreover, these studies are biased toward West Indian species and toward more terrestrial species. We have little idea of the biology of arboreal species. The mainland giant anoles, most of which are highly arboreal, are particularly poorly known. In summary, the population biology of almost all anole studies still requires detailed investigation, both to understand the biology of those species, and to establish more broadly based comparative patterns.

Probably the most intriguing generality emerging from this review is the differences between mainland and West Indian species. Clearly, more thorough sampling of species is needed to determine the extent to which differences truly represent an island-mainland dichotomy, as opposed to being a reflection of the species studied to date or regional differences in habitat use. In addition, Andrews' (1979) hypothesis needs to be explicitly tested to determine the extent to which predators, food, or other factors influence population biology. Experimental studies, though hard work, are probably the best way to test many of these hypotheses.

---

206. The greedy little buggers took as many as 11 mealworms—on average eating 14% of body mass in one sitting—in *A. lineatopus* (Licht, 1974). Food supplementation in these studies was accomplished by sprinkling insects around a lizard's home range, tossing mealworms at the base of a tree occupied by lizards, or presenting a mealworm attached to a two-meter fishing pole directly in front of a lizard.

207. One exception is that juvenile *A. aeneus* take supplemental food and increase their growth rates in the dry season, but not in the wet season (Stamps and Tanaka, 1981).

9

SOCIAL BEHAVIOR, SEXUAL SELECTION, AND SEXUAL DIMORPHISM

Sexual selection—"the advantage which certain individuals have over others of the same sex and species solely in respect of reproduction" (Darwin, 1871)—is a topic of great interest to behavioral and evolutionary biologists. The past 25 years have seen a tremendous upsurge in interest in sexual selection, and a concomitant documentation of its near ubiquity throughout the animal and even plant worlds (e.g., Andersson, 1994; Andersson and Simmons, 2006). In addition, we now have a greater appreciation of the many and varied ways in which sexual selection may occur, encompassing not only the traditional views of male combat and female mate choice, but many other mechanisms including sperm competition, alternative male mating strategies, cryptic female choice and sexual antagonism (Andersson, 1994; Eberhard, 1996; Hosken and Snook, 2005; Parker, 2006). The role that sexual selection may play in driving speciation and macroevolutionary trends is now widely discussed (Price, 1998; Seehausen and van Alphen, 1999; Gavrilets, 2000a; Panhuis et al., 2001; Mank, 2007).

Sexual selection certainly occurs in anoles. Males are often substantially larger than females and fight with each other to maintain territories providing sole—or at least primary—access to the females within them. Whether other types of sexual selection occur in these lizards is less certain. The traditional view is that female choice does not operate, but data are just beginning to emerge to challenge this assumption. The biology of anoles opens the door to the existence of other types of sexual selection, such as sperm competition and cryptic female choice, but few relevant data are currently available.

FIGURE 9.1
Anolis garmani from Jamaica in the survey posture.

This chapter reviews the behavior and reproductive biology of anoles, with an eye toward assessing the importance of sexual selection and the forms it takes, both within anole populations and potentially over evolutionary time spans.

TIME BUDGET

The first place to start in discussing anole behavior with regard to sexual selection is by considering what an anole does with its time: How much of their day do anoles allocate to different activities, and how do such allocations differ between the sexes and across the seasons?

Many anoles spend much—perhaps most—of their time in the "survey posture," head down, forequarters lifted slightly off the surface, and hindlegs extended backwards (Fig. 9.1). From this position, anoles can scan their surroundings for prey, predators, and conspecifics and can advertise their own presence. The importance of the survey posture for foraging was demonstrated by Stamps (1977a), who found that both male and female *A. aeneus* greatly decreased their use of this position after being fed to satiation. No comparative study of the use of the survey posture has been conducted, but most species seem to use it (e.g., Cooper, 2005a). Exceptions may include twig anoles, which are often found on surfaces that are too narrow for them to adopt this posture,[208] and the trunk-crown anoles, *A. carolinensis* (Jenssen et al., 1995) and *A. chlorocyanus* (Rand, 1962).

208. I base this statement on my own, unquantified impression that I have rarely seen a twig anole in the survey posture.

Time budgets—what individuals do throughout the course of a day—have only been measured for six species, *A. carolinensis* (Jenssen et al., 1995; Nunez et al., 1997), and five Central American species: *A. cupreus* (Fleming and Hooker, 1975), *A. polylepis* (Andrews, 1971), *A. nebulosus* (Lister and Aguayo, 1992), *A. humilis* and *A. limifrons* (Talbot, 1979). Comparisons among these studies are hampered slightly by different activity categorizations (Lister and Aguayo, 1992); nonetheless, the studies indicate many similarities among the species.

- In the breeding season, males spend more time in social interactions than females, much more in some species (twenty-fold difference in *A. humilis*; 49% of time versus zero percent in *A. nebulosus*). Most or all display behavior in both sexes is directed to conspecific males.
- Females spend most of their time foraging year round: most estimates are in the range of 80–100% of their time, much of which is spent in the survey posture scanning for prey.
- In comparison to females, males forage for comparable amounts of time in the non-breeding season, but for much less time in the breeding season, with concomitant lower food intake rate.[209]

The hypothesis that mainland and West Indian populations are regulated by different factors—predation and resource-limitation, respectively (Chapter 8)—would suggest that the daily activity patterns of species in the two regions might be very different. For example, mainland species might be expected to be less active, to behave in ways that are less conspicuous and to restrict their activities to areas in which they are less vulnerable to predation. However, in the absence of time budget data for any West Indian species, and with data only for five mainland species, this hypothesis cannot be evaluated.

SOCIAL STRUCTURE

TERRITORIAL BEHAVIOR

As a rule, male anoles are highly territorial.[210] They actively defend territories against other males and display frequently to proclaim their ownership. These "assertion" displays are often given when no other lizard is nearby and appear to serve as advertisements of the territory's ownership, just in case another lizard is watching (Carpenter, 1962; Jenssen, 1977, 1978). Appendix 9.1 briefly reviews anole display behavior. The rate

209. One interesting non-general finding is that *A. nebulosus* spends about a third of its time inactive and inconspicuous, often concealed under leaves or on the underside of branches ("resting/hiding" [Lister and Aguayo, 1992]). Moreover, in the non-breeding season, both sexes are apparently only active 1–2 days per week. Time budget studies for other species do not report comparable behavior.

210. Throughout this chapter, I will make broad statements about the behavior and ecology of anoles. These statements are meant to be taken as generalizations, with the realization that variation certainly exists among species—which I will mention in most cases—and that data are available for relatively few species.

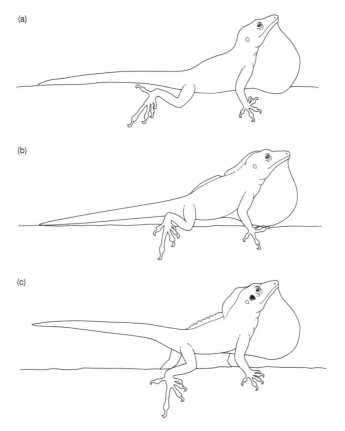

(a)

(b)

(c)

FIGURE 9.2
Aggressive behavior of
A. marcanoi showing
three stages of increasing
intensity. Figure modified
with permission from
Losos (1985).

at which male anoles display can be extraordinarily high in the breeding season. One *A. lineatopus* observed for 11 hours displayed 181 times (Rand, 1967b);[211] males in a population of *A. nebulosus* on a predator-free offshore island spent 95% of their time displaying and displayed 72 times per hour (Lister and Aguayo, 1992)! Other anoles display 2–100 times per hour and spend 1–10% of their time during the day displaying (Losos, 1990c; Lister and Aguayo, 1992; Jenssen et al., 1995; Bloch and Irschick, 2006).

Territory defense is accomplished by a series of displays and behaviors of increasing intensity.[212] My own work on *A. marcanoi* (Losos, 1985b)[213] can serve as a representative example. In this species, initial displays in an agonistic encounter often involve lifting the head and performing a dewlap and head-bobbing display, with the forequarters sometimes raised a little off the ground (Fig. 9.2). As an interaction escalates, lizards

211. In that period, the lizard also interacted agonistically with two males, copulated twice, and unsuccessfully courted several other times.

212. Darwin (1871) captured this progression of behaviors well, quoting Austen (1867, p.9) to describe the aggressive behavior of *A. cristatellus* (which Darwin incorrectly stated to be from South America): "During the spring and early part of the summer, two adult males rarely meet without a contest. On first seeing one another, they nod their heads up and down three or four times, and at the same time expanding the frill or pouch beneath the throat; their eyes glisten with rage, and after waving their tails from side to side for a few seconds, as if to gather energy, they dart at each other furiously, rolling over and over, and holding firmly with their teeth. The conflict generally ends in one of the combatants losing his tail, which is often devoured by the victor."

213. My first paper!

raise themselves off the substrate either by fully extending their forelimbs or by extending all four limbs to some extent, lifting the whole body, tail, and dewlap off the substrate. Finally, in the highest level of display, the entire body is elevated high off the ground and the head, tail, and dewlap are well above the ground.

A variety of other actions accompany this progression of display behavior. Nuchal and dorsal crests on the neck and back are erected at high levels of intensity. In addition, a dark spot on the side of the head posterior to the eye appears, usually in dominant individuals. Apparent body size is increased by engorging the head and lowering the hyoid apparatus. Lizards position themselves broadside and lean toward their opponent to maximize exposed surface area. High intensity displays also include hindlimb pushups, in which the lizard rapidly rocks the posterior of its body up and down or even jumps backwards. Competing lizards often align themselves in a "face-off" position, lined up side by side, often with heads pointing in opposite directions. Occasionally, one lizard will rapidly turn its head so that its snout points directly at the other lizard; sometimes this is followed by the lizards circling around, lunging at each other and maintaining their relative position; in other cases, pointing is followed by moving toward the other lizard and eventually, if one of the lizards does not retreat, to an attack, which takes the form of the lizards biting each other in the hindquarters or locking jaws (Fig. 9.3).

Most anoles studied show similar patterns, though differing in specific behaviors (e.g., Greenberg and Noble, 1944; Rand, 1967b; Gorman, 1968; Hover and Jenssen, 1976; Jenssen, 1979b; Jenssen et al., 2005; Ortiz and Jenssen, 1982; Scott, 1984). Among the most divergent anoles are the small Hispaniolan *A. bahorucoensis* (Orrell and Jenssen, 1998) and *A. Chamaelinorops barbouri* (Jenssen and Feely, 1991).

Jenssen's reviews (1977, 1978) are still the best overview of anole aggressive behavior. In addition to the highly stereotyped patterns of head-bobbing and dewlap extension, Jenssen identified a set of postures and body movements that are not stereotyped and are

FIGURE 9.3
Male *A. bimaculatus* from St. Kitts locking jaws in an aggressive encounter.

FIGURE 9.4

Male anoles in full battle mode. Note the raised nuchal and dorsal crests, the enlarged throat and postorbital black spot in *A. grahami* from Jamaica (a) and the extruded tongue in *A. pulchellus* from Puerto Rico (b).

only sometimes present in a lizard's display. Such "static modifiers" include the erection of crests, change in the shape of the head and body, color change, mouth opening, tongue bunching and other changes to the appearance of the lizard (Fig. 9.4).[214] "Dynamic modifiers" refer to actions such as snout pointing, head rolling, tail lifting, and non-stereotyped head movements added to the stereotypical head-bobbing display. A phylogenetic comparative analysis is needed to assess whether interspecific variation in display behavior is related to ecological or phylogenetic factors.

A great amount of work has been conducted on the neuroendocrinological basis of aggressive behavior in *A. carolinensis* (reviewed in Greenberg, 2003). For example, this work has identified how dominant and subordinate individuals differ in endocrinological and neurotransmitter activity and which parts of the brain are involved in stress and subordinate behavior (Greenberg and Crews, 1990; Tokarz, 1995; Baxter, 2003; Summers et al., 2005). Much of this work has taken advantage of the "split-brain" of anoles; i.e., because lizards lack a corpus callosum, each side of the brain regulates one side of the body (Jones et al., 1983)—even the function of the two hemipenes is independently controlled by the two sides of the brain (Ruiz and Wade, 2002; Holmes and Wade, 2004). Thus, investigators can make a lesion on one side of the brain and not on the other, using a lizard as its own control to compare the functioning and response to stimuli of bilaterally symmetrical parts of the body. Alternatively, by providing a stimulus to one side of the body (e.g., one eye, while covering the other eye), investigators can study the two sides of the brain to determine how the brain responds

214. See also Schwenk and Mayer (1991) on the derivation of different tongue displays and Myers (1971) and Milton and Jenssen (1979) on vocalizations.

(e.g., Baxter, 2003).[215] Comparative work of this sort on other anole species is just beginning (L. Baxter, pers. comm.).

TERRITORY SIZE AND OVERLAP

Most anoles are territorial in the sense that they defend part or all of their home range against other lizards. Males are generally territorial only to other males, but females are often territorial to all similar-sized individuals including males and occasionally heterospecific anoles; even hatchlings are territorial with regard to other hatchlings (Rand, 1967b; Stamps, 1977b).

Territories serve a number of purposes, providing territory holders with primary access to food, shelter, mates, and other resources. Multiple lines of evidence indicate that access to females is a major determinant of territory configuration in male anoles (reviewed in Stamps, 1977b):

- Males expand territory size as much as fifteen-fold in the breeding season (Stamps and Crews, 1976; Lister and Aguayo, 1992). In the non-breeding season, males have much smaller and often non-defended home ranges and perform fewer courtship and aggressive displays (Andrews, 1971; Fleming and Hooker, 1975; Stamps and Crews, 1976). In contrast, females maintain their territories year-round with constant levels of aggressive interactions and displays.

- Territories of males are much larger than the territories of females. Although males are often larger in body size, the difference in territory size is too great to be accounted for by differing energetic demands. Rather, this discrepancy has generally been interpreted to be a result of male territory size being determined by availability of potential mates, whereas female territories are determined by availability of food (Andrews, 1971; Fleming and Hooker, 1975; Schoener and Schoener, 1982a; Nunez et al., 1997; Jenssen et al., 1998).

- Male territories encompass the territories of 2–6 females (e.g., Andrews, 1971; Jenssen et al., 1998, 2005). The relationship between male territory configuration and the dispersion of females is particularly well documented for *A. aeneus* (Stamps, 1977b) in which the shape of male territories often is irregular, conforming to the shape of female territories; on several occasions, males changed their territorial boundaries to include a new female within the territory. Although female territories on average only covered 40% of a male's territory, males spent 89% of their time in that portion. Male *A. carolinensis* also tailor their territories to include as many females as possible (Jenssen et al., 2001).

- Males rarely are territorial with respect to heterospecifics, but females sometimes are territorial toward similar-sized individuals of other species (Rand 1967b,c).

---

215. Oddly, Deckel (1995, 1998) has reported that *A. carolinensis* seems to preferentially use its left eye in aggressive encounters. Although the statistics in these studies appear to suffer from pseudoreplication, a similar phenomenon has been reported in another lizard (Hews and Worthington, 2001).

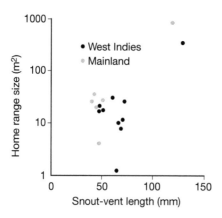

FIGURE 9.5

Relationship between body size and male home range size for mainland and West Indian anoles. An analyis of covariance controlling for body size on log-transformed variables finds a significant difference between mainland and island species (test for heterogeneity of slopes not significant; test for difference in intercepts, $F_{1,13} = 4.94$, p = 0.045). Caution should be taken in interpreting these results given the variability in both data quality and method of territory size estimation (discussed in Schoener and Schoener [1982a] and Perry and Garland [2002]). The analysis used the data presented in Perry and Garland (2002), taking averages for multiple estimates of male territory size, correcting the value for *A. lineatopus* to the mean of the range reported by Rand (1967b), adding the value for *A. nebulosus* reported by Lister and Aguayo (1992), using the combined sexes value for *A. auratus* because data for males were not presented, but excluding an estimate for *A. carolinensis* for combined sexes because estimates of male territory size were available, and considering *A. carolinensis* to be a West Indian species because of its phylogenetic heritage (the p value changes to 0.041 if *A. carolinensis* is removed from the analysis and to 0.037 if *A. carolinensis* is considered a mainland species). As with other mainland versus West Indies comparisons, this analysis does not account for phylogenetic relationships. It does, however, include a non-Norops mainland species, so all of the mainland species in this analysis do not form a single clade.

Male territory size varies both inter- and intraspecifically.[216] Territories of males range in size from 3 m² in *A. sagrei* to 806 m² in Panamanian *A. frenatus*;[217] across the 16 species for which data are available, territory size increases with body size and mainland anoles seem to have larger territories than comparable-sized island species (Fig. 9.5; Schoener and Schoener, 1982a).

216. Technically, most estimates of territory size refer to the home range of a lizard, i.e., the area it uses. Because most anoles behaviorally defend part or all of their home ranges, home range area may be a reasonable estimate of territory size; however, few studies have calculated the area of space actively defended by a lizard. For the purposes of this chapter, I refer to these estimates as territory size, with the recognition that some species may defend this space to a greater extent than others. Some definitions require that a territory be exclusive, with no overlap (reviewed in Stamps, 1994); in this sense, many anole species may not have territories, as discussed later in this chapter.

217. These figures probably underestimate the range of variation among species because species vary in the extent to which they utilize the third dimension, height. More terrestrial species, such as trunk-ground and grass-bush anoles, usually don't venture high into the habitat, but more arboreal species can occupy large swaths of vertical space (e.g., Hicks and Trivers, 1983; Jenssen et al., 1995). Consequently, calculation of home range volumes would doubtless show much greater interspecific variation. Only one species, *A. stratulus*, is known to stack territories vertically, with three-dimensional territories one on top of the next from near the ground to the canopy (Reagan, 1992).

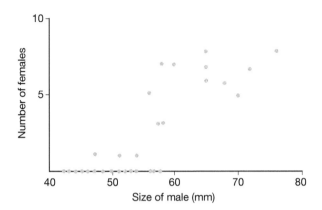

FIGURE 9.6

Relationship between the size of a male *A. aeneus* and the number of females within his territory. Modified with permission from Stamps (1977b).

Within a population, male body size correlates with both territory size (Jenssen, 1970b; Trivers, 1976; Jenssen and Nunez, 1998; but see Schoener and Schoener [1982a]) and the number of females residing within his territory (Fig. 9.6; Fleming and Hooker, 1975; Trivers, 1976; Stamps, 1977b; Ruby, 1984; Jenssen and Nunez, 1998; *A. nebulosus* is a rare exception and appears to be monogamous [Jenssen, 1970b]). In *A. sagrei*, mean male territory size across study sites is negatively related to density; this pattern may result because the higher the density, the greater the number of males competing for space (Schoener and Schoener, 1982a; see also Ruibal and Philibosian [1974b]) or because at lower densities, males must expand their territories to encompass the territories of the same number of females (Stamps, 1999).

The mechanistic underpinning for these intraspecific relationships is simple: to a large extent, body size determines dominance in anoles. For example, in nature, larger males won more than 85% of encounters in *A. carolinensis* and *A. lineatopus* (Rand, 1967b; Jenssen et al., 2005; also Greenberg and Noble [1944] and Tokarz [1985]). The only time a smaller *A. lineatopus* won an encounter was when it was the territory holder and the intruder was only slightly larger (Rand, 1967b).

The amount of overlap in male territories varies greatly among species (Johnson, 2007). In some species, territories are exclusive (Fleming and Hooker, 1975; Jenssen and Nunez, 1998), whereas in others overlap can be extensive. In the Bahamas, for example, *A. sagrei* and *A.distichus* generally (though not always) exhibit little overlap, *A. smaragdinus* exhibits more overlap, and the twig anole *A. angusticeps*, where abundant, has broadly overlapping territories (Schoener and Schoener, 1980a). The Jamaican twig anole, *A. valencienni*, also exhibits considerable territory overlap with as many as 11 males overlapping at one place (Hicks and Trivers, 1983).[218]

218. Overlap to this extent calls into question whether the home ranges of *A. valencienni* should be considered territories.

Although adult males are territorial toward other adult males, they may permit small males to live within their territory (Oliver, 1948; Rand, 1967b; Jenssen, 1970b; Stamps and Crews, 1976; Fleishman, 1988b), perhaps because they mistake them for females (Orrell and Jenssen, 2003).[219] Male *A. garmani*, which can attain an SVL of 131 mm, do not allow other males longer than 104 mm in their territory, but tolerate smaller males[220] and sometimes even mate with them (Trivers, 1976; see also Rand [1967b]).[221] When the smaller males outgrow this size, they have to disperse and find an empty tree (Trivers, 1976).[222] In *A. carolinensis*, each male territory on average contains 1.2 "covert" males (Passek, 2002). In many cases, these smaller males try to remain inconspicuous and display infrequently (Fleishman, 1988b; see also Orrell and Jenssen [2003]), although in some cases they do succeed in the occasional mating (see below).

Males are known to not be territorial in only three species: *A. agassizi* (Rand et al., 1975), *A. tropidolepis* (Fitch, 1972), and *A. taylori* (Fitch and Henderson, 1976).[223] The lack of territoriality in *A. agassizi* appears to be related to its unusual habitat on tiny Malpelo Island off the coast of Colombia. Essentially a large rock, this island hosts a large seabird colony and contains very little vegetation; the anoles feed on insects attracted to the colony. Rand et al. (1975) attributed the lack of territoriality and the very low levels of aggressive behavior to the scattered and (where they occur) superabundant resources, and to the fact that much of the island is very exposed and thus potentially uninhabitable for parts of the day, which requires lizards to move frequently over relatively large areas. Why the other two species are not territorial is not known.

Territoriality and aggressive behavior among adult females occurs in many anole species (Jenssen, 1970b, 1973; Andrews, 1971; Ruibal and Philibosian, 1974a; Fleming and Hooker, 1975; Nunez et al., 1997). In contrast, *A. valencienni* females are nonterritorial and nonaggressive toward each other; as many as 40 females may share a common feeding area (Hicks and Trivers, 1983). For those species in which it has been quantified, overlap in female territories is usually greater than in territories of males (Trivers, 1976; Jenssen and Nunez, 1998; but see Johnson [2007]).

The most extensive studies of spacing behavior in anoles have been conducted on hatchling and juvenile *A. aeneus* (reviewed in Stamps 1994, 2001). The patterns of territory overlap and behavior shown by these lizards differ little from those seen in adults.

219. In theory, small males resident in another male's territory could represent a genetically based alternative male morph (e.g., Gross, 1985; Ryan et al., 1990; Shuster and Wade, 1991). However, no evidence exists to support such a possibility; more likely, small males are simply young males which haven't grown large enough to contest for their own territories.

220. Males become reproductively mature at 85 mm SVL; females at this site grew to approximately 100 mm SVL (Trivers, 1976).

221. "An occasional buggery might be a small price to pay for the advantages of remaining within the large male's territory" (Trivers, 1976, p. 266).

222. In contrast, Fleishman (1988b) documented a small *A. auratus* eventually wresting away part of the territory of the larger male in whose territory he had lived.

223. Territorial behavior is also absent in laboratory studies of *Anolis* Chamaelinorops *barbouri* (Jenssen and Feely, 1991).

Unfortunately, almost nothing comparable is known about the social behavior and inter-actions of young lizards in other species.

MATING BEHAVIOR

Courting males approach females, headbobbing and dewlapping. Females respond by displaying, usually with headbobs. If the female is receptive, she will remain in place as the male approaches, bites her on the back of the neck, swings his tail underneath hers, and intromits one of his hemipenes (Fig. 9.7). In *A. carolinensis*, *A. sagrei* and *A. valenci-enni*, but not *A. aeneus*, the female arches her neck as the male approaches, inviting the neck-bite.[224] If the female is not receptive, she will display, but will not arch her neck and eventually will flee (Greenberg and Noble, 1944; Stamps and Barlow, 1973; Hicks and Trivers, 1983; Nunez et al., 1997; Tokarz, 1998).

Males tend to mate regularly with females within their territories. Seemingly, the more familiar a pair is, the shorter the preliminaries prior to copulation (Rand, 1967b). Stamps (1977b) quantified this relationship, finding that the amount of time a male *A. aeneus* spends courting a female is inversely related to how long they have had over-lapping territories: when females first meet a new territory owner, they usually flee, but after several weeks, females allow a male to approach and mate with little or no preced-ing courtship. This familiarity is taken to the extreme by *A. limifrons* pairs, which stay in close association, forming a "pair bond" that lasts 4–5 months (Talbot, 1979). To stay nearby, males follow females around; for example, when she climbs high up, he follows.[225]

FIGURE 9.7
Cuban *A. allisoni* mating.
Photo courtesy of
Richard Glor.

224. Scott (1984) did not observe neck arching in female *A. sagrei*.
225. Pair bonding has also been suggested for two Puerto Rican species, *A. occultus* and *A. cuvieri*, based on observations of male-female pairs seen sleeping in close proximity (Gorman, 1980; Rios-López and Puente-Colón, 2007).

In some species, females are only receptive for a limited period of time. In *A. aeneus*, female mating propensity changes with the egg laying cycle: females mate every 10–11 days, when their follicles are at an intermediate stage (Stamps, 1975). Both *A. aeneus* and *A. carolinensis*, which shows the same pattern, exhibit coition-induced inhibition of further mating—after mating, they are unreceptive until the next ovulatory cycle, unless the mating is brief (Crews, 1973; Stamps, 1975).[226] At the other extreme, female *A. valencienni* sometimes mate repeatedly on a single day (Hicks and Trivers, 1983), as do female *A. sagrei* (Tokarz and Kirkpatrick, 1991; Tokarz, 1998).

Many, but not all, anoles mate for extended periods. The average coupling of *A. carolinensis* takes 22 minutes (Jenssen et al., 1995; Jenssen and Nunez, 1998; Nunez et al., 1997),[227] whereas *A. garmani* copulations average 25 minutes (Trivers, 1976) and those of *A. nebulosus* 37 minutes (Jenssen, 1970b). By contrast, *A. valencienni* copulates for only two minutes (Hicks and Trivers, 1983) and *A. websteri* for less than one second (Jenssen, 1996)! In several species, copulation length differs consistently among males (Crews, 1973; Tokarz, 1988) and is greater later in the breeding season (Stamps, 1975; Tokarz, 1999).

The locations of anole mating have not been much studied. *Anolis valencienni* appears to mate randomly with respect to location in the habitat (Hicks and Trivers, 1983) and this may be true of most species (M.A. Johnson, pers. comm.). Female *A. garmani*, in contrast, position themselves in very conspicuous places—perhaps to make mating males highly visible to other nearby males or to solicit males so that the female may choose among them (Trivers, 1976; Hicks and Trivers, 1983). Conversely, when male *A. carolinensis* initiate copulation in the sunlight or in an open area, females often drag intromittent males into more sheltered spots, presumably to reduce their visibility to potential predators (Nunez et al., 1997).

SEXUAL SELECTION

INTRASEXUAL SELECTION

Intrasexual selection occurs when members of one sex compete with each other for the opportunity to mate with members of the other sex. To the extent that females mate with the male in whose territory they reside, then holding a territory should be a key to reproductive success for males. Moreover, given that territory size and number of females within a territory are related to male body size, intrasexual selection should favor large body size in males. In support of this hypothesis, the number of copulations, potentially a surrogate of fitness, increases with body size in *A. garmani*, *A. valencienni* (Trivers, 1976, 1985), and *A. carolinensis* (Trivers, 1976; Ruby, 1984); Ruby (1984) estimates that 15% of the males father most of the offspring.

226. Trivers (1976) suggested that *A. garmani* mates only once a month.
227. And increases with male body size.

Intersexual selection, or mate choice, occurs when members of one sex choose which members of the other sex with whom they will mate. The traditional view regarding anoles, and lizards in general, is that female mate choice occurs rarely (Stamps, 1983a; Tokarz, 1995). The basis for this view is that female territories usually lie within those of a single male, so females don't encounter other males, much less get an opportunity to choose among them. Moreover, females rarely switch territories, so even over time, they seem to have little opportunity to choose (Rand, 1967b; Stamps, 1983a; Jenssen et al., 2001).

On the other hand, the incredible amount of time—and presumably energy (cf. Bennett et al., 1981)—expended in displaying by males would suggest to some that displays function as an honest signal of a male's capabilities that might be the basis of mate choice for females (Sullivan and Kwiatkowski, 2007); indeed, the colorful dewlaps of anoles have been likened to the song and plumage of birds, traits that are often thought to be the subject of female mate choice (West-Eberhard, 1983).[228] Moreover, just as in anoles, female choice was thought to be absent in many monogamous and polygynous bird species prior to the advent of molecular paternity analyses, but we now know, in fact, that female birds are much more polyandrous and potentially choosy than previously recognized (Hughes, 1998). For these reasons, the possibility that female mate choice may occur in anoles deserves reconsideration.

Female mate choice could occur in several ways. First, females might mate with adjacent territory owners or males that wander through their territory. Second, the existence of smaller males residing within the territories of larger males (discussed above) provides an alternative means by which females may have the opportunity to mate with other males. Finally, third, at the time females settle into their adult territory, they may have the opportunity to choose among several vacancies (Stamps, 1983b), and thus among the male territory owners at those sites.

But do females actually show any evidence of being choosy and preferring some males over others? Several laboratory studies have been conducted on mate choice by female A. carolinensis (Andrews, 1985a; MacDonald and Echternacht, 1991; Lailvaux and Irschick, 2006). In all cases, females were allowed to view two males and to enter their compartments; preference was indicated when a female spent more time near one male than near the other. These studies found that only a minority of females showed a preference for one male over the other and that no traits distinguished chosen males from those that were not chosen.

Comparable data from natural populations are surprisingly scarce. In general, females not interested in mating simply run away from amorous males, who seem unable to

228. For example, recent work indicates that the yellow and red colors in the dewlaps of A. humilis and A. sagrei are produced by pterins and carotenoids, pigments that are used as condition-dependent signals in birds (Steffen and McGraw, 2007). Whether dewlap color varies among individuals in a condition-dependent manner, perhaps reflecting differences in ability to ingest carotenoids in their food, remains to be seen.

force the issue; in one case, a female *A. valencienni* cornered at the end of a branch jumped six meters to the ground to avoid an advancing male (Hicks and Trivers, 1983). However, whether and how often females choose to mate with a male other than the one in whose territory they reside is unclear.

Anolis carolinensis has been studied more extensively in the field than any other anole, and the results are somewhat contradictory. Following males intensively for eight-day periods, Jenssen and Nunez (1998) observed that males bypassed the opportunity to mate in 69% of their interactions with receptive females (receptive females defined as those that exhibited the neck arching posture discussed above). Why males would not mate given the opportunity is unclear, but possibilities are that frequent matings would prevent males from adequately patrolling and defending their territories against intruders, or that frequent matings would deplete sperm supplies and thus increase the risk that an interloping male's sperm might fertilize a female (Jenssen et al., 1995; Jenssen and Nunez, 1998).[229] During the eight-day time period, males mated on average once every 1.4 days and at least once with every female in their territory. Given that eight days is roughly the egg cycling period in female *A. carolinensis*, and that females exhibit coition-induced inhibition of receptivity, these data suggested that males should be fathering the eggs of most offspring produced in their territories.

However, the first study to use molecular means to quantify male reproductive success failed to support this conclusion. Working at the same study site several years later, Passek (2002) found that 52% of hatchling *A. carolinensis* were fathered by the territory owner, 15% by a neighboring male, and 21% by smaller males occurring within the larger male's territory.[230] These smaller males were inconspicuous and sedentary; Passek (2002) speculated that they were resident within the territory owner's territory, but was not sure. She also speculated that this "covert" approach was a temporary strategy used by growing males. Males with larger territories (presumably containing more females) suffered more paternity loss to extra-pair copulations. This study did not examine whether reproductive success was related to male body size or any other trait.

How to reconcile the inconsistency between the behavioral and genetic results of these two studies is not clear. Perhaps coition-induced inhibition is not as prevalent as laboratory studies suggest, or perhaps the population differed in its mating behavior in different years. A study that combines the detailed behavioral approach of Jenssen and Nunez (1998) with molecular parentage analyses is needed to distinguish among these possibilities.

---

229. Unexpected observations such as this demonstrate the utility of detailed studies of the natural behavior and ecology of organisms. The detailed study of *A. carolinensis* behavior by Jenssen and colleagues has provided a wealth of detail about the biology of this species, information that in many cases is otherwise unavailable, yet necessary to understand how and why species behave as they do. Comparable data on other species are desperately needed.

230. Another 12% were not fathered by the territory owner, but identity of the male could not be established.

Extra-pair copulations have been reported in several other species. *Anolis valencienni* females regularly mate with multiple males, sometimes on the same day (Hicks and Trivers, 1983). In *A. garmani*, five of 49 observed copulations were with males other than the territory owner: one with a neighboring male, two with small males resident within the male's territory, and two with males never seen before or after;[231] none of the these males was larger than the territory holder (Trivers, 1976). More generally, both *A. garmani* and *A. valencienni* show an interesting pattern in the distribution of copulations versus body size: some of the smallest males mate more than some larger males (Fig. 9.8). Trivers (1976, 1985) suggested that these small males may be subordinates that large males allow to stay in their territory.

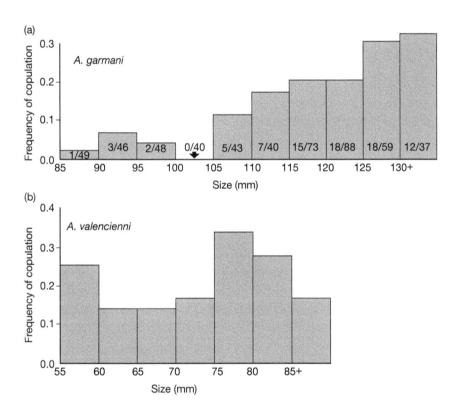

FIGURE 9.8

Relationship between body size and mating frequency (defined as the number of observed copulations by males in a size class divided by number of males in that size class) in (a) *A. garmani* and (b) *A. valencienni*. Figures modified from Trivers (1976, 1985) with permission.

231. Such "floaters"—territory-less males seemingly moving through the environment, looking for a place to settle—are known in anoles and other lizard species. Schoener and Schoener (1982b) noted a fairly high frequency of such individuals in four Bahamian species. Jenssen et al. (1998) observed three occasions in which an unknown small male entered a territory of one of their focal male *A. carolinensis*; in all cases, the lizard was quickly chased out and never seen again.

Two other studies have recently used molecular means to identify reproductive success resulting from extra-pair copulations, and both have found high levels of multiple mating among females. In *A. cristatellus*, 52% of wild-caught females brought into the laboratory produced eggs fathered by more than one male (Johnson et al., in review). Analysis of the males with whom a female mated showed similar trends to those observed for *A. carolinensis* (Passek, 2002). In particular, although most offspring were sired by the "primary" male (defined as the male who fathered the most offspring), a quarter of the offspring were sired by males living outside of a female's territory. Moreover, in a number of cases, offspring were sired by small males that may have been living within the territories of larger males. Overall, however, male size was correlated with reproductive success. *Anolis sagrei* also exhibits multiple paternity; 81% of females produced offspring fathered by more than one male (Calsbeek et al., 2007a).[232]

CRYPTIC FEMALE CHOICE

Intersexual selection could occur in at least one other way. Because female anoles can store sperm, the possibility exists that females could mate multiply over an extended period of time and subsequently choose which male's sperm to allow to fertilize an egg. Such "cryptic" mate choice has been an increasing subject of study in recent years (Eberhard, 1996; Birkhead and Pizzari, 2002).

Sperm storage has been best documented in *A. carolinensis*. Fox (1963) found sperm in a female seven months after she had last mated and Passek (2002) observed a female lay a fertile egg at least ten months after her most recent mating. Sperm are stored in special sperm storage tubules, 75–150 of which occur in the walls of each oviduct (Fox, 1963; Conner and Crews, 1980). Examination of mated individuals found as many as 633 sperm in a tubule, with an average of 319 (Conner and Crews, 1980). *Anolis sagrei* also has sperm storage tubules (Sever and Hamlett, 2002), and Calsbeek et al. (2007a) observed captive *A. sagrei* laying fertile eggs as long as 107 days after they last mated. In contrast, Stamps (1975) suggests that sperm storage does not occur in *A. aeneus* because females that fail to mate within their ten-day cycle resorb their eggs.

How sperm release occurs is not known; Conner and Crews (1980) suggest that muscular contractions accompanying ovulation might be responsible. If, in fact, sperm release is under the control of a female, then the possibility exists that females can determine the tubule that produces the sperm that fertilizes an egg (Passek, 2002). If females mate multiply and sequester sperm from different males in different tubules, and if females are able to keep track of which male's sperm resides in which tubules, then the possibility of cryptic female mate choice exists (Passek, 2002). These, of course, are big ifs for which currently there is strong evidence only for multiple mating. In a small experiment, Passek (2002) found that four females each mated by two males always had only one father for their offspring, which suggests the possibility that the female was controlling which male's sperm fertilized her eggs. More recently, Calsbeek and

232. Figures for both studies exclude females that only produced one offspring.

Bonneaud (2008) have presented evidence that female *A. sagrei* may choose the sire of their offspring depending on the sire's body size and sex of the offspring.

The existence of sperm storage could lead to sperm competition and cryptic mate choice in several ways. On one hand, females may mate multiply, with the male whose territory in which they reside as well as with others. A second possibility, however, is that a territory owner may be supplanted by a new male, in which case a female may harbor the sperm from both the old and the new male for some time. This possibility was clearly illustrated in a study of *A. sagrei* in which over a span of five weeks, most females mated with more than one male (Tokarz, 1998). In all cases but one, this was the result of one male taking over a territory from another. Females were almost always observed to mate only with the male in whose territory they resided; the exception was a case in which a neighboring male entered another's territory and mated with a female there.[233] No females were observed switching territories from one male to another, even when there was turnover in male territory holder.

Similarly short territory tenure is known for males in some populations of *A. aeneus* (Stamps, 1977b) and *A. carolinensis* (Ruby, 1984). On the other hand, males in other populations of these two species maintain territories for substantially longer periods (*A. carolinensis* [Jenssen et al., 1995]; *A. aeneus* [Stamps, pers. comm.]),[234] as do male *A. sagrei* in the Bahamas, whose territories remain approximately in the same place from one year to the next (Schoener and Schoener, 1982b).

Overall, the data on mate choice, and on sexual selection more generally, are pretty scant. There is good reason to believe that intrasexual selection operates strongly among males favoring large size. Females mate with multiple males to a much greater extent than previously recognized; whether, in fact, females are being choosy and favoring males with some characteristics over males with other characteristics—and thus producing intersexual selection—remains to be seen. This is clearly an area ripe for future work; anoles would seem to be a prime system for combining behavioral and molecular approaches to study mating systems and reproductive success.

SEXUAL DIFFERENCES IN ECOLOGY AND MORPHOLOGY

Anoles vary greatly in the extent of sexual dimorphism in both size and shape.[235] In the West Indies, males are almost always the larger sex, but in many mainland species, females are larger (Fitch, 1976, 1981). As discussed in Chapter 8, sexual dimorphism in

233. In five other cases, intruding males were unsuccessful in their courting. In three cases, the intruders were run off by the resident male, and in the other two cases the female fled from the Casanova.

234. These differences are a reminder that factors affecting social structure can vary substantially within a species, and thus it is not safe to generalize from a single study on a species, particularly because researchers often choose to work in areas in which a species is abundant and easily detectable. For example, survival, home range size, and density—factors which may have an important affect on social structure—all vary greatly among populations (Schoener and Schoener, 1980a, 1982a,b).

235. This section focuses on interspecific variation in sexual dimorphism. However, extensive variation in size dimorphism has been documented among populations of two species (Andrews and Stamps, 1994; Stamps, 1999). Further work on more species, incorporating shape as well as size dimorphism, could prove very interesting.

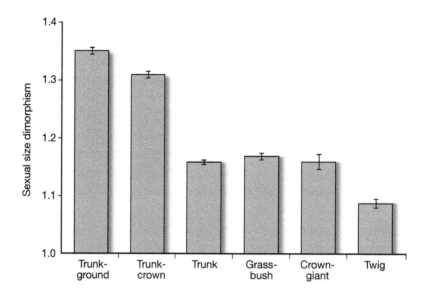

FIGURE 9.9

Variation in sexual size dimorphism (male SVL/female SVL) among the ecomorphs. Data from Butler et al. (2000).

body size is often, but not always, associated with differences in prey size. Among West Indian species, the degree of dimorphism differs among the ecomorphs: trunk-ground and trunk-crown species are highly dimorphic, whereas twig anoles and crown-giants exhibit relatively little dimorphism (Fig. 9.9; Butler et al., 2000).

Anole sexes also differ in body proportions.[236] For example, males usually have longer heads and hindlimbs than females, once the effects of differences in body size are statistically removed (Schoener, 1968; Butler and Losos, 2002). As with size dimorphism, the ecomorphs differ in the extent of shape dimorphism. The details of specific characters are complicated and somewhat inconsistent among studies, but in terms of overall shape, crown-giant anoles have the greatest and twig anoles the least dimorphism (Fig. 9.10; Butler and Losos, 2002; Losos et al., 2003a; Huyghe et al., 2007).[237]

Three factors could account for sexual dimorphism in anoles (Stamps, 1995; Butler et al., 2000; Butler and Losos, 2002; Butler, 2007; Losos et al., 2003a):

236. Such differences, however, are not great enough to obscure ecomorph differences. In a morphometric study including males and females from 15 species, most of the variation was explained by ecomorph class, with substantially less variation attributed to either differences between the sexes or to the sex-by-ecomorph interaction (Butler et al., 2007).

237. These analyses did not include head dimensions. In contrast to the patterns seen for other morphological characters, sexual dimorphism in head dimensions was greatest in the twig anole *A. angusticeps* in comparison to three other Bahamian species (Schoener, 1968). Examination of head dimorphism in other species would be interesting (e.g., Herrel et al., 2006).

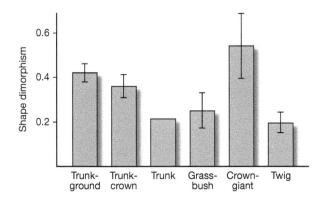

FIGURE 9.10
Variation in sexual shape dimorphism (quantified as the distance in morphological space between males and females of each species) among ecomorphs. Modified from Losos et al. (2003) with permission. In another study using different species and methods (Butler and Losos, 2002), trunk-ground and grass-bush anoles switched places in terms of their dimorphism relative to the other ecomorph classes.

SEXUAL SELECTION As discussed earlier in this chapter,[238] intrasexual selection may favor large body size in males to control territories and access to females. Why larger size in females would be favored in some species is unknown: in other lizard taxa, larger females have more offspring (Dunham et al., 1988), but anole clutch size is fixed at one. Perhaps larger anole females lay eggs more frequently or produce larger, and hence more fit (cf. Sinervo et al., 1992; Le Galliard et al., 2004), offspring? Few data are available to assess these possibilities (see Chapter 8).

Sexual selection could promote shape dimorphism in a number of ways. To the extent that some traits allow males to outcompete other males, those traits would be favored. Relatively great muscle mass and head length probably are advantageous in agonistic encounters; perhaps longer legs enhance fighting or displaying ability as well (cf. Husak et al., 2006a; see Chapter 13). Female mate choice might also promote shape dimorphism if females preferred males with particular traits.

INTERSEXUAL NICHE PARTITIONING Sexual dimorphism may be adaptive as a means of minimizing intersexual resource competition if the sexes use different microhabitats and resources. Size dimorphism could be favored if different-sized individuals eat different prey, which is often the case (Chapter 8). Shape dimorphism could be favored if the sexes evolve appropriate adaptations to use different parts of the environment (e.g., by being dimorphic in limb length or toepad size) or to eat different prey (e.g., by being dimorphic in head size [e.g., Herrel et al., 2006]). Intersexual microhabitat differences in anole species are pervasive (reviewed in Butler et al., 2007). Males, for example, often

238. And reviewed for lizards in general in Schoener (1977), Stamps (1983a) and Ord et al. (2001).

perch higher than females. For both sexes, a relationship exists among species between morphology and ecology, but these relationships are not the same, which suggests an adaptive basis to sexual shape dimorphism (Butler and Losos, 2002). In addition, the degree of sexual shape dimorphism among species is correlated with the degree of intersexual difference in microhabitat use (Butler et al., 2007), which further suggests that dimorphism is adaptive.

DIFFERENT REPRODUCTIVE ROLES Selection may differ between the sexes for reasons related to reproduction. For example, females have to carry developing eggs until they are laid, and displaying males are exposed to heightened predation risks. Moreover, different reproductive roles can cause the sexes to use different microhabitats; males may perch higher than females, for example, to display more effectively, rather than to partition resources (Andrews, 1971). These different reproductive roles can lead to selection for dimorphism. Females, for example, might require larger toepads for a given body mass to support the extra mass of an egg, and males might need longer limbs or other traits related to the greater need to be able to escape from predators (Chapter 13).

Differences in dimorphism among ecomorphs may result from differences in the extent to which these three factors are important in different microhabitats (Butler et al., 2000; Butler and Losos, 2002; Losos et al., 2003a). For example, intrasexual selection may be greatest in open microhabitats, where males can easily see and intercept intruding males (Schoener, 1977). Intersexual niche partitioning may depend on the range of resources available in different microhabitats: the greater the range, the greater the opportunity for divergence. Furthermore, differences in the number of species coexisting in a microhabitat may affect the opportunity for intersexual divergence; the greater the number of species, the less the opportunity for sexes to diverge. Lastly, differences in reproductive roles may be more consequential in some microhabitats than others. The added mass of eggs may affect arboreal females more than terrestrial ones; the risk to displaying males may be greater in some microhabitats than in others.

Unfortunately, the data are not available to test these hypotheses. Ideally, we would have information on whether and how sexual selection varies among microhabitats and the extent to which intersexual differences in resource use and reproductive roles select for different traits in males and females. However, these data are generally lacking. As a result, we cannot distinguish among these possibilities directly.[239]

In the absence of such data, I will speculate. Size dimorphism almost surely is affected by sexual selection; it is no coincidence that males are much larger than females

239. Of course, there may not be a single answer. The extent of dimorphism in different traits may be determined by different selective factors. In this light, it is notable that dimorphism in size and shape are not strongly correlated, which suggests that different processes are operating upon them (Butler and Losos, 2002; Losos et al., 2003a).

in the most territorial species (trunk-ground anoles) and that the least territorial species (twig anoles) are the least dimorphic.[240] But compelling evidence also suggests a role for niche differentiation in determining size dimorphism: among populations of widespread species, the degree of size dimorphism decreases with increasing number of sympatric congeners; moreover, when the geographic range of a species partly overlaps that of another ecologically similar species, size dimorphism is greater in the allopatric populations than in the sympatric ones (Schoener, 1977). These trends strongly implicate ecological factors as important determinants of size dimorphism.[241] As a result, it is hard to predict the relative importance of sexual selection, niche partitioning, and reproductive roles in causing variation among ecomorphs in size dimorphism. Predictions based on indirect measures of the opportunity for sexual selection and niche partitioning in the different ecomorph microhabitats are inconclusive (Butler et al., 2000); only direct measurement of these factors can answer this question.

By contrast, no evidence links sexual selection to most of the morphological "shape" characters that differ among the sexes, and differences in these characters have clear functional significance. My guess, and it is only that, is that variation in shape dimorphism among ecomorphs primarily reflects differences either in the extent of intersexual niche partitioning or in reproductive roles, rather than differences in sexual selection pressures (Butler et al., 2007). The one exception to this generalization is dimorphism in head size. In all eight West Indian species examined, males have larger heads than females (Schoener 1967, 1968; Schoener and Gorman, 1968; Herrel et al., 2006).[242] Although it is associated with intersexual differences in prey size, dimorphism in head size may also indicate that males have greater bite force (Herrel et al., 2006), which could be favored in intrasexual combat, as it is in *A. carolinensis* and other lizards (Lailvaux et al., 2004; Lappin and Husak, 2005; see Chapter 13).

As with so many other characteristics discussed in the last few chapters, patterns of size dimorphism differ between West Indian and mainland anoles.[243] In the West Indies, females are larger than males in only a few species, and in those cases, females are only slightly larger (Butler et al., 2000). In contrast, not only are females larger in 40% of mainland species, but the degree to which they are larger than males is much

240. I follow Schoener and Schoener (1980a) in my subjective ranking of most-to-least territorial ecomorph classes. This scale is generally consistent with a wide variety of data (e.g., display rates [Chapter 3]), but a recent comparative study on extent of overlap in male territories provided surprisingly inconsistent results (Johnson, 2007).

241. An alternative hypothesis, however, is that the presence of competing species reduces the density of a focal species, which in turn reduces the selective advantage of large size in males, thus leading to a negative relationship between the presence of sympatric species and sexual size dimorphism (Stamps et al., 1997).

242. A phenomenon seen in many other lizard species (e.g., Huyghe et al., 2005; Pinto et al., 2005), even among polygynous herbivores, where the larger heads of males are unlikely to be related to intersexual differences in diet (Carothers, 1984).

243. Dimorphism in shape has been studied in a number of mainland species (Vitt and Zani, 1996a, 2005; Vitt et al., 1995, 2001, 2002, 2003a,b, 2008). The only generality that emerges from these comparisons is that males often have relatively longer limbs than females. Because studies vary in how shape is measured, measurements cannot be compared across studies. Thus, a comparison of patterns of shape dimorphism between mainland and island species requires a study that includes species from both regions.

greater than that seen in the few female-larger West Indian species (Fitch, 1976, 1981). Even in the mainland, however, females are rarely substantially larger than males: in those cases in which the sexes differ in size by more than 10%, males are the larger sex in 88% of the species (Fitch, 1976).

Among mainland species, size dimorphism is related to habitat seasonality. The species in which males are substantially larger than females tend to live in seasonally dry habitats in which the breeding season is relatively short. Fitch (1976) and Fitch and Hillis (1984) speculated that intrasexual selection is particularly intense in these habitats. This is a hypothesis worth investigating. Most West Indian anoles also have breeding seasons (Chapter 8) which, by this logic, might explain the preponderance of male-larger dimorphism in the West Indies, although the existence of male-biased dimorphism even in those species that do breed year-round[244] would seem to argue against this hypothesis. Regardless, study of variation in size dimorphism in mainland anoles should prove interesting, particularly with regard to the occurrence of female-larger dimorphism, for which we currently have no hypotheses.

In addition to the morphometric characters used to measure shape dimorphism (e.g., limb and tail lengths, lamella number), sexual dimorphism occurs in two other interesting characters. The first is the dewlap. In many species, females have very small dewlaps, much smaller than those of males. In contrast, in a few species, the dewlap of the female is as large as that of the male (Fig. 9.11; Fitch and Hillis, 1984). In addition, in some species, mostly on the mainland, the sexes differ in dewlap color and patterning (Fig. 9.11; Savage, 2002; Köhler, 2003; Bartlett and Bartlett, 2003).

Unfortunately, little is known about how females use their dewlaps, and the little information that is available from three species permits few generalities. *Anolis carolinensis* females only rarely use their dewlaps in intersexual displays (Jenssen et al., 2000), whereas female *A. valencienni* use their dewlaps primarily to discourage courting males, including those of other species (Hicks and Trivers, 1983). Both *A. carolinensis* and *A. bahorucoensis* females use their dewlaps in intrasexual displays (Orrell and Jenssen, 1998, 2003); in *A. carolinensis*, females use the dewlap more at close range and less at long range in female-female interactions compared to dewlap use in male-male interactions (Jenssen et al., 2000; Orrell and Jenssen, 2003). Unfortunately, without more information on how females use their dewlaps, we will not be able to explain sexual dimorphism and dichromatism in anole dewlaps.

Anoles are also often dimorphic in their color and dorsal patterning. In many of the more brightly-colored species, adult females are drabber than adult males. Although this has not been studied, in some of these species, the male's coloration seems to make them more obvious in their environment (cf. Stuart-Fox et al., 2003; Husak et al., 2006b). I would think a saurophagous bird could detect a male *A. allisoni* from a great

---

244. E.g., *A. trinitatis* (Licht and Gorman, 1970), *A. opalinus* (Jenssen and Nunez, 1994).

FIGURE 9.11

Sexual differences in the dewlaps of two species in which the females have large dewlaps. *A. insignis* (a) male and (b) female. Photos of euthanized lizards being prepared as museum specimens courtesy of Steve Poe. (c) Male *A. transversalis*. Photo courtesy of Arthur Georges. (d) Female *A. transversalis*. Photo courtesy of Alexis Harrison.

distance (Figures 3.5a and 9.7)! This suggests that the male coloration may be favored by sexual selection (Macedonia, 2001), although an alternative possibility is that the sexes are adapted to being cryptic in different microhabitats.

The sexes commonly differ in dorsal patterning as well. In many species in which the male's back is uniformly patternless, females sport stripes, diamonds, speckles, or other designs (Fig. 9.12; e.g., Jenssen, 1970b; Schoener and Schoener, 1976; Stamps, 1977b; Calsbeek and Bonneaud, 2008). Less frequently, males are the more patterned sex (Fig. 9.13). Finally, in some species, both sexes are patterned, but with different styles (Fig. 9.14). The explanations for this dimorphism are not clear. Female patterning has been thought to increase crypsis, but this hypothesis has not been directly tested (Schoener and Schoener, 1976; Macedonia, 2001; for review, see Stamps and Gon [1983]). Several of the species with male-enhanced patterning have reduced or absent dewlaps, which may suggest that for communication, males use their body patterning, which is often colorful, in lieu of a well developed dewlap (Williams and Rand, 1977; Losos and Chu, 1998).

FIGURE 9.12
Dorsal patterning in
female (a) *A. porcatus*
from Cuba and (b) *A.
sagrei* from South Bimini,
Bahamas.

FIGURE 9.13
Male *A. bahorucoensis*.

Finally, no discussion of sexual dimorphism in anoles would be complete without mention of the aptly named *A. proboscis* and its close relative, *A. phyllorhinus*. Males of these species sport a long, laterally compressed, fleshy nasal appendage at the tip of their snout which can exceed 25% of snout-vent length (Fig. 9.15). Females of this species, which had not been collected until recently,[245] lack this protuberance. Several species of agamid lizards and chameleons exhibit somewhat similar nasal appendages that are sexually dimorphic in size and appearance, or absent from the female entirely (Manamendra-Arachchi and Liyanage, 1994; Manthey and Schuster, 1996; Nečas, 2004). In none of these species is the function of the nasal appendage known.[246]

245. Fewer than 20 specimens of these two species exist in museum collections (Rodrigues et al., 2002).
246. Some other chameleons species have nasal horns (the difference being that horns are made of bone, rather than being flesly appendages) that are sexually demorphic in size of presence. These horns are used in male-male combat and, perhaps, female mate choice (Martin, 1992; Nečas, 2004).

FIGURE 9.14

A. transversalis. (a) Male. Photo courtesy of Arthur Georges. (b) Female.

FIGURE 9.15

A. proboscis from Ecuador. Photo courtesy of Wanda Parrott.

HOW SMART ARE THESE GUYS?

I'll end this chapter by speculating on what goes on inside the head of an anole. Although many view reptiles as nothing more than robotic automatons, anyone who has looked into the eyes of an anole knows that they are so much more than that. Anoles are crafty and cunning; herpetologists have been known to refer to them as "clever" and even "witty." And it's always humbling, when trying to catch one, to be outsmarted by a creature with a brain the size of a pea (or smaller). Those are my impressions, any way, but what does science have to say on the subject?

Early studies concluded that lizards had limited learning capabilities. However, when investigators conducted experiments designed with the natural history of lizards in mind, they found that lizards can learn readily (Burghardt, 1977; Brattstrom, 1978). For example, *A. cristatellus*, when rewarded by receiving a mealworm, can learn to approach a sphere that pops out of a box near its perch and to respond differently to spheres of different colors (Shafir and Roughgarden, 1994); in the laboratory, *A. grahami* can learn to respond to a pulse of sound when the sound is always followed by the investigator prodding the lizard with a rod (Rothblum et al., 1979).

Anoles also can distinguish familiar from unfamiliar lizards and alter their behavior accordingly (Tokarz, 1992; Paterson and McMann, 2004): males are less aggressive toward neighbors than they are to novel males (the "dear enemy" effect [Qualls and Jaeger, 1991; Paterson, 2002]) and males court and mate with novel females more than they do with familiar ones (Rand, 1967b; Tokarz, 2002, 2007; Jenssen et al., 1995; Orrell and Jenssen, 2002).

As a rule, anoles exhibit extensive behavioral flexibility in response to environmental conditions. Other than the work just described, little research has focused on anole cognitive capabilities, much less the role such capabilities play in anole behavioral, population, and community ecology.

FUTURE DIRECTIONS

To evaluate all manner of hypotheses, basic time budget data and other information on other aspects of behavioral ecology are sorely needed. Such data are not difficult to collect; they require only time and patience. Particularly needed are time budget data for West Indian taxa; more time budgets for mainland species would also be useful (all five mainland species studied to date are small). These data would be particularly useful to test the hypothesis that mainland and West Indian populations are regulated in fundamentally different ways.

Head-bobbing patterns and other aspects of anole displays were extensively studied in the 1970s and 1980s, but little work has been done since then. Given the importance of display behavior for understanding not only anole behavioral ecology, but also species recognition and speciation (discussed in Chapters 2 and 14), a resumption of this work would be worthwhile. In particular, analysis of evolutionary diversification in display behavior, conducted in a phylogenetic context and investigating potential causal factors, such as shifts in habitat, could prove very interesting (Ord and Martins' [2006] study is a start in this direction, but more data are needed for species for which good ecological data are also available).

The twig anole, *A. valencienni*, is an outlier in many respects (e.g., high levels of territory overlap and female mating frequency). Detailed studies of other twig anoles, where feasible (*A. angusticeps* in the Bahamas would probably be the best candidate), would be valuable to see whether the unusual features of *A. valencienni* are characteristic of twig anoles in general. Given the many ways that twig anoles differ from the other ecomorphs, finding similar patterns in other twig species would not be surprising.

The study of sexual selection requires determination of reproductive fitness. The advent of molecular methods has revolutionized our understanding of the social systems of birds and many other organisms, in many cases revealing surprising findings. Such work should be relatively straightforward with anoles, with the complication that anoles only lay one egg at a time. Such data will allow determination of the extent and manner in which sexual selection operates; in turn, these data should provide great insight on the genesis of variation in sexual dimorphism among species.

ANOLE DISPLAY BEHAVIOR: CONTEXT, VARIATION, AND STEREOTYPY

Many anole species have multiple, distinct stereotyped patterns of headbobbing and dewlap extension (Fig. 9.16). Traditionally, these different displays have been labeled by the context in which they were thought to be given (e.g., assertion, courtship). However, in recent years, it has become clear that this practice confuses the form and the function of displays. For example, although only one display type is performed in the assertion context in some species (e.g., Stamps and Barlow, 1973), in other species the same display type is used in multiple contexts, and in still others, multiple display types are given in the same context (DeCourcy and Jenssen, 1994; Lovern et al., 1999; Bloch and Irschick, 2006).

Anole display patterns were documented in detail in the 1970s and 1980s (reviewed in Jenssen, 1977, 1978). After a lull, researchers are again examining anole display behavior, focusing primarily on how displays vary among age classes and between sexes and in different situations (e.g., DeCourcy and Jenssen, 1994; McMann, 2000; Lovern and Jenssen, 2003; Orrell and Jenssen, 2003). Relatively little comparative work is being conducted (e.g., Orrell and Jenssen, 1998), which is unfortunate; Ord and Martins (2006) have conducted a nice comparative analysis of the data collected to date, but data for more species, conducted within a uniform methodological framework, would be useful.

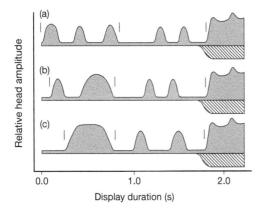

FIGURE 9.16

Three stereotyped head-bobbing patterns of *A. carolinensis*. The black area represents the amplitude of the head as it moves up and down. Although amplitude can vary within and among individuals, the cadence remains constant for each of the display types. The hatched area refers to times when the dewlap is displayed. Length and number of dewlap displays and associated headbobs at the end of the display can be quite variable. Modified with permission from Jenssen et al. (2000).

10

HABITAT USE

A key factor in understanding anole biological diversity is habitat use. Within localities, coexisting species invariably differ in some aspect of habitat use. Across the landscape, species replace each other as the environment changes. Through time, habitat use evolves within clades in predictable ways. These will be important themes throughout the remainder of the book. In this chapter, I will discuss the various aspects of the environment that are important to anoles, as well as the extent to which habitat use shifts through time.

I have already discussed how sympatric anole species partition the environment by using different structural microhabitats: trunks, twigs, ground, grass and so on. However, there are other aspects of the environment that vary within and among localities and to which anoles specialize. This specialization allows anole species to adapt to extreme habitats, and permits sympatric species to coexist while occupying the same structural microhabitat.

Two important environmental factors are temperature and moisture: anoles can be found across the broad range of habitats that occur throughout their range, from deserts to cool mountaintops and rainforest interiors. The latter half of the twentieth century saw a flowering of the field of physiological ecology and work on reptiles, and particularly anoles, played an important role. I will begin this chapter by discussing the extensive knowledge of anole thermal biology, and then will move on to other aspects of habitat use.

FIGURE 10.1

A male *A. valencienni* having its body temperature taken with a cloacal thermometer.
In the heyday of anole thermal ecological studies, researchers measured the temperatures
of hundreds of lizards throughout the course
of the day. According to a—perhaps apoc-
ryphal—story (R.B. Huey, pers. comm.), Stan
Rand was once engaged in such research
and attracted a crowd of curious onlookers.
Finally, one gentleman stepped forward and
asked "Excuse me, sir; are the lizards sick?"
Photo courtesy of Luke Mahler.

TEMPERATURE

THERMOREGULATION

ANOLIS *AND THE HISTORY OF THERMAL BIOLOGY*

The 1960s and 1970s might well be termed the "noose 'em and goose 'em" decades in herpetology. The invention of the rapid-reading cloacal thermometer, combined with a growing appreciation of the importance of thermal biology to ectotherms, led every able-bodied herpetologist to head out to the field, Schultheis® thermometer in hand, to measure the body temperature of unsuspecting reptiles (Fig. 10.1). The result was a golden era in the study of reptile thermal biology and a wealth of data on how reptiles regulate their body temperature. Perhaps no group of reptiles was studied as intensively or was as important for the development of the field as were anoles (reviewed in Huey, 1982).

Following Cowles and Bogert's (1944) pioneering work, it was widely believed that all lizards bask in the sun to regulate their body temperature precisely. Ruibal's (1961) study on several Cuban anoles was the first to contradict that idea by showing that some species do not bask or otherwise attempt to regulate their body temperature. Initially Ruibal's study was treated as an exception, but subsequent work by Rand (1964a; Rand and Humphrey, 1968) and Ruibal and Philibosian (1970) confirmed that a variety of tropical forest lizards, and not just anoles, are thermoconformers. Subsequent work on *A. cristatellus* by Huey (1974) led to the development of a conceptual framework for understanding when thermoregulatory behavior should be favored. In sum, early anole studies played an important role in the history of thermal biology, as these studies forced a reinterpretation not only of the complexities of the thermoregulatory behavior of lizards, but also of the paradigm of homeostasis as being central to an animal's ecology (Huey, 1982).

COSTS AND BENEFITS OF THERMOREGULATION

Physiologists have long believed that thermoregulation is adaptive because it allows animals to regulate their body temperature within the range in which they function best (Cowles and Bogert, 1944; Huey, 1982). A wealth of data for *Anolis* (summarized in Chapter 13), as well as for other lizards and ectotherms, generally supports this conclusion (Huey, 1982). Why, then, do some lizards not thermoregulate?

The reason is that thermoregulation has costs, such as the energy required to move into and out of the sun and the concomitant increased exposure to predators. Huey and Slatkin (1976) pointed out that, given these costs, thermoregulation is only beneficial in some circumstances. In particular, in situations in which thermoregulation is costly because the distance between different environmental patches (e.g., shaded areas versus sunny areas) is too great, lizards should not attempt to thermoregulate and instead should passively adopt the temperature determined by their surroundings.[247] As predicted by this theory, populations that live in deep forest tend not to bask and instead are thermoconformers, whereas those that occur in open or edge habitats tend to bask frequently (Fig. 10.2; reviewed in Huey and Slatkin, 1976).

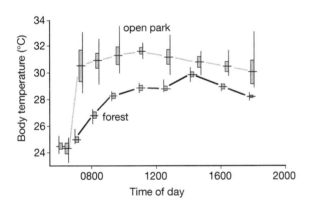

FIGURE 10.2

Anole thermoregulation. Body temperature of *A. cristatellus* in open and forest habitats. Note that in the open area, temperature rises rapidly early and then is maintained at a fairly constant rate throughout the course of the day; this pattern is commonly seen in lizards with ready access to basking sites. By contrast, body temperatures in the forest vary more through the course of the day as air temperatures rise and fall. Figure modified with permission from Huey (1983).

247. A related issue is the difference in temperature between different patches. In thermally extremely heterogeneous environments, the difference may be so great that the benefit of thermoregulating (and thus avoiding much lower body temperatures) may outweigh the costs even when distances between patches are great (cf. Blouin-Demers and Nadeau, 2005).

Although the thermal biology of a large number of anole species has been studied (see below), only recently has a conceptual framework been developed to quantitatively investigate the extent to which individuals within a population are thermoregulating (Hertz et al., 1993; see discussion in Christian and Weavers [1996]; Currin and Alexander [1999]; Hertz et al., [1999]). An important idea is that of the "operative environmental temperature" (T_e), which is the temperature to which a non-thermoregulating animal would equilibrate in a particular environment (see Appendix 10.1 regarding methods in thermal biology). By comparing the temperatures of real lizards in an environment to the distribution of T_e values that a non-thermoregulating lizard would attain in that environment, the extent of thermoregulation can be quantified.

Hertz's (1992b) study of *A. cristatellus* and *A. gundlachi* in Puerto Rico illustrates this approach. In many habitats, *A. cristatellus* spent more time in direct sunlight than lizard models randomly placed in the environment, and as a result had higher body temperatures than the mean T_e measured for the models. Moreover, *A. cristatellus* basked more often in January than in August and at higher compared to lower elevations, with the result that mean body temperature varied little betweens seasons and elevations. By contrast, *A. gundlachi* did not bask more frequently than expected at random, and its body temperature did not differ significantly from the T_e of randomly placed models in any season or at any elevation. *Anolis cristatellus* is a thermoregulator and *A. gundlachi* is a thermoconformer.

As yet, few comparable studies have been performed on anoles (see also Hertz [1992a]). One exception is a study of the leaf-litter dwelling South American species *A. nitens* (Vitt et al., 2001). These lizards avoid basking and maintain a body temperature that does not differ from air or substrate temperature at the particular sites they occupy. However, by choosing relatively warm sites, they are able to maintain body temperatures approximately 1–3° C higher than the T_e that lizards randomly placed at the study site would attain (Fig. 10.3).[248] Anoles in the Lesser Antilles also use perches with warmer T_e values than random and the extent of this non-random habitat selection varies by species and elevation (Buckley and Roughgarden, 2005b).

A related question concerns the effectiveness of thermoregulatory behavior. Laboratory studies—in which lizards are placed in a thermally heterogeneous chamber or trackway with homogeneous illumination—confirm that, given a choice, anoles (and many other types of ectotherms) regulate their temperature within a particular range (reviewed in Huey, 1982; Hertz et al., 1993). But are they able to do so in nature?

The effectiveness of microhabitat selection for regulating body temperature can be seen in the thermoregulating species *A. cooki* and *A. cristatellus*: body temperatures attained in the field by these species are closer to the preferred temperature range selected in the lab than would be expected if they selected sites randomly (Hertz et al.,

248. Similarly, at montane sites in Hispaniola, *A. shrevei* often rested under logs and planks in warm, decomposing sawdust and attained body temperatures higher than air temperature (P.E. Hertz, pers. comm.; see Hertz and Huey [1981]).

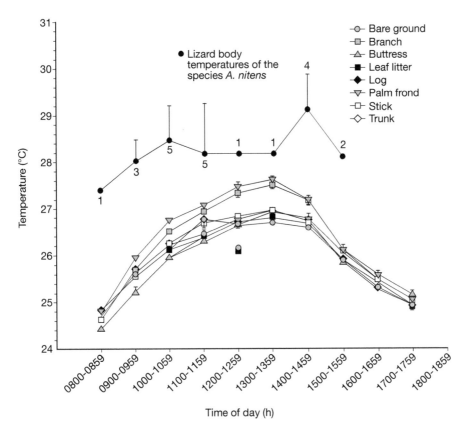

FIGURE 10.3

Body temperature of *A. nitens* through the course of the day, relative to body temperatures that would be attained by lizards randomly selecting perch sites on a variety of substrates. Figure modified with permission from Vitt et al. (2001).

1993). Of course, habitat selection is not always required to attain body temperatures within the preferred range. Lowland populations of *A. gundlachi* occur in forests where the ambient temperature is often within their preferred range, and thus body temperatures fall within this range even though the lizards use the habitat randomly with respect to T_e. However, this thermoconforming behavior causes populations of *A. gundlachi* in high elevation forests to experience body temperatures substantially below those they select in the lab (Hertz et al., 1993).

INTERSPECIFIC AND INTERPOPULATIONAL VARIATION

Given that anoles occur in many habitats, elevations, and latitudes and that they differ in extent of basking, we might expect anole species and populations to vary in the body temperatures they attain. On the other hand, most clades of lizards show relatively little variation in body temperatures (Huey, 1982; Hertz et al., 1983; but see Castilla et al. [1999] for

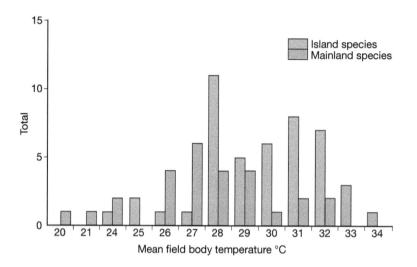

FIGURE 10.4

Mean field body temperature of anoles. Island anoles have higher temperatures than mainland species ($F_{1,71}$ = 12.75, p = 0.001). As in Chapters 8 and 9, mainland versus island analyses do not incorporate phylogenetic information, although in this case, *A. agassizi* from Malpelo Island represents an additional island colonization event. Data from Hertz et al. (in prep.).

one exception); *Sceloporus* is a particularly good comparison to *Anolis*, being a species-rich clade that occurs in many habitats and elevations in North and Central America, yet demonstrates little variation in field body temperatures (Bogert, 1949; Andrews, 1998).

In contrast to *Sceloporus*, anole species exhibit considerable variation in field body temperature, with specieś mean values ranging from 20.5–34.2°C (Fig. 10.4). To a large extent, this variation reflects differences in both macro- and microhabitats: species living at high elevations tend to have lower body temperatures than species in the lowlands, and species which live in deep shade have lower temperatures than species out in the open (Clark and Kroll, 1974). The predominance of higher temperatures in island species reflects the fact that most island species occur in open, lowland habitats (Fig. 10.4).

Field body temperature does not vary by ecomorph type (Fig. 10.5), which makes sense given that all ecomorph classes are represented in just about all habitats (e.g., open versus deep forest) and elevations. By the same token, because closely related species often occur in different habitats and elevations, field body temperature is an evolutionarily labile trait with no detectable phylogenetic signal (Hertz et al., in prep).

These phenomena are clearly exhibited by the trunk-ground anoles of Cuba. At Soroa in western Cuba, four species of the *sagrei* Series co-occur (Fig. 2.8). *Anolis sagrei* is found out in the open in the sun, with an average field temperature of 30.6°C; at the other extreme, *A. allogus* in the deep shade of the forest interior—some times a mere

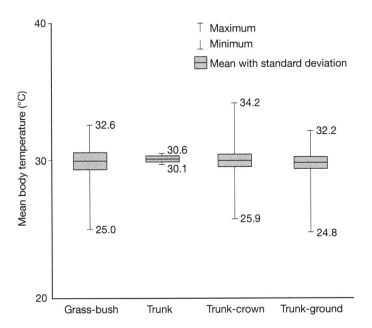

FIGURE 10.5

Body temperatures of ecomorphs. Ecomorphs vary little in mean body temperatures of constituent species; the range of variation is also comparable for most ecomorphs. Data from Hertz et al. (in prep.).

stone's throw away—maintains a temperature of 27.0°C (data from Losos et al. [2003b]; see also Ruibal [1961]).[249]

Intraspecific studies also demonstrate the lability of anole thermal biology. All anole species show a decrease in mean body temperature among populations with increasing altitude (reviewed in Huey and Webster, 1976; Hertz, 1981, 1992a). Most species bask more at higher elevations, thus behaviorally minimizing the decrease in body temperature that would otherwise result from decreasing air temperatures; however, the thermo-conforming deep forest *A. gundlachi* bucks this trend and does not increase its basking rate at higher elevations (Huey and Webster, 1976; Hertz, 1981; Hertz and Huey, 1981; Sifers et al., 2001). Anoles also generally alter their activity times elevationally, being inactive at midday in xeric, lowland areas and restricting activity to midday at high elevations (Hertz and Huey, 1981).

On an evolutionary time scale, interspecific comparisons show a strong match between the temperature a species selects in the lab and the temperature that the species attains in the field: for the nine West Indian species for which data are available,

249. Unfortunately, after a proliferation of thermal studies in the 1970s and early 1980s, relatively little research has been conducted on anole thermal ecology, particularly in the West Indies.

preferred temperatures, which range from 25.1–34.0°C, correlated strongly with field body temperatures (Hertz et al., in prep.). This correlation may suggest that species are generally quite good at attaining the body temperatures which they prefer; alternatively, however, the data might suggest that preferred temperatures evolve to adapt to the thermal environment in which a species occurs, a topic which will be revisited in Chapter 13.

MOISTURE

As with the thermal environment, anoles occupy a wide variety of hydric environments from xeric deserts to mesic rainforests. Elevationally, water stress should be greatest at lower elevations where conditions of high temperature and low rainfall often prevail (Hertz, 1980b). The organismal consequences of living in habitats differing in aridity are straightforward; particularly for small organisms, the risk of dehydration increases with decreasing moisture content of the air. Thus, one would expect that in the lab, species that live in xeric habitats should have lower rates of water loss than species from more mesic areas. For the most part, this prediction is confirmed, as will be discussed in Chapter 13.

No precise analogue to T_e exists to measure variation in hydric environment among sites within a habitat. Whether the hydric environment is more homogeneous than the thermal environment is unclear; nonetheless, variation in moisture probably exists in most habitats, and anoles may alter their microhabitat use to hydroregulate (Hertz, 1992b). For example, the small Central American species *A. limifrons*, which loses water at high rates (Sexton and Heatwole, 1968), basked less and maintained a lower body temperature in the dry season, presumably staying in cooler, moister sites to limit water loss (Ballinger et al., 1970).[250] Similarly, *A. gundlachi* does not use open habitats at high elevations, even though the thermal environment is suitable; Hertz (1992b) attributes this species' confinement to closed habitats at high elevations to the risk of dehydration, to which it is vulnerable (Hertz et al., 1979).

As with thermal biology, the hydric ecology of anoles shows no phylogenetic or ecomorphic signal: closely related species can occur in very different environments. For example, the trunk-ground anole clades on Cuba, Puerto Rico, and Hispaniola all include representatives living in desert environments and others occurring at high elevations, and the clades on Cuba and Puerto Rico also contain deep forest shade species.

LIGHT

Recently, Leal and Fleishman (2002) have suggested that microhabitats in close proximity may differ in their light intensity and spectral qualities, providing the opportunity for species to partition these microhabitats. In particular, they showed that the two sympatric trunk-ground species in southwestern Puerto Rico, *A. cristatellus* and *A. cooki*, use perches

250. An alternative possibility is that insect abundance is reduced during the dry season and that lizards consequently reduced their body temperature to minimize metabolic energy expenditures (Huey, 1982; Christian and Bedford, 1995).

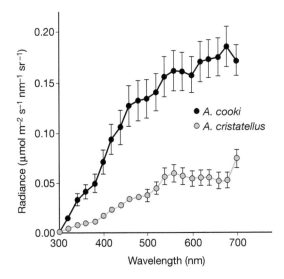

FIGURE 10.6

Differences in light environment for *A. cooki* and *A. cristatellus*. In locations occupied by *A. cristatellus*, the light spectrum peaks at 550 nm, which is green, whereas *A. cooki* sites have a more broad-based light spectrum with much higher levels in the low spectra, where ultraviolet occurs. Modified with permission from Leal and Fleishman (2002).

differing in light environment. *Anolis cooki* uses perches that are more open, with less vegetation, whereas *A. cristatellus* is found in more vegetated areas. In the vegetated areas used by *A. cristatellus*, plants tend to absorb short and long wavelengths, producing a light environment which peaks in the green region of the spectrum. By contrast, the areas used by *A. cooki* are more open to blue sky and thus not only have greater light intensity, but also a broader light spectrum, including ample light in the UV region (Fig. 10.6).

Of course, light and thermal environments will be correlated in many cases, so disentangling their effects on anole habitat use will be difficult. For example, in sympatry, *A. cristatellus* and *A. cooki* exhibit different body temperatures (Huey and Webster, 1976; Hertz, 1992a), and at a more mesic study site, light intensity and T_e were strongly correlated among perch sites of *A. cristatellus* (Hertz et al., 1994). On the other hand, light intensity and thermal environment are not always related. In a closed forest site near to the *A. cristatellus* mesic site, no relationship existed between light intensity and T_e for perch sites used by *A. gundlachi* (Hertz et al., 1994).

The role of the light environment in driving evolutionary divergence in signaling behavior and structures is potentially very important and will be discussed in Chapter 14. The possibility that species can diverge to adapt to different light environments as a means of partitioning the habitat is an exciting new possibility, the generality of which remains to be investigated.

REMOTE SENSING APPROACHES TO INVESTIGATION OF SPECIES' HABITAT REQUIREMENTS

The integration of satellite data and distributional records to understand the habitat factors shaping a species' distribution has taken off in recent years (e.g., Guisan and Zimmerman, 2000; Peterson, 2001). To a large extent, these Geographic Information Systems (GIS) approaches are useful in elucidating the role of temperature and moisture (e.g., maximum, minimum, seasonality) in determining where species occur. However, these approaches are too coarse-grained to shed light on the factors that affect microhabitat partitioning within local communities.

Although rich in potential for understanding ecological and evolutionary aspects of anole distribution, GIS approaches are just beginning to be applied to anole data. Knouft et al. (2006) studied the *A. sagrei* group on Cuba and found that ambient temperature, precipitation, and seasonality all were important determinants of species' distributions. As for thermal and hydric biology (discussed above), no phylogenetic effect is apparent in the environmental niches of different species: some closely related species have similar environmental niches and some have highly divergent niches; distantly related species also can be very similar or very divergent.

ONTOGENETIC AND SEASONAL SHIFTS IN HABITAT USE

Anoles change their habitat use both as they grow and across seasons. Ontogenetic habitat shifts have been reported in many species; anoles generally shift to higher and wider perches as they get older (reviewed in Stamps, 1983b). These size-related shifts probably have a variety of causes (Huey and Webster, 1975; Scott et al., 1976; Moemond, 1979a; Stamps, 1983b; Jenssen et al., 1998; Ramírez-Bautista and Benabib, 2001). Larger lizards have greater locomotor capabilities—including the ability to jump across larger gaps and to capture prey and escape to a refuge from a greater distance (Chapter 13)—and require broader surfaces to support their mass; in addition, larger lizards need to use wider surfaces to minimize their visibility to predators approaching from the opposite side of the object upon which they are perching.[251] In addition, larger lizards are dominant over smaller ones (Chapter 9) and thus able to secure the most desirable microhabitats.

Ontogenetic habitat shifts have been particularly well documented in *A. aeneus* (Stamps, 1983b). In this species, juveniles move into open clearings, then return to shady areas when they reach subadult size. Presumably, the juvenile shift is to avoid predation by the larger *A. richardii*, which is not found in open clearings and which poses a threat particularly to smaller *A. aeneus* (Stamps, 1983b).

Seasonal changes in habitat use have received relatively little attention. Not surprisingly, many species bask more in the winter to compensate for lower air temperature (e.g., Hertz, 1992a,b) and as mentioned above, *A. limifrons* basked less in the dry season, perhaps to minimize water loss (Sexton and Heatwole, 1968).

---

251. Lizards have a blind spot behind and underneath their heads, and the size of this blind spot is a function of head size. As a result, a predator approaching from the other side of a tree may not be visible to a lizard. Consequently, lizards should choose surfaces broad enough that they can't be seen by a potential predator located in their blind spot on the other side of the surface. Larger species, being wider, require broader surfaces.

Upward shifts in perch height in the non-breeding season occurred in three Puerto Rican rainforest anoles and *A. carolinensis*.[252] Two mainland species, *A. nebulosus* and *A. cupreus*, also exhibited shifts in perch height, but in the opposite direction, from near the ground in the non-breeding dry season to much higher in the vegetation in the wet season. In both species, the upward shift was substantially greater for males, which at the same time greatly increased their territorial behavior, than for females (Fleming and Hooker, 1975; Lister and Aguayo, 1992).[253]

Seasonal shifts in perch height also lead to shifts in foraging location. *Anolis nebulosus* changed from foraging almost entirely on the ground in the dry season to foraging mostly in arboreal situations in the wet season. Comparable shifts in foraging location occurred in male, but not female, *A. cupreus* (Fleming and Hooker, 1975) and in *A. stratulus* (sexes not differentiated [Reagan, 1986]).

Puerto Rican rainforest anoles also shifted their perch diameter use across seasons, but the direction of changes differed among species, and even among sexes (Lister, 1981; Jenssen et al., 1995; Dial and Roughgarden, 2004).

Habitat shifts as a result of the presence of other species have been commonly reported and are discussed in Chapter 11.

HABITAT SELECTION

The segregation of species into different microhabitats suggests that species can select the appropriate microhabitat, but little work has investigated how this selection occurs (Sexton and Heatwole, 1968; Kiester et al., 1975; Talbot, 1977). Several studies suggest that anoles may use conspecifics as cues when settling into new habitats (Kiester, 1979; Stamps, 1987, 1988).

That anoles use temperature in habitat selection is suggested by the data on thermoregulation discussed above. The physiological mechanisms underlying temperature detection and response in ectotherms are an area of active research and have not received much attention in anoles (reviewed in Seebacher and Franklin, 2005). *Anolis cristatellus* may use light intensity as a cue for habitat selection. In the open habitats that it uses, warmer sites are more brightly illuminated, and in a laboratory experiment, lizards of this species use light as a cue when attempting to thermoregulate (Hertz et al., 1994).

252. Thermoregulating lizards also shift perch height over the course of the day to avoid hotter temperatures near the ground at midday (e.g., Huey, 1974), as well as to avoid predators that are most active at midday (Chapter 11).

253. In a study on *A. nebulosus* that commenced just as Lister and Aguayo's (1992) study at the same site was ending, Ramírez-Bautista and Benabib (2001) found somewhat different patterns of seasonal change in perch height.

Anoles are almost entirely diurnal. The only exception is that many species have been reported active after dark on walls near electric lights, feeding on the insects attracted to the light (e.g., Rand, 1967b; reviewed in Perry et al., 2008).

Many anoles sleep on leaves or on the ends of branches (Fig. 10.7).[254] The presumed function of this behavior is that any potential predator approaching the lizard will cause the branch or leaf to vibrate, alerting the lizard in time to escape by jumping into the void. This behavior may work well against such predators (as far as I am aware, no one has ever studied the efficacy of this behavior), but at least some arboreal snakes have thwarted this defense by adopting an airborne approach, stretching across from another branch to pluck the unsuspecting lizard while it still slumbers (Fig. 10.8; Henderson and Nickerson, 1976; Yorks et al., 2004). The presence of anole remains in owl pellets suggests the existence of another threat to sleeping anoles, although another possibility is that crepuscular owls nabbed still-active anoles just as they were preparing for bed (Hecht, 1951; Etheridge, 1965; Buden, 1974; McFarlane and Garrett, 1989; Gerhardt, 1994; Debrot et al., 2001).[255]

For many years, just about every field biologist I knew who worked on anole ecology or behavior contemplated the idea of studying whether sympatric anoles partition their sleeping sites as they do their diurnal haunts. Many workers, myself included, set out to collect the relevant data, only to discover that this was a full time project in itself. Finally, such a study has been conducted. For three Jamaican species, sleeping perches are generally higher, narrower and more horizontal than diurnal perches (Singhal et al.,

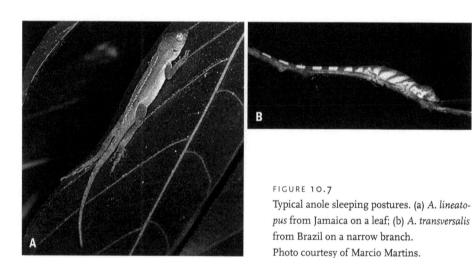

FIGURE 10.7
Typical anole sleeping postures. (a) *A. lineato-pus* from Jamaica on a leaf; (b) *A. transversalis* from Brazil on a narrow branch.
Photo courtesy of Marcio Martins.

254. But not all. Jenssen (1970b) described a population of *A. nebulosus* that slept in the leaf litter.
255. Sleeping on leaves and at the end of branches also makes sleeping anoles vulnerable to nocturnal biologists and other rapscallions: many anoles blanch in color at night and stand out quite vividly against the background in the beam of a flashlight.

FIGURE 10.8
A Brazilian blunt-headed vinesnake
(*Imantodes cenchoa*) eating an anole captured
while sleeping. Photo courtesy of Marcio
Martins.

2007).[256] Females increased height at night substantially more than males and day and nighttime habitat use was significantly different for each sex within all three species (except for *A. grahami* males). Despite these shifts in habitat use, interspecific differences in habitat use occurred at night, just as they did during the day (Fig. 10.9).

The implication of these findings is that community and functional biologists should consider the potential importance of sleeping sites. Could species be partitioning sleeping sites as a resource? Perhaps more importantly, could the morphological differences among species represent adaptations for using different microhabitats at night, as well as during the day? The narrowness of nighttime perches is particularly notable and might make strong biomechanical demands on lizards snoozing on such perches (see discussion of competition and adaptation in Chapters 11 and 13). Another question concerns whether perch sites are chosen for their thermal properties, either at night or early in the morning, when lizards may need to raise their body temperature quickly. Finally, could nocturnal predation exclude anoles from some microhabitats, thus affecting their diurnal microhabitat use (Chandler and Tolson, 1990)?

Anolis lineatopus individuals use sleeping sites that are within their diurnal home ranges (Singhal et al., 2007).[257] Some anoles appear to use the same perch repeatedly,

---

256. Some or all of these patterns have been reported for many other species (e.g., Ruibal and Philibosian, 1974b; Vitt et al., 2002; Vitt et al., 2003b; Poche et al., 2003).

257. Comparable data are not available for the two other species in this study.

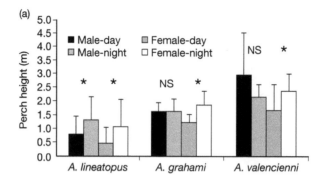

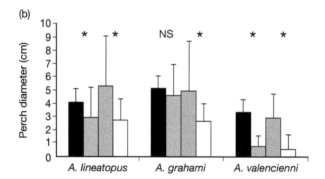

FIGURE 10.9
Shifts in (a) perch height and
(b) perch diameter between
day and night for three
Jamaican species. Values are
means + 1 standard error.
Asterisks indicate significant
differences between daytime
and nighttime habitat use.
Modified with permission
from Singhal et al. (2007).

but others do not (Rand, 1967b; Kattan, 1984; Clark and Gillingham, 1990; Shew et al., 2002; Poche et al., 2005; Singhal et al., 2007).

FUTURE DIRECTIONS

As with the data on population biology reported in Chapter 8, it is remarkable how little we know about habitat use of most species. Even for thermal biology, which has been extensively studied, few clades or communities have been well sampled. Moreover, most of the work on anole thermoregulation was conducted prior to the development of modern methods for assessing the extent to which anoles nonrandomly regulate their body temperature. Thus, despite a great number of studies on many species, we cannot quantitatively assess the extent of thermoregulation in most species. Now that the tools are in hand, a re-blossoming of anole thermal biology—mostly dormant since the early 1980s other than work on several Puerto Rican species—would be welcome.

The situation for other aspects of habitat use is much worse. The hydric and light ecology of only a few species have been studied. Now that remote sensing methods are available, such data will be critical to cross-validate the conclusions concerning the environmental factors that determine anole distributions (e.g., Kearney and Porter, 2004).

The history of the study of thermoregulation in ectotherms is a long one, and studies on lizards—especially on anoles—have played an important role (reviewed in Avery, 1982; Huey, 1982; Angilleta et al., 2006). A variety of early ideas concerning how to study the extent of thermoregulation have proven too simplistic: for example, neither the slope of the regression line between air temperature and body temperature nor the variance in body temperature among individuals in a population are good indicators of the extent of thermoregulation (Huey, 1982; Hertz et al., 1993).

The body temperature of a small ectotherm is a function of air temperature, wind speed, whether the animal is in the sun, the temperature of the surface on which it is sitting and a variety of other factors. Sophisticated biophysical models have been developed to calculate what the equilibrium temperature of a lizard occupying a particular spot with particular parameter values should be (Porter et al., 1973; Roughgarden et al., 1981; Waldschmidt and Tracy, 1983). However, a much easier approach is simply to build a model lizard, of appropriate size and with appropriate reflectance, conductance and other thermal properties, and place it in the environment (Fig. 10.10; Bakken and Gates, 1975; Bakken, 1992; Grant and Dunham, 1988; Hertz, 1992b; Dzialowski, 2005; see comparison of approaches in Huey [1991]). The temperature to which the model equilibrates is an estimate of the temperature a live lizard would attain if sitting in the same spot and not using any behavioral or physiological means to alter its body temperature (See Hertz [1992b] for review).

This approach can be taken one step further. By randomly placing many such models in the environment and monitoring them, one can estimate both the mean and the variance in body temperature that a population of lizards would attain if they were using the environment randomly and thus not behaviorally thermoregulating. By comparing real lizard temperature data to those generated by models, we can determine the extent to which lizards are actively thermoregulating (Figure 10.3).

FIGURE 10.10
Photo of lizard models. Instead of models cast from a real lizard, other researchers have used cylindrical tubes plugged at either end or small temperature sensors (e.g., Van Berkum et al., 1986; Vitt et al., 2001). Photo courtesy of Kevin de Queiroz.

Moreover, from such data we can also estimate the extent to which lizards might need to thermoregulate in a particular environment. Laboratory choice experiments (usually conducted by placing a lizard in a thermally heterogeneous gradient and seeing what temperature its selects) can determine the preferred temperature range of a species. Comparisons with the temperature that models attain in the field can indicate how far non-thermoregulating lizards would be from their preferred temperature (that is, how much thermoregulation is needed). The precision of thermoregulation can then be defined as the extent to which real lizards are closer to their preferred temperature than they would be if they were randomly sampling the environment (Hertz et al., 1993; Blouin-Demers and Nadeau, 2005).

11

ECOLOGY AND ADAPTIVE RADIATION

Adaptive radiation is the evolutionary divergence of members of a clade to adapt to the environment in a variety of different ways (Simpson, 1953; Givnish, 1997; Schluter, 2000).[258] Some of the most spectacular case studies in evolutionary biology are adaptive radiations. Consider Darwin's finches which, in the absence of many other types of landbirds in the Galápagos, have diversified to adapt to a wide variety of niches usually occupied elsewhere by members of different families (Grant, 1986; Grant and Grant, 2008). Similarly, African Rift Lake cichlids fill an enormous number of ecological roles—from grazers and molluscivores to scale-raspers, eye-pluckers and fish-eaters— with a corresponding diversity in morphological form (Fryer and Iles, 1972; Kornfield and Smith, 2000).

The concept of adaptive radiation is important to evolutionary biology and biodiversity studies for two reasons. First, the great differences between species and, in many cases, the great species richness of these radiations makes them focal cases for the study of adaptation, speciation, and other evolutionary phenomena. Second, many workers suggest that much of the diversity of life may be the result of adaptive radiation (e.g., Givnish, 1997; Schluter, 2000).

258. Other workers include the timing of diversification as part of the definition of adaptive radiation. In agreement with Givnish (1997), I believe that the important aspect of adaptive radiation is the extent of ecological disparity exhibited by a clade; whether the evolution of this disparity arises as part of an early burst of speciation or gradually through time is an empirical question to be tested, rather than subsumed within the definition itself (cf. Schluter, 2000). Givnish (1997) provides an interesting list of definitions of adaptive radiation from different authors.

Two questions are central to the study of adaptive radiation:

1. Why do some clades and not others experience adaptive radiation?

2. What is the process by which adaptive radiation occurs?

The first question—and the important related point concerning how adaptive radiations are identified—will be discussed in Chapters 15 and 17. In this and the next chapter, I focus on the second question.

Although many ideas have been presented about how adaptive radiation occurs, what I consider the classic idea—following Simpson (1953) and Schluter (2000)—has the following steps (Fig. 11.1):

1. A species finds itself in an environment in which resources are plentiful. This may occur due to colonization of a new area, extinction of other species, or evolution of a trait that provides access to previously unattainable resources.

2. Speciation occurs, leading to sympatric co-occurrence of several to many species. This sympatry may ensue either directly if speciation is sympatric or may be the result of non-sympatric speciation followed by range expansion.

3. The abundance of individuals leads to resource depletion (this step could occur prior to step number 2 above).

4. Species alter their behavior and/or habitat use to partition resources and minimize interspecific interactions.[259]

5. Species evolve adaptations to their new regime of resource use.

The end result is a set of species specialized to use different parts of the resource base; i.e., an adaptive radiation.[260]

The hypothesis of resource competition as the driver of adaptive radiation makes three testable predictions:[261]

- Sympatric species interact ecologically, primarily by competing for resources.

- As a result of these interactions, species alter their resource use.

- As a result of shifts in resource use, species evolve appropriate adaptations.

259. If speciation occurred in allopatry, the species may have diverged to some extent prior to sympatry as they adapted to different local circumstances. Indeed, without such differences, competitive exclusion may prevent coexistence if resources are limiting in sympatry (Grant and Grant, 2008). Nonetheless, in this scenario for the development of an adaptive radiation, the differences are not envisioned as being great enough to prevent resource competition and subsequent resource partitioning in sympatry.

260. Adaptive radiation could also result from sympatric speciation driven by disruptive or frequency-dependent selection, a type of "ecological speciation" (Schluter, 2001; Rundle and Nosil, 2005). In such a scenario, Step 1 would be followed by Step 3 prior to speciation. Subsequently, as the individuals deplete the resources, disruptive selection would lead to speciation during Steps 4 and 5, with the same outcome: an adaptive radiation. Some workers consider Greater Antillean anoles to be an example of this phenomenon (to be discussed in Chapter 14).

261. These predictions hold for the sympatric speciation model as well; just substitute "subpopulations" for "species."

(a) A species finds itself in a
    resource-rich environment

(b) Speciation occurs

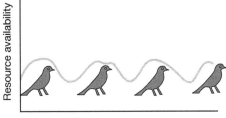

(c) Resources become scarce

(d) Species partition resources to minimize
    interactions

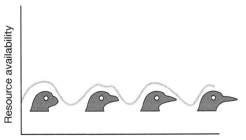

(e) Species adapt to new "niches"

FIGURE 11.1

The progression of an adaptive radiation, using the evolution of Darwin's finches as inspiration (Grant
and Grant, 2008). (a) A species finds itself in a resource-rich environment. In this case, the resource
spectrum might be thought of as seeds of different sizes. For anoles, it might be structural microhabitats. (b) Speciation occurs leading to sympatry of ecologically similar species (causes and geographical
context of speciation unspecified). (c) As species populations grow, resources are depleted (alternatively,
the initial species could deplete the resources prior to the arrival or origin of additional species).
(d) Species behaviorally partition resources to minimize interspecific competition. (e) Species adapt
to the different resources they are using.

In this chapter, I will review the evidence from anoles for the first two of these predictions, while deferring the more evolutionary considerations of the third prediction to the next chapter.

SYMPATRIC *ANOLIS* INTERACT ECOLOGICALLY

Caribbean anoles have been workhorses of community ecology. An enormous amount of research, primarily in the 1960s through 1980s, was devoted to studying the ecological relationships among coexisting anole species. Indeed, work on *Anolis* played a central role in the development of modern community ecology theory (e.g., Schoener, 1968, 1974; Roughgarden, 1974).

This work has taken almost every form imaginable: behavioral studies of individual lizards, comparative studies of populations in different areas, examination of resource use of coexisting species, null model analyses of patterns of species co-occurrences and experimental manipulations. The conclusion that shines through is clear: evidence for the importance of ecological interactions in structuring communities of Caribbean anoles is pervasive.[262]

THE STRUCTURE OF ANOLE COMMUNITIES

Early workers noted that sympatric anoles use different microhabitats and exhibit many behavioral differences (e.g., Oliver, 1948; Collette, 1961; Ruibal, 1961; Rand, 1962; Rand and Williams, 1969). Rand (1964b, 1967c) pioneered a quantitative approach to anole community ecology by walking through the environment and noting the location of each lizard observed.[263] This work, in turn, was followed by Schoener's sophisticated statistical analyses of similar data collected in both the Greater and Lesser Antilles (Schoener, 1968, 1970; Schoener and Gorman, 1968; Schoener and Schoener, 1971a,b), and subsequently by a large number of other studies, many of which are cited throughout this chapter.

262. Ecologists differ on how to refer to a group of closely related species that co-occur at a given locality. The term "community" is often used, although some point out that "community" refers to *all* co-occurring species, and that a taxonomic subset of these species should be referred to as an "assemblage" or some other term (Fauth et al. [1996] is a good entrée to this literature). Although mindful of these semantic distinctions, I use the term "anole community" to refer to species of *Anolis* sympatric at a given locality.

263. What has subsequently become known among the cognoscenti as a "Rand Census." Rand censuses are usually conducted by walking through suitable habitat and collecting data on every lizard observed. If a lizard is observed moving apparently in response to the investigator, then data are not collected for that individual; otherwise data are collected at the spot at which the lizard was first observed. Data taken usually include species, sex, size, perch type, height, diameter, and, if the sun is shining, whether all, some, or none of the the lizard is in the shade. Other data sometimes recorded include lizard orientation, response to being approached, distance to nearest object to which it could jump, and an index of habitat visibility. Sometimes the lizard is captured and its body temperature recorded as well.

Two main conclusions have emerged from these studies. First, sympatric anole species exhibit ecological differences. This resource partitioning[264] generally involves differences along one of three axes of resource use: structural microhabitat, thermal microhabitat, or prey size. Second, as a corollary, the more similar two species or size/sex classes are along one resource axis, the less similar they will be upon another, a phenomenon termed "niche complementarity."[265]

A great deal of research on anole community ecology conducted over the past 30 years has demonstrated the near ubiquity of the first phenomenon, and has also revealed that patterns of resource partitioning differ between the Greater and Lesser Antilles. By contrast, relatively little further work has examined niche complementarity.

RESOURCE PARTITIONING

Greater Antilles

As discussed in Chapter 3, sympatric species in the Greater Antilles always differ in either structural or thermal microhabitat or in prey size (with a very few exceptions discussed below). By definition, sympatric members of different ecomorph classes differ in structural microhabitat, except to some extent crown-giants versus trunk-crown anoles, which differ in size (Chapter 3). The resource axis partitioned when members of the same ecomorph class are sympatric differs as a function of perch height. Within the more terrestrial ecomorphs (e.g., trunk-ground, grass-bush), sympatric species tend to be similar in body size (which generally correlates with prey size: see Chapter 8), but differ in thermal microhabitat: one species will occur in the sun, for example, and another in the shade. These differences can lead to the entertaining situation that the species found at a particular spot—say, a tree trunk—will differ through the course of the day. In the morning when the sun hits the trunk, the heliothermic species will be found there, whereas at midday, when the trunk is shaded, the shade-loving species will be present (Schoener, 1970a).

Within the more arboreal ecomorph classes, sympatric species also exhibit differences in thermal preferences, but the differences are less extreme, with the consequence that these species tend to occur syntopically to a much greater extent.[266] However, sympatric members of the same arboreal ecomorph class almost invariably differ in body size to a substantial extent. This phenomenon is most evident in trunk-crown anoles: on all four islands of the Greater Antilles, two species of trunk-crown anoles occur

264. The term "resource partitioning" (which produced 16,900 hits on a Google Scholar search in January 2009) was coined by Schoener in his 1968 study of the anole community of South Bimini Island, Bahamas.

265. In recent years, the term "niche complementarity" has acquired a different meaning, referring to the idea that communities composed of functionally different species (hence, the "complementarity" of their niches) may be more stable, resilient, productive or otherwise different from less diverse communities (e.g., Hector et al., 1999; Tilman et al., 2001).

266. That is, sympatric arboreal anoles can be found together at the same spot, at the same time, much more frequently than more terrestrial ecomorphs.

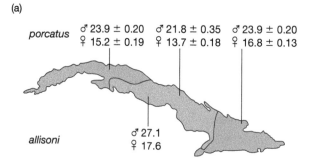

(a)

porcatus   ♂ 23.9 ± 0.20   ♂ 21.8 ± 0.35   ♂ 23.9 ± 0.20
          ♀ 15.2 ± 0.19   ♀ 13.7 ± 0.18   ♀ 16.8 ± 0.13

allisoni   ♂ 27.1
             ♀ 17.6

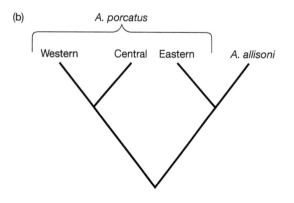

(b)

A. porcatus

Western   Central   Eastern   A. allisoni

FIGURE 11.2

Size differences between *A. allisoni* and *A. porcatus* in sympatry in Cuba. (a) Where sympatric, the two species differ greatly in size (represented by Jaw length), whereas lizards in the allopatric populations on either end of the island are about the same size. Figure modified with permission from Schoener (1977). (b) Central *A. porcatus* populations are related to western *A. porcatus*, whereas eastern *A. porcatus* populations are actually more closely related to *A. allisoni*, which only occurs in the center of the island (Glor et al., 2004). As a result, this constitutes a classic example of character displacement, in which the sympatric forms are greatly divergent in size, while the allopatric forms, each more closely related to one of the central forms, are about the same size (Brown and Wilson, 1956; Schluter, 2000). Compared to allopatric populations, centrally located *A. porcatus* also show character displacement in color and pattern, being duller in color with dark reticulations and white specklings, which makes them more distinct from sympatric *A. allisoni* (Ruibal and Williams, 1961).

sympatrically in some areas—one large (maximum SVL 70–84 mm) and the other small (44–57 mm). Furthermore, size differences are also exhibited by the two large trunk-crown anoles of Cuba, *A. allisoni* and *A. porcatus*, which co-occur in the center of the island. In sympatry, they differ substantially in size, whereas allopatric populations on the eastern and western side of the islands exhibit similar, intermediate body sizes, which makes this an excellent example of character displacement (Figure 11.2).

Size differences are also seen among sympatric members of other arboreal ecomorph classes. Among crown-giants, one case of sympatry is known from the Sabana Archipelago on the north coast of Cuba, where *A. pigmaequestris* co-occurs with *A. equestris*. As the names suggest, the two species differ substantially in size (Garrido, 1975). If Chamaeleolis is considered a twig anole (Chapter 4), then sympatric twig anoles differing vastly in size are also known.[267]

Lesser Antilles

The situation is slightly different in the Lesser Antilles, where sympatric species differ in not one, but two of the three resources axes (Schoener and Gorman, 1968; Roughgarden et al., 1981, 1983; Harris et al., 2004; Buckley and Roughgarden, 2005b; Hite et al., 2008). On 16 out of 17 two-species islands in the Lesser Antilles, one of the species is large and the other small, and, where studied, the species eat correspondingly different sized prey.[268] However, the identity of the other partitioned ecological axis varies geographically and phylogenetically. In the northern Lesser Antilles, sympatric anoles of the *bimaculatus* Series differ substantially in perch height, but to a limited extent in thermal microhabitat; in contrast, the situation is reversed in the southern Lesser Antilles, occupied by the *roquet* Series, where sympatric species differ relatively little in perch height, but substantially in thermal microhabitat (Figure 11.3). A corollary to the difference in extent of thermal microhabitat partitioning is that species in the south, which are adapted to use different thermal microhabitats, tend to segregate by habitat type (e.g., open scrub versus mature forest) to a much greater extent than species in the north.[269]

Mainland

Mainland anole communities exhibit the same general patterns as seen in the West Indies, though in general they have been studied much less quantitatively. Sympatric mainland anoles tend to differ along the same three niche axes as in West Indian communities (e.g., Rand and Humphrey, 1968; Fitch, 1975; Duellman, 1978, 1987, 2005; Castro-Herrera, 1988; Pounds, 1988; Vitt and Zani, 1996b; Vitt et al., 1999; D'Cruze, 2005). In addition, like some West Indian anoles, mainland anoles occurring in the same region sometimes segregate by habitat type (e.g., upland forest versus seasonally flooded forest).

267. There are no other known cases of sympatry of twig or crown-giant species.
268. The one exception being St. Martin, where A. gingivinus is medium-sized and A. pogus is small. The 17 islands are those reported by Schoener (1970b) and do not include the Grenadines or small islets offshore from some larger islands (e.g., near Antigua).
269. The same phenomenon occurs among the Greater Antillean ecomorphs. The terrestrial ecomorphs, which partition thermal microhabitat within a locality, also segregate across habitat types to a much greater extent than the more arboreal ecomorphs. The reason for this phenomenon is that species that use different thermal microhabitats differ in their thermal physiology (discussed in Chapter 13). Hot and open or closed and cool habitats may be suitable for only one species, although intermediate habitats may provide appropriate thermal microhabitats for multiple species. By contrast, species that partition the environment by perch height, as in the northern Lesser Antilles, or by prey size, as in the arboreal ecomorphs, can co-occur in almost types of habitat except those that have nothing but low vegetation or a limited range in prey size.

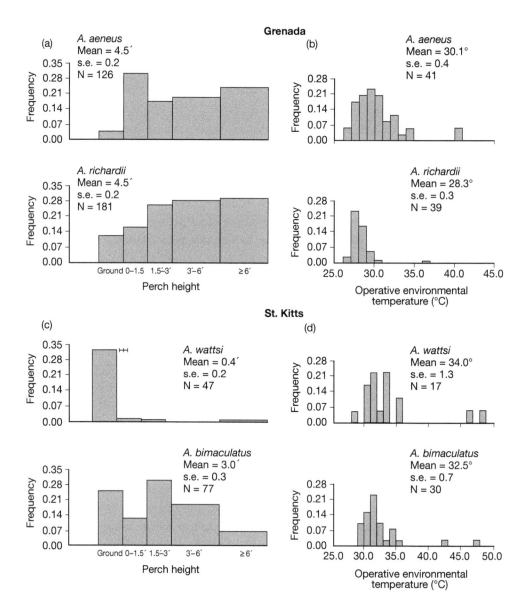

FIGURE 11.3

Resource partitioning between large and small species in the Lesser Antilles. (a) *A. aeneus* and *A. richardii* on Grenada in the southern Lesser Antilles; (b) *A. wattsi* and *A. bimaculatus* on St. Kitts in the northern Lesser Antilles. Operative environmental temperature is the equilibrium temperature a lizard would attain at the point at which it is located (Chapter 10). Figure modified with permission from Roughgarden et al. (1983).

Mainland communities contain many more non-anole lizard species than do West Indian communities, which might suggest that an anolocentric focus in mainland studies would paint a misleading or incomplete picture. In the West Indies, exclusive focus on interactions among anoles makes sense: when it comes to arboreal insectivores, anoles are just about the only game in town because the diversity and abundance of other types of lizards that are likely to interact with anoles is low in most habitats.[270] By contrast, in most mainland communities, anoles are a much smaller component of the insectivorous lizard community, both in terms of species richness and abundance. For this reason, consideration of other lizard species might seem more important in studies of mainland anoles.[271] However, detailed studies of several mainland lizard communities reveal that most species overlap little in habitat use and diet (e.g., Rand and Humphrey, 1968; Duellman, 1978, 1987, 2005); moreover, ecological variation among species has a strong phylogenetic component such that those species that do exhibit high levels of ecological overlap tend to be closely related (Vitt, 1995; Vitt and Morato de Carvalho, 1995; Vitt and Zani, 1996b; Vitt et al., 1999). As a result, in terms of habitat use and diet, few coexisting mainland lizard species are ecologically similar to anoles.[272]

Cases in Which Coexisting Species Are not Ecologically Differentiated

The primary exceptions to the rule that sympatric anoles differ ecologically are cases in which species, usually close relatives, meet where their geographic range boundaries

270. Another highly diverse lizard group in the West Indies is *Sphaerodactylus*, of which there are 19 species in Cuba and 35 in Hispaniola with a maximum of four occurring in sympatry (M. Leal, pers. comm.; species lists can be viewed at http://evo.bio.psu.edu/caribherp/). However, these diurnal geckos, which are usually found in the leaf litter, are very small (the largest species reaches 41 mm SVL, and most are substantially smaller [Schwartz and Henderson, 1991]) and probably overlap little in diet with all but the smallest anoles. Twelve species of curly-tailed lizards (*Leiocephalus*) occur in Hispaniola, six in Cuba and six in the Bahamas and nearby islands. These lizards can be very abundant in open and hot habitats, but are absent or rare in most other habitats, as are the teid lizards of the genus *Ameiva* that are found throughout the West Indies and have their maximum species richness (three) on Hispaniola. Most species of both *Leiocephalus* and *Ameiva* are larger—in many cases substantially larger—than most anoles and probably interact with them more as predators than as competitors (see Chapter 8). However, smaller curly-tailed species, such as *L. punctatus* in the southern Bahamas, may compete with sympatric anoles (Schoener, 1975). Other West Indian lizards (e.g., *Amphisbaena*, *Aristelliger*, *Celestus*, *Cyclura*, *Thecadactylus*) probably interact little with anoles.

271. The question also arises whether lizards form an exclusive guild (i.e., all the species in a community that utilize the same resources in a similar manner [Root, 1967]) of insectivores, or whether lizards are likely to be sharing resources with other taxa. Although frogs, spiders, and many insects may interact ecologically with anoles, the most obvious competitors are insectivorous birds. In the West Indies, the diversity and abundance of insectivorous birds is much lower than in the mainland. Some have argued that this is a result of the great abundance of anoles in the West Indies and that, conversely, the lower abundance of mainland anoles is attributable to competition from much more abundant insectivorous birds (e.g., Lister, 1976a; Wright, 1981; Wright et al., 1984; Waide and Reagan, 1983; Moermond, 1983; Buckley and Jetz, 2007). The extent and consequences of anole-bird interactions is crying out for experimental investigation, but such studies, which probably would take the form of excluding birds from some enclosures and not others, would not be easy.

Frogs are also extremely abundant in many places in the neotropics and being insectivorous and about the size of smaller anoles (and sometimes of not-so-small ones), they might also compete with anoles for food (Waide and Reagan, 1983). However, detailed diet analysis in the Luquillo Mountains in Puerto Rico indicated that anoles and the extremely abundant frog *Eleutherodactylus coqui* overlapped little in prey, probably because anoles are diurnal and coquis are nocturnal, and thus the two groups of insectivores ate prey that differ in their activity times (Reagan et al., 1996). A study in Amazonia also found little overlap between frogs and anoles, even though some mainland frogs are diurnal; much of the difference in diet resulted because frogs tended to eat smaller prey, often including large numbers of ants (Caldwell and Vitt, 1999), which most mainland anoles avoid (Chapter 8). In contrast, a study in Costa Rica found greater similarity in the diet of small anoles and frogs, although overlap was not quantified (Whitfield and Donnelly, 2006).

272. Two possible exceptions come to mind. First, the small, diurnal gecko *Gonatodes humeralis* occurs on tree trunks near the ground and is a sit-and-wait predator that overlaps substantially in diet with some anole species (Vitt et al., 1997, 1999). Second, aquatic anoles may interact with other lizards in their streamside habitats. Castro-Herrera (1988) suggested that at a site in the Chocó region of Colombia, an aquatic anole, *A. macrolepis*, is replaced in open, sunny areas by the basilisk, *Basiliscus galeritus*; he further noted that although adults basilisks are substantially larger, juveniles are about the same size as adult *A. macrolepis*.

overlap (e.g., Garcea and Gorman, 1968; Arnold, 1980; Pounds, 1988). In most cases, the zone of sympatry is relatively small; these parapatric distributions seem to reflect competitive exclusion, though this hypothesis remains to be tested.

In addition, in a few cases sympatric members of the same ecomorph class are only slightly differentiated ecologically (e.g., Hertz, 1980a; Arnold, 1980). The best documented case involves the trunk-ground anoles *A. cristatellus* and *A. cooki* which are broadly sympatric throughout most of the latter species' range in southwestern Puerto Rico (Jenssen et al., 1984). Although the species differ slightly in the thermal and light environments of the perches they occupy (reviewed in Chapter 10), territories of the two species are interdigitated throughout the landscape; given their similarity in size and perch characteristics, the species probably compete for food (Hertz, 1992a). Perhaps as a result, the two species are among the few anoles to exhibit interspecific territoriality (Jenssen et al., 1984).

If resources are limiting, how do such species manage to coexist? Indeed, several workers have suggested that *A. cristatellus* is pushing *A. cooki* to extinction (Williams, 1972; Jenssen et al., 1984; Marcellini et al., 1985). Hertz (1992a), however, pointed out the numerical dominance and longer daily activity period of *A. cooki*, and argued that *A. cooki* may be better adapted to the extreme conditions in southwestern Puerto Rico, suggesting that reports of its impending demise may be premature. Detailed population level studies on the interactions occurring in this and similar situations (e.g., *A. marcanoi* and *A. cybotes* in southern Hispaniola [Hertz, 1980a]) are needed to understand the factors which mediate these species' coexistence.[273]

Null Models

The observation that sympatric anole species almost invariably differ along one of three niche axes strongly suggests that a deterministic process—such as interspecific competition—is operating to prevent the coexistence of ecologically similar species. Nonetheless, as Morin (1999, p. 58) and others have pointed out, if sympatric species are examined closely, some ecological difference is likely to be found; they are different species, after all. The question, then, is whether sympatric species are more different than would be expected from any set of randomly assembled species.

For this reason, ecologists in the late 1970s and early 1980s developed null model approaches to investigate whether community level patterns are non-random; the basic idea is to create a null hypothesis about how the data—such as similarity between

273. As noted in Chapter 3, multiple grass-bush anoles appear to occur sympatrically in eastern Cuba; they possibly may constitute another example of sympatry of ecologically similar members of the same ecomorph class, but too little is known about their natural history to assess whether and how they partition resources (Garrido and Hedges, 1992, 2001). In addition, whether any exceptional cases of sympatry of ecologically similar species occur in the mainland is unclear because the ecology of many species, much less their ecological interactions, is poorly understood, and few quantitative community studies have been conducted. For example, Fitch (1975) reported a number of instances of ecological similarity in species with overlapping ranges, but more detailed studies are needed; some of the species mentioned by Fitch (1975) have been found to differ ecologically when examined more closely (e.g., Corn, 1981).

species—might be distributed if deterministic processes were not operating. Null models have always been controversial and one clear conclusion is that the results of null models are often critically dependent on assumptions made at the outset of the analysis (reviewed in Gotelli and Graves [1996] and for anoles specifically in Schoener [1988]).

Several null model studies have been conducted on West Indian anole communities. In the Lesser Antilles, size differences between sympatric species are greater than expected by chance (Schoener, 1988; Losos, 1990a). In the Greater Antilles, the ecomorph composition of small landbridge islands (Chapter 4) is non-random, with fewer instances of multiple occurrence of members of the same ecomorph on an island than would be expected by chance (Schoener, 1988). The ecomorph composition of multiple localities on Puerto Rico and Jamaica is similarly non-random (Haefner, 1988), although the results in this case are particularly dependent on model assumptions.

No general null model has asked whether the patterns of differentiation along all three resource axes described above are non-random.[274] Such a null model would be very difficult to construct and would probably entail many debatable assumptions. Nonetheless, the clear differences within communities, as well as the repeatable ecomorph-specific patterns of differentiation repeated across islands, causes me to suspect that the null hypothesis in such a study would be resoundingly rejected, leading to the conclusion that deterministic processes are at work in structuring anole communities.

NICHE COMPLEMENTARITY

In communities in which multiple resource axes are important, the more similar two species are along one resource axis, the less similar they should be along another (Schoener, 1974).[275] The best examples of niche complementarity in anoles are studies by Schoener in which size classes within each species (i.e., adult male; female-size, including sub-adult males, which often are difficult to distinguish from females from afar; and juveniles) were treated as separate entities. In the four-species community of South Bimini, Bahamas, size classes of different species that overlap most in microhabitat are least similar in prey size (Schoener, 1968; Fig. 11.4). Likewise, in Jamaica, those size classes most similar in perch diameter are most different in body size (Schoener and Schoener, 1971a).[276] Similar comparisons of size classes have found evidence for niche complementarity among size clases in some communities (e.g., Rand, 1962, 1967c), but

274. However, the pattern of niche complementarity in an anole community in Cuba, discussed in the next section, was demonstrated to be statistically non-random relative to a null model (Losos et al., 2003b).

275. Two species may be dissimilar along all resource axes. Hence, in the two-axis case, the relationship is expected to be triangular rather than linear, with only the space corresponding to similarity on both axes unoccupied.

276. This relationship occurred because body size and perch diameter were inversely correlated among the four common species at the study site, but positively correlated within species. Consequently, with regard to perch diameter, the most similar size classes would be the juveniles of the smaller species and the adult males of the larger species. The Jamaican crown-giant, A. garmani, was uncommon at the study site and was not included in the study. If it had been included, the results of this analysis might have been different because A. garmani is both larger and uses broader perches than all of the other species (Losos, 1990c).

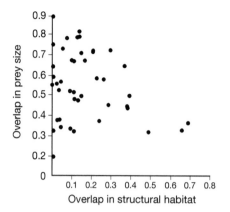

FIGURE 11.4

Niche complementarity among the anoles of South Bimini, Bahamas. Each point is an interspecific comparison between two size classes (e.g., female-sized *A. sagrei* versus adult male *A. distichus*). Overlap measures the extent of difference in frequency distributions of prey size or structural microhabitat use between the two groups: the greater the value, the more similar the groups are. Figure modified with permission from Schoener (1968).

not others (e.g., Schoener and Schoener, 1971b). In addition, an analysis using species rather than size classes as the ecological units found evidence for niche complementarity in an eleven-species community in Soroa, Cuba (Losos et al., 2003b).

Evidence for niche complementarity is mixed in mainland anole communities. Vitt et al. (1999) showed that for all lizard species in an Amazonian community, pairwise overlap in microhabitat use and diet type are positively related, rather than negatively as the niche complementarity hypothesis would predict; this finding also holds when considering only the four anoles in the community.[277] On the other hand, Corn (1981) found that overlap in structural microhabitat and diet were inversely correlated in a community of eight anole species in Costa Rica (see also Fitch [1975] and D'Cruze [2005]).

EXPERIMENTAL STUDIES OF SPECIES INTERACTIONS

The observation that sympatric anoles almost always differ ecologically suggests that ecological interactions occur among anoles. The most direct way to test such a hypothesis is to conduct an experiment: change the abundance of one species and see how other species respond.[278]

277. This study considered microhabitat and diet data. Consideration of data on the thermal microhabitat of these species, collected from populations in different areas (Vitt et al., 2002; 2003a,b), indicates that niche complementarity also does not occur when this resource axis is included.

Lack of niche complementarity could have several explanations, including: resource axes are not independent such that, for example, lizards in different microhabitats encounter different prey; species are greatly distinct on all resource axes so that, even if the species compete, complementarity would not be expected; or resource partitioning occurs for some reason other than to minimize interspecific resource competition.

278. Studies of how the abundance of species covary in undisturbed populations of anoles are surprisingly rare. One interesting example concerns the density of the Puerto Rican trunk-crown anoles, *A. evermanni* and *A. stratulus*, in the Luquillo Mountains of Puerto Rico after Hurricane Hugo. Dial and Roughgarden (2004) examined 14 trees that had been isolated from each other by the hurricane's destructive force. They found that no tree had high densities of both species: if one species was at high density, the other was invariably at low density. They attributed this negative relationship to recolonization of the trees in the hurricane's aftermath; whichever species had been able to become established first in a tree was postulated to have been able to keep the other from attaining high density. See also Buckley and Roughgarden (2006) for a different approach to investigating interspecific interactions using population size covariance data, and Schoener and Schoener (1980a) on population density covariation both across habitats and years in the Bahamas.

The late 1970s ushered in the era of experimental community ecology, and studies on anoles were in the first wave of manipulative field studies (reviewed in Connell, 1983; Schoener, 1983). Since that time, a number of other studies have taken an experimental approach to studying the interactions that occur among anoles.

A variety of approaches have been taken: experimental removal of lizards from single trees (Heatwole, 1977), continual removal of lizards from unenclosed plots (Salzburg, 1984; Leal et al., 1998), construction of enclosures within forests into which one or two species were stocked (Pacala and Roughgarden, 1982, 1985; Rummel and Roughgarden, 1985), and introduction of one or two species onto tiny islands (Roughgarden et al., 1984; Losos and Spiller, 1999; Campbell, 2000). Ecological effects have been quantified in a variety of ways, including perch height, activity times, growth rates, and population density.

In general, these studies have consistently revealed strong evidence of interspecific interactions. For example, on St. Martin, *A. gingivinus* kept in two 12 × 12 m enclosures for three months with *A. pogus* had lower growth rates, perched higher, ate fewer and smaller prey, and reproduced more slowly than conspecific lizards maintained for the same period in the absence of *A. pogus* (Pacala and Roughgarden, 1985).

Longer experiments are less common, but two experiments have been conducted on the interaction between *A. sagrei* and members of the *carolinensis* Species Group on small islands over the course of several years.[279] In the Bahamas, David Spiller and I found that two years after introduction, *A. smaragdinus* populations were more than five times denser on islands lacking *A. sagrei* than on islands onto which both species had been introduced (Losos and Spiller, 1999). Similarly, in Florida, Campbell (2000) found that *A. carolinensis* populations were reduced to 25% of their initial population size 3.5 years after the introduction of *A. sagrei*. In the Bahamas experiment, an effect on the density of sympatric *A. sagrei* was also suggested in the first year of the experiment, but disappeared thereafter, probably because of the declining density of *A. smaragdinus*; effects of *A. carolinensis* on *A. sagrei* were not investigated in the Florida study.

Only one experimental study has failed to find strong evidence of interspecific interactions among anoles. On St. Eustatius, enclosures were stocked with *A. schwartzi*, *A. bimaculatus*, or both species.[280] Allopatric and sympatric populations of the larger *A. bimaculatus* did not differ in any measure (Pacala and Roughgarden, 1985); although the smaller *A. schwartzi* was active at different times and moved to lower and hotter microclimates in the presence of *A. bimaculatus*, these shifts did not result in differences in food capture, growth, or reproductive rates between the sympatric and allopatric

279. The advantage of using islands as experimental replicates is that populations are self contained, with little or no immigration and emigration. By contrast, enclosures need to be constantly inspected and repaired to maintain their integrity.

280. Populations now referred to as *A. schwartzi* were called *A. wattsi schwartzi* by Pacala and Roughgarden (1985) and *A. wattsi* by Rummel and Roughgarden (1985).

populations (Rummel and Roughgarden, 1985). Overall, these results suggest that the two species, which differ greatly in body size (*A. bimaculatus*: 86 mm SVL; *A. schwartzi*: 51 mm SVL [Rummel and Roughgarden, 1985]), compete little, if at all.

THE FATE OF INTRODUCED POPULATIONS

In addition to carefully planned experimental manipulations, introductions of anoles to islands on which they don't occur naturally can serve as quasi-experimental tests of the interaction hypothesis. Although introductions of alien species have become a global scourge (Wilcove et al., 1998; Mooney and Hobbs, 2000), they do have one side benefit: they create conditions (which for ethical reasons never could be set up intentionally) that can provide insights concerning ecological and evolutionary hypotheses (e.g., Carroll et al., 1998; Phillips and Shine, 2004; Sax et al., 2007; Vellend et al., 2007).[281]

Anoles have been widely transported around the Caribbean. Although the means of introduction is in most cases unknown, a likely candidate is the nursery trade: lizards or eggs stow away in plants and can be transported great distances (Norval et al., 2002; Meshaka et al., 2004). Some introductions may represent escapes from the exotic pet trade (Meshaka et al., 2004). In addition, some introductions have occurred intentionally, both for biological control purposes[282] and more commonly by anole aficionados (Wilson and Porras, 1983; Meshaka et al., 2004).[283]

Ecological interactions between native and introduced anoles have been studied in detail in two cases. In the Florida experiment already discussed (Campbell, 2000), the native *A. carolinensis* not only declined in number, but also shifted to higher perches and different habitat types in the presence of the exotic *A. sagrei*. Similarly, on Grand Cayman, another native trunk-crown anole, *A. conspersus*, shifted to higher perch locations in the presence of the introduced *A. sagrei* (Fig. 11.5; Losos et al., 1993a).

281. Which is not to say either that introductions should be condoned or that such studies are as informative as well designed experiments. Nonetheless, taking advantage of introductions when they occur is worthwhile, especially because introductions occur on a scale and in ways that cannot be duplicated experimentally.

282. The story of the introduction of *A. grahami* to Bermuda is a classic in the annals of biological control disasters (Wingate, 1965; Lever, 1987). This Jamaican species was introduced in 1905 in an effort to control the medfly, *Ceratitis capitata*, itself an introduced and destructive pest on fruit trees. The lizard population boomed and quickly covered the entire archipelago, but failed to control the medfly, which was finally eradicated in the early 1960s by chemical means (Hilburn and Dow, 1990). Meanwhile, in the late 1950s coccinellid beetles and parasitic hymenopterans were introduced in an attempt to control an introduced scale insect that was damaging the native cedar trees. This effort was not particularly successful. In an attempt to eliminate the introduced anole, which was thought to be hindering the establishment of the insects which were hoped to prey upon the introduced plant pest, Simmonds (1958) recommended introducing a predatory Central American bird, the great Kiskadee, *Pitangus sulphuratus*. Unfortunately, this species has a broad diet of which lizards make up a small part. The introduction did little to affect the lizard population (and thus presumably little to help eradicate the scale insect), but the now extremely abundant kiskadees may have played an important role in the decline of the native Bermudian avifauna, as well as causing considerable damage to fruit orchards.

283. Who wouldn't want some attractive new anole species in his or her backyard? Alas, such anoliphily cannot be condoned.

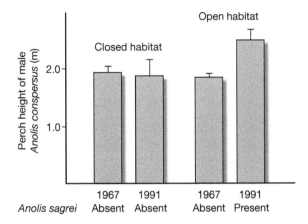

FIGURE 11.5

Habitat shift in *A. conspersus* on Grand Cayman in the presence of an introduced species. Perch height data for adult males were compared for *A. conspersus* before and ten years after *A. sagrei* was introduced. In closed habitats, where *A. sagrei* did not occur, *A. conspersus* habitat use was unchanged. By contrast, in open habitats where the more terrestrial *A. sagrei* was abundant, the more arboreal *A. conspersus* shifted to using higher perches. Data from Losos et al. (1993a).

A more general survey of the outcome of anole introductions also supports the hypothesis that anole species interact ecologically: on islands on which the introduced species is not ecologically similar to resident anoles, many introduced species have spread widely;[284] by contrast, on islands on which a resident species is ecologically similar, no introduced species has become widespread, and several have perished (Fig. 11.6; Losos et al., 1993a).

Many additional anole introductions have occurred since our 1993 review (summarized in Powell and Henderson [2008b]). Although in most cases too little time has passed to assess the outcome of these introductions, we can predict that anoles introduced to areas lacking ecologically similar species will prosper, such as *A. sagrei* on Grenada, St. Vincent, and the Grenadines (Greene et al., 2002; Henderson and Powell, 2005; Treglia et al., 2008),[285] and *A. equestris* on Oahu (Lazell and McKeown, 1998). On the other hand, several arboreal species recently introduced to Florida—*A. chlorocyanus*, *A. extremus*, and *A. ferreus*—are not thriving (Meshaka et al., 2004), as might be expected

284. As they do when introduced to islands lacking anoles entirely. For example, *A. carolinensis* and *A. sagrei* introduced to previously anole-free islands in the Pacific generally are doing quite well (Hasegawa et al., 1988; Rodda et al., 1991; McKeown, 1996; Suzuki and Nagoshi, 1999; Norval et al., 2002), the one exception being the decline of *A. carolinensis* on Guam, which probably results from predation by the brown tree snake (Fritts and Rodda, 1998). Similarly, *A. grahami* spread rapidly after being introduced to then anole-free Bermuda (Wingate, 1965; Losos, 1996a).

285. Although Simmons et al. (2005) suggested that predation by the terrestrial lizard *Ameiva ameiva* (see Chapter 8) may limit *A. sagrei*'s abundance and range expansion.

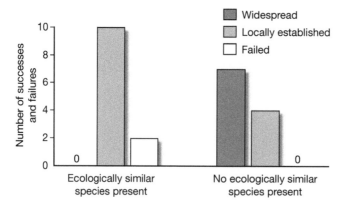

FIGURE 11.6

The success of introduced *Anolis* species as a function of their ecological similarity to resident anole species. Residents include both native species and previously introduced species that have become well established. Data from Losos et al. (1993a).

from the presence of ecologically similar species (*A. carolinensis* is similar to the two smaller species and *A. equestris* and *A. garmani* are similar to the larger *A. ferreus*). For the same reason, *A. carolinensis* on Anguilla may not fare well in the face of the native *A. gingivinus* (Eaton et al., 2001).

ECOLOGICAL INTERACTIONS LEAD TO SHIFTS IN RESOURCE USE

The previous section summarized the evidence that sympatric anoles interact ecologically. In this section, I review the data that indicate that, as a result of such interactions, sympatric species alter their behavior and ecology to minimize the extent of overlap in resource use.

HABITAT SHIFTS

A wide variety of studies demonstrate that anoles alter their structural microhabitat use in the presence of other anole species. These alterations can take two, sometimes related, forms: populations can shift their mean habitat use or they can contract the range of habitats used. Habitat shifts and changes in the breadth of habitat use can be related if expansion or contraction of habitat use is asymmetric (e.g., increased or decreased use of habitat on only one side of the mean), but populations can also shift their mean habitat use while maintaining the same degree of variation around the mean. This section will discuss shifts in habitat use, and the next will review changes in habitat niche breadth.

Behavioral studies reveal that individuals alter their habitat use in the presence of other species. The Jamaican trunk-crown *A. opalinus*, for example, moved higher in the tree in the presence of the larger trunk-ground *A. lineatopus* (Jenssen, 1973). More generally, although most species are not interspecifically territorial, agonistic interactions can occur when lizards of different species encounter each other: usually the smaller individual retreats (e.g., Rand, 1967b; Jenssen, 1973; Jenssen et al., 1984; Talbot, 1979; Losos, 1990b; Brown and Echternacht, 1991).

COMPARISONS AMONG POPULATIONS

So-called "natural experiments" (Diamond, 1986), comparisons among populations in the presence or absence of a second species, also provide evidence of habitat shifts. For example, *A. cooki* occupied lower and thinner perches on scrubbier vegetation in the presence of *A. cristatellus* in comparison to its microhabitat use at an allopatric site (Jenssen et al., 1984). Similarly, among offshore islets near Antigua, *A. wattsi* occurred higher in the vegetation and used trees more often on an island lacking the more arboreal *A. leachii* than on islands on which *A. leachii* was present (Kolbe et al., 2008a).

Far and away the most exhaustive comparative study of habitat use is Schoener's (1975) comparison of the perch height of populations of four widespread Caribbean taxa—*A. distichus*, *A. sagrei* and members of the *A. carolinensis* and *A. grahami* Species Groups—across 20 islands. By thoroughly measuring vegetation availability at each site, Schoener was able to account for the effect of habitat differences among sites and to demonstrate that perch height is frequently altered by the presence of other species. Habitat shifts were more frequent when the other species was similar in climatic microhabitat and effects were greatest between lizards that were similar in size; for a given size difference, the effect of presence of another species was greater when the other species was larger.

EXPERIMENTAL MANIPULATIONS

As mentioned above, experimental studies frequently reveal shifts in structural microhabitat in the presence of other species (Heatwole, 1977; Salzburg, 1984; Pacala and Roughgarden, 1985; Losos and Spiller, 1999; Campbell, 2000). Not surprisingly, these shifts serve to move individuals of one species away from members of the other species.

Shifts in other resource axes such as diet, thermal microhabitat or activity time have been examined much less often than shifts in structural microhabitat. Nonetheless, where looked for, these shifts are often found. For example, *A. gingivinus* altered its diet in the presence of *A. pogus* (Fig. 11.7; Pacala and Roughgarden, 1985) and *A. schwartzi* shifted its activity time in the presence of *A. bimaculatus* (Rummel and Roughgarden, 1985). Both experimental and comparative studies provide evidence of thermal microhabitat shifts (Salzburg, 1984; Rummel and Roughgarden, 1985; Kolbe et al., 2008a).

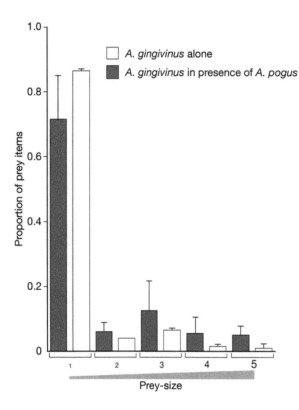

FIGURE 11.7
Effect of the presence of
A. pogus on arthropod prey size
of *A. gingivinus* on St. Martin.
Values are proportions of prey
items in five size classes,
from smallest to largest (means
±1 s.e. for two enclosures of
each treatment). Data from
Pacala and Roughgarden (1985).

NICHE BREADTH AND ECOLOGICAL RELEASE

In addition to shifts in resource use, species often use a narrower range of resources when in the presence of other species (reviewed in Schluter, 2000). Conversely, for the opposite reason, populations may expand their resource use in the absence of other species, a phenomenon termed "ecological release" (Schoener, 1986b).

Ecological release and its converse can clearly be seen in the *carolinensis* Species Group. On Cuba, and wherever else they occur with other species, members of this group are highly arboreal, usually seen on tree trunks, branches, and vegetation from eye level to high in the canopy. By contrast, when members of the group are the only anoles at a locality—e.g., *A. longiceps* on Navassa, *A. brunneus* on Acklins in the Bahamas, *A. carolinensis* in the southeastern U.S.—they tend to occur throughout the habitat, from the leaf litter to the canopy (Schoener, 1975; Jenssen et al., 1995; Powell, 1999; Campbell, 2000). Not surprisingly, the introduction of *A. sagrei* to Florida has forced *A. carolinensis* back into the trees, reversing the ecological release and returning the green anole to its ancestral habitat niche (Campbell, 2000).

The most comprehensive analysis of anole niche breadth was conducted by Lister (1976a), who found that the diversity of structural microhabitat use of *A. sagrei* was negatively related to the number of co-occurring species (Fig. 11.8). In two other comparisons of pairs of related species, Lister (1976a) also found that the allopatric species had

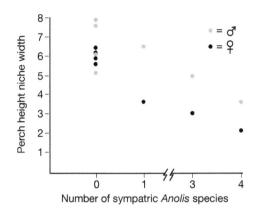

FIGURE 11.8

Relationship between number of sympatric species and habitat niche breadth in *A. sagrei*. Niche breadth was calculated using Simpson's Diversity Index (a measure of variation in frequencies among categories) on perch height data. Modified with permission from Lister (1976a).

a greater niche breadth than its close relative that occurs in sympatry with other species.[286] Some (Rand and Rand, 1967; Laska, 1970; Ruibal and Philibosian, 1974b), but not all (Hertz, 1983; Losos and de Queiroz, 1997) other solitary anoles also exhibit broad structural microhabitat use.

Lister (1976a) also showed that allopatric *A. sagrei* populations had greater variance in body temperature relative to populations in sympatry with other species, a result that holds for interspecific comparisons as well (Lister, 1976a, Hertz, 1983). Hertz (1983) suggested that the high variance of allopatric taxa stems from their behavioral flexibility: in open habitats, they thermoregulate carefully, whereas in closed habitats where thermoregulation becomes more costly (Chapter 10), they become thermoconformers. Hertz (1983) further argued that this thermal versatility is a characteristic of colonizing species (Williams, 1969), and hence the large thermal niche breadth of anoles on one-species islands may be an ancestral trait, rather than a derived response resulting from the lack of congeners on those islands.

In contrast to the trends for structural and thermal niches, dietary niche breadth, measured in terms of prey size, is not particularly large in allopatric anoles (Roughgarden, 1974; Lister, 1976a; Schoener, 1977).

SUMMARY: SYMPATRIC ANOLES INTERACT STRONGLY

Taken as a whole, these studies provide strong evidence that sympatric anoles interact ecologically—almost every study that has looked for evidence of such interactions has found it. Although habitat shifts are the most frequently detected effect, this trend may be more indicative of the types of data that have been collected than of a real distinction. The few studies that have looked for other effects, such as shifts in thermal microhabitat or food type, have also found supporting evidence.

286. Although Gorman and Stamm (1975) and Losos and de Queiroz (1997) did not find evidence for ecological release in structural microhabitat in these two comparisons.

The only study that found little evidence for interspecific effects is the comparison between *A. bimaculatus* and *A. schwartzi* on St. Eustatius. These species differ vastly in size and structural microhabitat use, so the lack of ecological interaction is not surprising.[287] Most studies in the Greater Antilles have focused primarily on pairs of species that overlap much more in size and structural microhabitat than these two species and have found evidence for interspecific interactions. However, examination of ecomorphs that occur in different parts of the habitat (e.g., in the canopy versus in the grass) and differ in size—e.g., crown-giant versus grass-bush anoles—might also detect non-interaction among pairs of species. Nonetheless, I would predict that these same species would interact with other sympatric species (e.g., grass-bush with trunk-ground anoles; crown-giant with trunk-crown anoles). Although Greater Antillean anoles probably don't interact with all other species in their community, almost all anoles probably interact with some of their sympatric congeners (e.g., Schoener, 1975).

HOW DO ANOLES INTERACT?

The traditional view has been that anole interactions take the form of interspecific competition. Nonetheless, over the course of the last two decades, ecologists have increasingly recognized that other types of interactions—such as predation, parasitism and mutualism—can be as or more important than interspecific competition in structuring communities. Moreover, in many cases, patterns that have been attributed to competition—e.g., decreases in population size or habitat shifts in the presence of another species—can result from other processes.

PREDATION

Predation, in particular, has been put forward as an alternative explanation for many patterns attributed to interspecific competition. One predation-based explanation for community structuring is intraguild predation, the idea that competing species may also prey upon each other. Such predation could explain why a species has lower population sizes and shifts its habitat use in the presence of a second species (Polis et al., 1989; Arim and Marquet, 2004).

Certainly, intra-guild predation occurs among anoles. Most species of anoles will eat other vertebrates when given a chance (Chapter 8), and predation by one anole on another is widespread (Gerber, 1999). In most cases, this intra-guild predation takes the form of adults of one species eating juveniles of another (Fig. 11.9). However, larger species, such as crown-giants and large species in the Lesser Antilles, prey on adults of

---

287. Although Kolbe et al.'s (2008a) comparative study of the closely related *A. leachii* and *A. wattsi*, which differ to an even greater extent in size, did find evidence for interspecific interactions.

FIGURE 11.9

Female *A. cristatellus* eating a juvenile *A. krugi* in Puerto Rico. Photo courtesy of Alejandro Sanchez.

many species, and adults of smaller species are vulnerable to predation by many other species (e.g., Fitch and Henderson, 1987; see review in Gerber [1999]).

Despite its widespread occurrence, the community effects of intra-guild predation have not been studied in anoles. Nonetheless, intra-guild predation is unlikely to be a general explanation for the pervasive evidence of anole interactions discussed in this chapter. The reason is simple: intra-guild predation is only possible between individuals that are substantially different in size—Naganuma and Rougharden (1990) estimated that an anole is capable of eating another anole up to 1/3 of its body length.[288] Yet, a wide variety of comparative and experimental data indicate that the strength of ecological interactions decreases with dissimilarity in size (e.g., Pacala and Roughgarden, 1985): interactions are strongest among species and age/sex classes most similar in size (e.g., Schoener, 1975). Nonetheless, to the extent that smaller species (or size classes of a species) are affected by larger ones (e.g., Kolbe et al., 2008a), the cause may be intraguild predation instead of, or in combination with, interspecific competition.[289]

Although intraguild predation is probably not a general explanation for anole interactions, predation in theory could be responsible in another way. The "apparent competition" or "competition for enemy-free space" hypothesis asserts that changes in species

288. This fraction is probably a bit on the low side. Although precise data are rare, perusal of Gerber's (1999) review—as well as Fig. 8.10—suggests that 1/2 to 2/3 may be closer to the mark (e.g., Fitch [1975] on an *A. capito* eating an adult *A. polylepis* and Fitch and Henderson [1987]; see also Gerber and Echternacht [2000]). Nonetheless, the general point remains true: anoles can only eat prey items of substantially smaller size.

289. Effects of intraguild predation might have been predicted to be especially strong in Rummel and Roughgarden's (1985) study of the effect of *A. bimaculatus* on *A. schwartzi*, given the size difference between the species. However, although *A. schwartzi* exhibited habitat and activity shifts in the presence of the *A. bimaculatus*, no direct evidence of predation was detected, either in the form of stomach contents or in differences in its population sizes in cages with or without *A. bimaculatus*.

abundance and habitat shifts in the presence of a second species may result from both species sharing a common predator (Holt, 1977, 1984; Holt and Lawton, 1994; see review in Vamosi [2005]). The abundance effect occurs because the second species may attract and sustain predators, which then feed on the first species as well. In turn, habitat shifts may result as the first species moves away from the parts of the habitat occupied by the second species, and thus away from predators of the second species. In contrast to intraguild predation, and mirroring the predictions of interspecific competition, we might expect that the more similar in size two species are, the more likely they would be to be affected by the same predator.

For this reason, we cannot dismiss the shared predator hypothesis as readily as the intraguild predation hypothesis. Unfortunately very little is known about the role of predation in anole communities, as discussed in Chapter 8. One prediction the shared predator hypothesis might make is that habitat divergence should lead to species being preyed on by different predators. This is not a particularly strong test, because even if habitat divergence is driven by some other process such as competition, species in different microhabitats might be subject to predation by predators that differ in microhabitat use. In any case, the hypothesis is difficult to test, because many studies that report the diet of anole predators do not identify the anoles eaten to species (e.g., Wetmore, 1916). However, some species level data on snake diets is available, and these data do not provide evidence for species-specific predation. In Hispaniola, the racer *Uromacer frenatus* eats at least four different ecomorphs, including both more terrestrial and more arboreal ecomorph types; different sets of three of these ecomorph classes are eaten by several other West Indian snake species (Franz and Gicca, 1982; Henderson et al., 1987; Rodríguez-Robles and Leal, 1993; Wiley, 2003). Similarly, the blunt-headed vinesnake (*Imantodes cenchoa*) in Brazil eats at least three anole species that differ in habitat use (Martins and Oliveira, 1998). The frog *Eleutherodactylus coqui* also eats both arboreal and terrestrial anole species in Puerto Rico (Leal and Thomas, 1992; Reagan et al., 1996).

PARASITISM

The effects of parasitism on communities are conceptually similar to those of predation. Indeed, parasitism in a sense is just a type of predation in which the predator is much smaller and its effects are sometimes only debilitating, rather than lethal. As they have with predation, community ecologists increasingly are recognizing that parasitism can play an important role in structuring communities (Settle and Wilson, 1990; Grosholz, 1992; Hatcher et al., 2006).

The best and almost only study of the community effects of parasitism on anoles concerns malaria and the two species of anoles on St. Martin (Schall, 1992), which is the only Lesser Antillean island on which a small and a medium-sized species coexist (see Chapters 4 and 7). The smaller species, *A. pogus*, has an unusually patchy and restricted range on the island as compared to its close relatives on other islands. *Anolis gingivinus* is more vulnerable to malaria than *A. pogus*, and *A. pogus* is only found in places in which malaria has been found, whereas *A. gingivinus* occurs throughout the island. Moreover, the abundance of *A. pogus* correlated positively with the prevalence of infection in *A. gingivinus*. Schall's (1992) conclusion based on these data is that in the absence of malaria, *A. gingivinus* excludes A. *pogus*, but that the malaria parasite's presence alters the competitive balance enough to permit coexistence. This hypothesis may be tested quite soon, because recent surveys have failed to find the malaria parasite on St. Martin, suggesting that it may have disappeared (Perkins, 2001). If the parasite-mediated coexistence theory is correct, then *A. pogus* populations ought to be in decline.

What role, if any, malaria plays in structuring anole communities elsewhere is unclear. On other islands in the West Indies, malaria occurs in some anole species, but not others: in one of five species in Puerto Rico (Schall and Vogt, 1993)[290] and in 12 of 22 populations of 12 species in the Lesser Antilles and Virgin Islands (Staats and Schaal, 1996). In addition, the physiological consequences of infection are variable, with little detectable effect in two of the three species studied to date (Chapter 8).

Our knowledge of the importance of other types of parasite is similarly scant. The presence of macroparasites has been reported for many anole species (Chapter 8), but the effect that they have on individual lizard fitness, much less on population or community biology, is unknown (Chapter 8).

INTERSPECIFIC COMPETITION

Despite uncertainties about the prevalence and importance of predation and parasitism, the data as a whole make a strong case that interspecific competition is a major, probably the predominant, force structuring anole communities, at least in the West Indies. Four lines of evidence, discussed in the last four chapters, bolster this claim.

1. West Indian anoles, as a generality, appear to be resource limited. Anoles can have a large affect on prey populations (Chapter 8), and supplementation experiments indicate that anoles will respond readily when food is provided. Such resource limitation can lead to population effects and shifts in resource use in the presence of competing species.

2. Interspecific aggression occurs between some anole species, particularly those that are ecologically similar.

3. Anoles respond immediately to the presence of potential competitors by shifting microhabitat use.

4. Effects of the presence of other anole species include decreases in body condition, feeding rate and egg production.

---

290. The other four species were infected at levels of less than 1%.

All of these observations are consistent with—indeed, predicted by—interspecific competition. Of course, one could imagine scenarios in which these observations also conform to an apparent competition or parasitism hypothesis. For example, microhabitat shifts in behavioral time could be adaptive as a means by which a lizard could immediately remove itself from the predators that might be attracted to a lizard of a second species; decreased feeding and reproduction might be a consequence of greater inactivity caused by the presence of predators attracted by a second species, and so on. These explanations, however, are less parsimonious than the more straightforward explanation that anoles compete with each other, producing the effects documented in this chapter. Nonetheless, parsimonious explanations are not necessarily correct; what is urgently needed are detailed studies examining the role that predators play in structuring anole communities.

More generally, competition, predation and parasitism are not mutually exclusive processes. All may occur within a community, and their effects may be synergistic (reviewed in Chase et al., 2002). For example, predators may alter competitive interactions in almost any way imaginable, at least in theory (reviewed in Chase et al., 2002). The presence of predators may permit co-occurrence of, and thus interaction between, species that otherwise could not coexist by reducing the density of the competitively dominant species, or they may prevent coexistence by forcing prey to overlap more in resource use. In theory, one example might involve *A. sagrei*, *A. smaragdinus*, and *Leiocephalus carinatus*. On small islands in the Central Bahamas, *A. sagrei* has a large, negative and mostly unreciprocated effect on *A. smaragdinus*. This result is not surprising because the habitat on these scrubby islands seems better suited for the more terrestrial *A. sagrei* (Losos and Spiller, 1999). On similar islands in the northern Bahamas, the terrestrial, lizard-eating *L. carinatus* not only has a large effect on *A. sagrei* density, but forces these anoles up into the vegetation, where they often must use narrow vegetation to which they seem poorly adapted (Losos et al., 2004, 2006; see Chapters 8 and 12). We can only speculate that in an experiment involving all three species on similar islands, the interaction between *A. sagrei* and *A. smaragdinus* might be very different: *A. sagrei*, forced into *A. smaragdinus*'s element, conceivably could end up on the short end of the competitive stick. Similarly, in the Greater Antilles, the presence of crown-giants may force other anole species away from tree trunks either toward the ground or the periphery of trees, potentially altering the interactions that occur among them.

Assuming that anole species compete, the obvious next question is: for what limiting resources are they competing? One obvious answer is food. Coexisting species differ in body size, structural microhabitat, and thermal microhabitat. Differences in body size among sympatric species clearly have the effect of partitioning prey resources by size: larger species generally eat larger prey. Moreover, food supplementation experiments support the supposition that food is often a limiting resource, and other experiments show that the presence of another species decreases body condition and feeding rate. Food clearly is an axis of resource competition among anoles.

By contrast, the resources affected by divergence in microhabitat are less clear. By partitioning these microhabitats, anole species occupy different physical parts of the habitat and thus are partitioning space. Some workers contend that space is a limited resource and that anoles minimize competition for space by partitioning thermal or structural microhabitats (Roughgarden et al., 1981). Certainly, space can be a limiting resource for organisms such as bivalves or trees. But lizards are not large enough that their physical presence prevents another lizard from being in more or less the same place, especially given that most anole species are not interspecifically territorial (Chapter 9); in other words, lizards do not consume space. In theory, partitioning the use of space could be a means of partitioning a resource in short supply, perhaps basking or egg-laying spots, or refuges into which a lizard could hide from predators. But nothing in the behavior or ecology of anoles suggests a focus on any of these possibilities as an important and potentially limited resource.

More likely, by partitioning space, anoles are partitioning prey. Many arthropods and other anole prey species have their own habitat requirements, and thus are not found uniformly throughout the environment (reviewed in Brown et al., 1997). Thus, by using different parts of the habitat, anoles are likely to be partitioning food resources: put simply, arboreal lizards eat arboreal prey, and more terrestrial lizards eat more terrestrial prey (Schoener, 1968; Campbell, 2000).

Admittedly, few relevant data are available. Sympatric species using different microhabitats usually exhibit dietary differences even when they are similar in body size (e.g., Cast et al., 2000; Sifers et al., 2001), and competitor-induced shifts in structural microhabitat are associated with shifts in diet (Pacala and Roughgarden, 1985). One experimental study has attempted to disentangle the effects of body size and spatial partitioning. Rummel and Roughgarden (1985) prevented *A. bimaculatus* from using perches higher than 1 m above the ground in two experimental enclosures, using two others as controls; the smaller and more terrestrial *A. schwartzi* was present in all enclosures. In the "lowered perch" treatment, *A. bimaculatus* continued to take larger prey than *A. schwartzi*, but its feeding rate was considerably diminished; surprisingly, *A schwartzi* was little affected by the greater microhabitat overlap with its larger cousin. Unfortunately, the prey eaten by the species were not identified, so we don't know whether the increased spatial overlap led to increased overlap in prey type. The lower feeding rate of *A. bimaculatus* is consistent with the hypothesis that spatial partitioning provides anole species with access to different food resources, although other explanations are always possible: for example, in theory shortage of suitable basking or escape sites resulting from lowered perch heights and increased spatial overlap could have resulted in less time available for foraging.

Overall, the most likely explanation is that resource partitioning in anoles occurs primarily to minimize competition for food. Nonetheless, more directed investigation of this hypothesis would be useful.

Chapter 8 presented the hypothesis that differences between mainland and West Indian anoles occur because the former are regulated by predation, whereas the latter are most affected by food limitation and interspecific competition. If this hypothesis is correct, we might expect that mainland and West Indian communities would be structured differently. Admittedly, the data for mainland communities is not extensive, but the information we have reveals the same general pattern in both areas: in particular, resource partitioning is the rule, with very few instances of sympatric species occupying the same niche.

This seeming contradiction could be explained in three ways.

1. More data, particularly for mainland anoles, may reveal undiscovered differences in how communities are structured between the two areas. The lack of niche complementarity in some, but not all, mainland communities is suggestive in this regard.

2. The premise is wrong: possibly the factors affecting communities in the two areas are the same, and the differences in life history, population biology and other aspects of anole biology detailed in Chapter 8 result from some other cause.

3. Different processes lead to the same end result: as discussed above, both competition and predation could lead to resource partitioning.

A recurring theme in this book is that more data are needed, especially concerning mainland anoles. Nowhere is this more evident than in the study of the factors affecting community structure and population biology.

ECOLOGICAL INTERACTIONS AND ADAPTIVE RADIATION

The model of adaptive radiation posits that interactions occur between entities—different species, or in sympatric speciation models, subpopulations—that initially are ecologically similar. These interactions drive divergence and, ultimately, adaptive radiation. This scenario is rarely directly tested in adaptive radiations for the simple reason that ecologically similar species almost never co-occur. We have seen that this is the case for anoles, and it is often true in many other adaptive radiations as well. As a result, most studies of interspecific interactions are between species that are ecologically different. For example, many of the studies in the Greater Antilles examine interactions between anole species that are members of different ecomorph classes, and studies in the Lesser Antilles have no choice but to examine ecologically differentiated species.[291] The fact

291. Even *A. gingivinus* and *A. pogus* on St. Martin are ecologically different, if less so than on any other two-species island in the Lesser Antilles.

that these species almost always have negative effects on each other suggests that ecological divergence has not been great enough to eliminate ecological interactions. Sensu Connell (1980), the ghost of competition past is still alive and kicking! By extrapolation, this finding suggests that when ecologically more similar species come into contact, as might happen at a contact zone between closely related species or in the putative first stage in an adaptive radiation, the two species probably would interact particularly strongly.

The data reviewed in this chapter thus support the first two tenets of the adaptive radiation hypothesis outlined at the beginning of the chapter. In the next chapter, I will review the evidence that habitat shifts resulting from interspecific interactions lead to evolutionary divergence and adaptation.

Before going there, however, I would like to address one additional topic. The conclusion of this chapter is that interspecific competition is the primary force responsible for ecological interactions among anole species. More generally, Schluter (2000) commented that previous studies of adaptive radiation have generally assumed that it is interspecific competition that drives resource divergence and subsequent evolutionary diversification. However, as noted in this chapter, processes other than competition can also lead to resource partitioning, and thus in theory could be instrumental in driving adaptive radiation. Schluter (2000) further noted that members of some adaptive radiations, such as African lake cichlids, occupy multiple trophic levels, which indicates not only that predation is an ongoing interaction among members of the radiation, but that in the process of evolutionary diversification, species that were initially ecologically similar diverged to the extent that some now prey upon others. How such divergence occurs and the role that predation may play in driving adaptive radiation are unanswered—indeed, mostly unaddressed—questions (Vamosi, 2005).

Anoles may be just the group for taking up this line of inquiry. Anole radiations have in all cases (Greater Antilles, Lesser Antilles, mainland) led to the coexistence of species differing in size, and thus in the capacity for predation by larger species on smaller species. Gerber's (1999) survey shows that such predation occurs in many places. Although there isn't a lot of evidence that predation is a significant factor in communities today, that by no means rules out the possibility that predation drove the divergence of species to use different microhabitats and resources, so much so that predation is no longer as potent an ecological force today. In other words, it is possible that the ecological diversity of forms we see in anole communities today represents (with apologies to Connell) "the ghost of predation past."

FUTURE DIRECTIONS

Much still remains to be learned about anole community ecology. More studies of interactions, both observational and experimental, need to be conducted. The relative merits of experimental versus observational studies have been extensively debated, and the consensus is clear: each approach has its advantages, and the best framework is one in which extensive comparative work is bolstered by smaller scale experimental studies (Case and Diamond, 1986). Work on anoles is an excellent example: a voluminous body of non-experimental work strongly indicates the occurrence of interspecific interactions, the existence of which is readily apparent in experimental manipulations. Nonetheless, given the suitability of anoles for experimental studies—great abundance, ease of observation and data collection, ability to conduct manipulations—it is surprising that more work of this sort has not been conducted.

Two types of experimental studies would be particularly useful. First is a study of the role of predation in structuring anole communities. Our experimental work with *Leiocephalus* in the Bahamas (Schoener et al., 2002, 2005; Losos et al., 2004) suggests that predation can be a very important factor affecting the ecology of populations, as do observational studies on *Ameiva* elsewhere (Simmons et al., 2005; Kolbe et al., 2008a). Whether predation can alter interspecific interactions among anoles remains to be studied. The Cuban crown-giant, *A. equestris*, is now abundant in parts of Florida and would seem to provide an excellent opportunity to explore this question.

The second area for future study concerns the ecology of mainland anole communities. No community level experiments have been performed, and these could be quite instructive concerning the relative importance of interspecific competition and predation.

An additional question that might prove interesting concerns why some ecomorph types are absent from some Greater Antillean communities. Trunk-ground and trunk-crown anoles are generally present almost everywhere, except in some high elevation localities, but other ecomorph types are sometimes absent from some localities (Moermond, 1979a; Schoener and Schoener, 1980a). Whether this is the result of biotic interactions, either between anole species or between anoles and other taxa, or whether it is a function of habitat, climate, or some other factor is unknown and these questions have received little study (Haefner, 1988; Schoener, 1988).

Finally, introduced species provide unparalleled opportunities to study ecological interactions and their evolutionary effects. As a result of introductions, ecologically similar species with no prior experience of evolutionary interaction have been brought together in many places. Opportunities are particularly rich in Florida, where communities of as many as five species representing four ecomorph types can be found in Miami; overall, four different pairs of species in the same ecomorph class from different islands occur in sympatry somewhere in Florida (Wilson and Porras, 1983; Meshaka et al., 2004). If ever there were a place to test the idea that interspecific competition should lead to shifts in resource use and subsequent evolutionary change, this is it!

12

NATURAL SELECTION AND MICROEVOLUTION

The anole radiation is characterized by divergence of closely related species into different ecological niches, producing communities composed of ecologically differentiated species. The theory of adaptive radiation presented in the last chapter posits that this diversity is the evolutionary result of ecological interactions between initially similar species.

Three predictions stem from this theory. Two of these predictions—that sympatric anole species interact and that these interactions lead to shifts in resource use—have been amply documented, as the last chapter attests. The third prediction—that resource shifts lead to evolutionary adaptation—is much more difficult to investigate.

Until recently, the standard wisdom was that evolution generally proceeds too slowly to be observable over the course of a human lifetime, much less that of a scientific study (Gould, 2002). Thus, the idea seemed far-fetched that scientists could actually document evolutionary change occurring over the course of a few years. Recent years, however, have shown that when natural selection is strong and directional, evolutionary response can be rapid and large (Hendry and Kinnison, 2001).[292] This accords with

292. It is amazing how prescient Charles Darwin was. Based on logic, analogy, and a scant amount of data, Darwin correctly deduced much of the basic outline of evolutionary biology. Moreover, many of the detailed observations he made a century and a half ago are still pertinent today. For this reason, more than a century later, Darwin's writings are still widely cited in the primary literature, a phenomenon that rarely occurs in most other fields of science.

However, Darwin did not get everything right, and the field has sometimes been led astray by his influence. One such point is the rapidity with which evolution by natural selection occurs. Darwin argued that such change

laboratory research on *Drosophila* and other organisms that demonstrates that selection on almost any sort of trait—e.g., morphological, physiological or behavioral—results in rapid and detectable evolutionary response (Charlesworth et al., 1982).

Evolutionary change in natural populations has been documented in a number of ways. Some studies measure selection within a generation and the resulting evolutionary response in the next generation; in a few cases, such studies have been continued over many generations, such as the elegant long-term studies on Galápagos finches (Grant and Grant, 2002, 2006a, 2008). In most studies, however, evolutionary change is documented by comparison of population attributes before and after some environmental change (Hendry and Kinnison, 2001). Many of these studies are opportunistic in the sense that scientists take advantage of natural or human-caused environmental change to investigate how species respond evolutionarily. Alternatively, in some cases researchers have taken an experimental approach by directly manipulating the environment to examine how a population responds. Certainly the best known and most thorough experimental study of evolutionary change in natural populations is work on Trinidadian guppies (*Poecilia reticula*) which has shown that the experimental addition of predatory fish leads to a wide variety of evolutionary responses in guppy populations, including changes in escape ability, coloration, body size, age at maturity, and reproductive rates (Endler, 1980; O'Steen et al., 2002; Reznick and Ghalambor, 2005). This work is particularly elegant because the experimental and control conditions are replicated several times, allowing for assessment of the statistical significance of differences between experimental and control populations.

To date, no study on anoles has documented that resource shifts lead to changes in patterns of natural selection with subsequent evolutionary change. Nonetheless, the individual pieces have each been documented separately, making a cumulatively strong case in support of this prediction of the adaptive radiation hypothesis. More importantly, the situation is changing rapidly as increasing research is focusing on microevolutionary aspects of anole adaptation.

In this chapter, I will first review evidence regarding natural selection and microevolutionary change in anole populations. I will then increase the spatial scale of my examination to consider what patterns of geographic variation may reveal about anole microevolution. Finally, I will conclude with a discussion of the heritability of anole traits and the significance and evolutionary role of phenotypic plasticity in anole evolution.

---

was so gradual as to be almost imperceptible. Only over eons would such change accumulate to a detectable level. This view was in accord with Lyellian ideas about the pace of geological change, to which Darwin wholeheartedly subscribed (Browne, 1995, 2002). Gould (1984) has argued that these views were based not so much on evidence—as there was very little—as on a Victorian worldview of the pace of change. For more than a century after the publication of *On the Origin of Species*, the idea that evolution generally proceeds slowly was the standard view in the field. However, recent work has conclusively showed that this viewpoint is not correct: given strong selection, evolutionary change can occur quite rapidly.

NATURAL SELECTION

At first glance, it might seem surprising that anoles were virtually unrepresented during the boom in studies of natural selection that occurred in the last quarter of the 20th century (Hoekstra et al., 2001; Kingsolver et al., 2001). In many respects, anoles would seem to be a perfect organism for studies of selection: abundant, easy to capture and measure, many well-understood potential targets of selection, and an interesting evolutionary ecological context in which such studies could be placed.

But there's a big catch: the most powerful way to study natural selection is to measure the survival and reproductive success of individual organisms (Endler, 1986), and that requires a means of identifying individuals. Unfortunately, this is harder than it seems: because lizards shed their skin, superficial marks are short lived, and other marking methods are problematic. Only recently has this problem been solved, allowing researchers to follow the fates of individual lizards (for details on previous and new methods, see Appendix 12.1). The result is an awakening of studies of natural selection that is just beginning.

SELECTION IN NATURAL POPULATIONS

Researchers have taken two approaches to measuring selection on anoles. The first is simply to document patterns of selection in natural populations.[293] For example, in two populations of *A. sagrei* on scrubby islands in the Bahamas, individuals with fewer labial scales (the scales around the mouth) had a slight survival advantage, whereas on a more lushly vegetated peninsula, no selection on scale number was detected (Calsbeek et al., 2007b; see Chapter 13 for a discussion of the adaptive basis of scale variation with regard to aridity).

In one of the few comparative studies of natural selection ever conducted, Irschick et al. (in prep.) measured selection on sprint speed and bite force, as judged by survival of marked lizards over a one-year span, in the four sympatric anoles of South Bimini, Bahamas (these and other measures of functional capabilities are discussed in Chapter 13). They found that in the fastest species, *A. sagrei*, selection favored the fastest individuals, whereas in the slowest species, *A. angusticeps*, selection was stabilizing. Directional selection favoring greater bite force was found in the same two species, which ranked first and third in bite force. By contrast, measures of selection were not significant for the other species for either measure, nor for any of the species for any measured aspect of morphology (for additional studies of selection on morphology and performance in

---

293. Selection has usually (except when noted below) been measured by calculating selection gradients, which are a way to measure the extent to which fitness is related to a phenotypic character; values of zero mean that fitness is not related to that trait, whereas high values indicate that individuals with positive values (e.g., the largest individuals) have highest fitness. Negative values imply the converse (Lande and Arnold, 1983; Brodie et al., 1995).

A. sagrei in the Bahamas, see Calsbeek and Irschick [2007], Calsbeek and Smith [2007], Calsbeek and Bonneaud [2008], and Calsbeek et al. [2008]).

Another means of studying selection is to compare trait distributions among age classes. If one assumes that each age cohort had the same trait distribution at the time of hatching, then changes in distribution among age classes may indicate survival differences (reviewed in Endler, 1986). Collette (1961) found that the number of toepad lamellae increases with age in both *A. porcatus* and *A. sagrei* and attributed this to selection favoring more lamellae as the lizards get older and shift to more arboreal habitats.[294] Similarly, juvenile male *A. sagrei* have fewer dorsal scales than adults from the same population, which suggests that selection through ontogeny favors individuals with more scales (Lister, 1976b).

EXPERIMENTAL STUDIES OF SELECTION

Several studies have experimentally tested the prediction of the adaptive radiation hypothesis that changing environmental conditions should alter the selection pressures operating on a population. The first experimental approach taken with anoles was to study ecotypic variation in *A. oculatus* of Dominica in the Lesser Antilles. As discussed in greater detail later in this chapter, populations of *A. oculatus* differ in a variety of body dimensions, scalation and color characters, and interpopulation variation is correlated with aspects of the environment. Malhotra and Thorpe (1991) and Thorpe et al. (2005b) raised individuals from different populations in replicate enclosures at one site. For each source locality, they estimated selection by comparing the morphology of survivors with that of non-survivors, with the prediction that the magnitude of selection on a population would be correlated with the extent to which the population's natural environment was different from the environment at the study site. This prediction was confirmed (Fig. 12.1), and selection was detected on a wide variety of characters, including body size, scalation, color, and limb length.

Several other studies have experimentally altered conditions on some, but not all, small islands and then compared selection gradients between experimental and control populations. Schoener, Spiller, and I conducted a replicated experiment in the Bahamas to examine whether the presence of the predatory curly-tailed lizard changed the form of selection on *A. sagrei* (Losos et al., 2004, 2006; see Chapter 8). Mortality was high on all experimental islands, but was correlated with the vegetation structure on control islands,

---

294. One oddity with Collette's data, however, is that in both sexes of both species, the range of variation in number of lamellae exhibited by the smallest individuals does not include the maximum number seen among all individuals. An increase in lamella number with body size has been observed in several other anoles (Lazell, 1966, 1972) and a gecko (Hecht, 1952). Hecht (1952), like Collete, attributed the pattern to selection favoring individuals with greater numbers of lamellae, but Lazell (1966, 1972) proposed that the number of lamellae may actually increase ontogenetically, which would explain the lack of the maximum lamella numbers among the smaller individuals. The only relevant data on whether lamella number is fixed at birth or increases with age is Hecht's (1952) statement that lamella number did not increase after each shedding cycle in a Jamaican gecko.

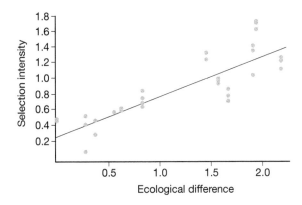

Intensity of selection on experimentally translocated populations as a function of ecological difference between the source area and the experimental site. As predicted, the greater the difference (the most extreme being lizards moved from a mesic, montane locality to the xeric, lowland study site), the stronger the selection. Modified with permission from Thorpe et al. (2005).

being low on islands covered with scrubby bushes and high on well-vegetated islands (Schoener et al., 2005).

We predicted that selection would favor larger lizards because they should be faster, and thus better able to elude predators (see Chapter 13) and also more difficult to subdue and ingest (Fig. 8.3; Leal and Rodríguez-Robles, 1995). In addition, we predicted that lizards with legs that were long relative to their body size should also have greater survival due to their increased speed (Chapter 13; we did not measure limb length on females because this measurement is difficult to take accurately on gravid females).

To test these hypotheses, we measured selection gradients on populations on each island.[295] Six months after the introduction of the curly-tails, we found significant differences in the form of selection on both males and females. All selection gradients on body mass in females were positive, indicating that larger females were favored on these islands; by contrast, on most control islands, selection gradients were negative (Fig. 12.2a). These results were in accord with our a priori prediction.

Selection results were more complicated for males. First, no difference in selection gradients for body size were found between experimental and control islands, perhaps

295. Our study differed from most investigations of selection in natural populations. Most studies, though often elegant and detailed, involve only a single population and thus have no generality—conclusions only pertain to selection operating on individuals within that population. By contrast, in our study experimental and control treatments were each replicated six times and the selection gradients measured in each population were the data used in statistical analyses. Such replication of studies of selection is rare (e.g. Bolnick, 2004; Calsbeek and Smith, 2007) and allows general conclusions about how selection may differ in different circumstances. The downside to our approach was that our sample sizes per island were considerably smaller than those used in most selection studies; consequently, errors around the estimated selection gradients were likely to be large. Nonetheless, such errors should be random in direction and should reduce statistical power, but not lead to increased chance of Type I error (Harmon and Losos, 2005).

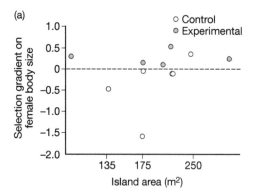

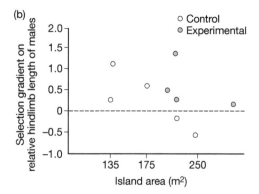

FIGURE 12.2

Selection gradients in relation to island area for (a) female body size and (b) male relative hindlimb length on islands with and without the predatory curly-tailed lizard, *Leiocephalus carinatus*. Selection gradients compare the phenotype of survivors versus non-survivors. Although curly-tailed lizards were introduced to six islands, with six others serving as controls, selection gradients could not be calculated on all 12 because all of the members of one sex either survived or perished on some islands, thus preventing comparison of survivors versus non-survivors. Modified with permission from Losos et al. (2004).

because the advantage of large size for escaping predators may be countered by increased vulnerability due to the conspicuous signaling and perching behavior of territorial males, which tend to be larger than non-territorial males (Chapter 9; Schoener and Schoener, 1982b). Second, across all islands, selection on hindlimb length was related to island area: the larger the island, the less relatively long hindlimbs are favored, regardless of the presence of predators (Fig. 12.2b). Why this trend occurs is unknown; an obvious correlate of island area is species diversity (Schoener and Schoener, 1983a,b), but why increased species richness should select for shorter legs is not clear.[296] Third, once the effect of island size is removed, a clear effect of predators was present: for islands of a given size, selection more strongly favors individuals with longer limbs on islands with predators than on islands without predators (Fig. 12.2b). Thus, though convoluted, our second prediction was also confirmed; the presence of predators was altering the direction of selection in the predicted manner.

However, at the same time that curly-tails were presumably catching the shorter-legged lizards, the anoles were becoming increasingly arboreal. The previous round of this experiment, wiped out by Hurricane Floyd in 1999, revealed that over the course of three years, *A. sagrei* progressively increased its perch height and decreased its perch diameter in the presence of curly-tailed lizards (Fig. 8.5; Schoener et al., 2002). From studies of ecomorph evolution, we know what happens when anoles start using narrow surfaces: they evolve shorter legs. So, we predicted that over time, selection would reverse

296. Although mortality rate in control islands is related to vegetation height of an island, vegetation height and island area are only moderately correlated ($r = 0.58$); some large islands are scrubby and some small ones thickly vegetated (Fig. 8.4). Mortality rate was not related to island area, nor selection gradients to vegetation height.

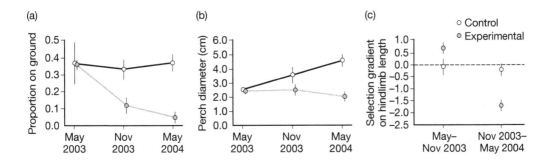

FIGURE 12.3

Shifts in habitat use and selection gradients in the predator introduction experiment. Selection gradients correspond to those in Fig. 12.1 with the effect of island area statistically removed. Modified with permission from Losos et al. (2006b).

direction and populations on islands with curly-tailed lizards would experience selection for shorter legs, relative to control islands (Losos et al., 2004).

To our surprise, this prediction was borne out much sooner than we expected. In the second six-month period of our study (November–May), not only did the difference in perch height increase between experimental and control islands, but selection did the expected 180° turn (Fig. 12.3; Losos et al., 2006).

This experiment thus directly confirms the predictions of adaptive radiation theory. Species interactions—in this case, predation—lead to habitat shifts and predictable changes in patterns of natural selection.

Calsbeek and Smith (2007) also conducted an experimental test of natural selection using Bahamian *A. sagrei*, but their study related more to social behavior than ecological morphology (for a more recent experimental study of selection more closely related to ecological morphology, see Calsbeek and Smith [2008]). In particular, they predicted that as lizard density increases, large size should be increasingly favored (a prediction that also has been made for other lizards [Harmon and Gibson, 2006]; see also Case [1978]). This prediction stems from the observation that larger males generally have an advantage in territorial interactions (Chapter 9); though few data exist, the same relationship plausibly exists for females. To test this prediction, Calsbeek and Smith experimentally changed lizard densities on small islands[297] and measured survival selection. They found that in populations with natural densities, mid-size individuals were favored. By contrast, as density was experimentally increased, selection increasingly favored larger size in both sexes. One population in which density was experimentally lessened displayed disruptive selection favoring both small and large individuals. Selection on hindlimb length was not detected on any island.

297. And on one island-like peninsula. Lizards in these experiments ranged in size from subadult to adult, and an attempt was made to match natural size distributions, though the introductions did have the effect of changing the variance in size from that seen in natural populations.

Although few studies of selection have been completed, general patterns are already emerging. Selection is often detectable in natural populations. Moreover, when the environment is altered, selection usually can be detected in the manner predicted based on prior studies of anole ecological morphology or population biology.

To date, all selection studies have focused on survival. Fitness, of course, is composed of three components: survival, mating success (sexual selection), and fecundity. For species in which sexual selection is strong, selection gradients for sexually selected traits may be substantially greater than those for traits related to other aspects of fitness (Hoekstra et al., 2001; Svensson et al., 2006). As discussed in Chapter 9, molecular approaches for assessing reproductive success in male anoles are just beginning, and are already indicating that estimates of reproductive success based on observed matings may be inaccurate. Assessments of fitness in females are difficult because of the fixed clutch size of one egg; fecundity differences, if they exist, must result from differences in the rate at which eggs are laid or from maternal effecs on the size or composition of the eggs, phenomena for which few data are available (Jenssen and Nunez, 1994). An alternative approach for measuring reproductive success of both sexes, genotyping wild offspring to determine reproductive success of both parents, has not yet been attempted, although some such studies are currently underway. The near future is likely to see many studies take these approaches, and thus provide a more complete picture of selection pressures affecting anole populations.

MICROEVOLUTIONARY CHANGE

Measurements of natural selection allow a test of the prediction that populations experiencing new environmental conditions will experience selective pressure to adapt to these conditions. Continued over generations, such studies can examine whether selection will lead to evolutionary change. But measuring natural selection in a single generation, much less in multiple generations, is difficult and time-consuming. An alternative approach is to examine whether a population presumed to be experiencing new selective pressures changes through time.

The traditional approach to detecting evolutionary change was paleontological, by measuring fossil populations and how they changed through strata. In the last few decades, with the advent of long-term field studies, scientists have documented change in extant populations (e.g., Kruuk et al., 2001; Grant and Grant, 2002, 2006a). Although this approach is usually correlational, examining whether evolution occurs subsequent to an environmental change, an experimental approach is possible when intentional introductions like those just described are conducted. Non-intentional introductions also present particularly good opportunities; such populations often face new environments and thus may be more likely to experience strong directional selection. In addition, introduced populations may exert selection pressure on native species. All of these approaches have been taken with anoles.

Subfossil *Anolis* remains are known from several islands in the Lesser Antilles. In at least six species, body size has decreased since the late Pleistocene (reviewed in Pregill, 1986; Roughgarden and Pacala, 1989; Losos, 1992b). For example, on Antigua, *A. leachii* attained an SVL more than 60% greater than its maximum today. Similarly, *A. sagrei* on New Providence was larger than any population of *A. sagrei* today, and more than 25% larger than any modern population in the Bahamas (Etheridge, 1965; Pregill, 1986). On the other hand, fossil data for other taxa failed to find evidence of increased size in fossils, including *A. wattsi* on Antigua, *A. wattsi* and *A. leachii* on Barbuda (Etheridge, 1964; Lazell, 1972), and *A. distichus* and *A. smaragdinus* on New Providence (Etheridge, 1965).

INTRODUCED POPULATIONS

The great number of anole introductions (Chapter 11) would seem to be a gold mine for the ambitious microevolutionist, but to date few prospectors have staked a claim. The lack of study of introduced anoles is particularly surprising because many introductions have occurred in areas convenient for study, such as the ten introduced anole species in the vicinity of Miami.[298]

The only documented cases of post-introduction change in non-experimentally introduced populations involve *A. leachii*, which was introduced from Antigua to Bermuda in 1940, and *A. sagrei*, which has been introduced to many islands as well as to the southeastern United States. In both species, individuals in introduced populations are larger than those from their natural range (Pregill, 1986; Campbell and Echternacht, 2003).[299]

EXPERIMENTAL INTRODUCTIONS

I have previously mentioned several experimental introductions of anoles to small islands in the Bahamas (Chapters 8 and 11). This line of research was initiated in the late 1970s by Tom and Amy Schoener. They started with the observation that anoles generally do not occur on very small (less than 10,000 m²) islands in the Bahamas. Wanting to learn something about the process of population extinction, a topic about which we are still too ignorant, the Schoeners introduced either five or ten *A. sagrei* to small islands in the vicinity of Staniel Cay, Bahamas.

298. Partly this is bad luck. In the last 20 years, several graduate students in Florida have begun doctoral work on such questions, but for various reasons, none of these studies was completed and published.

299. The cause of these size increases is uncertain, but we can always speculate. Perhaps relevant to the *A. leachii* introduction is the fact that *A. grahami*, which had already become widely established in Bermuda (Wingate, 1965), is substantially larger than *A. wattsi*, the species with which *A. leachii* co-occurs on Antigua. In theory, Bermudian *A. leachii* may have been forced to evolve larger size to minimize resource competition with *A. grahami*. A lack of sympatric anoles, particularly larger ones, may also have played a role in the size increases of *A. sagrei* (discussed in Campbell, 2003). Of course, other explanations for size differences are possible, as discussed later in this chapter. Data are needed to test these hypotheses.

To their surprise, the *A. sagrei* populations survived on all but the tiniest of islands, and in some cases a population explosion quickly ensued (Schoener and Schoener, 1983c).[300] The Schoeners concluded that it was the periodic occurrence of hurricanes, rather than any inhospitability of the environment, that accounts for the lack of anoles on small islands, a prediction that has been repeatedly confirmed in the last few years (Spiller et al., 1998; Schoener et al., 2004, 2005).[301]

In retrospect, these introductions constituted a nicely-designed experiment examining the determinants of population differentiation. All of the introduced lizards came from Staniel Cay, and the islands onto which the lizards were introduced varied in area, vegetation and other characteristics.

In 1991, Schoener and I returned to Staniel Cay to determine if the populations had differentiated. We found that not only had island populations diverged non-randomly from the source population, but that the degree of morphological divergence was related to the extent to which each island's vegetation structure differed from that on Staniel (Fig. 12.4a,b). This morphology-environment relationship was probably driven by a correlation across islands between limb length and perch diameter (Fig. 12.4c; Losos et al., 1997, 2001).

Schoener and Schoener also introduced *A. smaragdinus* to three of these islands. By 1991, the population on the smallest of these islands was reduced to a single individual (which subsequently perished without issue), but the other two populations had not only thrived, but had differentiated morphologically. In fact, the populations differed so greatly in body size that in a sample of 23 adult males, individuals from the two populations were completely non-overlapping: the smallest individual on one island was larger than the largest from the other island. In addition, the populations differed in size-corrected body length, toepad width, and limb dimensions. The population with smaller maximum SVL was indistinguishable from the Staniel population in body size, but was greatly divergent in body proportions; the other introduced population attained much greater SVL than the Staniel population, but was much less differentiated in body proportions. The explanation for these differences was not clear because the two introduced populations differed little in habitat use, probably because the islands themselves had similar vegetational structure (Losos et al., 2001).

Size increase was also seen in two populations of *A. sagrei* introduced to small dredge spoil islands in Florida (Campbell, 2003). In this study, all offspring in the first generation grew to larger sizes than their parents, which suggests that the size increase was not genetically based.

300. The record was set by the population on White Bay Cay, which grew from ten founder individuals to an estimated population of 98 lizards within a year. Similar rapid increases have been seen in subsequent introductions (Losos and Spiller, 1999). I should note that islands are censused many times prior to introduction to make sure that no lizards are on the island. Although it is always possible that a small population existed and repeatedly escaped detection, such a population could not consist of more than a handful of lizards.

301. The reason that hurricanes remove lizards only from small islands is that only small islands are overwashed by the storm surge that accompanies a hurricane. Larger islands are usually also higher, ensuring that lizards above the high-water mark will survive the storm (Spiller et al., 1998; Schoener et al., 2004).

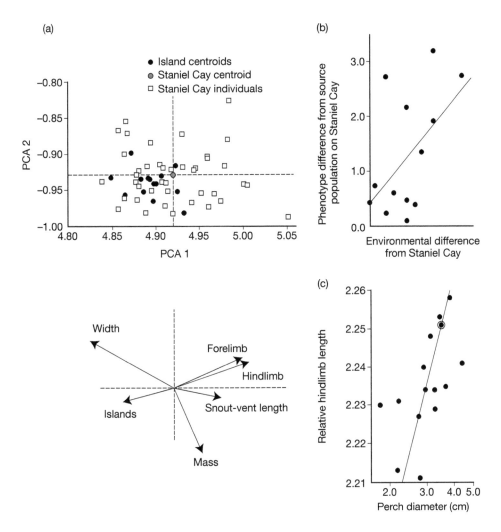

FIGURE 12.4

Phenotypic differentiation of *A. sagrei* populations introduced onto small islands near Staniel Cay, Bahamas. (a) Divergence of introduced populations is nonrandom relative to the source population. All lizards were measured in 1991; lizards on Staniel Cay were not measured at the outset because the experiment was designed to test very different questions. The principal components analysis was conducted on size-adjusted data; arrows indicate the loadings of variables on the PC axes. Divergence in the direction of the introduced populations (the arrow labelled "islands") corresponds primarily to decreased relative limb length. (b) Relationship between the difference in environment (as indicated by maximum vegetation height) between an introduced island and Staniel Cay and degree of phenotypic divergence of an introduced population from the Staniel population. Each point represents the mean value for an introduced population. (c) Relationship between hindlimb length relative to body size and perch diameter. Each point represents the mean value for a population. Circled point is Staniel Cay, others are introduced populations. Modified with permission from Losos et al. (1997).

GEOGRAPHIC VARIATION

An alternative means of testing the hypothesis that populations adapt to environmental differences is to examine variation in ecologically relevant traits among populations that occur in different environments, with the prediction that morphological variation will map onto ecological variation among localities. A number of such studies have been conducted, most of which find the predicted correlation between environmental and phenotypic variation.

LESSER ANTILLES

The relationship between environmental variation and interpopulational phenotypic variation has been studied most intensively in the Lesser Antilles by Thorpe, Malhotra,

FIGURE 12.5

Geographic variation in
(a) *A. roquet* from Martinique and
(b) *A. oculatus* from Dominica.
Photos of *A. oculatus* courtesy
of Roger Thorpe and Anita
Malhotra.

and colleagues at the University of Wales. As mentioned in Chapter 4, anole species in the
Lesser Antilles show substantial amounts of within-island geographic variation in body
size, scalation, limb proportions and, most notably, color patterns (Figures 4.8 and 12.5).
This has been most studied in the three large islands of the central Lesser Antilles—
Dominica (*A. oculatus*), Martinique (*A. roquet*), and Guadeloupe (*A. marmoratus*)—where
four, six and twelve subspecies have been described (Lazell, 1972). However, the detailed
studies by the Wales group have shown that similar patterns, though in some respects
more muted, occur on many other Lesser Antillean islands.

The Wales group's approach is to examine a large number of populations, measuring
phenotypic characters such as coloration, patterning, scalation, and body proportions.
Matrix correlation tests are then used to ask whether between-population similarity in
phenotype is related to similarity in environment. Matrices representing geographic

distance and phylogenetic relationship are also used to partial out effects of similarity due to geographic and phylogenetic proximity.

The results have been consistent in the six species studied: phenotypic variation is strongly related to environmental variation. Details vary somewhat from one species to another: for example, scalation is always related to environmental variation, but sometimes rainfall is the best predictor, whereas in other cases it is vegetation, humidity or temperature (summarized in Thorpe et al., 2004). Body proportions usually are related to altitude and sometimes also to vegetation type, and color patterns always are correlated with vegetation type, with brighter, usually greener, colors in lusher areas, and drabber colors in more xeric localities.

Anolis oculatus, the best studied of these species, illustrates the sometimes complex patterns. Geographic variation in overall body shape, limb length and head width in males is related to vegetation type, and body size is related to altitude, whereas in females, limb length is also related to vegetation type (longer legs in rainforests for both sexes), toe width is related to altitude and body size to rainfall (Malhotra and Thorpe, 1997b). Various aspects of scalation are related to a number of different environmental variables in both sexes; total scale number is negatively related to rainfall in both sexes (Malhotra and Thorpe, 1997a).

GREATER ANTILLES

Comparisons of populations of *A. sagrei* and *A. carolinensis* Species Group anoles[302] in the Bahamas reveal that in both species a correlation exists between relative hindlimb length and mean perch diameter (Fig. 12.6; Losos et al., 2004; Calsbeek et al., 2007).[303] The same relationship occurs among the introduced populations of Bahamian *A. sagrei* discussed earlier in this chapter, and among different species in the Greater Antillean anole radiation (Chapter 13). Similarly, *A. sagrei* populations also mirror Greater Antillean anoles in exhibiting a relationship between perch height and lamella number (Lister, 1976b).

As in the Lesser Antilles, geographic variation in scale number in *A. sagrei* is related to climatic conditions. Both in comparisons among islands and between populations in different habitats within an island, a relationship exists between temperature and scalation: higher temperatures are associated with fewer, but larger, scales (Lister, 1976b; Calsbeek et al., 2006). In addition, scale number is positively related to precipitation levels (Calsbeek et al., 2006).

302. Losos et al. (1994) included populations of the *A. carolinensis* Species Group that are now recognized as *A. smaragdinus* and *A. brunneus* (Glor et al., 2005).

303. Note that this analysis includes perch height and diameter data for *A. occultus*. However, contrary to the statements in that paper and in Calsbeek et al. (2006), I have never collected ecological data for this species, and the values used are instead the published values for a different twig anole, *A. valencienni* (Losos, 1990c). In fact, the habitat use of *A. occultus* is little known and no quantitative data exist (Webster, 1969).

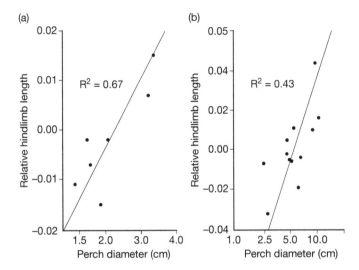

FIGURE 12.6
Relationship between perch diameter and relative hindlimb length among Bahamian populations of
(a) the *A. carolinensis* Species Group and (b) *A. sagrei*. Modified with permission from Losos et al. (1994).

A number of Greater Antillean species exhibit patterns of geographic variation in coloration associated with environmental moisture levels similar to those in Lesser Antillean anoles. For example, in *A. lineatopus* and *A. grahami* in Jamaica, and in *A. conspersus* on Grand Cayman, populations in wetter regions have greater abilities to change color and are greener than populations in more xeric areas (Fig. 12.7; see also Fig. 13.15; Underwood and

FIGURE 12.7
Geographic variation in *A. grahami* from Jamaica (see also Figures 3.5d and 4.4b). In species varying geographically in color, the greenest lizards are from the lushest parts of the islands, whereas the grayer lizards are from the most xeric areas. Photo in (b) courtesy of Richard Glor.

Williams, 1959; Macedonia, 2001). Similar patterns are seen in at least some other widespread Greater Antillean species (e.g., *A. distichus* [Schwartz, 1968]), but the phenomenon has not been examined in detail.

Geographic variation in body size occurs in many species in both the Greater and Lesser Antilles (e.g., Roughgarden, 1974), but its causal basis has not received much attention. In several species, the size of males decreases with the number of sympatric anoles on an island, perhaps as a result of resource limitation resulting from interspecific competition (Schoener, 1969). In *A. oculatus* and *A. aeneus*, male body size is correlated with insect abundance; in some cases, large changes in size can occur over distances as small as 100 m (Roughgarden and Fuentes, 1977).

THE GENETIC BASIS OF TRAIT VARIATION

The 800-pound gorilla in the room is the issue of whether variation—either among individuals within a population or between populations—is the result of genetic differences. Selection on variation within a population will only lead to evolutionary change if that variation has a genetic basis. By extension, differences among populations may represent evolutionary differentiation, perhaps as a result of selection, but they may also be the expression of phenotypic plasticity in genetically undifferentiated populations exposed to different environmental conditions. In the latter case, of course, a phenotype-environment correlation among populations would be evidence neither of natural selection nor of evolutionary change.[304]

PHENOTYPIC PLASTICITY

Variation in body size among individuals and populations is particularly likely to result from phenotypic plasticity. This is not to say that genetic variation in body size does not exist: quite the contrary, such variation has been amply documented in many taxa (Mousseau and Roff, 1987; Falconer and Mackay, 1996). Nonetheless, many non-genetic factors can produce variation in body size as well. Observations of substantial differences in body size in populations separated by very short distances (Roughgarden and Fuentes, 1977) or between parents and their offspring introduced to an empty island (Campbell and Echternacht, 2003) confirm that environmental differences can have a large effect on body size and caution against interpreting size differences among populations in different localities as the result of genetic change. Differences in prey availability, which could affect growth rates, and in predation rates, which could affect average lifespan, are two factors that could lead to non-genetically-based differences in the average size of individuals among populations (Roughgarden and Fuentes, 1977).[305]

---

304. Of course, like any other trait, the extent and form of plasticity has a genetic basis and can evolve by natural selection. Thus, an interpopulational phenotype-environment correlation produced by plasticity may be adaptive if the phenotype produced in each environment has the highest fitness. Such adaptive plasticity has been increasingly documented in a wide range of taxa (reviewed in Ghalambor et al., 2007).

305. Differences in lifespan, of course, are not an example of phenotypic plasticity.

By contrast, the potential extent of phenotypic plasticity for other traits is less clear. Scale characters are thought to be fixed at birth (Hecht, 1952; see footnote 294), so any environmental effect would have to occur during development prior to hatching (which has been documented to occur in garter snakes [Fox, 1948; Fox et al., 1961]). On the other hand, quantitative characters—such as limb and tail lengths—grow through ontogeny and thus potentially could be influenced by post-hatching environmental conditions.

Two approaches have been taken to investigate the existence of phenotypic plasticity in anoles. The first involved rearing lizards in different environments. A number of colleagues and I have shown that in both *A. sagrei* and *A. carolinensis*, baby lizards that are raised in cages with broad surfaces develop longer hindlimbs, relative to their body size, than individuals raised in cages with only narrow surfaces (Fig. 12.8; Losos et al., 2000, 2001; Kolbe and Losos, 2005). The conclusion that hindlimb length in anoles is a phenotypically plastic trait was a surprise to most zoologists with whom I've spoken, but not to many botanists.[306] Given the existence of this plasticity in a pair of distantly related anoles that are members of different ecomorph classes, the most parsimonious

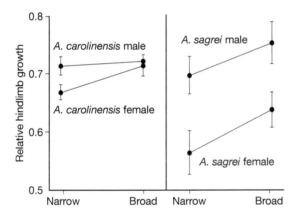

FIGURE 12.8

Phenotypic plasticity in hindlimb length resulting from surface diameter. The hindlimbs of *A. sagrei* and *A. carolinensis* that were raised in cages with broad surfaces grew faster relative to body size than those of lizards in cages with narrow surfaces. Some of the differences between the two species may have resulted from slight differences in experimental conditions (see Kolbe and Losos, 2005). Modified with permission from Kolbe and Losos (2005).

306. It was a surprise to me, too. In fact, we only performed the *A. sagrei* experiment so that I could answer the questions asked by pesky botanists in the Q-and-A sessions after seminars and tell them that plasticity didn't occur.

Bone is a dynamic substance that is constantly being remodeled in response to biomechanical forces. Most studies show that bones alter their width, rather than their length, in response to changes in such forces, but there are a few exceptions (reviewed in Losos et al., 2000). The most interesting is a study of professional tennis players that showed that the bones of the serving arm were longer than those of the other arm (Buskirk et al., 1956), suggesting that through years of smacking balls during ontogeny, the serving arm had grown longer (though, of course, we can't rule out the possibility that the causality runs in the other direction, and that asymmetrically-armed people make particularly good tennis players).

conclusion would be that plasticity is ancestral for the entire clade (Kolbe and Losos, 2005); more data should be collected to test this hypothesis.[307]

These findings suggest the possibility that observed differences among recently introduced populations, or even among naturally occurring populations, may be the result of plasticity, rather than genetic differentiation. On the other hand, these studies were conducted in the artificial confines of the laboratory, where no heterogeneity in substrates was available and where the lizards were not able to make the ontogenetic shift to broader perches that occurs routinely in anoles (Chapter 10). Thus, the extent to which plasticity accounts for variation in limb length within or between natural populations remains to be investigated; the only study to date suggests that interpopulational variation in limb length in *A. oculatus* is genetically based (see below).[308]

At a larger evolutionary scale, the differences in limb length between individuals in the two treatments of these plasticity experiments are much less than those observed between trunk-ground and twig anoles, even though the substrates on which the lizards were raised approximated twigs and trunks. Consequently, these results should not be generalized to suggest that differences between ecomorph classes are the result of plasticity (Losos et al., 2000).

An alternative means of testing for the existence of phenotypic plasticity is to raise lizards from phenotypically differentiated populations in the same environment. If differences observed in nature are the result of phenotypic plasticity, then individuals raised in a common garden should develop indistinguishable phenotypes.[309] Thorpe et al. (2005b) conducted a common garden experiment, raising *A. oculatus* from nine populations that occur naturally in a variety of habitats that differ greatly in elevation, moisture and vegetation. They examined limb, toe, and head dimensions, lamella number, and several scalation characters and found that interpopulational differences persisted among individuals raised in the common environment, suggesting that genetic differentiation, rather than phenotypic plasticity, was responsible for interpopulational differences.

HERITABILITY

The flip side to examining phenotypic plasticity is investigating the genetic basis of trait variation. Anole genetics has been until recently the most glaring shortcoming in our

307. Although I should point out that these studies are extraordinarily labor-intensive and the differences that resulted are small, raising the possibility that they could be obscured by experimental noise in a study. Students whose primary interest is in phenotypic plasticity, rather than anoles, might be better served choosing another organism. This may be the only place in this book (or elsewhere) in which I discourage the study of anoles!

308. Also relevant in this respect is a recent study on the western fence lizard, *Sceloporus occidentalis*, which showed that plastic differences in limb lengths that resulted from egg incubation at different temperatures disappeared by seven weeks of age (Buckley et al., 2007).

309. Non-genetic inheritance, such as maternal effects, can complicate such experiments. For this reason, studies, particularly of plants, are often conducted on multiple generations raised in the common garden (Roach and Wulff, 1987).

understanding of anole biodiversity. At the beginning of this century, next to nothing was known about the genetic basis underlying the incredible phenotypic diversity seen within and among anole species. The reason so little of this sort of work has been conducted was the perception that raising and breeding anoles for classical genetic studies would be labor intensive, expensive, and of uncertain success.[310] However, hobbyists, particularly in Europe, have been breeding and raising anoles for years, so this concern was probably unwarranted. Now, several laboratories are breeding and raising anoles, and this number is likely to increase in the near future.[311]

The first fruits of these efforts are now being published. Work by Calsbeek and colleagues has found significant heritability for an unspecified scale measurement ($h^2 = 0.76$ [Calsbeek et al., 2006]), for body size ($h^2 = 0.55$ [Calsbeek and Smith, 2007]), female dorsal pattern ($h^2 = 0.78$ [Calsbeek and Bonneaud, 2008]), and hindlimb length ($h^2 = 0.77$ [Calsbeek and Bonneaud unpublished data, cited in Calsbeek and Smith, 2008]).[312] No doubt this is just the tip of the iceberg; within a few years, quantitative genetic estimates of trait heritability in anoles should be common.

Even more exciting, however, are the possibilities presented by the availability of the genome sequence of *A. carolinensis*. With a genome sequence in hand, a wide variety of approaches can be deployed to identify the genes affecting particular phenotypic characters, as well as the particular allelic differences responsible for variation within populations and, potentially, between species (Feder and Mitchell-Olds, 2003). Because many developmental genes are phylogenetically conservative in their effects (Carroll et al., 2005), anole researchers may be able to make particularly rapid progress when working on genes known to have important effects on relevant traits in other taxa. Indeed, the genetics of some of the most interesting traits from an anole perspective, such as limb and head dimensions, have been extensively studied in other vertebrates (e.g., Niswander, 2002; Tickle, 2002; Abzhanov et al., 2004, 2006; Stopper and Wagner, 2005).

A lot of hard work stands between obtaining a genome sequence and actually finding the alleles responsible for phenotypic variation, but already the impending publication

310. I can't speak for others, but this was certainly the reason I didn't undertake such studies for many years.

311. As an example of the serendipity and contingencies of scientific research, here's how my colleagues and I started down this path: In 1999, Hurricane Floyd traveled over our Abaco, Bahamas study site. When my colleagues visited the islands two months later, they were surprised to find the island devoid of adult lizards, but covered with small juvenile lizards (Schoener et al., 2001). This led us to surmise that even though all the lizards had been washed away or drowned by the high water that accompanies a hurricane, eggs in the ground must have been able to survive immersion in saltwater for up to six hours. To test that hypothesis, we obtained gravid female *A. sagrei* and used their eggs in studies of the effect of saltwater immersion on hatching rates (there was no effect for eggs less than 10 days old [Losos et al., 2003c]). We found that incubating and hatching eggs was not difficult, and since we had females in the lab anyway, we decided to see how difficult it would be to breed them, so we obtained several males. To our surprise, this turned out to be easy as well, and thus our lizard breeding activities were underway (Sanger et al., 2008a).

312. The existence of both high heritability and extensive phenotypic plasticity for body size and hindlimb length is not contradictory. Heritability is a measure of the relationship between parent and offspring in the same environment, whereas phenotypic plasticity measures the extent that the same genotype produces different phenotypes in different environments, termed a "norm of reaction" (Malhotra and Thorpe, 1997b).

of the *A. carolinensis* sequence is generating widespread interest (see Schneider [2008] for a review of how the anole genome may be useful in studying anole evolution). My guess is that this interest will translate in short order to exciting new discoveries about anole genetics.

FUTURE DIRECTIONS

In the last two chapters, I have examined the three predictions of the adaptive radiation hypothesis: species interact; species shift their resource use in the presence of congeners; and these shifts in resource use lead to evolutionary change. The conclusions of these two chapters are quite similar. On one hand, each of the three predictions is strongly supported by a wealth and variety of data.

On the other hand, both chapters highlight the same call for more basic data relevant to these three predictions. In my mind, the biggest message from this chapter is how few studies of natural selection and microevolutionary change have been conducted on anoles. Given that current methods are now in place, such studies are relatively easy (if labor intensive) and present the opportunity to examine many important questions in evolutionary biology. Introduced populations would seem to be a particularly good opportunity, but examination of selection in undisturbed communities also could prove enlightening. Moreover, intraspecific comparative studies—examining how and whether selection differs among localities to produce patterns of geographic variation (Thompson, 2006)—would also be interesting.

In Chapter 11, I noted that more manipulative ecological studies are needed. Coupling such studies with measurements of selection would be particularly useful. The addition or subtraction of a species to a community should lead to changes in habitat use by other species; coupling such studies with measurements of the selective and evolutionary consequences of such shifts would constitute a direct test of the adaptive radiation hypothesis, combining the approaches outlined in the last two chapters.

This suggestion is relevant to the phenomenon of ecological release discussed in the previous chapter. One species in which ecological release undoubtedly occurred was *A. carolinensis* in the southeastern United States. In the absence of other anoles, *A. carolinensis* broadened its habitat use to include more terrestrial parts of the environment. Whether this shift was accompanied by morphological change from its Cuban ancestors has received little attention. Conversely, the introduction of *A. sagrei* and other species to Florida has reacquainted *A. carolinensis* with the species with which it evolved in Cuba, with the result that it appears to have retreated to its ancestral habitat niche. The evolutionary consequences of the *A. sagrei* invasion have not been studied, although Collette (1961) suggested the idea nearly five decades ago. One could easily imagine comparing present-day populations to pre-*sagrei* museum specimens. Moreover, an exciting study would involve examining what happens as *A. sagrei* invades new areas; nearby sites ahead of the advancing wave could serve as controls. Given that the spread of *A. sagrei* seems unstoppable (now in Georgia and introduced to New Orleans and elsewhere), many opportunities for such studies should be available.

Although long considered unimportant and an impediment to evolutionary change, the study of phenotypic plasticity has experienced a revival in recent years and now many workers emphasize the ways in which phenotypic plasticity can promote evolutionary diversification (reviewed in Schlichting and Smith, 2002; West-Eberhard, 2003; Ghalambor et al., 2007). Previously, I have suggested the possibility that plasticity in limb length may facilitate adaptive evolution. By allowing a population to utilize an environment that otherwise might not be inhabitable, plasticity may permit occupation of new environments, setting the stage for the incorporation of genetic changes that greatly amplify the initial plastic response (Losos et al. [2000, 2001], echoing ideas originally put forth by Schmalhausen [1949] and Waddington [1975]). Testing such ideas, however, will not be easy, as discussed above.

Genomics has revolutionized the study of natural selection (Eyre-Walker, 2006; *Drosophila* 12 Genomes Consortium, 2007; Stinchcombe and Hoekstra, 2008), and many genomic techniques can be brought to bear to study anole evolution. For example, field studies might focus not only on measuring selection on phenotypic characters, but also on candidate genes that are likely to underlie variation in those traits. Genes related to limb length might be particularly good candidates given the extensive body of research on limb developmental genetics (Niswander, 2002; Tickle, 2002; Stopper and Wagner, 2005). Another approach would compare genomic data among populations or species to attempt to identify particular genes or regions exhibiting the signature of selection, such as unusually high or low rates of nucleotide substitution (Vasemägi and Primmer, 2005; Voight et al., 2006; Stinchcombe and Hoekstra, 2008).

Until recently, the standard method for marking lizards and other small vertebrates was to cut off the tips of their toes—by cutting off several toes in different combinations, each lizard could get a unique mark. However, not only is this method not for the squeamish, but one has to wonder about the effect toe-clipping would have on the operation of an anole's toepads; indeed, a recent study suggested that the effect may be severely detrimental to clinging performance (Bloch and Irschick, 2004).[313] These concerns are particularly relevant to studies involving locomotion and movement patterns and consequently deterred workers from investigating natural selection on locomotor behavior and morphology.

In defense of those who have used toe-clipping as a means of identification, several studies on other types of lizards have shown that toe-clipping does not affect sprint speed (Huey et al., 1990; Dodd, 1993; Paulissen and Meyer, 2000; Borges-Landáez and Shine, 2003); no study of which I'm aware has directly assessed the effect of toe clipping on survival. Moreover, lizards in many (but not all) species routinely lose toes naturally (reviewed in Hudson [1996] and Schoener and Schoener [1980b]). Among Bahamian anoles, frequently more than 10% of the individuals in a population have sustained an injury to at least one toe (Schoener and Schoener, 1980b). I have caught fat and sassy-looking anoles, seemingly in the prime of life, that were missing most or all of the toes on a foot.

Other common methods of marking lizards are no more successful. Branding has been attempted and, horrible as it seems, it apparently may work on some lizards (Clark, 1971), but attempts on anoles have not been successful (J. Lazell, pers. comm.). PIT tags, now commonly used in studies of small vertebrates, are too big for all but the largest males of most anole species (Gibbons and Andrews, 2004).

At least in part for this reason, only two studies of natural selection in anoles were conducted before 2000, and they did not require individual identification of lizards (Malhotra and Thorpe, 1991; Thorpe et al., 2005b).

In recent years, two new methods for marking lizards have been developed. The first is the equivalent of bird banding: using surgical wire, colored beads are sewn into the dorsal tail musculature. Such bead tags seem to last for several years (Fisher and Muth, 1989) and have been used in behavioral studies of anoles (e.g., Tokarz, 1998; McMann and Paterson, 2003), but not, as far as I know, to measure natural selection.

---

313. It should be noted that Bloch and Irschick appear to have clipped off a larger portion of the toe than in previous studies of pad-bearing lizards, which clipped only a small portion of the toe attached to the claw (e.g., Bustard, 1968; Paulissen and Meyer, 2000; T.W. Schoener, pers. comm.; R.S. Thorpe, pers. comm.).

FIGURE 12.9
Anolis sagrei with colored elastomers injected
subdermally into the ventral side of the
limbs. Like bird bands, different positions
and colors of elastomers allow identification
of individual lizards.

The second method involves the injection of a non-toxic elastomeric substance
(Northwest Marine Technology, Shaw Island, WA, USA) subdermally into different locations on the underside of an anole. The liquid elastomer subsequently solidifies into a
rubbery strand that can easily be seen when the lizard is in hand (Fig. 12.9). By injecting
into different locations and with different colors, researchers can create a large number
of unique combinations.[314]

314. In the name of giving credit where credit is due, Duncan Irschick was the first person to realize the
possibility of applying this technique, developed for fish, to lizards.

13

FORM, FUNCTION, AND ADAPTIVE RADIATION

The previous two chapters have focused on the ecological side of adaptive radiation, discussing how interspecific interactions drive ecological shifts and how natural selection subsequently leads to evolutionary change. In this chapter, I take the macroevolutionary perspective: faced with a clade composed of species that are phenotypically differentiated and that occupy distinct ecological niches, how do we test the hypothesis that the phenotypic differences represent adaptations to different ecological circumstances?

In the case of Greater Antillean anoles, the approach is straightforward. Since the dawn of evolutionary biology, convergence—evolution of the same phenotypic feature by taxa that have independently occupied the same ecological situation—has been taken as *prima facie* evidence of adaptation driven by natural selection (Darwin, 1859; Conway Morris, 2003). Anoles, of course, exhibit such convergence in spades. Still, the question remains: Why, in a mechanistic sense, does the convergence occur? Why are particular traits favored in particular structural microhabitats?

To get at questions like these, we need information on function: what does a trait do and how does its performance vary in different circumstances? In a landmark paper, Arnold (1983) pointed out that if we want to know why a trait is favored by natural selection in a particular situation, we need to know what performance advantage that trait provides relative to other character states for that trait (e.g., absence, smaller, different color). In turn, we then need to investigate how differences in performance are related to fitness: does greater running ability, bite force, or any other measure of function translate into greater survival or reproductive success? This phenotype → performance

→ fitness paradigm has been widely adopted (Wainwright and Reilly, 1994; Irschick and Garland, 2001).

Extended to the macroevolutionary level, this approach predicts that phenotypic differences among species will lead to differences in functional capabilities; in turn, these functional capabilities will result in species performing best in the environment (or niche) in which they occur. Such comparative studies, of course, must be conducted in an explicitly historical framework, and a large body of literature now exists that relies on studying the evolution of performance capability in a phylogenetic context to elucidate evolutionary adaptation (Greene, 1986; Coddington, 1988, 1990; Arnold, 1994).[315]

THE STUDY OF ECOLOGICALLY RELEVANT FUNCTIONAL CAPABILITIES

Physiologists and functional morphologists have studied organismal function for more than a century, but Huey and Stevenson (1979) had the key insight that the traits examined must have clear and direct ecological relevance. That is, measuring the contractile speed of a muscle or the oxygen uptake of a cell may provide considerable insight into the workings of that muscle or cell, but extrapolating from sub-organismal traits to the performance of an entire organism is not always straightforward, and in some cases may be positively misleading (e.g., Licht, 1967; Marsh and Bennett, 1985, 1986a,b). Rather, Huey and Stevenson (1979) argued that measures of whole organism function, at tasks of potential ecological significance, are what is needed to investigate the potentially adaptive relationship between organismal traits and environmental features.

In the past 25 years, a vast array of measures of organismal function has been investigated, including locomotor speed and acceleration (running, swimming, flying, and crawling), jumping, biting, digesting, clinging, climbing, maneuvering, endurance, and many others (Wainwright, 1994; Irschick and Garland, 2001; Irschick, 2003; Biewener, 2003). This work has provided great insights into the relationship between phenotype and function, but less well studied is the ecological significance of variation in functional capacity. A simple first question is whether organisms actually make use of their maximal capabilities in nature: fast speed or great jumping ability is only ecologically relevant, for example, if and when animals run or jump at top speed in nature. Relatively few studies have actually examined the extent to which organisms use their maximal capacities, no doubt because such work is difficult and time consuming (Hertz et al., 1988; Jayne and Irschick, 2000; Irschick and Garland, 2001).

A complementary approach to understanding the adaptive significance of measures of organismal function is investigation of whether natural selection operates on variation in functional capability—i.e., do individuals with greater capabilities (faster runners,

315. In addition, in recent years, an armada of statistical approaches—each more sophisticated than the last—has been developed to analyze such questions (see Garland et al. [2005] for a recent review).

better jumpers) have higher survival or reproductive success? Such studies have shown that variation in performance is sometimes, but not always, associated with differences in fitness (e.g., Bennett and Huey, 1990; Jayne and Bennett, 1990; Le Galliard et al., 2004; Miles, 2004; Lappin and Husak, 2005; Husak et al., 2006a).

At the interspecific level, a hypothesis of adaptive diversification would predict that interspecific differences in functional capabilities should correlate with differences in habitat use and behavior; species should perform best at those tasks most relevant to where they live and what they do (reviewed in Garland and Losos, 1994; Wainwright, 1994; Irschick and Garland, 2001). For example, bats that forage in cluttered habitats have greater maneuverability than bats that forage in the open (Stockwell, 2001). Similarly, bite force in turtles is correlated with the hardness of prey consumed (Herrel et al., 2002).

ANOLES AND THE STUDY OF ADAPTATION

Anoles have proven to be exemplary subjects for studies of the relationship between phenotype, function, and environment: they exhibit substantial interspecific variation in all three, they usually perform well in laboratory settings, and they often can be observed readily in the field. Throw in extensive convergence and a well-corroborated phylogenetic hypothesis, and *Anolis* could be the poster child for the study of adaptation.

In this chapter, I will examine a variety of traits for which variation among species may have an adaptive basis. Some of these traits, such as limb length, have been extensively studied, whereas for others, the most we can do is hypothesize about potential adaptive significance and suggest future lines of investigation. This examination will focus on two questions:

1. How does variation in phenotype relate to variation in functional capabilities?
2. And how does variation in functional capabilities relate to variation in ecology and behavior?

These questions will be addressed primarily at the interspecific level, but with some comparisons among and within populations.

THE PERVASIVE EFFECTS OF BODY SIZE

Before launching into a discussion of the relationship between variation in morphology, performance capabilities, and ecology, I need to address the issue of allometry. The ubiquity of body size effects (Peters, 1983; Calder, 1984) is readily apparent in anoles. For example, among anole species, body size, as represented by SVL, accounts for most of the variation in a variety of morphological, ecological, and functional traits (Table 13.1).

A consequence of the near-universal effect of size is that almost all morphological characters are highly correlated with performance capabilities (Table 13.2). As a result,

TABLE 13.1   Interspecific Correlation of Body
Size with Morphological, Ecological
and Functional Variables

Variable	Correlation
Forelimb length	0.95
Hindlimb length	0.93
Tail length	0.85
Mass	0.98
Lamella number on 4th hindtoe	0.83
Perch height	0.57
Perch diameter	0.43
Clinging ability	0.97
Sprint speed	0.73
Jumping ability	0.75

NOTE: Sample sizes: morphology, 53 species; habitat use, 55 species; functional performance, 14–18 species. Data from Losos (1990c), Losos et al. (1991), Autumn and Losos (1997). See Beuttell and Losos (1999) for a more extensive list of morphological characters. These are non-phylogenetic correlations, but correlations from analyses incorporating phylogenetic information are equally high.

TABLE 13.2   Morphology-Performance Correlations

	Sprint speed	Jump distance	Clinging ability
Forelimb length	0.79	0.82	0.91
Hindlimb length	0.90	0.89	0.89
Tail length	0.73	0.81	0.86
Mass	0.84	0.82	0.94
Lamella number on 4th hindtoe	0.47	0.39	0.83

NOTE: See Table 13.1 for data sources.

untangling the web of causality is extremely difficult. For example, one theory holds that sprint speed should be a function of limb length, because longer limbs can take longer strides and, all else equal, longer strides translate into greater speed (Losos, 1990c; Irschick and Jayne, 1999; Vanhooydonck et al., 2002). However, species with longer legs are also bigger, and have larger everything: muscles, hearts, brains and almost every other quantitative morphological character. Consequently, determining which trait or traits is responsible for the increase in speed with body size is difficult. A variety of methods have been developed to examine situations such as this; however, these methods are often impotent in the face of high variable intercorrelations, termed 'multicollinearity' (Slinker and Glantz, 1985; Graham, 2003).

For this reason, morphometric and functional analyses often statistically remove the effect of size on traits before examining their interrelationships.[316] In this way, one can ask, for example, "Do species with long legs for their body size run faster than would be predicted given their size?" The advantage of this approach is that the effect of variation in a trait can be examined without the confounding effects of correlations with every other variable that scales strongly with body size. The disadvantage, however, is that absolute measures, rather than size-corrected ones, are sometimes the ones that matter in nature. For example, when escaping a predator, absolute sprint speed, not sprint speed relative to body size, probably determines whether prey will escape (but see Van Damme and Van Dooren, 1999). Thus, decisions about whether to remove the effect of size in an analysis depend on the questions being asked; in some cases, no right answer will exist and we must accept that all options have limitations.

THE ADAPTIVE SIGNIFICANCE OF INTERSPECIFIC VARIATION IN LIMB LENGTH

Perhaps the most obvious way that anole species differ—other than in size or color—is in the relative length of their limbs; species of the same total body length can differ by a factor of two in length of their hindlimbs (Fig. 3.3). This variation clearly is relevant to the ecology and behavior of anoles in two respects.

First, among Greater Antillean ecomorph species, a relationship exists between hindlimb length and the diameter of the surface each species usually uses, with the effects of body size removed from both variables (Fig. 13.1; Losos, 1990c; Irschick et al., 1997). This limb length–diameter correlation is remarkably consistent in West Indian anoles, also being observed among populations of both *A. sagrei* and *A. carolinensis*

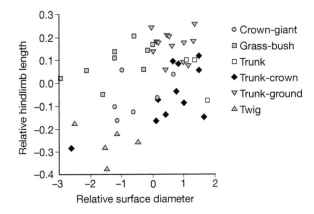

FIGURE 13.1

Relationship between surface diameter and hindlimb length in ecomorph species. Both variables have had the effect of size removed by calculation of residuals in a regression against SVL (variables log-transformed). Data from Losos (1990c), Losos (1992b), Irschick and Losos (1996) and Losos (unpubl.).

---

316. Many methods, some extremely complicated, have been developed to remove body size effects, and this is an area of great controversy (e.g., Rohlf and Bookstein, 1990; Marcus et al., 1996). In my own studies, I have usually opted for the simpler, and more intuitive, approaches which are adequate for unidimensional traits such as limb lengths (Beuttell and Losos, 1999). However, more sophisticated approaches are needed for traits that are related as parts of two- or three-dimensional shapes such as skulls (e.g., Harmon et al., 2005).

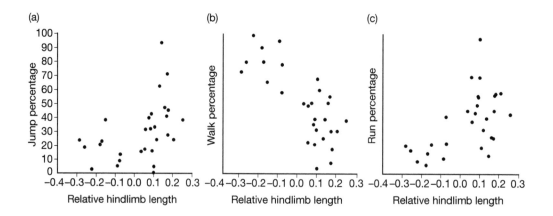

FIGURE 13.2

Relationship between relative hindlimb length and frequency of movements of different types among Greater Antillean anoles. All movements were categorized as runs, jumps, or walks. Y-axes are the percentage of all movements in each category. Movement proportions were arcsin-transformed prior to statistical analysis. All correlations are statistically significant based on independent contrasts analysis (p < 0.001 for walk and run percentage, p < 0.03 for jump percentage). Movement percentages are not related to body size, and thus are not size-corrected. Data sources as in Fig. 13.1.

Species Group anoles in the Bahamas and among recently established populations of *A. sagrei* in the Bahamas (Fig. 12.6).

Second, differences in relative limb length are also related to behavioral variation among West Indian anoles. Long-limbed species tend to run and jump more frequently, whereas shorter-limbed species walk more often (Fig. 13.2; Moermond, 1979a,b; Estrada and Silva Rodriguez, 1984; Pounds, 1988; Losos, 1990b). Moreover, the relationships of relative hindlimb length to habitat use and locomotor behavior are independent; for example, jumping frequency and perch diameter are not related and simultaneous consideration of both variables explains as much variation in limb length as the two variables do separately (Fig. 13.3). This is evident in Figure 13.1 in which the biggest outliers in the perch diameter-limb length relationship are the grass-bush anoles, which have moderately long legs, use narrow surfaces, and jump frequently, and trunk-crown anoles, which have relatively short legs, often use broad surfaces,[317] and don't jump all that much.

317. Although the data here can be a bit misleading. Trunk-crown anoles use a wide range of surfaces from narrow branches to broad tree trunks. Consequently, one observation of a trunk-crown anole on a 50 cm diameter tree trunk has a disproportionate effect on the calculation of mean perch diameter. Other ways of representing the data, such as looking at medians or log-transforming the data prior to calculating means, might be more representative (cf. Schoener [1968], who points out that for a small lizard, the difference between perches 0.5 and 1.0 cm in diameter may be much more functionally significant than the differences between, say, 5- and 10-cm diameter perches).

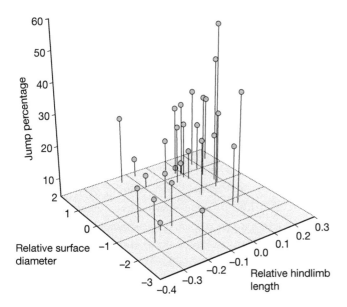

FIGURE 13.3

Relationship between relative hindlimb length, relative perch diameter, and percentage of movements that are jumps. Percentage jumps and perch diameter have independent effects on relative hindlimb length, explaining 31% and 35% of the variation respectively. Both variables are significant ($p < 0.005$) in a multiple regression conducted on the independent contrasts for these variables and together explain 64% of the variation. Data sources as in Fig. 13.1.

To investigate the adaptive basis of limb length variation, we need to ask several questions:

1. How do functional capabilities change in relation to limb length?
2. Do limb length and surface diameter have interactive effects on functional capabilities?
3. How do differences in functional capabilities affect the way species utilize the environment?

INVESTIGATING THE FUNCTIONAL CONSEQUENCES OF VARIATION IN LIMB LENGTH

SPRINTING

This is where the fun comes in. In 1981, Ray Huey and colleagues invented a portable lizard racetrack (Huey et al., 1981).[318] The concept is simple: construct a trackway through which a lizard (or other small animal) will run. Position sensors at regular intervals that the animal will trip as it runs by, feeding information into a timing device which calculates the time elapsed between each station (Fig. 13.4). Make the whole apparatus modular, so that it can be easily transported to field sites.[319]

With a lizard racetrack, investigation of maximal lizard sprinting speed became a snap: go to the field, catch a bunch of lizards, run them a few times on the track, then release them back into their homes. And what a joy it could be! Few things in life are as impressive as an anole hurtling down a trackway in excess of two meters per second, front legs barely touching the ground or maybe not at all, hellbent for the dark bag placed at the end of the track into which it can seek refuge.[320] In recent years, the advent of affordable high speed video cameras has allowed greater precision, which in turn has facilitated the more difficult measurement of acceleration from a standstill, as well as speed throughout the duration of a run (Vanhooydonck et al., 2006c).

318. Bennett (1980) independently developed the idea of a lizard racetrack, but Huey et al. (1981) emphasized the utility of portability for field work.

319. Perhaps a bit of an overstatement, because in the early days of lizard racetrack work, a portable computer was the size of a sewing machine. True, it was "portable," but a far cry from today's slender and light laptops.

320. Truth in advertising: lizard racing can also be a big drag when, for example, the lizard refuses to run and instead bites the investigator's fingers, or escapes from the track and leads the investigator on a 20-minute pursuit throughout the field lab room, managing to find and squeeze into every nook and cranny.

Also, I'll never forget my first lizard racing field season in 1988 at the El Verde Field Station in Puerto Rico. It was the summer after my fourth year in graduate school, and I was desperate to jump-start my thesis research, which had been stuck in low gear. I arrived in mid-June, only to be met with a prolonged stretch of extremely rainy weather. Not only did the continual downpours make field data collection difficult, but the resulting extremely high humidity had an adverse effect on my Compaq® portable computer. In particular, it rendered the "r" key inoperable. Unfortunately, to initiate the lizard racing sequence, I had to type "run" on the computer—"un" did not do the trick and I did not know how to re-write the program (Lesson to all graduate students: computer programming is an indispensable skill). Looking disaster in the face, I contemplated law school as the rain continued unabated for two weeks. Finally, just as all seemed lost, the rain stopped and the sun came out, the computer dried and the "r" resumed its assigned function, the data poured in, and the legal profession was forgotten.

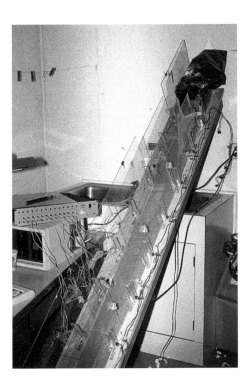

FIGURE 13.4

A lizard racetrack. Sensors are positioned
every 0.25 m; when the lizard breaks the in-
frared beam, the information is transmitted
to a computer, which calculates time elapsed
from one point to the next. A black bag is
placed at the top of the track to give the lizard
a refuge toward which it runs (when all goes
well). When in operation, dark paper is
placed on the side walls to minimize distrac-
tions and encourage the lizards to run up the
track. The track is angled for anoles because
on horizontal surfaces they often hop instead
of running. Regardless of the incline, lizards
often foil the experimenter's efforts by moving onto the walls and running along them,
instead of on the substrate.

The results of these studies are clear: in *Anolis*, body size is strongly related to maximal sprint speed (Table 13.1); with the effects of size statistically removed, hindlimb
length and sprint speed are still strongly correlated (Losos, 1990d; Vanhooydonck et al.,
2006c). This finding is readily explainable in biomechanical terms. All else equal,
lizards with longer limbs will take longer strides, which should in turn translate into
greater speed, unless compensated by reduced stride frequency (Losos, 1990d; Irschick
and Jayne, 1999). Detailed analysis suggests that interspecific variation in the length of
the tibia is of primary importance in determining differences in sprint speed (Fig. 13.5;
Vanhooydonck et al., 2006c).[321]

What is not necessarily predicted by biomechanical theory is a relationship between
sprint speed and acceleration (reviewed in Vanhooydonck et al., 2006c). Nonetheless,
such a relationship does occur in anoles: species that run quickly for their size also have
great relative acceleration ability (Fig. 13.6; Vanhooydonck et al., 2006c). The reason is
that ecomorphs with longer limbs also have larger knee extensor muscles. The relative
mass of these muscles is positively associated with both increased relative sprint speed
and acceleration, thus explaining the positive relationship between these variables.

321. A comparative study of phrynosomatid lizards also revealed a large effect of tibia length, although
more distal limb elements were even more strongly correlated with variation in sprint speed (Irschick and Jayne,
1999). Interspecific variation in sprint speed in mammals also is primarily determined by variation in distal
limb elements (Hildebrand, 1985).

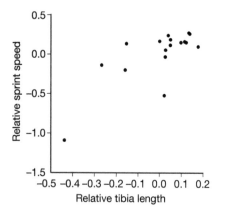

FIGURE 13.5

Relationship between relative tibia length and relative sprint speed. In a multivariate analysis incorporating phylogenetic information, relative tibia length was the morphological feature most strongly correlated with relative sprint speed. Data kindly provided by B. Vanhooydonck from Vanhooydonck et al. (2006c).

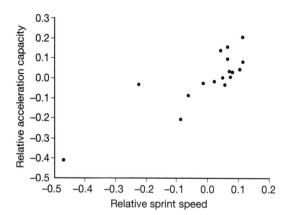

FIGURE 13.6

Relationship between relative sprint speed and relative acceleration capability among West Indian anoles. Modified with permission from Vanhooydonck et al. (2006c).

The finding that relative sprint speed is affected independently by both relative muscle mass and relative limb length indicates that a "many-to-one" relationship exists between morphology and performance (Wainwright et al., 2005; Collar and Wainwright, 2006; Wainwright, 2007)—multiple phenotypes may be equivalent in their functional capabilities (e.g., individuals with long legs and small muscles or with short legs and big muscles may have the same capabilities). Among Greater Antillean species, this functional equivalence is probably of secondary importance because relative hindlimb length by itself explains such a large proportion of the variation in relative sprint speed (more than 60% [Losos, 1990d]). However, an intriguing possibility, at least in theory, is that muscle mass could explain more of the variation in sprint performance among Greater Antillean unique anoles and mainland species, in which cases these species may have convergently evolved the same functional capabilities as some ecomorph species, but by different morphological means. Unfortunately, sprint performance data are available for few mainland species and no Greater Antillean unique anoles, so the extent and significance of the many-to-one phenomenon in anole evolution cannot be assessed.[322]

JUMPING

In the early days of performance studies, measuring jumping ability was a decidedly less sophisticated enterprise (e.g., Losos, 1990d). Lizards were placed on a jumping platform and cajoled to jump by threats, invectives, and taps to the base of the tail. Performance was determined by measuring the length of the jump with a tape measure. More recently, high speed video has allowed the quantification of many other aspects of the jump, such as acceleration and angle of takeoff (Fig. 13.7), and as a result, the kinematics of anole jumping are now well understood (Toro et al., 2006).

The biomechanical prediction is straightforward. All else equal, the longer a lizard's hindlegs, the greater the length of time it will take to straighten them during the launch phase of a jump, and thus the greater the time through which it will accelerate, resulting in greater flight speed and longer distance traversed. The data strongly support this prediction (reviewed in Losos, 1990d; Toro et al., 2004).

Jumping ability is also influenced by the speed with which a lizard accelerates as it jumps, which is related to knee extensor muscle mass, but not to hindlimb length (Toro et al., 2004; James et al., 2007). Thus, as with sprinting, two independent paths may exist to increasing jump distance, either by evolving longer limbs or larger muscles.[323]

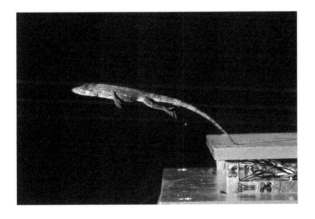

FIGURE 13.7
Measurements of jumping ability of *A. valencienni*. High speed video analysis allows calculation of factors relevant to jumping kinematics, such as angle of takeoff and acceleration at point of liftoff. Photo courtesy of Anthony Herrel.

322. Before getting too carried away with the many-to-one phenomenon, I should point out that the sprint speed measurements reported in Vanhooydonck et al. (2006c) show some discrepancies from speeds recorded for the same species in other studies (Losos and Sinervo, 1989; Losos, 1990c; Irschick and Losos, 1999). In particular, some of the faster species in previous studies did not run as quickly, whereas some slower species ran more quickly. The cause of these discrepancies is unknown, but may result at least in part from differences in experimental protocols between the studies. The species with relatively slower speeds in Vanhooydonck et al. (2006c) are trunk-ground anoles, which have the relatively longest legs and most massive leg muscles. The effect that the relatively slow speeds of these species had on the outcome of the study is not obvious.

323. A third factor that affects jump distance is takeoff angle. Surprisingly, however, this effect is relatively slight. Toro et al. (2004) demonstrated that lower takeoff angles substantially shorten the time in flight, while only slightly decreasing jump distance: they predicted that this would be a worthwhile trade-off when attempting to escape a predator; however, no data are available on the jump angles of lizards in nature.

In summary, interspecific variation in limb length in anoles is strongly related to variation in two ecologically relevant aspects of locomotor performance, sprinting and jumping.

INTERACTIVE EFFECTS OF LIMB LENGTH AND SURFACE DIMENSIONS ON FUNCTIONAL CAPABILITIES

The studies just summarized examined functional variation on broad, straight surfaces in the laboratory. However, not all species use such surfaces in nature. To investigate how substrate diameter affects anole performance, the sprinting and jumping capability of anoles has been measured on round surfaces ranging in diameter from very narrow (emulating twigs, 0.7–1.2 cm wide) to very broad (similar to large branches, 4.6–5.1 cm wide [Losos and Sinervo, 1989; Macrini and Irschick, 1998; Irschick and Losos, 1999; Vanhooydonck et al., 2006b]).

The initial hypothesis in this line of research was that each species would perform best when running on the diameter corresponding most closely to what it uses in nature. As expected, the species that use the broadest surfaces—and that have the longest legs— ran the fastest on broad surfaces and showed a steady decline in sprint speed with decreasing surface diameter (Irschick and Losos, 1999). But on the narrow surfaces, the hypothesis was not supported: the short-legged twig denizens not only were slower than longer-legged species, but they did not even run faster on narrower surfaces than they did on broader surfaces (Fig. 13.8).

The reason our hypothesis was not supported was that sprint speed declined with decreasing perch diameter for all species, but to a greater extent in some species than in others. Overall, the degree to which speed declines is related to relative limb length: long-legged species are affected much more than short-legged species (Fig. 13.8; Losos and Irschick, 1996). Subsequent detailed kinematic studies on one species, A. sagrei, explained why this occurs. As diameter decreased, lizards had to change the position of their feet from the sprawling stance adopted on a flat surface to a position in which the feet were more directly underneath the body. To accomplish this, the lizards flexed their knees more, lowered their bodies, and oriented their toes more perpendicularly relative to the long axis of the body. The result was shorter strides and, probably, less force in the direction of movement, resulting in slower speeds (Fig. 13.9; Spezzano and Jayne, 2004). Presumably, these effects are not as great in shorter-legged species, which even on a narrow surface can retain a more typical sprawling stance to a greater extent (Mattingly and Jayne, 2004); data are needed to test this hypothesis.[324]

324. Surprisingly, the effect of perch diameter on acceleration ability is exactly opposite: the long-legged *A. sagrei*'s acceleration capabilities were unimpaired on narrow substrates, even though overall speed declined sharply, whereas the twig anole *A. valencienni* ran at the same speed, but had much slower acceleration, on the narrow surface (Vanhooydonck et al., 2006b). The biomechanical explanation for these contrasting trends is unclear.

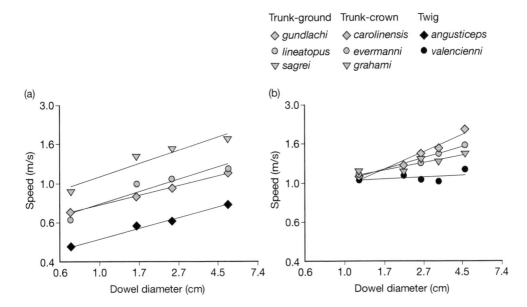

FIGURE 13.8

The effect of surface diameter on sprint speed in a variety of *Anolis* species. Measurements were taken in two studies that differed slightly in experimental methods, such as the diameters of the rods and the angle at which the rods inclined (Losos and Sinervo, 1989; Irschick and Losos, 1999). Overall, the effect of decreasing surface diameter on sprint speed (the slope of the line) increases with relative hindlimb length (p < 0.01, analysis on independent contrasts; sprint speed data from Losos and Irschick [1996] and Irschick and Losos [1999]; relative hindlimb length data for species in Irschick and Losos [1999] from Losos [1990 and unpubl.]). Figure modified with permission from Irschick and Losos (1999).

These results were at first perplexing from an evolutionary perspective. What is the advantage to a twig anole of evolving short legs if they do not provide a locomotor advantage on the narrow surfaces which twig anoles use?[325] Indeed, such short legs seem to come at a cost: the longest-legged species can run twice as fast on broad surfaces and at least as fast on the narrow surfaces as the short-legged species.

The answer lies in a different measure of locomotor performance. During the sprint trials, we recorded the number of times a lizard stumbled or fell as it ran along the dowel, a measure we later termed "surefootedness" (Sinervo and Losos, 1991). On the broad surfaces, the species did not vary substantially in surefootedness, but running on the narrowest surface was a different story: the longest-legged species had difficulty in more than 3/4 of the trials, whereas the shortest-legged species were unaffected (Losos and Sinervo, 1989).[326] In other words, short-legged lizards may have been

325. Given the nested phylogenetic placement of most twig anoles and the fact that all potential outgroups have longer legs, extremely short legs in twig anoles are almost certainly a derived trait.

326. Subsequently, videos revealed why long-legged species had such difficulty. On the narrowest surfaces, the lizards had to grab a narrow dowel directly under their body. Longer-legged species had great difficulty doing so and sometimes missed entirely, pitching forward and occasionally even falling off the surface entirely (Losos and Irschick, unpubl.).

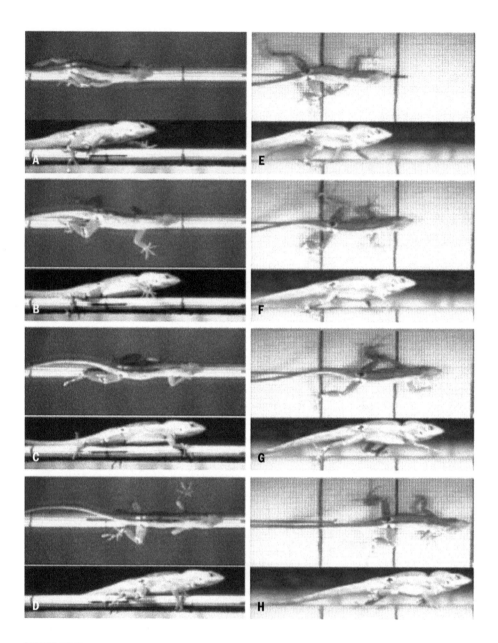

FIGURE 13.9

High speed video of *A. sagrei* running on narrow and broad surfaces. Each panel shows the same image shot laterally and above. On the narrow surface, the lizard holds its body lower to the surface and takes shorter strides; the toes are also to a greater extent oriented perpendicular to the long axis of the lizard's body. Figure from Spezzano and Jayne (2004) with permission.

slow, but they were surefooted wherever they went, especially on narrow perches. This result emphasizes the importance of examining multiple functional aspects of a trait.

In contrast to these results, surface diameter appears to have minimal effect on jumping ability regardless of limb length, at least in the laboratory, where lizards were given plenty of time to orient themselves before jumping from stable surfaces (Losos and Irschick, 1996). In nature, by contrast, lizards must position themselves quickly, usually by turning perpendicular to the branch on a surface that may not be stable, particularly when it is narrow (Cartmill, 1985; Pounds, 1988; Bonser, 1999). Laboratory studies that simulate more natural conditions might reveal limb length effects that are not yet apparent.[327]

Surfaces upon which anoles move not only vary in diameter, but also in straightness. Higham et al. (2001) and Mattingly and Jayne (2005) investigated the extent to which a 90° horizontal turn affected sprint speed in three Jamaican and four Bahamian species. Trials were conducted on a 4.8-cm wide surface, equivalent to the broadest surfaces in the studies just mentioned above.

As expected, all species ran more slowly when they had to make a turn than when they ran on the straightaway. In addition, the Jamaican species were tested on turns of 30° and 90° and, predictably, the decrease in speed was greater on the larger turn. In both studies, the trunk-crown species were affected by turning much less than were the longer-legged species. However, one oddity was that the Jamaican twig anole, *A. valencienni*, suffered a decline comparable to the Jamaican trunk-ground *A. lineatopus*, whereas in the study of the Bahamian species, the decrease in speed of the twig *A. angusticeps* was much less, and was comparable to that of the trunk-crown, *A. smaragdinus*.

In summary, anole sprinting capabilities are strongly habitat-dependent, and the extent of this dependence is a function of relative limb length.

FUNCTIONAL CAPABILITIES, BEHAVIOR AND MICROHABITAT USE

The studies just reviewed reveal the functional consequences of variation in limb length, but variation in capabilities is only relevant if it translates into differences in the way organisms interact with their environment (Greene, 1986). Overall, performance capabilities are correlated with what anoles do in nature: good jumpers tend to jump more, and poor runners get around more by walking (Fig. 13.10).

These correlations are a good first step toward demonstrating the adaptive significance of performance ability, but a key question is whether maximal capabilities are actually used in nature (Hertz et al., 1988; Garland and Losos, 1994; Jayne and Irschick, 2000; Irschick and Garland, 2001). In fact, most anoles spend most of their time moving around very slowly, well below their maximal abilities (Irschick and Losos, 1998; Irschick, 2000; Mattingly and Jayne, 2004). If anoles never run at top speed or jump as far as they can, then interspecific variation in maximal sprinting and jumping capabilities is unlikely to explain why species differ in how they move through their environment.

---

327. The other end of a jump might also be a profitable area for future research. Almost all work on the biomechanics of jumping in lizards so far has focused on takeoff and has measured the distance to landing on a flat surface (usually the floor). In nature, lizards jump to other arboreal surfaces, and the biomechanics of landing have yet to be examined extensively (Bels et al., 1992; cf. Bonser, 1999).

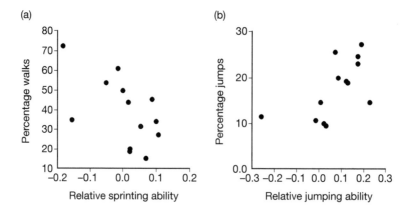

FIGURE 13.10

Relationship between maximal sprinting and jumping ability and the proportion of movements that are walks and jumps (statistics: jumping, $p < 0.03$; sprinting: $p < 0.10$; tests one-tailed and based on independent contrasts; non-phylogenetic analyses, $p < 0.025$ for both tests). Data from Losos (1990c).

To address this issue, Duncan Irschick and I measured sprinting and jumping performance in the field to determine whether species use their maximal capabilities and, if so, in what contexts (Irschick and Losos, 1998). We found that anoles ran at near top speeds when escaping a simulated predator.[328] These data suggest that the adaptive significance of sprint speed may be sought by studying variation among species in escape behavior (see Chapter 8). As this hypothesis would predict, slower anoles seem to rely less on rapid escape than faster species, although this generality has not been quantified. Twig anoles, in particular, rely on crypsis and stealth. In nature, when a predator appears, twig anoles usually squirrel around to the other side of a branch, remain pressed tightly to the surface and sometimes slowly walk away, no doubt in many cases eluding detection (Fig. 13.11; Schoener, 1968; Rand and Williams, 1969; Myers, 1971; Huyghe et al., 2007; J.B. Losos, pers. obs.) Only as a last resort, twig anoles will run as fast as they can to try to get away.[329] If anything, the ability to move on narrow surfaces without stumbling, and thus drawing the attention of a predator, seems more likely than speed to be the key locomotor adaptation of twig anoles.

Most species also run near their maximal capability when chasing a simulated prey item, though slightly less rapidly than during escape behavior (Irschick and Losos, 1998). Two species—one of them the twig anole *A. angusticeps*—bucked this trend, running much slower than they could in pursuit of the faux fly.

328. The same is true for collared lizards, *Crotaphytus collaris* (Husak, 2006).
329. In fact, slower species and age classes tend to use a slightly higher proportion of their maximal sprinting capability when attempting to escape a predator, perhaps because they need to do so to compensate for their slower maximal speed (Irschick and Losos, 1998; Irschick, 2000).

FIGURE 13.11

Escape behavior of a twig anole. When detecting a predator, the first response of a twig anole, such as this *A. valencienni*, is to press its body against the surface and sidle around to the other side, while keeping an eye on the predator. Then, the lizard often slowly creeps away without ever being noticed.

In contrast, the jumps of all species in all contexts averaged substantially shorter than maximal capabilities (Irschick and Losos, 1998).[330] One possible explanation is that jump distance may be more constrained by the surrounding environment than running speed; a jumping lizard has to avoid obstacles and have a suitable landing place. For this reason, lizards may jump maximally only occasionally. The alternative possibility, however, is that maximum jumping ability is rarely relevant to anoles and not of adaptive significance. Because jumping and sprinting ability are tightly linked to hindlimb length, the potentially excessive jumping capabilities of anoles may have evolved as an incidental consequence of selection for increased sprinting ability.

Overall, these results support the hypothesis that maximal capabilities are ecologically relevant and potentially adaptive. However, the preceding discussion has focused on the maximal sprinting and jumping capabilities of species as measured on optimal surfaces. As noted above, sprint speed, but not jumping ability, declines with decreasing surface diameter. Moreover, some species are more sensitive than others to narrow surface diameters. If performance capabilities are important, we would expect to see differences both within and between species in habitat use and behavior related to surface diameter. This hypothesis has been tested and confirmed in three different contexts:

- Species that experience a great decline in sprint speed on narrow surfaces tend to avoid such surfaces and use a more restricted range of substrates than species whose abilities are less affected by surface diameter (Irschick and Losos, 1999).[331]

330. Notably, this study did not include grass-bush anoles, the ecomorph type which jumps most frequently.
331. Perversely, this leads to the result that in nature, diameter affects escape speed more in short-legged species than long-legged ones because the latter avoid narrow surfaces (Mattingly and Jayne, 2005).

- Because sprinting but not jumping ability declines sharply with decreasing perch diameter in most species, anoles should increasingly jump to escape a threat as diameter decreases. This prediction was confirmed in four of the five species examined (Losos and Irschick, 1996).[332]
- When escaping a simulated predator and coming to a branching point on the surface on which they were moving, four Bahamian species tended to choose the larger-diameter branch and the branch that deviated least from the direction in which they were moving. Given that speed declines both on narrower surfaces and when an anole turns, this result is not surprising. When forced to choose in laboratory trials, all four species preferred staying on a broad diameter surface, even if it required a larger turn, than moving onto a very narrow surface, presumably because the decrease in speed caused by turning was less than the decrease that would have been caused by using the narrower surface (Mattingly and Jayne, 2005).

THE ADAPTIVE BASIS OF LIMB LENGTH VARIATION: CONCLUSIONS

Overall, the data support the hypothesis that variation in limb length among Greater Antillean ecomorphs is adaptive: variation in limb length affects sprinting and jumping ability and agility on narrow surfaces. Species utilize at least some of their maximal capabilities and avoid microhabitats in which sprint performance and agility are impaired.

As a generality, three locomotor strategies are seen in anoles. Long-legged species run quickly and only rarely use microhabitats in which their abilities are impaired. Short-legged species often do not rely on sprinting or jumping to capture prey or escape predators. Their ability to move without difficulty on narrow and irregular surfaces allows them to creep up on prey and avoid detection by predators. Grass-bush anoles take a different approach, using narrow surfaces even though they have moderately long limbs. These species tend to jump much more than any other ecomorph type. Because jumping ability isn't affected by perch diameter,[333] they are able to function effectively even on narrow surfaces. Future work should determine whether they take longer jumps relative to their maximal capabilities than other ecomorphs, as their great propensity to jump might suggest.

TOEPADS

The functional and adaptive significance of other traits have not been nearly as well studied as has limb length. Probably the next best studied and most interesting trait of anoles is their expanded subdigital toepads. Anoles differ tremendously in the relative size and

332. As well as subsequently in *A. gingivinus* (Larimer et al., 2006). In addition, the fifth species in our study showed a non-significant trend in the same direction.
333. At least on stable surfaces (the bushier side of "grass-bush" anole); grass blades might be a different matter.

composition of the pads. In some species, the pads are large and expansive, whereas in others, they are much smaller and narrower; one species (*A. onca*) has lost its pad entirely (Peterson and Williams, 1981; Nicholson et al., 2006).

The structure of the pad also is highly variable among species: some are composed of many lamellae (the laterally expanded scales that comprise the pad; see Chapter 2), whereas others have few (Figure 2.4). The microstructure of the pad also exhibits great variation, the functional significance of which is unclear (reviewed in Peterson, 1983; see Chapter 2).

Pad area increases with body size, both ontogenetically and interspecifically (Macrini et al., 2003). Number of lamellae also increases with size among species ($r^2 > 0.62$ [Glossip and Losos, 1997]), but not ontogenetically because scale number is believed to be fixed at birth, as discussed in the previous chapter. Absolute pad area and lamella number are strongly related among species: with body size effects removed, the correlation is much weaker (Losos, unpubl., based on data in Beuttell and Losos, 1999).

A correlation between toepad structure and arboreality has long been noted (e.g., Collette, 1961); terrestrial species tend to have narrow pads with few lamellae, whereas pads of species that occur higher in the trees are larger and better developed. Quantitative analysis reveals a significant, albeit fairly weak ($r^2 \leq 0.20$) relationship between lamella number and perch height and diameter. With size effects removed,[334] a relationship still exists with perch height, but perch diameter effects are reduced and non-significant in some analyses (Glossip and Losos, 1997; Macrini et al., 2003).

To date, only one aspect of pad function has been measured, the ability to cling to a smooth surface. Initially, clinging ability was measured in as low tech a way imaginable: by placing a loop around the lizard's waist and pulling backwards with a small scale to register the force needed to displace the lizard from its position on a horizontal sheet of plexiglas. More recently, the procedure has become considerably more sophisticated. By using a force plate covered by an acetate sheet, extremely accurate measures of the clinging force generated by a lizard can now be obtained (Fig. 13.12; Irschick et al., 1996; Elstrott and Irschick, 2004).

FIGURE 13.12

Measuring clinging ability of *A. equestris*. The investigator slowly pulls the lizard backwards and the force plate measures the force generated by the lizard's forefoot as it clings to the smooth surface. Tape around the mouth is for obvious reasons, as an *A. equestris* bite packs quite a wallop. Photo courtesy of Anthony Herrel.

334. The outlying *A.* (Chamaelinorops) *barbouri*, which is known to have aberrant pad microstructure relative to other anoles (Peterson, 1983), was also removed from the analysis.

These studies reveal that clinging ability is related to pad size. Both variables increase with body size and, with such effects removed, relative pad area is strongly related to relative clinging ability.[335] In turn, clinging ability and perch height are significantly related, both in absolute and relative terms (Elstrott and Irschick, 2004).

Why species that live higher off the ground should require greater clinging ability is not clear (reviewed in Glossip and Losos, 1997; Elstrott and Irschick, 2004). Hypotheses fall into two groups: either falls from a greater height have more severe consequences, so that more arboreal species need a greater margin of safety to avoid falling,[336] or the habitat use and behavior of more arboreal species require greater clinging ability. Specifically:

- Falls from greater heights may be more dangerous (although most anoles are so small that they would not attain a high enough velocity to hurt themselves, regardless of height).

- Energetic costs and predation risk may increase the further a lizard has to climb back up into the vegetation.

- More arboreal species more often use smooth surfaces on which claws are ineffectual (e.g., leaves; see Chapter 3 on differences in use of leaves among ecomorphs) and thus require greater adhesion ability.[337]

- Arboreal species engage in activities that require greater clinging ability.

No data are available to test these ideas.

The functional significance of lamellae is even less clear. The number of lamellae should affect the flexibility of the pad: the greater the number of lamellae for a pad of a given size, the greater the ability of the pad to mold itself to narrow or irregular surfaces. No relationship exists between lamella number and clinging ability on a smooth, flat surface (based on a combination of data in Losos [1990c], Irschick et al. [1996] and Ellstrott and Irschick [2004]), but a more appropriate test would examine the ability to cling to other types of surfaces (as in Losos et al., 1993b). In turn, further field work is required to understand how lamella function relates to habitat use.

Anoles, of course, have a second means of clinging, one that is better suited for rough and irregular surfaces: claws. Anyone who has tried to pull an anole off of a branch can attest that anoles can generate substantial clinging force with their claws, but

335. Clinging ability increases with pad area with a slope less than 1.0 (based on data in Irschick et al. [1996] and Elstrott and Irschick [2004]), which means that, per unit area of pad, species with larger pads have relatively less clinging ability. Potentially, this scaling may relate to pad microstructure. Although setal density appears relatively constant among anoles, the distribution of the setae over the pad and digits varies among species. In addition, variation in setal microstructure exists, which might also have consequences for clinging capabilities (Peterson, 1983). Peterson (1983) made a nice start at cataloguing and understanding setal variation, but this area of research has lain fallow since then.

336. Or they require greater ability to catch themselves when they do fall. Falling geckos can rescue themselves by adhering to a leaf with a single foot (Autumn, 2006).

337. In the laboratory, the relatively small-padded trunk-ground *A. cybotes* often fell off smooth leaves, whereas the larger-padded trunk-crown *A. chlorocyanus* used the same surfaces with little difficulty (Rand, 1962).

such clinging has not been studied in anoles.[338] Moreover, this clinging ability is readily used; for example, when in the survey posture, anoles hang head downward on a tree, gripping the trunk by the claws on their hindlimbs (Fig. 9.1). The claws of anoles, and hence probably their clinging ability, differ among anole species, but this variation has never been examined.[339]

HEAD SHAPE

Anole head shapes vary in a complex, multivariate way (Harmon et al., 2005). Much of the variation is captured in the first axis of a principal components analysis which distinguishes species whose heads are short from front to back, broad, and high versus heads that are long, narrow, and low (Fig. 13.13). This variation correlates with habitat use: ecomorphs that use broad surfaces have short, broad and high heads, whereas ecomorphs that use narrow structures exhibit the reverse.

One possible explanation involves locomotion. Species using narrow surfaces may need slender, low heads to maintain their balance and to move through a cluttered environment. Alternatively, variation in head shape may correspond to bite force. Species with shorter, higher and broader heads would be expected to be able to bite harder

FIGURE 13.13

Differences in head shape among anoles. Long and low, *A. brunneus* from Acklins, Bahamas, versus short and high, *A. gundlachi* from Puerto Rico.

338. Although it has been studied in other lizards (Zani, 2000, 2001).
Crown-giant anoles seem to have particularly well developed claws, which is painfully evident when an anole grabs onto a hand or arm. The great clinging ability provided by these claws may compensate for the relatively small toepads (in proportion to their body mass) of large anoles (Elstrott and Irschick, 2004).
339. Actually, this is not technically correct. Years ago, an undergraduate in my laboratory spent most of a year measuring anole claws (hence my unreferenced statement that they vary in shape). However, with 95% of the data complete and analyses already underway, all the data were lost in a computer crash, and no backup had been kept. Since then, I haven't had the heart to try to get someone else to repeat the endeavor (and the student went on to get a Ph.D. in Forestry).
Note, also, that Gans (1974, p. 19) stated, without attribution, that anole ecomorphs differ in claw size. I do not know the basis of this claim.

(Herrel et al., 2001; Verwaijen et al., 2002), which could be useful in eating harder prey or in fighting with conspecifics or predators. In addition, differences in head size may be related to the dimensions of the prey being subdued and swallowed: lizards with longer, wider heads may be able to eat bigger prey (Schoener, 1968; DeMarco et al., 1985). Lastly, the slender head shape of some ecomorphs could enhance crypsis on narrow surfaces.

Anoles vary in head shape in other ways beside the "short-broad-high" versus "long-narrow-low" continuum. For example, the second axis of the principal components analysis in Harmon et al. (2005) described the shape of the head, ranging from relatively flat to particularly low in the anterior and high in the posterior. Functional and ecological explanation for this variation awaits investigation.

Research on the functional ramifications of variation in head size has only just begun. Bite force in *A. lineatopus* increases ontogenetically, such that larger individuals can bite disproportionately harder. This trend can only partially be explained by ontogenetic changes in head shape (Herrel et al., 2006). In contrast, the relatively greater bite force of male *A. carolinensis* compared to that of females (Fig. 13.14) is readily explainable; males have relatively larger jaw adductor muscles, as well as differences in skull shape to accommodate the greater muscle mass (Herrel et al., 2007). In both species, differences in biting ability have ecological significance, because individuals with larger bite force eat larger, harder prey (Herrel et al., 2006, 2007).

Biting ability also is related to success in intraspecific interactions. As with *A. lineatopus*, bite force also increases with body size in seven out of eight other species (Lailvaux et al., 2004; Lailvaux and Irschick, 2007b; Vanhooydonck et al., 2005).[340] However, in

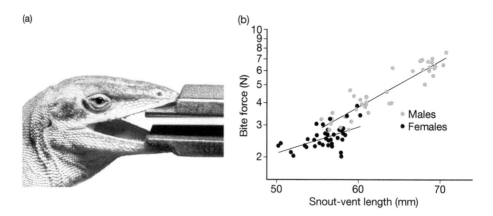

(a)

(b)

FIGURE 13.14

Bite force of *A. carolinensis*. (a) Bite force is measured by inducing a lizard to bite onto a specially-built force plate. Photo courtesy of Anthony Herrel. (b) Males bite harder than females, even when matched for size. Figure modified with permission from Herrel et al. (2007).

340. In most of the seven species, these papers reported a relationship between bite force and dewlap size, rather than body size. However, because dewlap size increases with body size (Echelle et al., 1978), I assume that bite force is also related to overall size. The twig anole *A. angusticeps* is the exceptional species that does not exhibit a relationship between body size and bite force.

size-matched trials within species, males with greater biting ability generally won in trunk-ground and trunk-crown anoles, but not in twig or trunk anoles (Lailvaux et al., 2004; 1Lailvaux and Irschick, 2007b).

COLOR AND PATTERN

Anoles vary in both color and pattern; this variation is not only interspecific, but also interpopulational, intersexual, and ontogenetic (Chapter 9). The most obvious difference in color is between those species which have the ability to change color from a bright green to a dark brown versus those species whose ability to change colors is limited to shades of brown or gray. The explanation for this variation is almost surely crypsis: green species are usually arboreal and thus more often occur on or near green surfaces (Collette, 1961). Moreover, geographic variation in many species takes the form of brighter, greener coloration in more humid areas and duller, browner color in more xeric areas (Chapter 12).[341]

Nonetheless, the crypsis hypothesis, though plausible, has not been investigated extensively. The field of visual ecology has taken off in the last decade, and methodologies now exist to test the hypothesis that a particular color or pattern is cryptic with regard to the environment in which it occurs (in this case, degree of crypsis is the measure of functional capability [Macedonia et al., 2005; Wilson et al., 2007]). The only applications of this approach to anoles have been studies of A. carolinensis in Florida and A. conspersus on Grand Cayman, which showed that anoles in most populations are relatively inconspicuous when judged against the color of their surrounding habitat (Fig. 13.15; Macedonia, 2001).

In theory, we might expect green anoles to match their background, turning green when in vegetation and brown when against a woody surface. Although widely believed, this idea is not strongly supported (reviewed in Jenssen et al., 1995).[342] In one study, male A. carolinensis mismatched the surface upon which they sat (green on brown substrate or vice versa) more often than would be expected by chance (Jenssen et al., 1995; but see Medvin [1990] for an opposite result). Indeed, males of green species often adopt a bright green coloration when in the survey posture, although a darker appearance would almost surely be more cryptic against a woody background; this tendency suggests the possibility that skin color is being used to make the lizards more, rather than

341. This pattern is also seen among closely related species that differ in habitat (Schoener, 1975).
342. This point was first made by Reverend Lockwood (1876), who noted that his captive anoles were usually brown during the day, even when on a green leaf, and were green at night, even when sleeping on brown surfaces. He concluded (p.13): "The belief that the color of the contiguous object is mimicked for the sake of protection is, I think, not confirmed by the observed facts. The truth is that in this matter of animals enjoying life there is a higher law than that of mere intention. I shall call it the law of spontaneous expression, which has its base in another law, to wit, that a joy unuttered is a sense repressed. Why should green be the favorite night-gown of our sleeping Anolis? I timidly venture the suggestion that it is because the animal is disposing itself for the luxury of sleep, its color changes being the utterances of its emotions . . . Whether it be the expression of enjoyment of repose, comfort, or emotional joy, the highest manifestation is its display of green."

FIGURE 13.15
Geographic variation in
A. conspersus on Grand
Cayman. Each color form
is cryptic in the environ-
ment in which it occurs
(Macedonia et al., 2001).

less, conspicuous (Macedonia, 2001; see also Trivers [1976] on the green color of mating
A. garmani).[343]

Although many anole species are uniform in color and patternless, others exhibit a
great variety of colors and patterns (Figures 9.13 and 9.14). Many of these species occur
relatively low to the ground in densely vegetated areas, and thus the patterns may provide crypsis. The same explanation has been put forth to account for the patterning of
females in sexually dimorphic species (see discussion of dimorphism in Chapter 9).

343. Color change is under hormonal control in anoles and often occurs in social encounters (reviewed in
Greenberg [2002, 2003]; the physical mechanism by which color change is produced is reviewed in Cooper and
Greenberg [1992]. For example, almost half of the instances in which *A. carolinensis* males changed from green
to brown occurred in the context of aggressive encounters (see also Trivers [1976] on *A. garmani*). Many of these

No study has directly tested these hypotheses, though female patterning correlates with perch diameter among Bahamian populations of *A. sagrei* (striped females are more frequently found on narrow surfaces, unstriped females on broad surfaces [Schoener and Schoener, 1976]).

In other species, males are boldly colored and patterned. At first glance, it is hard to believe that the brilliant blue of male *A. allisoni* (Fig. 3.5) or the red head of males in some populations of *A. marmoratus* (Fig. 4.8) could be for camouflage—attracting attention would seem a more likely explanation. These species, too, have received little attention; the possible significance of such a seemingly ostentatious wardrobe for sexual selection and species recognition will be discussed in Chapter 14.

In summary, substantial variation in color and patterning exists within and between species. These colors and patterns probably in many cases promote crypsis, but in some cases may serve to make their bearers more conspicuous. The tools are now available to test these functional hypotheses in sophisticated ways, but few such studies have been conducted on anoles.

THERMAL AND HYDRIC PHYSIOLOGY

Anoles vary in the microclimates they occupy, both across geographical gradients and even at particular sites. As I reviewed in Chapter 10, anoles often alter their behavior and microhabitat use to occupy suitable thermal and hydric microhabitats. However, varying conditions also require anoles to evolve physiological adaptations to function effectively.

THERMAL PHYSIOLOGY

Anoles live in different thermal environments and exhibit average body temperatures ranging from 20.5–34.2°C (Fig. 10.4). Given this diversity, we can ask whether species exhibit physiological adaptations to living at different temperatures. A wide variety of approaches have been taken to studying the physiological ecology of anoles, including measurements of preferred temperatures, maximum and minimum temperatures beyond which anoles cannot function, and optimal temperatures for physiological performance. A recent review tabulated studies on 28 species (Hertz et al., unpubl.), although most studies have only measured one or a few physiological aspects.

As discussed in Chapter 10, many species behaviorally thermoregulate and are able to maintain their body temperature within a narrow range. Not surprisingly, then, a

---

occurred as males were approaching the boundary of their territory, but before an opponent was visible (e.g., the male which owned the adjacent territory was on the other side of a tree trunk), which suggests that the male in some sense anticipated an agonistic encounter (Jenssen et al., 1995). In general, dark color is a response to heightened stress, although a variety of other factors—including predation attempts, temperature, and light levels—also affect color in *A. carolinensis* (reviewed in Jenssen et al., 1995; Greenberg, 2003). During male-male interactions, lizards will change color frequently; by the end of the encounter, the winner is usually green and the loser brown (Greenberg, 2003).

correlation exists between the temperatures which anoles select in the laboratory and the mean temperature for that species in the field. These data support the conclusion that anoles use the environment nonrandomly, but do not in themselves illustrate physiological adaptation.

Early approaches to studying thermal physiological adaptation focused on the extremes and investigated whether species that lived in warmer environments could tolerate higher temperatures and, conversely, whether species in cooler environments could survive at lower temperatures (reviewed in Huey, 1982). Many of the approaches taken to addressing these questions—such as increasing the temperature in a chamber by 1°C every five minutes and recording the onset of panting, spasms, uncoordinated movements, and other variables until the animal died—are no longer permitted, for good reason. Other methods, which examine, for example, the temperature at which a lizard tries to escape an experimental chamber as temperatures are increased or decreased (termed the Experimental Voluntary Maximum [EVM]), are more humane and cause no long-term effects.

These analyses indicate that the preferred temperature a species selects in the laboratory is related both to the EVM and to the temperature at which it loses the ability to right itself (Critical Thermal Maximum [CTMax]; Hertz et al., unpubl.). Moreover, among populations environmental variation is also related to tolerance of high temperatures. CTMax decreases with altitude in all six species studied, although only in one, *A. gundlachi*, is the magnitude of the decrease large (Hertz and Huey, 1981). The relatively small decrease in CTMax probably results because most species (but not *A. gundlachi*) bask more often at higher elevations, which results in mean body temperature declining only slightly with elevation (Huey, 1981; Huey et al., 2003; see Chapter 10).

Fewer data are available for cold tolerance. Among Puerto Rican anoles, upland species can survive longer at low temperature than lowland species (Heatwole et al., 1969; Gorman and Hillman, 1977). Data are available for too few other species to make broader, statistically substantiated statements, though the general trend is that species that experience cooler temperatures in nature are better able to withstand them.[344]

An alternative approach to studying thermal adaptation is to examine how physiological function varies in relation to body temperature, with the specific hypothesis that species are adapted to perform best at temperatures they most frequently experience (Huey and Stevenson, 1979; Bennett, 1980; Huey, 1982; Hertz et al., 1983). Most physiological traits in reptiles show a hump-shaped relationship with body temperature (Huey, 1982), and anoles are no exception. For example, sprint speed of *A. cristatellus* from a lowland population in Puerto Rico was very low at 10°C, increased to a maximum at around 30°C and then declined sharply at higher temperatures; the optimal temperature for sprinting corresponded almost exactly to the temperature these lizards selected in

344. Although laboratory physiological data exist for 22 species, different types of data have been collected for different species. As a result, insufficient data exist for many comparisons; for example, for only two species do we have data for both preferred temperature and critical thermal minimum temperature.

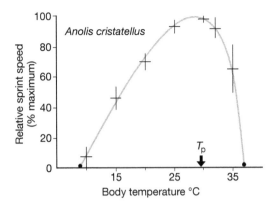

FIGURE 13.16
Thermal dependence of sprint speed in
A. cristatellus. The preferred temperature
(T_p) of this species in the lab corresponds
almost perfectly to the optimal performance temperature. Modified with permission from Huey et al. (1983).

the laboratory (Fig. 13.16). Similarly, among seven Costa Rican species, a relationship existed between the temperature at which each species sprinted most quickly and the mean body temperature for that species in the field (van Berkum, 1986).[345]

These data indicate that, as a generality, species are adapted to perform best at body temperatures they most frequently experience. Nonetheless, the thermal environment changes over the course of the day and from one locality to another; consequently, lizards may not always be able to maintain body temperatures that produce maximal capabilities. At Huey's (1983) lowland study site, *A. cristatellus* in a forested area had low body temperatures for the first few hours of the day, with the result that they could not run at maximal speed for much of the morning. By contrast, lizards in a more open area attained their preferred temperature before 8 a.m., but by midday the heat was so great that even in the shade lizards exhibited body temperatures above the optimal level, leading to submaximal sprint capabilities at that time. On the other hand, the non-thermoregulating forest-dwelling *A. gundlachi* often experiences body temperatures that restrict functional capabilities. At a cool high elevation site, individuals could only sprint at 80% or more of their maximal capacity 32% of the time in the summer, and three percent of the time in the winter; corresponding numbers at a lowland site were 95% in the summer and 74% in the winter (Hertz, 1992b).

This temperature dependence of sprint performance suggests the possibility that lizards may alter their behavior as a function of body temperature. Rand (1964b) found that when he approached *A. lineatopus,* the distance at which they would flee was inversely correlated with their body temperature—the warmer they were, the closer they would let him approach. Presumably they fled sooner at low temperatures to compensate for their diminished sprinting capabilities. Subsequent studies on a wide variety of other reptiles have shown similar patterns of temperature-dependent shifts in

---

345. Thermal dependence in anoles has only been examined for one other trait in one species. Maximal jumping performance of *A. carolinensis* occurs at about 34°C (Lailvaux and Irschick, 2007a), identical to the temperature that this species selects in the laboratory (Corn, 1971).

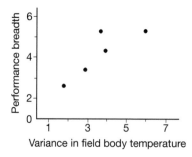

FIGURE 13.17

Relationship among Costa Rican anoles between variance in field body temperature and span of temperatures over which a species could sprint at 95% or more
of maximum speed. Figured modified from van
Berkum (1986) with permission.

anti-predator behavior (reviewed in Garland and Losos, 1994), but no subsequent work has been conducted on anoles.[346]

If performance-temperature curves are evolutionarily labile, species that experience greater variability in body temperatures—perhaps because of lessened opportunity to thermoregulate (Chapter 10)—might be expected to evolve the capability to sprint at near maximal speeds over a broader range of body temperatures than species that are able to maintain their body temperature over a narrower range (Huey, 1982; Hertz et al., 1983). This hypothesis was confirmed for the same seven Costa Rican species mentioned above; van Berkum (1986) calculated the range of temperatures over which a species could sprint at 95% or more of its maximal speed and found that this measure of thermal performance breadth was correlated among species with variability in field body temperatures (Fig. 13.17).

HYDRIC PHYSIOLOGY

Less research has been conducted on physiological adaptation of anoles to different hydric environments. As with studies of thermal biology, early approaches were cruel by today's standards, dehydrating lizards until they died; with increasing sophistication, researchers have been able to more humanely measure rates of water loss in dry condi-tions. These studies have all shown that rate of water loss is directly related to the moisture level in a species' environment (Sexton and Heatwole, 1968; Hillman and Gorman, 1977; Hertz et al., 1979). The same relationship holds true within populations of several species (Hillman et al., 1979; Dmi'el et al., 1997; but see Hertz, 1980b). No studies have looked

346. One interesting twist is that some aspects of performance are more sensitive to temperature than others (Huey, 1982). For example, maximum force generation by muscles is often less sensitive to temperature than is the rate at which muscles contract (Marsh and Bennett, 1985). Consequently, the shift from locomotion to aggressive defense seen at low temperatures in a number of lizards (e.g., Hertz et al., 1982; Crowley and Pietruszka, 1983), though not reported for anoles, may result because the ability to bite is much less affected by low temperature than the ability to run quickly (Herrel et al., 2007). In a similar vein, studies on the clinging ability of geckos suggest that the effect of temperature differs between clinging and sprinting, with clinging either unaffected by temperature (Bergmann and Irschick, 2005) or having an optimal temperature much lower than the optimal temperature for sprinting (Losos, 1990e). If these results are general, then they may have interesting implications for how lizards change their behavior as a function of temperature.

at the effect of hydric physiology on locomotor function or any other similar performance measure.

A number of studies on a variety of reptiles, including several on anoles, have reported a relationship among populations between scale number and temperature or moisture (see Chapter 12). However, in some cases the relationship is positive, whereas in other cases it is negative (reviewed in Malhotra and Thorpe, 1997a; Calsbeek et al., 2006). Adaptive explanations have been put forward to explain both trends (Malhotra and Thorpe, 1997a). The important question is whether water loss occurs primarily through the scale or through the interstitial skin that occurs at scale edges. If it occurs through the scale, then a greater number of small scales would be expected in arid environments. Alternatively, if water loss occurs through the skin between the scales, then the relationship should run in the opposite direction, with fewer but larger scales expected in more arid environments. Surprisingly, almost no physiological research has been conducted on this topic; the only study to compare rates of water loss and scale number among populations found no relationship (Dmi'el et al., 1997).

Overall, the data indicate that anoles are adapted to function at the temperatures and moisture levels they experience in nature. Data from more species are needed, as is investigation of the biochemical bases of these adaptive changes. The only work of the latter sort is research that indicated that decreased rates of water loss in *A. carolinensis* acclimated to dry conditions are the result of increased lipid deposition in the skin (Kattan and Lillywhite, 1989).

THE MYSTERY OF THE MAINLAND

For the most part, anoles on the mainland use the same range of environments as those in the West Indies: near the ground, up in the canopy, on branches, trunks, and twigs, and in the grass. However, as discussed in Chapter 4, the morphological variety exhibited by mainland anoles differs from that seen in the Greater Antilles. Moreover, the relationship between morphology and habitat use is fundamentally different in the two areas (Table 13.3). For example, relative forelimb length is negatively correlated with perch height in mainland anoles, but the variables are unrelated in Greater Antillean species (Irschick et al., 1997). Similarly, with the effects of body size removed, Greater Antillean anoles have larger and wider toepads for a given perch diameter compared to mainland anoles (Fig. 13.18; Macrini et al., 2003).

Two hypotheses could explain these differences. First, the many-to-one morphology-performance relationship suggests the possibility that morphologically very different species may have the same functional capabilities. Consequently, the relationship between functional capability and microhabitat use may be the same between the mainland and the Caribbean, even though morphology-microhabitat relationships differ. One possible example of a many-to-one relationship could involve the structure of the toepad and clinging ability. Differences in clinging ability could result, in theory, from

TABLE 13.3   Differences in Ecomorphological Relationships Between Mainland (ML)
and Greater Antillean (GA) Anoles

Comparison	Difference
Lamella number versus perch height	GA species have more lamellae
Lamella number versus perch diameter	GA species have more lamellae
Toepad width versus perch height	Pads narrower at low heights in ML; equal at great heights
Toepad width versus perch diameter	Greater width in GA
Toepad area versus perch height	Greater area in GA
Toepad area versus perch diameter	Greater area in GA
Forelimb length versus perch height	No relation in GA; decreases with height in ML
Forelimb length versus perch diameter	No relation in ML; increases with diameter in GA
Mass versus perch diameter	Slope of increase greater in GA
Tail length versus perch height	No relation in ML; decreases with height in GA

NOTE: Analyses based on 12 mainland and 27 Greater Antillean species. Toepad analyses in Macrini et al. (2003); all other analyses from Irschick et al. (1997). All variables size-adjusted.

differences in the structure of the microscopic setal hairs that affect clinging ability (reviewed in Chapter 2). Setae could differ in many ways among species, including setal structure, density, or distribution across the toe (in some species, setae occur on scales on the toe, as well as on the toepad [Peterson, 1983]). As a result of these differences, species with different toepad areas might be able to cling equally well, and species with the same toepad area might differ greatly in clinging ability. More detailed functional studies, particularly involving the significance of forelimb and tail length and relative body mass, as well as toepad microstructure, are needed to assess the frequency and importance of many-to-one relationships.[347]

If the existence of many-to-one morphology-performance relationships accounts for ecomorphological differences between mainland and Greater Antillean anoles, then alternative morphological means of producing the same functional outcome must have evolved only when evolutionary transitions occurred between the two areas: that is, early in anole phylogeny, when the basal mainland anoles colonized the Caribbean, and within the Norops group when the mainland was recolonized from the islands (see Chapter 6). Although such coincidence is possible, my intuition is that it is unlikely: ecomorphological variation occurs throughout both radiations and little of it seems to map phylogenetically to those two branches of the phylogeny. The only way to find out, however, is for someone to collect the relevant data.

The alternative hypothesis is that the relationship between functional capability and habitat use differs between mainland and island anoles. This hypothesis is plausible

347. Ironically, hindlimb length, which is one of the only documented many-to-one relationship in anoles, is the one trait for which ecomorphological relationships don't appear to differ among mainland and Greater Antillean anoles (Irschick et al., 1997).

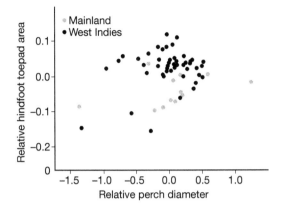

FIGURE 13.18

Relationship between perch diameter and toepad area. Variables relative to body size, as represented by SVL. Modified with permission from Macrini et al. (2003).

given the many differences between mainland and Greater Antillean anoles discussed in Chapter 8 and elsewhere. In particular, if predation plays a much greater role in regulating the population biology and guiding the life history of mainland anoles, then patterns of ecomorphological selection may be very different in the two areas. How differences in predation pressure and intraspecific interactions would lead to selection for different functional capabilities is not clear. In the Greater Antilles, anoles are highly visible and active, with high rates of intraspecific interactions. By contrast, mainland anoles tend to be less active and visible. Greater Antillean anoles might require greater maximal sprinting capabilities to successfully engage in intraspecific interactions, as has been shown in collared lizards (Husak et al., 2006a), as well as to escape predators. Greater clinging ability might be important in the context of intraspecific aggression, in which individuals often lock jaws and try to keep from being being thrown or falling to the ground (e.g., Trivers, 1976; Passek, 2002; Fig. 9.3). By contrast, mainland anoles may rely more on crypsis, rather than high speed, to avoid predators, and intraspecific interactions may be less common. These possibilities are, of course, purely speculative, but could explain why mainland and Greater Antillean anoles occupying the same structural microhabitat nonetheless differ morphologically.

Obviously, more detailed functional and ecological studies are needed on mainland species. With data comparable to those we have in hand for West Indian species, we will be able distinguish between these hypotheses and attempt to understand the cause of the divergent evolutionary paths taken in these regions.

FUTURE DIRECTIONS

Phenotype-environment correlations, especially when they recur repeatedly in evolutionarily independent clades, strongly suggest an adaptive basis for phenotypic variation. Nonetheless, without understanding the functional consequences of phenotypic variation and in turn the ecological significance of functional variation, we cannot fully understand how adaptation has occurred, much less the role that interspecific interactions have played in driving adaptive diversification.

FIGURE 13.19
Anolis pulchellus preparing to jump from a dried grass stem. The effect of unsteady surfaces on locomotion in anoles remains to be studied.

In anoles, the best case study concerns the significance of variation in hindlimb length. The data strongly support the conclusion that differences in hindlimb length often represent adaptive responses to changes in habitat use. Even for hindlimb length, however, many functional and ecological questions remain. I have suggested a number already in this chapter. One area that is ripe for investigation is the interaction of habitat and morphology. The effect of surface diameter on sprint speed and jump distance has been investigated, but these studies have used stiff supports. How performance changes when surfaces are more pliant and, if so whether the magnitude of change is a function of morphology remains to be seen (Fig. 13.19). In addition, in some microhabitats animals must move through a maze of vegetation. For example, small diameter branches commonly have more obstructions that can impede the movement of an animal. Consequently, short legs may be advantageous to enhance the clearance of anoles moving through such cluttered environments (Spezzano and Jayne, 2004). The biomechanics of movement in such conditions has not been studied.

The functional significance of toepad design also requires more study. We know that the ability to cling to a smooth surface is related to toepad size, but little more. How toepads function on narrow, irregular, and rough surfaces, and the functional consequences of variation in number of lamellae is mostly unknown (cf. Russell and Johnson [2007] on the clinging ability of geckos on rough surfaces), as is the extent and significance of toepad microstructure. Moreover, the functional capacities of toepads seem excessive; even *A. sagrei*, with relatively poorly developed toepads, can hang from a single toe (Fig. 2.5). Why anoles—and geckos—should have such great clinging ability is not clear: most likely, the extraordinary clinging ability of these lizards represents a combination of adaptation to extreme circumstances plus a safety factor to prevent failure (Autumn, 2006; Irschick et al., 2006b).

Many other traits vary among anoles and may have functional and ecological significance. Perhaps most obvious is variation in tail length, which ranges from slightly greater than body length in twig anoles to as much as four times the length of the body in grass-bush anoles. Lizard tails serve as counterbalances in running and jumping, are used for maneuvering in midair during a jump, and act as props while moving through vegetation (Collette, 1961; Ballinger, 1973; Arnold, 1988; Higham et al., 2001), but these roles would not seem to be able to explain the huge range in tail size. Anole tails can regenerate, so a role in predator escape is also possible.

The shape of the pectoral and pelvic girdle also varies among ecomorphs (Peterson, 1972; Beuttell and Losos, 1999). Arboreal lizards require greater mobility of the pectoral girdle to facilitate placing the limbs in the many different positions required to move through the three-dimensional vegetation matrix. Peterson's (1974) preliminary studies revealed a trade-off between girdle structures that maximize relatively low speed maneuverability in arboreal species, especially twig and crown-giant anoles, versus structures that provide joint stability and muscle leverage, thus increasing the ability to withstand the forces generated during rapid movements, as exemplified by the trunk anole, *A. distichus*. Grass-bush and trunk-ground anoles, which often move rapidly, but also commonly maneuver through three dimensions, exhibit interesting functional compromises. Unfortunately, Peterson only published the conclusions of her work and not the supporting data. Further work on this topic could prove very illuminating.

Anole teeth and claws also vary, but this variation has largely been ignored. Adult Chamaeleolis develop broad molariform teeth that are used in crushing snails and other hard prey items; other lizards develop similar teeth for the same reason (Fig. 13.20; Estes and Williams, 1984). No other study has examined variation in anole tooth structure, much less how it relates to diet (cf. Hotton, 1955). Similarly, little comparative work has been conducted on the musculature of anoles (but see Herrel et al., 2008). Surely, functionally important variation in muscle type, composition and placement exists among anole species; recent work has demonstrated that ecomorphs differ in the mass of several muscles, but much more detailed investigation is needed (Vanhooydonck et al., 2006c; Herrel et al., 2008).

Understanding the functional significance of trait variation will require more sophisticated investigation of performance capabilities. Anoles often move on irregular surfaces and through complicated three-dimensional environments. Agility and maneuverability may be as important as, or more so than, raw speed. Other aspects of

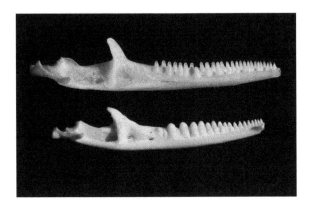

FIGURE 13.20
Molariform teeth at the back of the jaw of *A. Chamaeleolis porcus* (below) are used to crush snails and other hard prey (Leal and Losos, 2000). By comparison, the teeth of *A. equestris* (above) are more typical of anoles, conical toward the anterior of the jaw and tricuspid toward the posterior. Photo courtesy of Luke Mahler.

locomotion—even swimming[348] and gliding—may be important to some species. Vision, hearing, digestive abilities and many other aspects of biological function may represent additional avenues of adaptive differentiation.

Sexual differences in functional capabilities of anoles have received little attention (e.g., Lailvaux and Irschick, 2007a; Calsbeek, 2008). As discussed in Chapter 9, many anole species exhibit sexual differences in both morphology and ecology. The role that functional capabilities play in mediating these differences remains to be studied, as well as the extent to which the morphology-performance and performance-ecology relationships differ between the sexes due to sex-specific roles, such as egg-bearing in females or territoriality in males (e.g., Scales and Butler, 2007).

348. In addition to aquatic species, several other anoles are known to escape predators by jumping into water (e.g., Heatwole et al., 1962; Heatwole and Torres, 1963; Franz and Cordier, 1986).

14

SPECIATION AND GEOGRAPHIC DIFFERENTIATION

Adaptive radiation involves both multiplication of species from a single ancestor and ecological and phenotypic diversification of these species, with the end result that communities are composed of multiple species adapted to different niches. The focus of the last several chapters has been on the second of these two aspects, but the first, the manner in which one ancestral species gives rise to many descendant species, is equally important. Anoles have speciated prolifically, and in the Greater Antilles most of this speciation has occurred within islands, rather than resulting from cross-island colonization and subsequent divergence (Chapter 6). Despite the great extent of this within-island speciation, surprisingly little research has addressed the means by which it occurs, much less the role that speciation plays in anole adaptive radiation.

SPECIATION AND ADAPTIVE DIVERGENCE

At the extreme, two views could be taken on the relationship between speciation and adaptive divergence (reviewed in Schluter, 2000, 2001). On one hand, the two could be completely unrelated. During the process of speciation,[349] differentiating populations might not diverge adaptively, with the result that speciation would produce the raw

---

349. By "speciation process," I refer to cladogenetic speciation in which one ancestral species gives rises to two descendant species, rather than anagenetic speciation, in which an ancestral species transforms into a different descendant species.

material—multiple reproductively isolated entities—for adaptive radiation, but the adaptive component would come later in the process, perhaps as the result of ecological interactions as discussed in previous chapters. This scenario might be particularly likely if speciation occurs in allopatry, followed by ecological differentiation of descendant species when they secondarily come into sympatry.

Selection and adaptive divergence are not necessary in several types of speciation, such as speciation resulting from founder effects, polyploidy, or genetic drift in large populations. For example, allopatric speciation in many groups (e.g., salamanders, snails) often produces geographically isolated species that exhibit few or no adaptive differences (Gittenberger, 1991; Kozak and Wiens, 2006; Wake, 2006); such speciation might result from adaptively neutral processes such as genetic drift or some forms of sexual selection (reviewed in Schluter, 2000; Rundle and Nosil, 2005).

On the other hand, at the opposite extreme, the processes of speciation and ecological divergence might be intimately interrelated. Disruptive selection could lead to ecological differentiation occurring within an ancestral species at a single locality. As the subpopulations differentiated ecologically, selection might favor the evolution of reproductive isolation to prevent interbreeding, thus avoiding the production of offspring ecologically intermediate and unfit for either niche; in turn, the evolution of reproductive isolation would avoid the homogenizing effect of genetic exchange, thus allowing further ecological differentiation. This of course is the highly controversial process of sympatric speciation, in which adaptive differentiation and speciation are causally related and occur simultaneously (reviewed in Coyne and Orr, 2004). Repeated instances of sympatric speciation could lead to a multitude of ecologically differentiated species—an adaptive radiation—all produced in situ.

Two intermediate possibilities exist between the extremes of sympatric speciation and nonadaptive speciation in allopatry: adaptively-driven divergence in allopatric populations resulting in speciation, and speciation among parapatric populations arrayed along an ecological gradient.

ADAPTIVE DIVERGENCE IN ALLOPATRY

Populations that speciate in allopatry may also diverge phenotypically, and some of this divergence may reflect adaptation to different environmental conditions. In fact, laboratory and field studies make clear that allopatric populations diverging to adapt to different environmental situations are much more likely to evolve barriers to successful interbreeding than are allopatric populations living in similar environments (Rice and Hostert, 1993). This result is in agreement with Dobzhansky's (1937) view that reproductive isolation often results as an incidental by-product of evolutionary change, rather than being selected for directly. Consequently, the divergence that occurs in allopatric speciation may contribute to the extent of phenotypic diversification that occurs in an adaptive radiation.

Assuming that populations diverge adaptively in allopatry, what happens when they come back into sympatry? Basically, three possibilities exist. If complete reproductive isolation has already evolved, the populations will interact as different species, and whether they can coexist will be determined by ecological mechanisms. If, on the other hand, no reproductive isolation has evolved at all (i.e., if the populations are completely interfertile), then the two gene pools are likely to meld back together as alleles flow between populations. Perhaps the most interesting situation is when some degree of reproductive isolation has occurred; perhaps individuals tend to mate with others from their own population, but do not always do so. In this case, an evolutionary race will ensue. On one hand, selection will favor the evolution of prezygotic isolating mechanisms because hybrid offspring would be at a disadvantage; parents that mate with individuals of their own population will have greater reproductive fitness. This is the evolutionary process termed "reinforcement." On the other hand, the genetic exchange that does occur will tend to homogenize the gene pools, and selection will favor genetic variation that maximizes the fitness of hybrids, thus decreasing the cost of hybridization.

Reinforcement has long been controversial because many workers predicted that the homogenizing effects of genetic exchange between incompletely isolated populations would usually swamp divergent natural selection and thus lead to the fusion of the two populations, rather than to completion of the speciation process. This is, of course, essentially the same reason that sympatric speciation is thought by many to be unlikely in most cases.[350] The circumstances under which reinforcement is likely to occur are still hotly debated, though in recent years proponents of reinforcement appear to be gaining the upper hand (reviewed in Servedio and Noor, 2003; see also Rundle and Nosil, 2005).

Regardless of whether reproductive isolation evolves before or after sympatry is attained, the end result—assuming speciation occurs—is sympatry of species that are ecologically differentiated to some extent, thus setting the stage for subsequent, greater ecological divergence resulting from interspecific interactions, as discussed in Chapter 11. Scenarios involving initial adaptive divergence in allopatry followed by much greater divergence in sympatry have been particularly well discussed with regard to the evolutionary radiation of Darwin's finches (Grant and Grant, 2002, 2006a, 2008).

SPECIATION ON ECOLOGICAL GRADIENTS

Another possibility intermediate between speciation without adaptive divergence and sympatric speciation is speciation along an ecological gradient, otherwise known as parapatric speciation. Although enjoying a resurgence of interest in recent years, the

---

350. The difference between reinforcement and sympatric speciation is that in the former, the populations may have evolved some degree of reproductive isolation prior to coming back into sympatry, thus increasing the possibility that complete reproductive isolation can evolve quickly enough to forestall genetic homogenization. Another difference is geographic context; the usual scenario for reinforcement is a hybrid zone, with the two species being mostly allopatric, whereas most sympatric speciation models envision extensive or total sympatry of the two species.

idea is a fairly old one, based on the observation that populations of a species distributed across a geographic landscape experience differing selection pressures. Even in the presence of gene flow, populations on opposite sides of an environmental gradient will diverge and adapt to local conditions (e.g., Schneider and Moritz, 1999; Schneider et al., 1999; Smith et al., 2001, 2005). As a result, selection may strongly favor the evolution of reproductive isolation (Endler, 1977; Gavrilets, 2004). This scenario is intermediate between sympatric speciation and reinforcement. As in sympatric speciation, the diverging populations begin with no reproductive isolation; however, the diverging populations are mostly not overlapping geographically, only being in contact at their range borders, rather than throughout their entire ranges.

Like the allopatric scenario just discussed, parapatric speciation would lead to the production of reproductively isolated species that are ecologically differentiated. However, because the reason they had diverged is that populations were adapting to conditions that varied across an ecological gradient, the species would not coexist; rather, their ranges would abut somewhere within the environmental gradient. Subsequent sympatry would require further evolutionary change to allow coexistence.

The preceding discussion indicates that speciation and adaptive divergence may be unrelated or intimately connected. Distinguishing these possibilities is not easy. An important first step is consideration of the mechanisms by which reproductive isolation evolves. Once these are understood, a potential link between the evolution of reproductive isolation and adaptive divergence can be examined (Schluter, 2000; Rundle and Nosil, 2005). In addition, as the preceding discussion also has made clear, the role of speciation in adaptive radiation depends, at least in part, on the geographic context in which speciation occurs. For these reasons, in the remainder of this chapter, I address two questions.

1. What is the mechanism of speciation in anoles and how does the evolution of this mechanism relate to adaptive evolution?
2. What is the geographical context of speciation?

MECHANISTIC APPROACHES TO THE STUDY OF SPECIATION

In the case of anoles, a mechanistic approach should start with those characteristics known to be involved in reproductive isolation: head-bobbing patterns and dewlap configuration. As discussed in Chapters 2 and 9, anoles use both the dewlap and the species-specific head-bobbing patterns to distinguish conspecifics from heterospecifics.

Consequently, as a first approximation, to understand speciation, we need to understand what causes dewlaps and head-bobbing patterns to diverge.[351] Natural selection

351. Much of the literature on speciation focuses on postzygotic reproductive barriers (Coyne and Orr, 2004). By contrast, I focus on prezygotic barriers for two reasons: first, very few cases of hybridization are known among naturally co-occuring anoles (Chapter 2). Thus, either postzygotic barriers between all species are complete, such that no hybrids result from interspecific matings, or coexisting anole species today are

could lead to evolutionary change in dewlaps or head-bobbing patterns in two ways: either selection directly favors evolution of these characters to minimize mating between two incipient species or the characters evolve in response to some other selective factors, with the incidental consequence of causing reproductive isolation. In addition, genetic drift could also lead to divergence in species recognition signals.

SELECTION FOR DIVERGENCE IN SPECIES-RECOGNITION SIGNALS

Almost no cases are known in which sympatric anole species have identical dewlaps—sympatric species always differ in the size, color, or patterning of their dewlaps (Rand and Williams, 1970; Nicholson et al., 2007; Fig. 2.8). This trend suggests that some process is at work that prevents coexistence of species with similar dewlaps.[352] Anole head-bobbing patterns have been studied in less detail (for the simple reason that it is much more laborious to quantify headbobbing pattern than to score dewlap appearance), but the results are much the same: each species seems to have its own, species-specific pattern (Jenssen, 1977, 1978; Appendix 9.1). Moreover, interspecific differences in head-bobbing patterns are related to the number of congeners with which a species is sympatric; species that co-occur with many other species tend to have displays with a greater number of distinct components, as might be necessary to facilitate species recognition, whereas species that co-occur with few other species have displays in which the display components are more homogeneous (Ord and Martins, 2006).[353]

One possibility, as suggested previously, is that species evolve differences in dewlap and head-bobbing patterns to avoid mating with an individual of the wrong species. If hybrid individuals have lower fitness, then natural selection should favor the evolution

---

characterized by prezygotic reproductive isolating mechanisms that prevent interspecific matings. The paucity of observations of interspecific matings, combined with the well understood pre-mating isolating mechanisms (Chapter 9), suggests that pre-mating isolation is probably responsible for the lack of hybridization among anoles. This is not to say that anole species couldn't first become isolated by postzygotic means and only subsequently evolve prezygotic isolation by reinforcement; nonetheless, no data exist to examine this idea. More generally, given the current lack of any data on postzygotic reproductive barriers in anoles, at this point there is nothing to say about how such isolating mechanisms might evolve.

Anoles do exhibit substantial inter- and intraspecific variation in chromosome number and morphology (reviewed in Gorman, 1973; Williams, 1977b), but whether these differences lead to reduced interspecific fertility is unknown. A particularly interesting question for future research might be an investigation of the extent to which post-zygotic isolation evolves in closely-related allopatric populations and thus contributes to the initial stages of allopatric speciation.

352. A recent analysis, however, found that the lack of co-occurrence of species with similar dewlaps is not statistically unexpected. The reason is that so much variety exists in anole dewlaps that even communities assembled by randomly choosing anole species would be expected to contain few or no species with identical dewlaps (Nicholson et al., 2007).

353. This analysis was hampered by difficulty in determining the number of species with which a given species co-occurs. Data were unavailable for many species, and for at least some species, the number of co-occuring species was understated (e.g., A. carpenteri, A. humilis, and A. limifrons all co-occur at La Selva, Costa Rica with five other species [Guyer and Donnelly, 2005]), some times greatly (e.g., A. rodriguezi, which occurs at Los Tuxtlas in sympatry with ten other species [Vogt et al., 1997]). As information on geographic ranges and overlaps become more accessible, a reanalysis of these data could prove interesting.

of means by which hybridization can be avoided. In anoles, this could be accomplished by evolving differences in the dewlap and in head-bobbing patterns.

Not many relevant data are available for anoles. However, one particularly suggestive case occurs in the *brevirostris* species complex in Haiti. Three species of these trunk anoles, nearly indistinguishable in appearance, occur contiguously along the western coast of Haiti (Figure 14.1). The southernmost of these species is *A. brevirostris* itself, which has a light colored pale dewlap. By contrast, the northernmost species, *A. websteri*, has a vivid orange dewlap. Most interesting, however, is the species sandwiched in

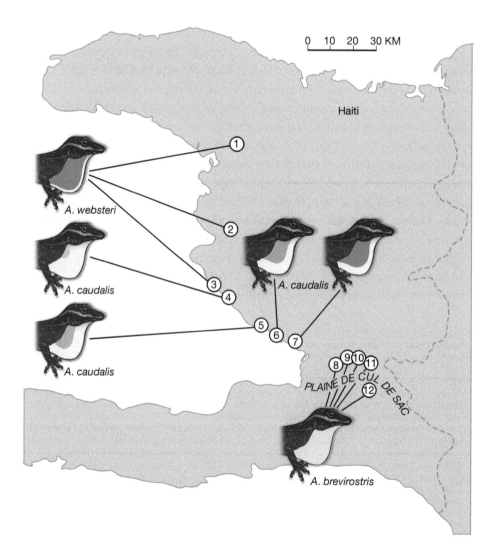

FIGURE 14.1

Dewlap evolution in the *A. brevirostris* species complex in Haiti. Figure modified with permission from Webster and Burns (1973).

between the other two, *A. caudalis*, whose dewlap color varies geographically: at the southern border of its range, near *A. brevirostris*, its dewlap is bright orange, and at the northern edge of its range, where it comes into contact with *A. websteri*, its dewlap is white. Interior populations exhibit variability in dewlap color with change occurring at least somewhat clinally from one end of the range to the other (Webster and Burns, 1973; Jenssen, 1996). Display behavior also differs among all three species, with the behavior of *A. caudalis* being the most distinct from the other two species (Jenssen and Gladson, 1984). The most parsimonious explanation for these differences—particularly the geographic variation in dewlap color in *A. caudalis*—is that they evolved to prevent hybridization between closely related species. Indeed, electrophoretic analyses confirm that levels of gene flow are high among populations within each species, but extremely low or nonexistent between species, including adjacent heterospecific populations (Webster and Burns, 1973).[354]

Of course, the lack of co-occurrence of species with similar species-recognition signals does not mean that differences in the signals among sympatric species evolved in situ. An alternative explanation is that species with similar signals are unable to co-occur in sympatry; if two similar species come into contact, mating mistakes may cause one species to become locally extinct too quickly for evolutionary divergence to occur. This scenario would explain why sympatric species always differ in species-recognition signals, but would not account for how such differences arise in the first place.

SPECIATION AS AN INCIDENTAL BY-PRODUCT OF ADAPTATION

An alternative, selection-based explanation for the evolution of species-recognition signals and the reproductive isolation they produce is that these signals evolve as a by-product of divergent adaptive evolution to some aspect of the environment, as discussed earlier in this chapter. A number of recent examples have highlighted that if species recognition is based on traits that have an ecological function, then adaptive evolution of these traits in response to differing ecological conditions can lead to reproductive isolation between two populations. For example, reproductive isolation between anadromous and stream-living sticklebacks results primarily from the size differences that repeatedly evolve between populations occupying these habitats (McKinnon et al., 2004). Similarly, the beaks of Darwin's finches evolve adaptively in response to variation in the availability of different sized-seeds (Grant and Grant, 2002); because mate choice in these birds is related to beak size, adaptive divergence in beak size among populations can lead to reproductive isolation (Podos, 2001; Grant and Grant, 2006b; for a similar example in crossbills, see Smith and Benkman [2007]).

354. However, one very small-scale study found no assortative mating by species in several mating trials between *A. caudalis* and *A. websteri* in semi-natural conditions, calling into question the role of behavior in preventing genetic exchange between these species (Jenssen, 1996). Follow-up studies are needed to confirm this result.

Reproductive isolation could evolve as an incidental by-product of adaptation to different environments for a second reason. For a signal to be useful in communication, it needs to overcome the background noise and be perceived by the sensory system of the intended receiver in that environment. If the effectiveness of a signal differs among environments, then a population that occupies a new habitat may have signals that initially are ineffective in that environment. As a result, selection may favor signal evolution to enhance the ability of individuals to communicate in their new environment. A consequence of such evolution is that the population may become reproductively isolated from its ancestral population, which retains the ancestral signal. This is the basis of the sensory drive theory of speciation (Endler, 1992; Boughman, 2002), which has long been applied to vocal communication signals (reviewed in Slabbekoorn and Smith, 2002; see also Boncoraglio and Saino, 2007), and more recently to visual signals.

The sensory drive theory likely applies to anoles. The signaling effectiveness of both the dewlap and head-bobbing almost certainly depends on the environmental context. A great deal of recent research has established that different colors are more detectable in different environments (Endler, 1993; Fleishman, 2000). Put simply, in closed forest, the little light that penetrates is primarily in the green and yellow parts of the electro-magnetic spectrum; consequently, highly reflective or transmissive dewlaps[355] (which tend to be white or yellow) are favored. Conversely, in open areas, full spectrum light is available and the optimal dewlap is dark, usually having low reflection and transmission properties and thus contrasting well against the bright background (Fig. 14.2; Fleishman,

FIGURE 14.2

Dewlap detectability in different light environments. (a) In open, sunny habitats, dark-colored dewlaps contrast well against the bright background (*A. sagrei* in the Bahamas), (b) whereas in dark habitats, lighter dewlaps best reflect the available light (*A. cybotes* in Hispaniola).

355. Highly reflective dewlaps reflect more light, whereas highly transmissive dewlaps allow more light to pass through and be seen by an observer on the other side of the dewlap.

1992, 2000).[356] Not surprisingly, forest anoles tend to have white or yellow dewlaps, whereas species found in open habitats tend to have dewlaps that are orange, red, blue, or black (Fleishman, 1992, 2000).[357]

Similarly, the ability to detect the up-and-down movements of a lizard's head should depend in part on what the background vegetation is doing (Fleishman, 1992). In fact, lizard headbobs seem designed to stand out: their jerkiness is very different from the less abrupt swaying of vegetation in the wind (Fleishman, 1988a,c). Although some components of anole head-bob patterns are species-specific (Chapter 2), other parts—such as the initial head bob, which is critical for catching the attention of onlookers (Fleishman, 1992)—are less stereotyped and are altered to increase communication effectiveness in different conditions. For example, as vegetation moves more quickly with increased wind speed, both *A. cristatellus* and *A. gundlachi* increase the rapidity of their head bobs, presumably to keep them distinguishable from the movements of the background (Fig. 14.3; Ord et al., 2007). Anoles also appear to change their display behavior depending on the location of the receiver. For example, when individuals are far apart,

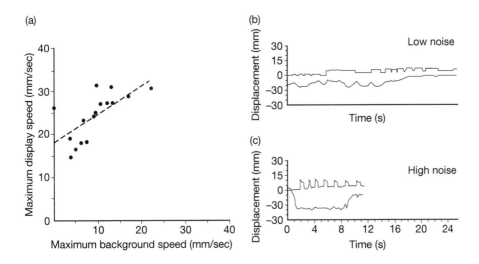

(a)

(b) Low noise

(c) High noise

FIGURE 14.3

Changes in display behavior of *A. cristatellus* as a function of background vegetation movement. (a) As the vegetation moves more quickly due to higher wind speeds, lizards increase the speed of their own movements during displays. (b) Representative display action pattern graphs (Fig. 2.9) illustrating head-bob movements (above the line) and dewlap extension (below the line) during times of low and high background vegetation movement. Modified with permission from Ord et al. (2007).

356. Note that a range of colors is effective in a given habitat. Consequently, multiple species can coexist by evolving different dewlap colors without compromising the ability to communicate (Fleishman, 2000).

357. This is the reason for the lack of consistent inter-ecomorph differences in dewlap color patterns. Because most ecomorph classes contain species that use a wide variety of different light environments, from deep shade to open sun, most variation in dewlap color occurs within, rather than between, ecomorph classes (Nicholson et al., 2007).

A. aeneus alters its display in ways that make the signal more detectable (Stamps and Barlow, 1973; Fleishman, 1992; see also Orrell and Jenssen, 2003).

In the same way, we might expect the stereotyped, species-specific aspects of anole displays to be tuned to environmental conditions. Interspecific microhabitat differences could affect display effectiveness in a variety of ways, such as lighting conditions, wind speed, and the spacing of conspecifics. Data on these variables are not at hand. Nonetheless, a comparative survey revealed differences in display structure between anoles that used primarily sunny versus shaded habitats, and among species occupying different structural microhabitats (Ord and Martins, 2006).

These environment-signal relationships provide a plausible hypothesis for how speciation may occur in anoles. Many widespread anole species have populations that occur in a variety of different habitats; divergent evolution of the signaling behavior and structures of these populations to adapt to differing circumstances may have the incidental consequence of leading to the evolution of pre-mating reproductive isolation between species.

To date, this hypothesis has only been tested in *A. cristatellus*, in which divergence among populations is in accord with the sensory drive predictions. Populations in xeric habitats tend to have dark dewlaps that stand out against the bright visual background, whereas populations in more mesic areas have more reflective and transmissive dewlaps that appear to be lighter against the darker backgrounds.[358] Quantitative analysis conducted in the context of the visual system of these lizards indicates that dewlaps of lizards from mesic populations are more detectable in mesic habitats, whereas xeric population dewlaps are more detectable in xeric habitats (Leal and Fleishman, 2004).

Detailed analyses have revealed substantial geographic variation in dewlap color in several other species (e.g., Case, 1990; Macedonia et al., 2005).[359] Coincident patterns of geographic variation in *A. trinitatis* on St. Vincent and *A. roquet* on Martinique suggest an adaptive basis for dewlap differentiation—in both species, populations occupying forests along the Atlantic coast have dewlaps with greater reflectance in the ultraviolet and blue parts of the spectrum (Thorpe, 2002; Thorpe and Stenson, 2003; see also Thorpe et al. [2008]). More work is needed to assess the causes of this divergence and its effect on genetic exchange between populations.

More generally, the sensory drive hypothesis suggests an explanation for why anoles are so much more species rich than most other lizard clades of comparable age. The combination of using visual signals both for species-recognition and intraspecific

358. See an online video clip in Leal and Fleishman, 2004.
359. The amazing variation of dewlap color in *A. distichus*, which is uniquely variable among anoles both in intra- and interpopulational variability, must be highlighted. This variation has received surprisingly little study (e.g., Case and Williams, 1984; Williams and Case, 1986). Whether it correlates with environmental variation is unknown; some genetically divergent populations (as revealed by electrophoretic protein analysis) differ in dewlap color and pattern, but much of the interpopulational variation is not correlated with genetic differentiation (Case, 1990).

communication, combined with the tendency to readily occupy habitats varying in light availability, may be a strong promoter of speciation, a point to which I will return in the next chapter.

NON-ADAPTIVE CAUSES OF SPECIATION

A wide variety of theories have been proposed to explain how speciation may occur by nonadaptive processes (reviewed in Coyne and Orr, 2004). Some, such as speciation by autopolyploidy, clearly do not apply to anoles.

GENETIC DRIFT AND FOUNDER EFFECTS

Both laboratory and field data indicate that reproductive isolation evolves much more readily when populations are adapting to different environments (Rice and Hostert, 1993), but this finding does not rule out the possibility of speciation resulting from genetic drift in isolated populations.

A more specific hypothesis that is relevant to adaptive radiation on islands is the role of founder effects.[360] Speciation resulting from a founder event is one of the most controversial topics in evolutionary biology; currently, the tide seems to have turned toward those who argue that the necessary requirements for founder effect speciation are so stringent that it probably rarely occurs (e.g., Coyne and Orr, 2004; Gavrilets, 2004; Price, 2007), but the debate certainly is not settled (Futuyma, 2005).

Among anoles, those species that frequently disperse over water and colonize islands—particularly A. carolinensis Species Group anoles and A. sagrei, but also A. distichus and A. grahami (Williams, 1969)—would seem to be most prone to founder effects. However, despite occurring across many islands in the Caribbean and elsewhere in the northern Caribbean, most populations in these clades are little differentiated phenotypically.[361]

The introduction experiments of Schoener and colleagues employed relatively small founder sizes, but have revealed no obvious examples of the sort of rapid and large divergence that would be expected by founder effect speciation.[362] In the one experiment in which founder population size was varied, populations founded by five individuals did not diverge to a greater extent than populations founded by ten individuals (Losos et al., 2001).[363] In these studies, females used to establish populations were probably gravid and perhaps carried sperm from more than one male (Chapter 9); thus, a substantial

---

360. A founder effect occurs when a new population is established by a small number of individuals whose genetic constitution (e.g., number of alleles at a genetic locus, allele frequencies) may differ greatly from the source population as a result of random sampling.

361. Although there are a few exceptions: e.g., A. cons/persus on Grand Cayman, descended from Jamaican A. grahami (Underwood and Williams, 1959; Jackman et al., 2002); A. maynardi (Little Cayman) and A. longiceps (Navassa), both of which arose from within the A. porcatus clade (carolinensis Species Group) in Cuba (Glor et al., 2005); A. sagrei luteosignifer on Cayman Brac and A. sagrei nelsoni on Swan Island (Lister, 1976b).

362. Small scale divergence in limb length and other traits has occurred (Chapter 12).

363. A 2:3 sex ratio was used in all introductions.

amount of genetic variation may have been retained in the founder populations; more-over, given the rapidity with which these populations often increased in number (Schoener and Schoener, 1983c; Losos and Spiller, 1999), the effects of the population bottleneck on genetic variation were probably not large (Nei et al., 1975).

SEXUAL SELECTION

Recently, the idea that sexual selection can promote speciation has gained traction (reviewed in Panhuis et al., 2001; Coyne and Orr, 2004). The idea is that if sexually-selected traits are the same traits that are used in species-recognition, then divergence driven by sexual selection may have the incidental consequence of leading to reproduc-tive isolation between populations. Although a number of theoretical mechanisms have been proposed (e.g., Lande, 1981; Holland and Rice, 1998), most actual evidence for this hypothesis is indirect, relying on among-clade correlations between putative proxies for the strength of sexual selection and species richness (reviewed in Coyne and Orr [2004]; see also Futuyma [2005]).

In the broadest sense, sexual selection refers to differential mating success among members of one sex. In this sense, the sensory drive mechanism outlined above would be an example of sexual selection: males with dewlaps or displays that were more easily detected by females in a given light environment would be more successful at attracting mates. However, much of the current interest in sexual selection refers to various mech-anisms by which mating preferences might evolve independent of the environmental context (Andersson, 1994; Andersson and Simmons, 2006).[364]

Certainly, any mechanism that caused females in two populations to evolve different preferences for dewlap colors or display patterns could lead to populations becoming re-productively isolated. Unfortunately, as discussed in Chapter 9, currently no data exist on female mate choice. The recent discovery that females mate multiply, and not only with the male in whose territory they reside (Chapter 9), indicates that the opportunity for female mate choice exists. Whether this mate choice, if it occurs, is based on signals that are also involved in species recognition, and thus might cause speciation, remains to be seen.

THE GEOGRAPHY OF *ANOLIS* SPECIATION

The geographic context of anole speciation has been little studied. Much of what has been published is fairly speculative. On the one hand, some authors (e.g., Lazell, 1996, 1999) have presented scenarios by which populations may have become geographically

364. It is for this reason that I discuss speciation by sexual selection under the "nonadaptive causes of speciation" heading. Certainly, some speciation driven by sexual selection involves adaptive evolution, as when signaling behavior evolves to adapt to environmental circumstances. Moreover, some theories of mate choice evolution—e.g., the good genes theory—might also be considered adaptive in the sense that females choose males that have the best genes for survival in the given environment (Schluter, 2000; Coyne and Orr, 2004). Nonetheless, many mechanisms of sexual selection do not involve changes that lead to greater adaptation to the environment, and the extent to which sexual selection promotes adaptive change is debated (e.g., Rundle et al., 2006).

isolated, leading to allopatric speciation. In a few cases, concordance of genetic and geological data supports this hypothesis (e.g., Glor et al., 2004; see also Ogden and Thorpe [2002], discussed below).

On the other hand, others have assumed that the initial stages of the anole radiations have been the result of sympatric speciation producing the different ecomorphs on each island (e.g., Doebeli and Dieckmann, 2000; Shaw et al., 2000; Thomas et al., 2003); these assertions are usually stated without explanation and probably result from authors placing their own worldviews about how speciation proceeds into what they read about anole diversity.

Nonetheless, both of these views have merit. The existence of closely related species that are allopatrically distributed—the endpoint of the intra-ecomorph radiations on Hispaniola and Cuba (see Chapters 3 and 7)—strongly suggests the occurrence of allopatric speciation.[365] Conversely, sympatric speciation scenarios have an attractive simplicity to explain the evolution of the ecomorphs. In the relatively simple one- and two-species communities of the Lesser Antilles, most species occur island-wide. Similarly, in the Greater Antilles, each ecomorph is usually represented by at least one species that occurs island-wide on each island. Thus, one might reasonably conclude that the ancestral anole on an island would have been similarly widespread. Assuming an initial island-wide distribution for the ancestral anole, it is not easy to envision a scenario by which this species might have become fragmented into geographically isolated populations on Puerto Rico and particularly on Jamaica, which has a central mountain range and very few offshore islands (but see Lazell, 1996). Assuming that the ancestral anole species on an island was widespread, sympatric speciation would seem to be a more parsimonious explanation for evolution of the ecomorphs on each island than the ad hoc invocation of geological or climatological events that fragmented an ancestral species' range, followed by allopatric speciation and subsequent range expansion of the newly-arisen species, producing the sympatry of the ecomorphs that occurs today. Of course, these two viewpoints are not incompatible: the predominance of allopatric speciation leading to ecologically little-differentiated forms at the end of a radiation does not preclude the occurrence of sympatric speciation producing major ecomorphological differences early in a radiation (cf. Price, 2007, pp. 33–34).

365. In species-rich ecomorph clades, many of the species have relatively small geographic ranges that are allopatric from most or all other members of the clade (e.g., *alutaceus, sagrei* and *carolinensis* Species Groups on Cuba; *cybotes* and *chlorocyanus* Species Groups on Hispaniola [Williams, 1965; Garrido and Hedges, 1992; Glor et al., 2003, 2005; Knouft et al., 2006]). More detailed intra-clade phylogenies are needed for these species-rich ecomorph clades to better identify the sister taxa of many of these species.

Along these lines, Glor and I have argued that inferences about the geographic mode of speciation based on phylogenetic examination of species' current geographic ranges are likely to be unreliable (Losos and Glor, 2003). Nonetheless, the widespread occurrence of allopatry among sister taxa is much more likely to result from allopatric speciation than from sympatric or parapatric speciation followed by shifts in geographic range that cause sister taxa to become secondarily allopatric (Barraclough and Vogler, 2000; Losos and Glor, 2003; Fitzpatrick and Turelli, 2006).

Consideration of anole species distributions in a phylogenetic context provides two additional points of importance in thinking about anole speciation. First, most anole cladogenetic speciation has occurred within Greater Antillean islands. Put another way, phylogenetic analysis reveals evidence for few instances of dispersal from one Greater Antillean island to another (Chapter 6). As a result, most speciation events in the Greater Antilles must have occurred by divergence entirely within a single Greater Antillean island. Allopatric speciation, if it did occur, must have resulted from isolation in different parts of these islands or on offshore islands.

Second, within-island cladogenesis appears to be limited almost entirely to the Greater Antilles. In almost no cases do sister species co-occur on any island smaller than Puerto Rico (Losos and Schluter, 2000).[366] Allopatric speciation of populations on different islands thus must be mostly or entirely responsible for the endemic species that occur in the Lesser Antilles and on small islands in the Greater Antilles.[367]

Three explanations could account for this area threshold, two of which may be readily dismissed. The first possibility is that species on small islands are relatively recent arrivals and haven't been there long enough to speciate. Many landbridge islands were connected to nearby Greater Antillean islands during the last ice age, when sea levels were lower, and thus populations on these islands have been isolated for only a few thousand years. However, many other small islands have never been connected to larger landmasses. Moreover, genetic data indicate that species on some of these islands have been genetically isolated for many millions of years (e.g., Malhotra and Thorpe, 1994, 2000; Schneider, 1996; Brandley and de Queiroz, 2004; Glor et al., 2005), which would seem long enough for speciation to occur given that many ecomorph clades have diversified greatly over roughly the same period.[368]

The second possibility is that oceanic islands smaller than Puerto Rico are not ecologically heterogeneous enough to allow the coexistence of multiple species, even if they should appear by speciation. This suggestion is not tenable. Many much smaller landbridge islands now contain 2–4 or more species (Chapter 4). For example, near Puerto Rico, three species co-occur on an island 0.15 km² in area and four species occur on South Bimini, Bahamas, which is 8 km² (Rand, 1969). Given that the anole ecomorph

366. Thorpe et al. (2004) suggested that more than one species may occur on Martinique. However, these putative species, which are parapatrically distributed, are thought to have arisen by allopatric speciation on distinct islands which then were subsequently merged by volcanic activity into the single island of Martinique that exists today. In addition, at the time Losos and Schluter (2000) was published, *A. trinitatis* and *A. griseus* on St. Vincent were not thought to be sister taxa, in contrast to current phylogenetic information (Chapter 7).

367. "Mostly" because it is always possible that sympatric speciation occurred on one island in the Lesser Antilles, and then the two species each sent out colonists to other islands which evolved into different species so that the original sympatric pair of species would no longer be sister taxa, but rather members of sister clades. This caveat does not apply to satellite islands in the Greater Antilles, whose species clearly originated from ancestors on the Greater Antilles.

368. For example, compare the branch lengths of Lesser Antillean species in the *roquet* and *bimaculatus* Series to those of species within the ecomorph radiations in the *alutaceus*, *cybotes*, and *sagrei* Series (Afterword Figure 1).

story is one of partitioning of structural microhabitat within a single locality, it is not surprising that small areas can maintain multiple species, as long as the islands have more than scrub vegetation and the species have a way of getting there. Lack of ecological opportunity cannot account for the speciation-area threshold.

This leads to the third possible explanation: lack of speciation on ecologically diverse islands suggests that sympatric speciation does not occur in anoles. Consider, for example, Guadeloupe, which is split into two parts that are connected by a narrow isthmus and has a total area of 1,628 km². Much of the island is covered with dense and tall rainforest; the highest peak, La Grande Soufrière, soars to 1,467 m. Moreover, the prevailing winds produce a wet side and a dry side, with sometimes sharp ecological gradients where they meet. Despite this environmental heterogeneity, only one species, *A. marmoratus*, occurs on Guadeloupe. Genetic data suggest that it has been there since at least the early Pliocene (Malhotra and Thorpe, 1994; Schneider, 1996), which is in accord with the geological history of the island (Komorowski et al., 2005).

To my mind, the environmental heterogeneity of Guadeloupe is comparable to that on Puerto Rico, an island five times greater in area.[369] My guess is that if one were to airlift the anole fauna of Puerto Rico to Guadeloupe, all ten species would be able to exist in appropriate areas. Certainly at first glance it would seem that appropriate habitat for all ten occurs somewhere on Guadeloupe.[370]

Consequently, I conclude that if the processes of sympatric or parapatric speciation were possible in anoles, then these processes should have occurred on Guadeloupe and similar islands (e.g., Dominica and Martinique, also large and environmentally heterogeneous islands occupied by only one species). Lack of speciation on these islands suggests that anole biology does not meet the exacting conditions needed for sympatric speciation to occur, nor even the less exacting requirements of parapatric speciation (Coyne and Orr, 2004; Gavrilets, 2004).[371]

Why, then, is there an area threshold for speciation in anoles? The answer would seem to be that smaller islands are not large enough to provide opportunities for geographic isolation.

On this admittedly somewhat flimsy evidence (the lack of speciation on large Lesser Antillean islands and the occurrence of allopatric speciation among closely related

369. I am unaware of any quantitative data to test this assertion. Ricklefs and Lovette (1999) used vegetation maps to quantify habitat diversity among islands in the Lesser Antilles and found that Guadeloupe was the most diverse. However, comparable vegetation maps were not available for Puerto Rico (R. Ricklefs, pers. comm.).

370. It is too bad that this hypothesis cannot be tested ethically. I have daydreamed about transplanting only males, which would at least test whether these species could survive on Guadeloupe. *Anolis marmoratus* is divergent enough from Puerto Rican anoles that it is extremely unlikely that they would hybridize. In any case, the recent invasion by the Puerto Rican trunk-ground *A. cristatellus* (Eales et al., 2008) of nearby Dominica, which also hosts only one native species, indicates that at least one Greater Antillean ecomorph can exist on large Lesser Antillean islands in sympatry with a native species.

371. Price (2007) draws the same conclusion from the lack of speciation by birds on isolated islands; the case of the single species of Darwin's finch on Cocos Island is a particularly apt parallel case (Grant and Grant, 2008).

members of the same ecomorph class on Cuba and Hispaniola), I conclude that the evidence to date provides no support for the occurrence for non-allopatric speciation in anoles, and moderate corroboration of an allopatric speciation model.

INTRASPECIFIC GENETIC DIVERGENCE

A time-honored approach to the study of speciation is to examine intraspecific divergence on the assumption that it is an early stage along the road to species-level differentiation (Mayr, 1963). Of course, this need not be the case: intraspecific divergence in many cases will not ultimately result in speciation, and some speciation mechanisms (e.g., founder effect speciation) might be expected to occur so quickly that intermediate, intraspecific stages are unlikely to be observed. Nonetheless, the study of intraspecific differentiation has a long history in speciation research that continues to this day (e.g., Knowles and Richards, 2005; Kozak et al., 2006).

Research on intraspecific divergence within anole species dates to the allozyme era. A number of studies beginning in the early 1970s examined among-population differentiation using gel electrophoresis methods. The results indicated that levels of interpopulation variation in *Anolis* are not exceptional (e.g., Webster et al., 1972; Gorman and Kim, 1975, 1976; Buth et al., 1980; Case and Williams, 1987; Case, 1990), although a few cases of high interpopulation divergence were reported (Wade et al., 1983; Case and Williams, 1984), including the discovery of the cryptic species in the *brevirostris* species complex discussed above (Webster and Burns, 1973).

MITOCHONDRIAL DNA STUDIES OF GENETIC DIFFERENTIATION AMONG POPULATIONS

This line of research then was quiescent until the DNA era was in full swing. Examination of anole phylogeography,[372] as the field came to be called, focused initially on the anoles of the central Lesser Antilles—Dominica, Guadeloupe, and Martinique—which exhibit such extensive among-population variation in morphology (Chapter 12). Examination of mtDNA across localities yielded a big surprise: immense differences were found among populations—as much as 10% or more sequence divergence (Malhotra and Thorpe, 1994, 2000; Schneider, 1996; Ogden and Thorpe, 2002; Thorpe and Stenson, 2003). Such extensive differentiation is as great as that often seen between species and suggests that populations have been diverging for millions of years (see Chapter 6 for discussion of dating using DNA data). Moreover, these differences were not related to geographic distances separating populations. Rather, in some cases neighboring populations exhibited great genetic differences (Fig. 14.4). Subsequent work on other species

---

372. The phylogenetic study of the geographic distribution of genetic variation within a species.

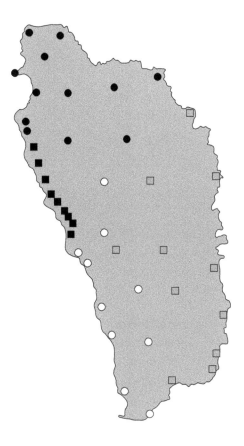

FIGURE 14.4
Phylogeography of *A. oculatus* on Dominica.
Four divergent groups are recognized on the
basis of variation in the cytochrome *b* gene. In
some cases, geographically proximate populations occur in different groups and thus are
genetically very different. Modified with permission from Malhotra and Thorpe (2000).

demonstrated that such differentiation in the Lesser Antilles seems to be the rule, rather than the exception (Thorpe, 2002; Thorpe et al., 2005a).

These results were surprising, but perhaps not completely unexpected. After all, Lesser Antillean anoles on one-species islands were known for their exceptional geographic variation in morphology (Lazell, 1972), so finding similarly great divergence in genetics just reinforced the conclusion that something unusual is going on in these species.

What was not expected was that similar patterns also characterize Greater Antillean species. The first study to look at the phylogeography of Greater Antillean taxa was Jackman et al.'s (2002) study of Jamaican anoles. As part of an examination of species-level relationships, DNA sequencing included geographic sampling of several species. Sample sizes were small (one individual per locality; maximum of four localities per species), but high levels of intraspecific divergence were clearly indicated: three species (the trunk-ground *A. lineatopus* and the trunk-crown *A. grahami* and *A. opalinus*) exhibited mitochondrial divergence exceeding 10% between some populations.[373]

---

373. The only species that did not show comparable divergence among geographically widely-separated populations was the crown-giant *A. garmani*.

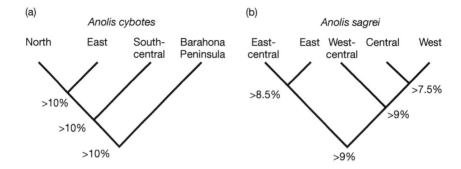

FIGURE 14.5

Interpopulational genetic differentiation in (a) *A. cybotes* in the Dominican Republic and (b) *A. sagrei* in Cuba. Clades are parapatrically-distributed with little or no geographic overlap. Numbers represent pairwise uncorrected mitochondrial sequence divergence between populations across clades. Substantial genetic differentiation also exists between geographically non-overlapping subclades within some of these clades. Data from Glor et al. (2003) and Kolbe et al. (2004).

Subsequent work indicated that deep interpopulational divergence is the norm for Greater Antillean anole species: every geographically widespread species that has been examined on Cuba, Hispaniola and Puerto Rico exhibits similar results, including the trunk-ground anoles *A. allogus*, *A. cristatellus*, *A. cybotes*, and *A. sagrei* (Fig. 14.5; Glor et al., 2003; Kolbe et al., 2004; Knouft et al., 2006; Rodríguez-Robles et al., 2007; Kolbe et al., 2007a) and the trunk-crown anoles *A. allisoni*, *A. chlorocyanus* and *A. porcatus* (Glor et al., 2004, 2005; Kolbe et al., 2007a). Deep mitochondrial divergence was also found in three out of four Amazonian species (Glor et al., 2001b).

These results were unexpected for two reasons. First, most of these species have broad and continuous distributions. Their ranges generally are not broken by geographic barriers such as mountains or large rivers, so impediments to gene flow would not seem to exist for these highly mobile animals. Moreover, most genetic discontinuities are not obviously associated with an area that may have served as a barrier in the past. Thus, a priori, no reason existed to expect to find deep genetic subdivisions. In addition, second, although some geographic variation in morphology has been noted in some of these species, it is generally neither striking in magnitude,[374] nor geographically abrupt, and thus gives little indication of such drastic underlying genetic fragmentation.[375]

374. With some exceptions, such as the transition between *A. grahami grahami* and *A. grahami aquarum* in eastern Jamaica (Underwood and Williams, 1959).

375. At this point, I should offer an apology to the memory of Albert Schwartz. Schwartz was one of the great herpetological taxonomists of the twentieth century. He made prodigious collections and described an enormous number of new taxa: 86 new species and 279 new subspecies (Schoener, 1996)! Many of these taxa were described on the basis of typical herpetological taxonomic characters, such as scale counts and shapes, coloration and patterning, and body size. I have to admit that I was of the opinion that his description of subspecies was, to put it mildly, excessive. However, I have been chastened to note that at least a few of his subspecies conform at least roughly to some of the clades discovered by genetic studies. It will be interesting to see whether future morphological work discovers that these genetically-identified clades differ morphologically in the types of characters upon which Schwartz relied.

Consequently, these results are tremendously exciting, as well as challenging. I personally am astonished that despite forty years of intensive study of anole evolutionary ecology in the Greater Antilles, based on just about every molecular method developed to study evolutionary processes, Richard Glor and Jason Kolbe (then both graduate students) and others have discovered a potentially important aspect of anole diversity that was previously completely unknown.

The challenge comes in figuring out what these findings mean. Given that in most of these species, extremely genetically differentiated populations come into contact without homogenizing their gene pools, one obvious hypothesis is that the populations are reproductively isolated and thus represent distinct species. This scenario would imply that what are currently recognized as geographically widespread species are actually comprised of a series of parapatrically distributed species, perhaps as many as 6–8 in some cases (e.g., *A. cybotes* [Glor et al., 2003] and *A. sagrei* [Kolbe et al., 2004]). Certainly, the amount of divergence in mtDNA is comparable to that separating taxa recognized as distinct species in other groups (Johns and Avise, 1998; Schulte et al., 2006). If every widespread anole species is actually a complex of several-to-many species, then the extent of anole species diversity has been greatly underestimated. This result may seem unlikely at first blush, but in fact researchers working on a wide variety of taxa have discovered similar examples of interpopulation genetic differentiation—based primarily, but not exclusively, on mtDNA—and have inferred the existence of extensive and previously unrecognized species-level diversity (García-Paris et al., 2000; Riddle et al., 2000; Yoder et al., 2000, 2005; Sinclair et al., 2004; Gübitz et al., 2005; Kozak et al., 2006; Bickford et al., 2007; Boumans et al., 2007).

ARE MITOCHONDRIALLY-DIFFERENTIATED CLADES REPRODUCTIVELY ISOLATED?

Whether these genetic discontinuities correspond to reproductively isolated units in anoles—and in most other taxa—has not been determined (but see Gibbs et al., 2006). Data on reproductive isolation are essential for understanding how such divergent clades can exist in parapatry without melding, regardless of whether one thinks that such information is necessary to evaluate the species-level status of clades (see discussion in Chapter 2). The hypothesis of reproductive isolation could be tested in two ways. First, reproductive isolation could be directly investigated by examining how lizards interact at the contact zone where divergent clades meet. Field and laboratory studies could determine whether populations interbreed and, if so, whether fertile offspring result. To my knowledge, such studies have not yet been conducted on anoles. If populations are reproductively isolated, then behavioral and ecological studies can investigate the ecological or behavioral mechanisms that prevent widespread coexistence.

Second, the reproductive isolation hypothesis could be investigated indirectly. If mitochondrially divergent populations are reproductively isolated, then little nuclear gene flow should occur between them. This hypothesis has not been widely tested, but several studies have used nuclear markers to examine whether gene flow occurs across mitochondrial boundaries. The results of these studies are mixed. In Cuba, little nuclear gene flow occurs among populations of large green anoles that are greatly divergent in mtDNA and that probably differentiated in allopatry in the Miocene, when Cuba was sundered into three paleoislands by high sea levels (Glor et al., 2004). On the other hand, in *A. roquet* on Martinique, also thought to be a case of secondary contact of divergent populations that arose on separate islands that are now joined, nuclear gene flow occurs unimpeded across contact zones where highly differentiated mitochondrial clades meet (Ogden and Thorpe, 2002; Thorpe et al., 2008). Similarly, in Dominica, levels of nuclear gene flow are generally high, even between populations with highly divergent mitochondrial haplotypes; however, at one locality (discussed below), gene flow is low where two populations that are both morphologically and genetically distinct come into contact (Stenson et al., 2002).

Studies of introduced populations can investigate how individuals descended from genetically differentiated populations interact, both behaviorally and genetically. The extreme genetic fragmentation of native populations facilitates investigation of whether multiple introductions have occurred from different parts of a species' native range. In fact, that is exactly what has happened: genetic data indicate that many introduced anole populations are derived from more than one native source population (Kolbe et al., 2007a); at the extreme, the occurrence of *A. sagrei* in Florida is the result of introductions from at least eight different areas in Cuba and the Bahamas (Kolbe et al., 2004).[376] The coexistence of individuals from genetically distinct populations sets the stage for testing the hypothesis that these populations are reproductively isolated. Preliminary studies based on nuclear microsatellite loci suggest that, quite the contrary, individuals descended from different regions of Cuba are interbreeding and homogenizing their initially distinct gene pools (Kolbe et al., 2008b).

It is too early to generalize from these studies of nuclear genes. Certainly, however, the results from Martinique, Dominica, and Florida are intriguing: if gene flow is occurring between highly divergent populations that exist in parapatry, why don't the mitochondrial clades intermix, producing populations that are polymorphic for greatly differentiated mitochondrial haplotypes, a phenomenon that has been rarely detected in anoles (e.g., Glor et al., 2004)?

One possibility is that the boundaries between highly distinctive mitochondrial clades are the result of secondary contact between populations that diverged in allopatry and that have not yet had time to intermix. Given the large population sizes of anoles, we might expect introgression of genotypes to occur relatively slowly. Nonetheless, given enough time and no barrier to gene flow, such intermixing should occur and should lead

---

376. Because descendants from as many as five different introductions may occur at a single locality (Kolbe et al., 2007b), introduced populations are often genetically much more variable than native range populations—whether this has anything to do with the success of anole invasions remains to be seen (Kolbe et al., 2007a).

to widespread mitochondrial polymorphism within populations, erasing the genetic differentiation that existed between populations. The rarity of such polymorphism suggests that this explanation likely is not general.

An alternative hypothesis is that the divergent mitochondrial clades are simply the result of in situ differentiation in which large genetic discontinuities arose through random genetic processes within a continuously distributed population. Irwin (2002) showed by simulation that if individuals are extremely philopatric, then given enough time, large genetic breaks in neutral markers may arise between adjacent populations. Essentially, this is the concept of isolation by distance applied to discrete genetic markers such as mitochondrial haplotypes.

Evaluating the hypothesis with respect to anoles is hard because Irwin's simulations were fairly simplistic, making comparison to actual data difficult. The model does make two relevant predictions, that the likelihood of developing large genetic discontinuities should increase with decreasing female dispersal (because mtDNA is maternally inherited) and with decreasing population size.

On the face of it, anoles would not seem to be likely candidates to develop such deep genetic breaks in situ. Trunk-ground and trunk-crown anoles have the physical ability to move large distances[377] and most species for which genetic data exist are not only wide-ranging, but also are generally very abundant.[378] Consequently, one would think that the prerequisites necessary to develop genetic fragmentation within a continuous population across a homogeneous landscape—low dispersal and small populations sizes—are unlikely to be met in widespread anole species (keeping in mind that few actual data are available on anole dispersal; see Chapter 8). On the other hand, discordance between multiple unlinked markers—as occurs in *A. roquet* and *A. oculatus* for mitochondrial and nuclear genes—would be an expected result of such neutral divergence across a homogeneous landscape (Irwin, 2002; Wilkins, 2004). Moreover, sex-based differences in dispersal could produce differences between patterns of genetic variation of maternally-inherited versus autosomal markers (Stenson et al., 2002). Clearly, collection of detailed data on dispersal and population sizes and development of a model tuned to anole biology are needed to better evaluate this isolation-by-distance hypothesis.

THE RELATIONSHIP BETWEEN MORPHOLOGICAL AND GENETIC DIFFERENTIATION

The great degree of geographic variation in both morphology and genetics in Lesser Antillean anoles suggests that the two phenomena are linked. This, of course, is a stepping stone to the idea of parapatric speciation, that natural selection across an environmental

---

377. For example, I have seen trunk-ground anoles sprint 10–15 m in a few seconds to get away from a conspecific or a potential predator (me).

378. Of course in population genetics terms, the important parameter is not population size per se, but the "effective population size" which is an estimate of the effect of nonrandom mating and other factors on the genetic variation in a population. Given the enormous population sizes of many West Indian anole species, effective population sizes are likely to be very large even if mating is highly nonrandom, but no actual estimates are available.

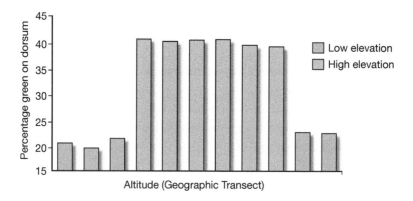

FIGURE 14.6

Abrupt morphological shift associated with environmental change in *A. roquet* on Martinique. Lizards from high elevation populations, which occur in montane rainforest, are much greener than lizards from the lowlands. Modified with permission from Thorpe and Stenson (2003).

gradient can produce both adaptive divergence and reproductive isolation, as discussed earlier in the chapter.

A hypothesis of parapatric speciation is very difficult to test. Its phylogenetic signal—populations of different species on either side of an ecological gradient being more closely related to each other than to members of their own species—is likely to be obliterated by gene flow once reproductive isolation is attained, unless levels of intraspecific genetic exchange are very low (Thorpe, 1984). Moreover, such a phylogenetic pattern might also be attributed to hybridization after secondary contact, rather than to parapatric speciation (Endler, 1977).

An alternative method to detect parapatric speciation is to catch populations in the act of diverging: populations in the process of speciating parapatrically should display a pattern in which morphological divergence across an environmental gradient is accompanied by decreased gene flow. Such a pattern is expected during the course of parapatric speciation, but the converse is not true: detection of such a pattern does not mean that parapatric speciation is occurring, because concordant clines in genes and morphology can be static, rather than representing an intermediate step in the speciation process. Given these difficulties, probably the only way to conclusively demonstrate the occurrence of parapatric speciation would be to document concordant clines steepening as reproductive isolation increased through time.

In the Lesser Antilles, strong relationships exist between variation in the environment and in presumably adaptive morphological characters (reviewed in Chapter 12). When the environment changes abruptly over short distances, steep morphological clines often are found (Fig. 14.6). These are the locations in which reduced gene flow would be expected in a parapatric speciation hypothesis.

Evaluating this hypothesis is complicated because patterns of mitochondrial and nuclear genetic variation are not concordant in the Lesser Antilles. The two best-studied cases are *A. oculatus* in Dominica and *A. roquet* in Martinique. In *A. oculatus*, morphological variation is for the most part independent of both mitochondrial and nuclear genetic variation, except at one location. This exceptional locality is not the site of an environmental transition, and the concordant morphological and genetic transition is interpreted as a case of secondary contact after allopatric differentiation which resulted from a volcanic lava flow. Although reduced nuclear gene flow is documented across this contact zone, the populations are still connected by genes flowing through adjacent populations elsewhere in the island (Stenson et al., 2002; Thorpe et al., 2004);[379] overall, *A. oculatus* would not seem to be a case of parapatric speciation in action.

In Martinique, the situation is different. Morphological and mitochondrial variation are not related (Fig. 14.7),[380] but where the environment changes greatly, sharp clines in morphology and reduction in nuclear gene flow occurs (Figs. 14.6 and 14.7; Ogden and Thorpe, 2002; Thorpe and Stenson, 2003; Thorpe et al., 2008). Whether this situation

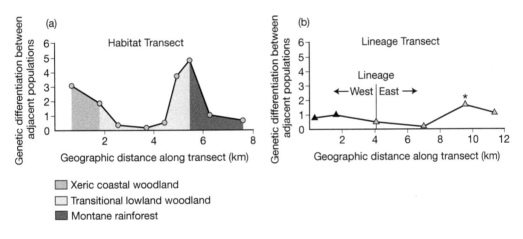

Xeric coastal woodland
Transitional lowland woodland
Montane rainforest

FIGURE 14.7
Reduced nuclear gene flow across environmental gradients along geographic transects in *A. roquet*.
(a) Genetic differentiation in nuclear microsatellite loci is significantly higher between populations on opposite sides of environmental transitions compared to comparisons between populations that both occur within the same habitat. By contrast, (b) in a transect within the montane rainforest habitat type across a contact zone between two mitochondrial clades, populations on either side of the contact are no more genetically differentiated for the nuclear genes than are populations within the mitochondrial clades.

379. A broadly similar situation exists on Guadeloupe, where morphological and mitochondrial variation are mostly unrelated, but concordant clines exist in several places (Malhotra and Thorpe, 1994; Schneider, 1996). Data on nuclear genes are needed to clarify what is going on there, and elsewhere, in the Lesser Antilles (e.g., *A. trinitatis* [Thorpe, 2002]).

380. Population differentiation in mtDNA evolved in allopatry, before volcanic eruptions joined several unconnected landmasses into a single island. However, these mitochondrial clades mostly come into contact in areas of environmental homogeneity, and neither divergence in morphology nor reduction in nuclear gene flow occurs across these contact zones (Ogden and Thorpe, 2002; Thorpe and Stenson, 2003).

represents a case of populations caught in the act of speciating is not clear (Thorpe et al., 2004, 2008). Given the rapidity with which parapatric speciation is thought to occur (Gavrilets, 2000b), one might have expected speciation to be completed by now, under the assumption that the environmental transition—from montane rainforest to xeric lowland forest in the rainshadow of the mountain—has been in existence for some time. Alternative possibilities are that a selection-gene flow equilibrium has been reached across the transition with divergence stopping short of speciation, or that gene flow occurring indirectly through other populations on the island (as occurs in *A. oculatus*), is countering selection for divergence directly across the transition.

Few studies have examined interpopulation divergence in morphology and gene flow in Greater Antillean species. Calsbeek et al. (2007) examined nearby populations of Bahamian *A. sagrei* that occurred in different habitats. In habitats with broader surfaces, lizards had longer legs than lizards in habitats with narrower structures, mirroring patterns seen in previous intra- and interspecific studies (Chapter 12). Nonetheless, levels of gene flow between these populations were high, providing no evidence of any incipient reproductive isolation.

SPECIES AND SPECIATION IN *ANOLIS*, AND FUTURE DIRECTIONS

Taking advantage of the lack of data necessary to resolve some of the issues discussed in this chapter, I will conclude by providing my take on anole speciation, fully recognizing that some of my positions are based more on intuition than on data.

1. Natural selection drives phenotypic divergence among populations. This has been amply demonstrated in Chapter 12. Adaptation may lead to the evolution of reproduction isolation of populations occurring in different environments.

2. Nonetheless, no evidence indicates that selection can lead to speciation among genetically connected populations. The environmental conditions for parapatric and sympatric speciation exist on many Lesser Antillean islands, but in situ speciation has not occurred. Gene flow may be reduced by divergent selection, as in *A. roquet*, but this phenomenon seems too localized to lead to the production of genetically isolated entities, even if some measure of reduced genetic exchange evolves across some environmental gradients. By default, I presume that most or all speciation in anoles occurs in allopatry. This does not mean that selection is not involved in speciation; quite the contrary, allopatric populations experiencing divergent selection are probably more likely to speciate than those in the same selective environment (e.g., Thorpe et al., 2008).

3. The unexpectedly high levels of mitochondrial divergence seen in many anole species raise more questions than answers. These data indicate either

that some population-level process specific to the mitochondrial genome is occurring or that these forms represent divergent gene pools, i.e., species. The data currently in hand do not provide evidence for restricted gene flow between mitochondrial clades in most species. Why, then, do those clades persist, and why are they found in almost all widespread anoles? My hunch is that many of these clades in the Greater Antilles actually are genetically isolated and that, as a consequence, anole species diversity has been greatly underestimated (for a different view, see Thorpe et al. [2008]).

Explaining how these clades originated in the first place, however, remains a mystery. Given my views on allopatric speciation, I would have to assume that even though these species—or species complexes, if that's what they are—are widespread and continuously distributed today, they must have been geographically isolated in the past. In most cases, no data support this prediction (see Glor et al. [2004] for an exception). Paleoclimatic modelling of species' niches,[381] could provide a test of the hypothesis that species had less continuous ranges in the past.

Much of the research needed to address these points is obvious. The most exciting questions, in my opinion, revolve around the unexpectedly large genetic subdivisions that occur within species. These subdivisions have been detected primarily with mtDNA; studies with other markers are needed to test whether the mtDNA is accurately tracking population history.

These studies need to be complemented with detailed behavioral and ecological studies to understand what goes on at contact zones. Assuming that the mtDNA divergence is confirmed by other markers, we need to know how such divergences are maintained. How do members of the populations interact at contact zones? Are they reproductively isolated? If so, are they isolated by behavior or through post-mating barriers? And are the clades ecologically differentiated?[382]

Alternatively, if the divergence in mtDNA is not echoed by divergence in other markers, then we will need to understand what aspect of anole natural history leads to divergence only in a female-inherited marker. Study of anole dispersal would be an obvious place to start. Selection directly on the mitochondrial genome has also been occasionally suggested, both in anoles (Malhotra and Thorpe, 1994) and in other taxa (e.g., Doiron et al., 2002).

---

381. Such modeling could be accomplished by combining remote sensing to model species' present-day niches with paleoclimatological estimates of environmental conditions in past times. This approach has been conducted for a number of taxa in the rainforests of northeastern Australia (e.g., Hugall et al., 2002; Schneider and Williams, 2005) and elsewhere (Martínez-Meyer and Peterson, 2006; Carstens and Richards, 2007). For a critique, see Pearman et al. (2008).

382. These questions and those further below, illustrate how a focus on genetic exchange and reproductive isolation, as proposed by adherents of the Biological Species Concept, is a tractable and productive framework in which to consider the origin and maintenance of anole species diversity (see Chapter 2).

More work is also needed to understand the basis of species recognition and how it evolves. The extent to which changes in dewlap color are adaptive to the local environment is currently being studied, but the other species-recognition signal, head-bobbing display patterns, has not received much work in recent years. Greater understanding of the extent to which the evolution of these displays is affected by environmental conditions could prove interesting (e.g., Ord et al., 2007); so could further work on the extent to which changes in head-bobbing patterns can lead to reproductive isolation. In addition, the hypothesis that both head-bobbing and dewlap patterns are used in species recognition could be examined in a study of geographic variation. Widespread species co-occur with different sets of species across their ranges; examination of whether changes in signaling behavior are related to patterns of species co-occurrences would be a good test of the hypothesis that these signals evolve to enhance species-recognition.

15

THE EVOLUTION OF AN
ADAPTIVE RADIATION

In Chapter 11, I defined adaptive radiation as the evolutionary divergence of members of a clade to adapt to the environment in a variety of different ways and presented three predictions made by a hypothesis of adaptive radiation:

- Species interact ecologically, primarily by competing for resources.
- As a result of these interactions, species alter their resource use.
- As a result of shifts in resource use, species evolve appropriate adaptations.

A combination of experimental, observational, and comparative data (reviewed in Chapters 11 and 12) strongly supports these predictions. Thus, Greater Antillean anoles meet the expectations of an adaptive radiation.

In this and the next chapter, I will shift the focus from how adaptation occurs within species to the nuts and bolts of how an entire radiation unfolds. Gould (1989, 2002) argued for the predominance of contingency in evolution, suggesting that if we "re-ran the evolutionary tape" and started again from the same point, the outcome likely would be very different. However, the replicate adaptive radiations of anoles in the Greater Antilles challenge this view, a point on which I'll focus in the next chapter. Before doing so, however, it's worth looking in close detail at how anole diversification has proceeded. Can we trace the actual course of evolution, and has it occurred in the same way on each island in the Greater Antilles? What factors, if any, have instigated anole diversification, and have rates and patterns of evolution changed through time?

In this chapter, I'll focus primarily on the anoles of the Greater Antilles, finishing by examining the concept of adaptive radiation itself. In Chapter 16, I will examine the other four anole faunas (Chapter 4) to understand why their evolutionary diversification for the most part has taken a different path than that followed by the Greater Antillean ecomorphs.

HISTORICAL INFERENCE OF PATTERNS OF ADAPTIVE RADIATION IN THE GREATER ANTILLES
ECOMORPHOLOGY OF THE ANCESTRAL ANOLE

We might wonder, for starters, what the progenitor of the anole adaptive radiations was like. Is there any way to infer the morphology and ecology of the species ancestral to the anole radiations on each of the Greater Antilles?

Several lines of evidence provide somewhat conflicting insights. Initially, I attempted to infer the morphology of ancestral species by reconstructing their character states using parsimony (Losos, 1992a). This exercise resulted in hypothetical ancestral species on Jamaica and Puerto Rico that were intermediate in morphology between existing species, leading to the conclusion that the ancestral anoles were ecological generalists that presumably had relatively broad niches in the absence of competing species.[383] However, as discussed in Chapter 5, reconstructing ancestral states for characters that evolve at relatively high rates is fraught with difficulty, and Schluter et al. (1997) showed that the confidence limits on reconstructions for the ancestral anole in Puerto Rico are enormous, overlapping the position of most ecomorphs in morphological space (Fig. 7.3; see also Fig. 7.4).[384] Consequently, character reconstruction appears unable to provide much insight into the morphology of the ancestral anole of the Greater Antilles.

An alternative approach to investigating the ancestral anole is to look at the species occupying one-species islands today (henceforth, "solitary anoles") with the assumption that those species approximate the state of the ancestral anole. Consideration of solitary anoles suggests the hypothesis, in contrast to the results of phylogenetic reconstruction, that the ancestral species to the anole radiations may have been a trunk-crown anole.

---

383. This is a fairly standard view for the ancestor of an adaptive radiation, implying that a generalist ancestor gives rises to specialized descendants. Schluter (2000) showed that the evidence for this pattern as a general scenario for adaptive radiation is not compelling.

384. Two notes on my analysis in the 1992 paper. First, the conclusion that the ancestral form was intermediate is not an artifactual outcome that resulted because the method for inferring ancestral traits averages phenotypes of descendant species. Although some algorithms to reconstruct ancestral states work in this manner (e.g., squared change parsimony [Huey and Bennett, 1987]), the method that I employed, linear parsimony, does not (Swofford and Maddison, 1987). Second, the phylogenies were patched together based on a variety of data; subsequent studies (e.g., Jackman et al., 1997; 2002; Brandley and de Queiroz, 2004) have resulted in somewhat different preferred phylogenetic hypotheses. Although it might be of interest to repeat these studies using this new phylogenetic information, my guess is that the results for the ancestral anole would not be much different, and, given the observed phylogenetic lability of ecomorphological characters, the confidence in the resulting reconstructions almost surely would be no higher.

Two sets of such species exist: in the Lesser Antilles, members of the *roquet* and *bimaculatus* Series occur by themselves on a number of islands. These species have never in their evolutionary history co-occurred with other anole species.[385] On the other hand, a variety of solitary anoles in the northern Caribbean are derived from within the ecomorph radiations of the Greater Antilles; some of these are still considered to be conspecific with the Greater Antillean species from which they are descended, whereas others have diverged to the extent that they are considered different species (Chapter 4).

Of the solitary Lesser Antillean species, almost all species are classified on the basis of morphology and microhabitat use as trunk-crown anoles (Losos and de Queiroz, 1997; Knox et al., 2001).[386] These species do not have exceptionally large breadths of habitat use as; rather, the coefficient of variation in habitat use is comparable to that seen in Greater Antillean ecomorph species.

Examination of nine northern Caribbean species derived from ecomorph ancestors (all trunk-ground or trunk-crown anoles) revealed that three of four trunk-crown descendants are, on morphological grounds, trunk-crown anoles, whereas only two of five trunk-ground descendants have retained their ancestral ecomorph status (Losos and de Queiroz, 1997). All of the species that do not correspond to their ancestral ecomorph class exhibit morphologies that are centrally located in morphological space and are intermediate between trunk-ground and trunk-crown anoles; one trunk-ground descendant[387] even approaches trunk-crown morphospace. Ecological data are more confusing, but generally consistent with the morphological results in that more of the species derived from trunk-ground anoles have become generalized in their structural microhabitat use (reviewed in Losos and de Queiroz, 1997).[388]

These solitary anole data suggest the hypothesis that the anole radiations were initiated by a species similar to a trunk-crown anole. The data from the Lesser Antilles species are particularly persuasive on this point, as those species would seem to have a historical and environmental setting similar to that experienced by the anole species ancestral to the Greater Antillean radiations—i.e., occupation of a large, environmentally heterogeneous island, and perhaps not descended from an ecomorph species.[389] The northern Caribbean species support this argument to some extent as well in that

385. I do not include some Lesser Antillean species that occur by themselves on some tiny offshore islets, but in sympatry with another species on a major Lesser Antillean island, nor *A. gingivinus*, which occurs with *A. pogus* on St. Martin and by itself on St. Bart's and Anguilla. On Anguilla, *A. pogus* apparently perished some time in the twentieth century (Lazell, 1972).

386. The exceptions are *A. ferreus*, which is too large to qualify as a trunk-crown anole, but is similar in other respects morphologically, and some populations of *A. oculatus*, which are classified as trunk-ground anoles. Several species from relatively inaccessible islands (e.g., Blanquilla, Redonda) have not been examined.

387. The little known *A. desechensis* from Isla Desecheo near Puerto Rico.

388. The confusion results because different studies come to contrasting conclusions about the extent of ecological generalization of some species.

389. The *roquet* Series is related to the Dactyloa clade anoles of the mainland; its ancestor was unlikely to be a member of an ecomorph class given that most mainland anoles are not; by contrast, the *bimaculatus* Series is embedded within the Greater Antillean radiation and thus its ancestral state is uncertain, given the difficulties in reconstructing ancestral ecomorph states (Chapter 5 and above).

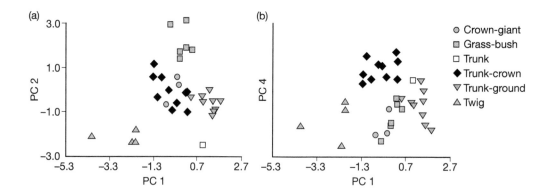

FIGURE 15.1

Position of ecomorphs in multivariate morphological space. Principal components scores are based on an analysis using SVL as a proxy for body size and a number of size-corrected variables. As in Fig. 7.3, PC III loads primarily for SVL, a trait for which trunk-crown anoles are also intermediate relative to other ecomorphs. Modified with permission from Losos and de Queiroz (1997).

descendants of trunk-ground anoles tend to evolve away from their ancestral condition and toward a trunk-crown-like state, whereas trunk-crown derivatives generally have remained as trunk-crown anoles.

Why might the ancestral anole have been a trunk-crown anole rather than a hypothetical, ecomorphologically intermediate, species capable of using a broad array of microhabitats? Actually, these two possibilities might not be all that different because trunk-crown anoles may approximate a generalist species in two respects. First, among the ecomorphs, trunk-crown anoles are the most ecologically and morphologically intermediate. In terms of morphology, they lie toward the center of multivariate ecomorphological space (Fig. 15.1) and exhibit average values for most characters (their well-developed toepads being the major exception [Beuttell and Losos, 1999]). Ecologically, they also seem intermediate between the various ecomorph structural microhabitats. Thus, to the extent that a generalized "jack-of-all-trades" (Huey and Hertz, 1984; Sultan, 1992) is expected to possess an intermediate ecology and morphology, trunk-crown anoles may fit the bill.

Second, trunk-crown anoles are the most generalized of the ecomorphs in that they use the broadest range of substrates, from leaves and twigs to tree trunks and from near the ground to the canopy. The habitat use of trunk-crown anoles doesn't completely encompass the habitat use of all of the ecomorphs—although they do use narrow surfaces, they rarely go to the extremes of twigginess exhibited by twig anoles and they are not often seen in the grassy microhabitats frequented by grass-bush anoles. But the question is whether any anole species could have a greater habitat breadth. It may be that conflicting functional demands make it impossible for any one species to use the full range of structural microhabitats used by all of the ecomorphs. Thus, the trunk-crown morphology may be the closest an anole can come to being a "jack-of-all-trades."

A related point concerns whether the greater range in structural microhabitat use of trunk-crown species is accomplished by all individuals using the same, broad range of structural microhabitats or by the population being composed of individuals each with an unexceptional range of habitat use, but differing from each other in which part of the habitat they use.[390] This was a topic of interest primarily in the 1970s and early 1980s; the consensus seems to be that most cases of niche expansion result from increased within-individual variation (Schoener, 1986b; Futuyma and Moreno, 1988; but see Bolnick et al., 2007). No comparable study has been conducted on anoles to investigate differences in niche breadth among ecomorphs. However, intraspecific comparisons among populations that differ in habitat breadth reveal the same result, that niche expansion is accomplished by individuals having a greater breadth of habitat use, rather than the population being composed of individuals morphologically specialized to use different parts of the habitat (Lister, 1976b).[391]

SEQUENCE OF ECOMORPH EVOLUTION

Discussion of the state of the ancestral anole leads to a logical next question: did the Greater Antillean ecomorph radiations progress in the same way? Put another way: was there a consistent sequence in which the ecomorph types evolved? Two approaches have been taken to this question, and neither has withstood close scrutiny. Indeed, this is the sort of historical question for which it may not be possible to get a decisive answer, as discussed in Chapter 5.

The first approach was presented by Williams (1972), who suggested that the Lesser Antilles might represent the first stages in faunal buildup. We have seen that one-species Lesser Antillean islands might be a reasonable approximation of the starting condition for the anole radiations. However, examination of two-species islands is less convincing: nine out of ten species on two-species islands do not correspond to any of the ecomorph types (the tenth species is a trunk-crown anole). Either the anoles of the Lesser Antilles are on a different trajectory than the Greater Antillean radiations, and thus cannot inform us about patterns of ecomorph evolution, or the intermediate stages of ecomorph radiation produced a suite of forms that later disappeared as the eco-morphs evolved. If this latter scenario is the case, however, then we also would have to

390. That is, is increased breadth accomplished by increasing the within- or between-individual component of niche variation (Roughgarden, 1972).

391. Roughgarden (1974) made a similar point concerning prey use. By contrast, Lister (1976b) found that increased dietary niche width in some populations of *A. sagrei* was accompanied by increased between-individual variation in the size of prey, which he attributed to an increase in the range of body sizes in those populations, combined with the relationship between prey size and body size reported in Chapter 8 (see also Bolnick et al., 2007).

postulate that trunk-crown anoles evolved in the ancestral stage, disappeared at the two-species stage on most islands, and then re-evolved later in the progression.

The second approach to examining the sequence of ecomorph evolution involved phylogenetic estimation of ancestral character states. As discussed above and in Chapter 7, I tried to reconstruct the evolution of the anole communities of Jamaica and Puerto Rico using parsimony (Losos, 1992a). The results were quite exciting, indicating that the sequence of ecomorph addition was nearly identical for the two islands (Fig. 7.2). Moreover, the grass-bush ecomorph was inferred to have been the last ecomorph to evolve on Puerto Rico. Hence, its absence in Jamaica might have been explained not as a result of some aspect of the grass-bush microhabitat on that island, but more generally as stemming from the failure of the Jamaican radiation to advance to the five-ecomorph stage.

Unfortunately, this finding has not held up. The phylogenies on which it was based have been revised, and although the formal analysis has not been re-done, it is unlikely that the same result—the different ecomorphs appearing in the same order on both islands—would be found. The reason is that my analysis inferred that twig anoles evolved early in anole evolution on both islands. However, the more recent phylogeny of Jamaica (Jackman et al., 2002) nests the twig anole A. *valencienni* high in the phylogeny, rather than in a basal position as sister taxon to the rest of the radiation (cf. Hedges and Burnell, 1990). As a result, the extreme morphology of twig anoles is almost certain to be reconstructed as a later addition in the Jamaican radiation, whereas twig anoles are still thought to have evolved early in Puerto Rico due to the basal phylogenetic position of the twig anole A. *occultus* (Nicholson et al., 2005).[392]

More generally, my confidence in the ability to reconstruct ancestral ecomorph character states has been shattered since I published my analysis in 1992. As I have discussed above and in Chapters 5 and 7, we can probably place little confidence in ancestral character reconstructions for anole ecomorphs, which renders this entire approach problematic.

An alternative approach that does not require reconstruction of ancestral phenotypes would be to ask whether the phylogenetic topology of ecomorphs[393] is the same across islands. This question is not the same as examining the sequence of ecomorph evolution because no necessary correspondence exists between phylogenetic topology and ancestral character states: the same phylogenetic outcome could be produced through very different intermediate ancestral stages and, conversely, identical ancestral forms could give rise to radically different phylogenetic arrangements among extant taxa (Fig. 15.2).

---

392. As a simplification, my analysis treated Puerto Rican anoles as a single clade, even though they were known not to be. In defense of this approach, A. *occultus* is the sole occupant of one of the most basal branches in the anole phylogeny. For this reason, a phylogenetic reconstruction of character states that used the entire anole phylogeny would also almost certainly identify a twig anole as one of the first ecomorphs to have been present on Puerto Rico.

393. By "phylogenetic topology," I mean the branching structure of the phylogeny and the position of the ecomorphs on the terminal tips.

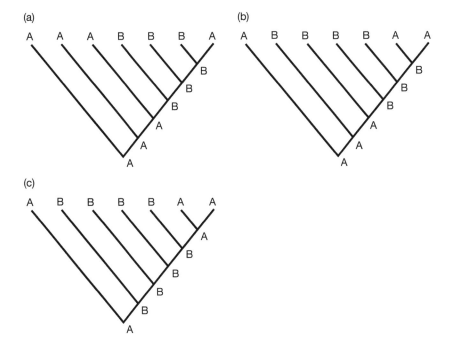

FIGURE 15.2

Phylogenetic topology does not necessarily predict ancestral character states unless no uncertainty exists in ancestor reconstruction. In this example, parsimony accurately reconstructs character evolution in (a) and (c), but in (b), evolution has occurred in a very non-parsimonious fashion. As a result, the sequence of ancestral states is identical in (a) and (b), even though their phylogenetic topology is very different. Conversely, the phylogenetic topologies of (b) and (c) are identical, but their ancestral sequences differ greatly.

Regardless of this interpretational ambiguity, one result of such an analysis is clear: no consistent pattern of phylogenetic relationship of ecomorphs is evident among the four Greater Antillean islands. In fact, in only one case are the same two ecomorphs sister taxa on more than one island (Fig. 7.1). This result might be interpreted to suggest that the convergent faunas have attained their similarity by taking different evolutionary trajectories; certainly, evolution would have had to occur very non-parsimoniously for the ecomorphs to have arisen in the same sequence on all four islands given these topologies.

We might always hope that the fossil record will enlighten our understanding of ecomorph evolution, but I wouldn't hold my breath waiting on this one. All pre-Holocene Greater Antillean anole fossils come from one place and point in time (the Dominican Republic, approximately 15–20 mya; Chapter 6). If *Anolis* originated more than 40 mya (Chapter 6), then the phylogeny suggests that all of the ecomorphs already were in place in the Dominican Republic by this time (Fig. 7.1). As a result, these fossils cannot provide insight into the early stages of anole radiation. However, the fossils can be used to test the hypothesis that if a clade today is composed of members of only one ecomorph class, then members of that clade that lived 15–20 mya also should also belong to that eco-morph class. This inference has been supported for one clade of trunk-crown anoles. The *chlorocyanus* Series is composed of four trunk-crown species that occur today on Hispan-iola and two Dominican amber specimens that appear to belong to this group based on os-teological characters. Morphometric analysis of these fossils clearly indicates that they were, at least morphologically, trunk-crown anoles (de Queiroz et al., 1998). Further exam-ination of additional amber specimens will be needed to test this hypothesis on other clades of Dominican ecomorphs.

In the absence of new fossil finds from deeper in time, we will have to rely on the ambiguous inferences we can take from phylogenetics. These inferences preclude con-fident assignment of specific ancestral types, but do suggest the possibility that the island radiations have taken different courses to attain their current convergent state.

STAGES OF RADIATION

Williams (1972) suggested that there were distinct stages of evolutionary diversification within Puerto Rico (Fig. 15.3). The first stage involved divergence in body size: an ances-tral anole that occurred in the shade in arboreal vegetation gave rise to three arboreal species differing in size (small, medium, and large). At that point, the canopy was full and the next stage of divergence involved change along structural microhabitat lines, produc-ing species using the trunk-ground and grass-bush niches, again in shaded microhabi-tats. Finally, the last stage of divergence was along the climatic axis, producing species of similar size and structural microhabitat to their ancestors, but moving from the shade to occupy hotter, more open microhabitats. This stage occurred in trunk-ground, grass-bush, and trunk-crown anoles.[394] As Figure 15.3 illustrates, Williams (1972) derived these conclusions from the phylogenetic relationships of the species. Although the analytical method he used bore no resemblance to the algorithms we use today, Williams clearly presaged by two decades the phylogenetic approach to evolutionary analysis.

Similar evolutionary scenarios have been proposed for other organisms. For example, Richman and Price (1992) showed that leaf warblers in the genus *Phylloscopus* diverged first in body size, then in foraging morphology and behavior, and finally in habitat use, and Streelman et al. (2002) hypothesized that parrotfish diversified first in habitat, then in diet, and finally in sexually selected traits (see reviews in Streelman and Danley [2003] and Ackerly et al. [2006]).

These hypotheses share a common theme: a clade first diversifies in one way, such as in habitat use.[395] Once that avenue is fully utilized, species stop diversifying along that axis, but begin subdividing or diverging along a different axis, such as prey size. A vari-ety of explanations could account for why divergence occurs first along one axis and only

---

394. The Puerto Rican trunk-crown anoles do vary substantially in size, but both might be considered "medium" in size in comparison to the size range of Puerto Rican anoles (the diminutive *A. occultus* to the giant *A. cuvieri*).

395. Usually the axes are some type of resource, though divergence in communication signals is implicated in some cases (Streelman and Danley, 2003).

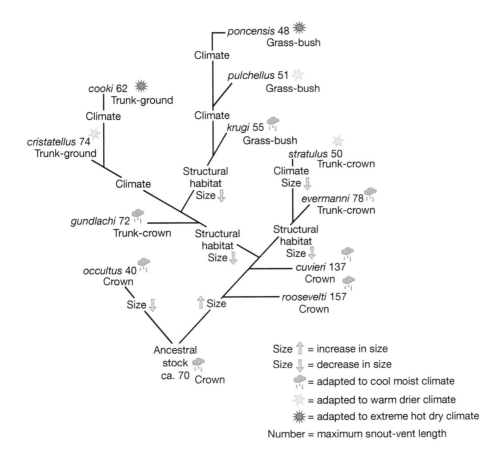

FIGURE 15.3

Stages of radiation in Puerto Rican anoles. Reprinted from Williams (1972) with permission.

subsequently on another, such as optimality considerations, genetic constraints or historical contingencies such as the ancestral starting point of a clade (although this point has actually been little addressed [e.g., Diamond, 1986; Schluter, 2000; Gavrilets, 2004; see also Schoener, 1977]).

The difficulty with these proposals is that they usually rely on reconstructing ancestral traits.[396] As I've argued repeatedly, we simply cannot have much confidence in

396. Diamond (1986) took an alternative, process-based approach. Under the assumption that speciation in New Guinea birds is always allopatric, he ordered species pairs by their degree of geographic overlap and differentiation, ranging from continuously distributed subspecies through allopatric populations and allospecies to fully sympatric congeners, with the assumption that the allopatric pairs represented the earliest stages in divergence and the pairs with progressively greater amounts of sympatry represented later stages in diversification. This approach has some logic to it, but also makes assumptions that at face value seem to run to counter to those employed in phylogenetic analyses. For example, Diamond's method concludes that traits that differ between the most closely related taxa represent those traits that diverge early in a clade's history, at the initial, allopatric stage; in contrast, phylogenetic analysis would suggest that traits that are invariant among close relatives, but differ among more distant relatives, are the ones that diverged early in a clade's history (see Ackerly et al. [2006]). A process-based analysis such as Diamond's, implemented in an explicitly phylogenetic framework, might be able to reconcile these two approaches and could prove interesting.

ancestral reconstructions for traits which evolve rapidly relative to the frequency of cladogenesis (see Chapter 5). For example, in the case of the leaf warblers (Richman and Price, 1992), the phylogeny exhibits a basal split between two clades, one containing three large species and the other containing five smaller species. This is the sort of situation in which ancestral reconstruction is likely to be most accurate, and thus the conclusion that these warblers diverged in body size early in their radiation is strongly supported. By contrast, both clades contain species with high and low values for habitat use, and some of the largest differences are between the most recently diverged sister taxa. Given the evolutionary lability of this trait, we can have little confidence in reconstructions of habitat use for ancestors deep in the phylogeny. Unfortunately, this means that one aspect of a stages-of-radiation hypothesis—that traits that are inferred to have diverged in later stages of a radiation did not also diverge early in the radiation—cannot be tested; an alternative to the hypothesis of discrete and distinct stages of evolution in leaf warblers would be that body size diverged early in the radiation without much subsequent change, but that habitat use has been diverging throughout the radiation (Ackerly et al., 2006).

This is exactly the situation seen in Greater Antillean anoles. Many islands contain clades composed of species all of which are members of the same ecomorph class. The *alutaceus* Series in Cuba, for example, contains 14 grass-bush anoles. Overall, there are 14 Greater Antillean clades with three or more species all of the same ecomorph (Fig. 7.1). For these clades, particularly the more species-rich ones, the conclusion that the ancestor of that clade was a member of the same ecomorph class as all of its descendant species seems reasonable.[397] By contrast, considerable diversity exists within these clades in thermal microhabitat (Chapters 3 and 10), so much, in fact, that no relationship exists between degree of phylogenetic relationship and degree of similarity in thermal biology (Chapter 10). Not surprisingly, then, phylogenetic methods are not able to reconstruct ancestral thermal microhabitat use with any degree of confidence (Glor et al., in prep.).

For this reason, the hypothesis that anole radiation has occurred by distinct stages of evolution, in which the type of diversification that occurred differed in each stage, is not supportable. The reason is that although it is clear that closely related species have diverged in thermal microhabitat, we simply cannot make reliable inferences about whether such divergence occurred deep in anole phylogeny, at the same time as structural microhabitat (e.g., ecomorph type) was diverging.[398]

What we can say, however, is still quite interesting. Phylogenetic analysis clearly indicates that anoles diverged into different ecomorph types early in their radiations on all

397. The alternative is that the ancestor was something else and the descendants evolved in parallel to attain the same ecomorph state.
398. Phylogenetic analysis also provides no indication that body size is more strongly conserved than microhabitat type, in contradiction to Williams' (1972) hypothesis (Glor et al., in prep.).

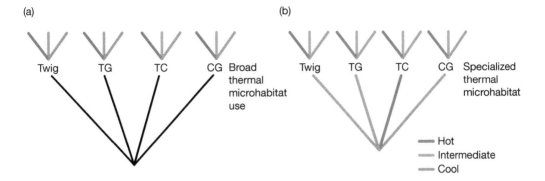

FIGURE 15.4

Two views on the evolution of microhabitat specialization in the Greater Antilles. For simplicity's sake, this cartoon only considers evolution on a single island and treats the species on the island as monophyletic, as occurs in Jamaica. In (a) the ecomorphs evolve early in an island's history, but these ancestral species have broad thermal niche use; subsequently, climatic microhabitat specialization occurs, either by adaptation in allopatry in different environments or by character displacement among initially similar species that came into sympatry after allopatric speciation. Alternatively, in (b) initial diversification produced species differing both in ecomorph and in thermal specialization. Subsequent evolutionary diversification produced a diversity of microclimate specialists within each ecomorph clade; as in the first scenario, this could occur by invasion of different thermal microhabitats and subsequent adaptation, or by character displacement within a thermally heterogeneous locality.

four islands (Chapter 7)—whether these species differed in thermal microhabitat is unknown. Although these ancestral ecomorphs proliferated, their descendants almost invariably remained within the same ecomorph class even as they diversified in thermal microhabitat, macrohabitat type (e.g., semi-desert, pine forest) and elevation (e.g., Ruibal, 1961; Glor et al., 2003).

Based on this information, I can envision two scenarios. In the first scenario (Fig. 15.4a), the earliest stage of anole radiation is marked by evolution of the different ecomorphs. Each ecomorph is hypothesized to have a broad thermal niche and thus is found throughout most habitats on the island, as occurs in most anole species in the northern Lesser Antilles and in solitary anoles in the southern Lesser Antilles. For this reason, an evolutionary priority effect (Chapter 7) would have existed so that as ecomorph clades diversified, new species in one clade could not evolve into other ecomorph types because those niches were already occupied by members of other clades. However, as populations became isolated in allopatry, they diverged into different species adapted to the different thermal microhabitats they occupied, such as semi-desert or cool montane rainforest, while still retaining their ancestral ecomorph type. These differences may have facilitated coexistence if two such species came back into contact at thermally heterogeneous localities (Chapters 3 and 10). Alternatively, microclimatic habitat special-ization could have arisen by character displacement in a thermally heterogeneous environment when initially similar species became sympatric and evolved physiological differences to use different climatic microhabitats, thus permitting coexistence.

The alternative scenario assumes that the early stage of ecomorph evolution led to species that differed in both structural and thermal microhabitat (Fig. 15.4b). Thus, for example, one species might be a trunk-crown anole restricted to shady, cool microhab-itats, whereas another might be a grass-bush anole occurring primarily in sunny micro-habitats. Few comparable situations exist today. Species on most two-species islands in the Lesser Antilles and elsewhere usually differ in either structural or thermal micro-habitat, but not both (of course, Lesser Antillean species also differ in body size, whereas most ecomorphs overlap in body size to a greater extent). One exception is the unusual case of *A. gingivinus* and *A. pogus* on St. Martin, where the species differ to some extent in both structural and thermal microhabitat use (Roughgarden and Pacala, 1989).

This alternative scenario is problematic for a second reason: it doesn't explain the evolutionary stasis in ecomorph clades. If the early radiation of anoles didn't result in a species using grass-bush microhabitats in the shade, why would this niche necessar-ily be filled by the clade that contained a grass-bush anole in the sun, rather than by the one that contained a trunk-crown anole in the shade? Of course, by the same token, one could ask why, in the first scenario, the initial stage of radiation should have led to a set of structural microhabitat specialists, rather than a set of thermal microhabitat specialists, each of which used all available structural microhabitats. One might sug-gest that it has something to do with the patchiness of thermal versus structural micro-habitats or the relative ability to evolve phenotypes able to function effectively across a range of different microhabitats—perhaps an all-purpose thermal physiology pheno-type would be more fit than a comparably broad, "jack-of-all-trades" morphological phenotype.

At this point, we have no data to address these speculations, and the fauna of the Lesser Antilles cautions against a simplistic answer, given that the two-species islands in the north contain species that have diverged in structural microhabitat, whereas those in the south have diverged to a greater extent in thermal microhabitat (Chapter 11).

These questions highlight the limitations of phylogenetic reconstruction of past events. The phylogeny strongly indicates that the ecomorphs evolved early, and that sub-sequent diversification in ecomorph clades only extremely rarely leads to a species that breaks out of the ancestral ecomorph mold. My intuition is that thermal microhabitat specialization did not occur until later in the anole radiation, but this is a weak inference made by comparison to existing species and communities. Unfortunately, more direct examination of the events that occurred millions of years ago may be impossible to obtain—these are the sorts of answers lost in the "fog of time" to which I referred in Chapter 5.

A standard idea concerning adaptive radiation—sometimes ensconced in the definition of the term (see Footnote 258)—is that phenotypic diversification is very rapid at the outset of a radiation and diminishes through time. On the other hand, a corollary of the stages-of-radiation hypothesis is that different phenotypic characters have had different rates of evolutionary change through clade diversification; consequently, some characters do not experience their highest rate of evolution early in the radiation.

Certainly, many clades do exhibit a burst of diversification early in their history (Simpson, 1953), and a trend toward decreased rates of evolution through time is seen in many cases (Harmon et al., in review). Nonetheless, Schluter (2000) has provided a variety of different lines of evidence that suggest that ecological specialization and niche occupation occurs throughout the course of a radiation, rather than being temporally clustered at its outset.

Anole morphological evolution certainly displays the classic expectation: early divergence into the different ecomorph types, followed by evolutionary stasis.[399] By contrast, closely related species differ greatly in thermal physiology; as Schluter (2000) anticipated, species have specialized to different thermal microhabitats late in the radiation, with no evidence for decreasing rates of change through the course of the radiation (Hertz et al., in review).

KEY INNOVATIONS

The term "key innovation" is now widely used in evolutionary biology. Unfortunately, it is used in a number of different ways. Initially, Miller (1949) defined a key innovation as a trait that allows members of a clade to interact with the environment in a novel way, potentially—but not necessarily—opening the door for evolutionary diversification and adaptive radiation (reviewed in Galis, 2001).[400]

In recent years, however, key innovation has taken on a second meaning: a trait that leads to increased species richness in a clade (reviewed in Heard and Hauser, 1995;

---

399. Quantitative assessments of rates of anole morphological diversification have been somewhat contradictory and have not found a strong and consistent signature of early bursts and declining rates through time (Harmon et al., 2003; Hertz et al., in review). However, this no doubt results from the fact that in these studies, all Greater Antillean anoles have been analyzed as a single group, confounding the different ages of the radiations on the different islands. If each island radiation were analyzed separately, they would no doubt confirm what is obvious from examination of the phylogeny (Fig. 7.1), that the rate of ecomorphological evolution slowed through time as ecomorph stasis kicked in.

400. A key innovation could allow members of a clade to interact with the environment in a novel way, but diversification might not ensue. For example, salamanders in the genus *Aneides* are characterized by a novel morphological structure of their feet which allows them to climb, in contrast to their terrestrial relatives. Although it interacts with the environment in a fundamentally different way compared to its ancestor, *Aneides* has diversified little, producing only six morphologically little-differentiated species (Baum and Larson, 1991). An even more extreme case would be the evolution of the features that have allowed the monotypic aardvark to adopt its termite-feeding ways (Hunter, 1998). Note that this point runs counter to one definition of key innovation, which is the evolution of a trait that facilitates an adaptive radiation (e.g., Anker et al., 2006).

Hunter, 1998). These two concepts can be related: taxa that diversify as a result of the opportunity provided by exploiting the environment in new ways may exhibit both great adaptive diversity (often termed "disparity" [Foote, 1999; Erwin, 2007]) and great species richness. Nonetheless, the two outcomes are not necessarily linked: great adaptive disparity can occur with low species richness, and great species richness can occur in adaptively homogeneous clades (Erwin, 1992; Foote, 1993; Losos and Miles, 2002).

For this reason, the two phenomena need to be distinguished. Being a traditionalist, I will retain "key innovation" to refer to its original meaning of a trait that allows members of a clade to interact with the environment in a new way, potentially promoting adaptive radiation. I'm sure alternative terms have been proposed for traits that promote species diversification, but I am unaware of them; so as to avoid proposing yet another term, I will just refer to the concept itself. I want to emphasize that a given trait can simultaneously promote increased species richness and function as a key innovation.

Testing the hypothesis that a trait constitutes a key innovation or leads to increased species richness is difficult. The traditional approach is to examine a group that exhibits great disparity or species richness and look for some trait that evolved at the base of the clade and that plausibly could be related to evolutionary diversification. For example, Salzburger et al. (2005) identified mouth-brooding, the possession of anal tail spots, and the existence of color morphs as ancestral traits in haplochromine cichlids that could have promoted species diversification in African Rift lakes.

This approach is fraught with peril for a number of reasons (Lauder, 1981; Lauder and Liem, 1989; Cracraft, 1990; Erwin, 1992; Heard and Hauser, 1995; Donoghue, 2005): first, any clade is characterized by more than one derived character. How can one know which character is responsible for the increased diversification? Second, the great diversity or disparity of a clade may be due to enhanced rates of diversification[401] of a subclade within that clade. The great species richness of mammals, for example, traces not to the base of the Mammalia, but to the subclade Theria, which is comprised of the Marsupialia and Eutheria; its sister taxon, the Monotremata, contains only three living species. Hence, examining characters that evolve at the base of a clade may miss the evolutionary events that were responsible for increased diversification of the subclade. Finally, third, many important adaptive complexes are built up incrementally over long evolutionary periods and thus cannot be localized to a single phylogenetic branch. In such cases, identifying a single evolutionary change occurring on a single branch of a phylogeny as being responsible for increased diversification may not be possible.

Given these difficulties, what is to be done? Three options exist. First, one can conduct detailed studies to understand the mechanistic implications of the evolution of a trait. Even though it is a single case study, one can put forward the strongest argument possible by combining phylogenetic, functional, selective, and ecological analyses to

401. Resisting the urge to coin the term "disparification," I use "diversification" to refer to evolutionary increase in either species richness or adaptive disparity.

bolster the case that evolution of a particular trait has been a key factor in subsequent diversification (Baum and Larson, 1991; cf. Coddington [1994] on the study of adaptation). In cases such as this, we need to remember that evolutionary biology is a historical science. Some times, the best we can do is make a plausible case, testing it in as many ways as possible (Chapter 1).

An alternative approach is to investigate whether the same trait has evolved in other clades and, if so, whether it has led to enhanced diversification in each case. This approach is attractive both for its generality and because it allows one to make statistical statements that are not possible when dealing with a unique event. For example, plant clades that have evolved secretory canals that conduct latex or resin have consistently greater species richness than their sister taxa which lack such defenses against herbivores (Farrell et al., 1991).

This approach can be taken one step further. Although many putative key innovations are uniquely evolved, the mechanism by which they are presumed to facilitate increased diversification may have evolved multiple times. For example, the evolutionary duplication of a feature—whether it be a gene or a morphological structure—may enhance the evolutionary potential of a clade by loosening selective constraints: one copy of the trait may be free to change while the second copy continues to perform the trait's original function. Although each particular trait may be duplicated evolutionarily only once, one can test the general hypothesis that trait duplications are correlated with increased rates of diversification (Lauder, 1981; Lauder and Liem, 1989).

The statistical multiple comparison approach is useful, but it should not take the place of detailed mechanistic analysis of the trait and how it is causally related to subsequent evolutionary diversification. Many statistical comparative studies are difficult to interpret because they are lacking this mechanistic understanding (Coddington, 1994). The strongest case that a trait, or a type of evolutionary change such as duplication, is a key innovation or a cause of species proliferation comes from a combination of these two approaches: a statistical association between the evolution of the trait and subsequent diversification and a detailed understanding about how and why this relationship exists in many or all of the cases (cf. Arnold [1994] and Larson and Losos [1996] on the study of adaptation).

As in any statistical analysis, failure to reject the null hypothesis in a multiple comparisons test does not mean that a key innovation hypothesis is not correct. In addition to standard considerations of statistical power and noise in the data, an important additional issue remains: just because the evolution of a trait enhances diversification in one clade does not mean that it will do so in all clades. Quite the contrary, for many reasons acquisition of a trait may lead one clade and not another to diversify, including possession of other relevant traits, environmental setting, and other historical contingencies (de Queiroz, 2002). These considerations emphasize the importance of detailed, mechanistic analysis across clades: if the evolution of a trait is associated with increased diversification in some clades, but not others, detailed study may help determine whether the heterogeneity results because the trait actually has no causal relationship with the extent of diversification versus the alternative that it does in some cases, but not in others (Donoghue, 2005).

With these considerations in mind, we may ask what traits, if any, are candidates to have promoted adaptive radiation and species diversification in anoles? To address this question, the first place to look is at traits that arose at the base of the anole radiation.[402] Two obvious candidates are traits that characterize anoles, the toepad and the dewlap. I'll consider each in turn.

EXPANDED SUBDIGITAL TOEPADS AS KEY INNOVATIONS

Anoles use a greater range of microhabitats, from leaf litter and grass stems to rainforest canopy and boulder-strewn streams, than other comparable clades of iguanid lizards. One feature that distinguishes anoles from other iguanids is the extent of their arboreality. Most iguanids are either terrestrial or, to the extent that they get off the ground, they use broad surfaces such as boulders, tree trunks, and large branches (Vitt and Pianka, 2003).

In this respect, the evolution of subdigital toepads may represent a key innovation that allowed anoles to interact with their environment in a new way. By allowing these lizards to use a variety of arboreal surfaces such as narrow twigs, leaves, and grass blades that lizards lacking pads have difficulty accessing, the evolution of toepads may have facilitated the radiation of anoles into a variety of ecological niches otherwise little explored by iguanids.

How might this hypothesis be tested? Certainly, toepads provide functional capabilities not available to padless lizards, in particular the ability to adhere to smooth surfaces (Chapter 13). On the other hand, iguanid species that climb on vertical or arboreal surfaces often have sharp, curved claws that provide clinging ability (Zani, 2000). Presumably, toepads allow anoles to use smooth or narrow surfaces upon which claws are ineffective, but this hypothesis has never been tested. One way of examining this idea might be to interfere with the action of the setal hairs on toepads to see whether anoles can still use these habitats with only their claws providing clinging capabilities.

---

402. Of course, as argued above, traits responsible for diversification in anoles might have arisen in subclades of *Anolis*. The basal split within *Anolis* is between the Dactyloa clade, found primarily on the mainland, and the Caribbean clade, within which Norops arose and re-colonized the mainland (Chapter 5). In Chapter 17, I show that both clades have great amounts of morphological disparity; consequently, if a key innovation is responsible for the great disparity of anoles, either it occurred at the base of *Anolis*, or different traits arose independently in both subclades (if the same trait evolved in both subclades, then phylogenetic analysis would infer a single origin at the base of the clade). With regard to species richness, Dactyloa (including Phenacosaurus) has 87 species and the Caribbean clade nearly 300. This difference is not statistically significant by at least some tests (Slowinski and Guyer, 1989), which suggests that the base of the tree is the appropriate place to investigate the existence of a trait responsible for the great species diversification of anoles.

No candidates are available for an alternative possibility, that a trait responsible for the great diversity of *Anolis* might have been constructed by sequential evolutionary changes spanning several ancestral nodes.

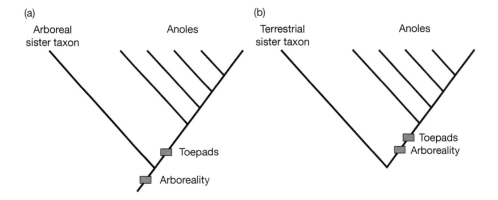

FIGURE 15.5

Evolution of toepads and *Anolis* diversification. (a) If the sister taxon to *Anolis* is arboreal, then the most parsimonious reconstruction is that the ancestral anole was already arboreal when toepads evolved. Alternatively, (b) if the sister taxon is terrestrial, then both the evolution of toepads and the transition to arboreality occurred on the same branch of the phylogeny. Yet another scenario, not illustrated, is possible if more distant outgroups are arboreal. In that event, even if the sister taxon to *Anolis* is terrestrial, the ancestor of *Anolis* may still have been arboreal, with terrestriality being the derived state in the sister taxon.

Phylogenetic evidence generally supports the toepads-as-key-innovation hypothesis. Toepads evolved at the base of the anole phylogeny: shortly thereafter, anoles radiated into a wide variety of ecological niches, just as a key innovation hypothesis would predict. What is not clear is whether anoles moved into the trees before evolving toepads. If, as traditionally believed, *Polychrus* (or some other arboreal clade) is the closest relative of anoles (Chapter 6), then the most parsimonious interpretation is that the ancestor of *Anolis* was also arboreal. In this scenario, the evolution of toepads would have evolved subsequent to the evolution of arboreality (Fig. 15.5a). Alternatively, if the sister group to *Anolis* is terrestrial, as some molecular data suggest (Schulte et al., 2003; see Chapter 6), then toepads may have arisen phylogenetically coincident with the movement of anoles into the arboreal realm (Fig. 15.5b). In this scenario, distinguishing which occurred first—moving into the trees or evolving toepads—is not possible (Arnold, 1994; Larson and Losos, 1996).

In summary, a strong case on functional and phylogenetic grounds has been made that the evolution of toepads permitted anole radiation by allowing the use of a wide variety of ecological habitats not previously accessible. This is about as far as a test of a key innovation can go in most cases when dealing with a single evolutionary event.

However, the evolution of toepads has not occurred just in anoles, but also in two other lizard clades, the Gekkonidae and the skink genus *Prasinohaema* (Fig. 15.6). In both cases, the toepads are covered with microscopic setal hairs and provide enhanced clinging capability (Ruibal and Ernst, 1965; Williams and Peterson, 1982; Irschick et al.,

1996, 2006b). The Gekkonidae is the second most species-rich family of lizards (Vitt and Pianka, 2003) and exhibits a remarkable extent of ecological and morphological diversity. *Prasinohaema*, by contrast, is not species-rich (five species), nor does it seem to be ecomorphologically diverse, although few ecological data are available. Overall, although this hypothesis has never been formally tested, my guess is that the great species richness and ecomorphological disparity of anoles and geckos would lead to a statistical association of both attributes with toepad evolution, the lackluster diversity of *Prasinohaema* notwithstanding. This hypothesis could most profitably be pursued by detailed studies within the Gekkonidae, in which toepads appear to have evolved independently many times (Han et al., 2004).

Thus, the key innovation hypothesis of toepad evolution seems well supported. The way in which the evolution of toepads leads to increased ecomorphological disparity is straightforward: pads give lizards the ability to move effectively on a variety of surfaces on which padless lizards are not competent. But the link between toepads and increased species richness is not so obvious.

Increased species richness can result either from increased speciation rates or decreased extinction rates (Dorit, 1990; Heard and Hauser, 1995). The evolution of features like toepads could plausibly be related to either. On one hand, the possession of toepads could indirectly increase rates of speciation through mechanisms of ecological speciation by opening evolutionary avenues down which populations could diverge. This could happen sympatrically, through disruptive selection, or in allopatry as populations in different localities diverged in different ways, with reproductive isolation evolving as a by-product of this divergence, as discussed in Chapter 14. Alternatively, the possession of toepads might decrease rates of extinction in several ways. For example, if two populations speciated in allopatry but did not diverge ecologically, the possession of toepads would give the two new species increased possibilities for resource partitioning and character displacement if they came into secondary contact, thus potentially decreasing the rate of extinction for young species. Extinction rate might also decrease if the possession of toepads simply made populations better adapted to the environment, and thus more likely to persist over long periods.

FIGURE 15.6

Other lizards with toepads. (a) Madagascar leaf-tailed gecko (*Uroplatus fimbriatus*); (b) the skink *Prasinohaema virens* from New Guinea and (c) its toes. Skink photos courtesy of Chris Austin.

These possibilities make clear why linking a putative key innovation to changes in species richness is so difficult and problematic. The possibilities just outlined are plausible, but that is about as far as it goes. Actual direct mechanistic evidence demonstrating a link between evolution of a trait and increased species diversification is rare in most cases, and nonexistent for anoles with regard to toepad evolution.

EVOLUTION OF THE DEWLAP AND SPECIES RICHNESS

The flip side of the key innovation coin is the dewlap, the second characteristic feature of anoles. The evolution of the dewlap probably did not open new ecological opportunities for anoles in contrast to the effect of toepad evolution. Thus, the great ecomorphological disparity of the anole clade is probably not a direct result of the evolution of the dewlap. Conversely, the dewlap may explain the great species richness of the clade.

The reason is simple: the use of a visual signal both for intraspecific communication and for species identification increases the possibility that shifts in habitat may lead to divergence in these signals, thus resulting in speciation. The evidence for this hypothesis in *Anolis*, as I reviewed in Chapter 14, is suggestive, but far from conclusive.

A further test of the hypothesis might involve those few anole clades that have greatly reduced dewlaps or none whatsoever. If possession of the dewlap enhances the rate of speciation, then dewlap-deficient clades should have relatively few species compared to other clades. This is exactly what is observed. The only anoles to completely lack a dewlap are *A. bartschi* and *A. vermiculatus*. These species comprise a clade that is very old (Fig. 5.6), but nonetheless only contains two species; other clades of comparable age have dozens of species. Other species with notably reduced dewlaps are *A. poncensis* (a Puerto Rican grass-bush anole), *A. ophiolepis* (the Cuban grass-bush anole that arose within the clade of trunk-ground anoles in the *sagrei* Series), *A. Chamaelinorops barbouri*, the two small trunk-crown anoles of Hispaniola (*A. singularis* and *A. aliniger*), *A. agassizi*, and the three species, all grass-bush anoles, in the *A. hendersoni* Series (Fig. 15.7; Losos and Chu, 1998). The low species richness of all of these clades of small-dewlapped

FIGURE 15.7

Small-dewlapped anoles. (a) A. Chamaelinorops *barbouri* has the second smallest dewlap relative to its body size among 49 West Indian species (not including the two Cuban species that do not have a dewlap [Losos and Chu, 1998]). (b) *A. agassizi* from Malpelo Island off the coast of Colombia also has a very small dewlap; large, reproductively active adult males have a permanently erected nuchal crest, unlike other anoles, in which crest erection is facultative (Rand et al., 1975). Photo courtesy of Margarita Ramos.

species—young or old—suggests that when small dewlaps evolved, for whatever reason,[403] the rate of species differentiation decreased.

Dewlaps or dewlap-like structures have evolved in a number of other lizard clades. The most similar are the dewlaps of several Asian agamid lizards which are strikingly like those of anoles (Fig. 2.3c). This clade, containing the seven species in *Sitana* and *Otocryptis*, is not particularly species rich. In contrast, the flying dragons of southeast Asia, genus *Draco*, sport a structure fairly similar to the anole dewlap (Fig. 2.3b) and are relatively diverse in both species number and ecomorphology (Lazell, 1992; McGuire and Alcala, 2000; McGuire et al., 2007a).[404] Interspecific variability of the *Draco* dewlap is reminiscent of that seen in *Anolis*, but flying dragons have another trick up their sleeve: during displays, they also extend their wings, which also exhibit interspecific variation in coloration and pattern (Fig. 2.3b; Lazell, 1992; Mori and Hikida, 1994; McGuire and Alcala, 2000). Further, like anoles, sympatric *Draco* tend to differ in the color of their display structures (Inger, 1983; Lazell, 1992). No research of which I am aware has directly tested the species-recognition role of *Draco* dewlap and wing coloration, much less a hypothesized role in *Draco* speciation. Nonetheless, the parallels are obvious.

403. Fitch and Henderson (1987) suggested that the small dewlap of *A. bahorucoensis*, a member of the *hendersoni* Series, evolved to make display less conspicuous to larger anoles which preyed upon them. Another possibility is that evolving a small dewlap could be another way to differentiate one species from another (Rand and Williams, 1970; Losos and Chu, 1998; Nicholson et al., 2005). No doubt other possibilities exist as well, but the evolution of dewlap size has received little attention.

404. With more than 20 described species, *Draco* is already one of the most species-rich genera of agamids (Stuart-Fox and Owens, 2003). However, many new species have been described recently, and by all indications the number of species may have been greatly underestimated (Lazell, 1987, 1992; McGuire and Alcala, 2000; McGuire et al., 2007a).

As with the evolution of toepads, some, but not all, clades characterized by the possession of a dewlap have high species richness. No statistical analysis has been conducted, but the great species richness of *Anolis* and *Draco* compared to their close relatives is highly suggestive of a causal relationship, even considering the modest diversity of the *Sitana + Otocryptis* clade.[405]

In summary, the evolution of both toepads and dewlaps may have played a role in anole evolutionary diversification. In both cases, a plausible mechanism exists, and comparative data are generally supportive. In addition, these observations suggest a further hypothesis: perhaps evolutionary radiations that combine both great species richness and great adaptive disparity may be the result of the evolution of multiple features that increase both ecological opportunity and rate of speciation. In the case of anoles, the hypothesis would be that the dewlap and the toepads have had an interactive effect: the dewlap has enhanced the production of new species, whereas toepads have increased the likelihood that species would diverge to explore new ecological areas. In Chapter 17, I will explore the extent to which similar scenarios may account for adaptive radiation in other groups.

The study of the factors sparking evolutionary diversification is both fascinating and frustrating. Fascinating, because this is what evolutionary biology is ultimately about, trying to explain the diversity around us. Frustrating because of the difficulty of actually testing the hypotheses that are so easily generated. The discussion in this section, to me, embodies that conundrum: the ideas are interesting, the data somewhat persuasive, but the ability to strongly test the hypotheses limited.

DETERMINANTS OF SPECIES DIVERSIFICATION WITHIN *ANOLIS*

The discussion of the effect of dewlap size on rates of species diversification highlights the fact that species richness varies among anole clades. This variation is evident simply by inspecting the phylogeny in Figure 5.6: clades that originated at approximately the same time vary greatly in species number, from one in the *occultus* Series to 151 in the Norops clade. Such variation is highly unlikely if diversification has occurred in a homogeneous fashion among clades ($p < 0.05$, methods following Ricklefs [2003], Ricklefs et al. [2007]). Moreover, examination of the phylogeny reveals that many anole clades originated in a short period early in anole history; statistical analysis confirms that the rate of species origination in the Greater Antilles has decreased with time (Harmon et al., 2003).

---

405. The appropriate statistical evaluation of this hypothesis might take the form of asking: What is the probability that if an investigator randomly selected three clades of iguanian lizards, at least two would have substantially higher species diversity than their sister taxa? Alternatively, one could test whether rates of species diversification on those three branches of the Iguania were significantly higher than on branches throughout the rest of the clade? This latter analysis would have the advantage of explicitly incorporating information on evolutionary age, which is always a potential problem when clades—or members of a taxonomic rank such as a genus—differ in age.

A pattern of explosive species diversification early in the history of a clade followed by decreasing rates of diversification later on is found in many radiations (e.g., Gould et al., 1987; MacFadden and Hulbert, 1988; Nee et al., 1992; Lovette and Bermingham, 1999; Rüber and Zardoya, 2005; Seehausen, 2006). Usually this pattern is attributed to the occupation of initially empty ecological space as a result of colonization, extinction of an ecologically dominant form, or evolution of a feature permitting access to previously unavailable resources (Simpson, 1953). This explanation fits anole history well: early on in the radiation, ecomorph types evolved repeatedly; subsequently, ecomorph stasis has been accompanied by lower rates of diversification. An interesting test of the "ecological opportunity" hypothesis might involve mainland Norops, which diversified in part of their range in the absence of other anoles, but in the other part in the presence of the Dactyloa clade;[406] if this hypothesis is correct, we might expect to see greater rates of diversification among Norops in the Dactyloa-free region.

Given that rates of diversification are not constant within *Anolis*, we can now ask what accounts for the heterogeneity in rates. I have already discussed the potential role of dewlap size; no other phenotypic characters is obviously linked mechanistically to rates of species diversification. However, species richness may be affected by extrinsic factors as well. I will consider two: island area and microhabitat.

THE SPECIES DIVERSIFICATION: AREA RELATIONSHIP

The effect of island size on the rate of species diversification is an obvious place to start. The species-area relationship is one of the most consistent findings in all of ecology—across almost any set of islands or island-like entities (e.g., lakes, mountaintops), species richness increases as a function of area (Schoener, 1976b; Lomolino, 2000). The species-area relationship could result purely from ecological processes of extinction and colonization, but recent work has illustrated an evolutionary component as well by demonstrating that rates of species diversification are also a function of island area (Steppan et al., 2003; Gillespie, 2004; Parent and Crespi, 2006). This relationship was first demonstrated for anoles in the Greater Antilles (Fig. 15.8; Losos and Schluter, 2000); statistical analysis indicates that the relationship between rate of diversification and area results primarily from an increase in the rate of speciation with area, rather than a decrease in the rate of extinction.

Why speciation rates should be a function of island area is not clear. One obvious possibility is that the potential for allopatric isolation increases with island area, a hypothesis which appears particularly plausible given the number of mountain ranges on Cuba and Hispaniola. In addition, island area is often correlated with vegetational diversity and the number of different habitats (reviewed in Ricklefs and Lovette, 1999;

---

406. This assumes that Dactyloa wasn't more widespread in the past. Dactyloa's range currently extends as far north as Costa Rica.

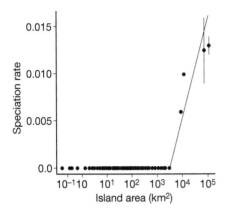

FIGURE 15.8

Speciation-area relationship in West Indian *Anolis*. Rates were calculated with the assumption that the occurrence of sister taxa on the same island is evidence of a cladogenetic speciation event on that island and were calculated relative to branch lengths; the y-axis has no units because branches weren't calibrated to time. Bars indicate ranges resulting from analyses based on different reconstructions of ancestral biogeography. This plot also reveals the threshold island size required for speciation discussed in Chapter 14. Modified with permission from Losos and Schluter (2000).

Whittaker and Fernández-Palacios, 2007), which potentially could increase the available niche space on larger islands. However, as mentioned in Chapter 14, islands the size of Guadeloupe and larger do not seem to differ greatly in habitat availability. Finally, a third possibility is that larger islands have a greater complement of other species—competitors, predators, parasites—which may drive ecological divergence and rates of diversification.

ECOMORPHS AND SPECIES RICHNESS

A second factor that may affect the rate of species diversification is microhabitat use. Many aspects of anole biology correlate with ecomorph class: is probability of speciation and extinction yet another? A priori, we can imagine a variety of ecomorph attributes that might have effects on speciation or extinction, such as population size or dispersal ability and its relationship to levels of gene flow.

Analysis of species richness across the Greater Antilles shows a strong effect of both island area and ecomorph (Fig. 15.9). In particular, trunk-ground and grass-bush anoles are particularly species rich, whereas twig and crown-giant anoles tend to be less diverse.

Why these differences exist is not obvious. Clearly, body size is not a factor because the smallest ecomorphs, the twig and grass-bush anoles, differ greatly in the number of species per island. If propensity for habitat fragmentation were responsible, we might predict, in contrast to Figure 15.9, that trunk-ground anoles should be the least

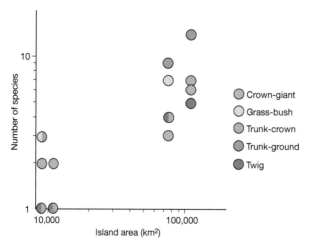

Differences among ecomorph classes in species richness (circles with more than one color indicate ecomorphs with the same number of species on an island). Ecomorphs differ in species richness across islands (analysis of covariance, heterogeneity of slopes non-significant; ecomorph effect, $F_{1,4} = 5.74$, $p = 0.007$; island area as covariate). Trunk anoles were not included in the analysis because they are only found on two islands; trunk anoles do buck the trend, however, being more species-rich on Hispaniola than on Cuba. Results are qualitatively unchanged if grass-bush anoles, absent from Jamaica, are excluded (Ancova, heterogeneity of slopes non-significant; ecomorph effect: $F_{1,3} = 4.33$, $p = 0.030$). Because ecomorphs are independently derived on each island, statistical significance of the ecomorph effect is not confounded by phylogenetic relationships.

likely to speciate because their populations seem least likely to be isolated by habitat disruptions.[407]

On the other hand, differences in species richness might be a function of extinction, rather than speciation, rates. Both on landbridge islands and throughout the Greater Antilles, trunk-ground anoles are nearly ubiquitous: if anole species are present, trunk-ground anoles are there. On landbridge islands, this pattern results because trunk-ground anoles survive even after other species have perished (see discussion of faunal relaxation in Chapter 4). Although this decreased rate of extinction results because trunk-ground microhabitats are present on even very small islands, it may indicate a general hardiness and resistance to extinction of trunk-ground anoles. Conversely, crown-giants and twig anoles often appear to have low population densities (but see Hicks and Trivers [1983]) and are rarely found on landbridge islands, perhaps bespeaking a high vulnerability to extinction. This line of reasoning, however, breaks down in a comparison of trunk-crown and grass-bush anoles, where the prediction of lower extinction rates of trunk-crown anoles based on patterns of occurrence on landbridge islands and in species-poor sites on the Greater Antilles[408] does not square with the higher species richness of grass-bush anoles.

407. This prediction assumes that trunk-ground anoles are more likely to cross open ground from one habitat patch to another than are more arboreal species.

408. In both situations, trunk-crown anoles are often in places where grass-bush anoles do not occur (see Chapter 4 on landbridge islands; no quantitative data exist for species-poor sites, but my impression is that trunk-crown anoles are usually more likely to be present than grass-bush anoles).

A variety of other aspects of anole biology could, in theory, affect rates of species diversification. Other factors such as environmental stability and seasonality or trophic position might plausibly have an effect. As discussed in Chapter 14, degree of sexual selection has been suggested recently as one factor that may affect rate of species diversification. If ecomorphs differ in extent of sexual selection (which remains to be determined [Chapter 9]), then this hypothesis would be worth investigating.

SEXUAL DIMORPHISM AND ADAPTIVE RADIATION

Despite the tremendous amount of research over the past several decades on both sexual dimorphism and adaptive radiation, little attention has been paid to the relationship between these two topics. Most research on sexual dimorphism has focused on its causes and consequences within single species and has considered neither the role that sexual dimorphism may play in adaptive radiation, nor how dimorphism might evolve during the course of a radiation.

THE EVOLUTION OF SEXUAL DIMORPHISM DURING AN ADAPTIVE RADIATION

Imagine the first anole species occupying a Greater Antillean island. Presumably, resources would be abundant and many different ways of making a living—corresponding to the different ecomorph types—would be available. What's a species to do? One possibility is that disruptive selection could drive adaptive radiation as all of the ecomorph types evolve *in situ*. I've already argued in Chapter 14 that sympatric speciation doesn't seem to occur in anoles, so—for whatever reason—this option appears to be out.

Another possibility is niche expansion. As discussed in Chapter 11, anole populations in species-poor localities tend to have broad resource use. An evolutionary response to such wide niche breadth is the evolution of increased intra-population phenotypic variation in which individuals are adapted to use different parts of the resource spectrum. At the extreme, these differences could take the form of discrete morphs, as in the African fire-cracker finch (*Pyrenestes ostrinus*), in which large- and small-billed morphs are adapted to eat seeds of different sizes (Smith, 1993). However, as discussed earlier in this chapter, quantitative analysis indicates that broad resource use is not generally accompanied by increased phenotypic variation within a population, but rather by phenotypically similar individuals with broader resource use (Lister, 1976b); moreover, few examples of ecologically relevant, non-sex-linked polymorphisms exist in anoles.

An alternative response is for populations to evolve sexual dimorphisms in which the sexes use different parts of the ecological spectrum (Schoener, 1986b). Such sexual dimorphism in both size and shape is rampant in anoles and varies by ecomorph (Chapter 9). Consequently, we might predict that the hypothetical initial Greater Antillean anole population would be comprised of individuals with broad resource use and that substantial ecological differentiation would occur between the sexes leading to the evolution of sexual dimorphism in morphology.

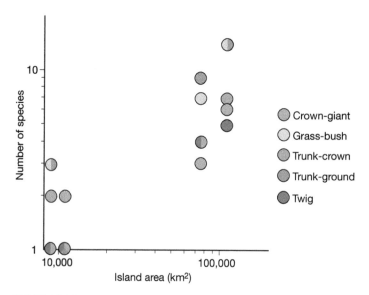

FIGURE 15.10

Sexual size dimorphism as a function of number of coexisting species on an island. Each point represents the median value of sexual size dimorphism for all of the species on one island. Values on the x-axis represent number of described species per island in the mid-1970s. Many species have been discovered since then, particularly on the larger islands. Modified with permission from Schoener (1977).

Eventually, however, more anole species evolve, probably in allopatry, and then become sympatric. As a result, ecological contraction—the opposite of ecological release—should occur, leading to diminished sexual dimorphism. Moreover, as more and more species join the community, this decrease should continue and the extent of sexual dimorphism should get steadily smaller.

This prediction has been tested most thoroughly with regard to size dimorphism. In comparisons both among species and among populations within species, the degree of sexual size dimorphism is negatively correlated with the number of sympatric species (Fig. 15.10; Schoener, 1977). This inverse correlation has several components:

1. Species in depauperate communities on landbridge islands have high levels of dimorphism due to ecological sorting. As landbridge islands decrease in size, ecomorphs drop out in a predictable sequence, and the ecomorphs that tend to persist, trunk-ground and trunk-crown anoles, tend to have high dimorphism (Chapter 4). One possibility is that these ecomorphs are successful in persisting on depauperate islands because of their high dimorphism; however, an alternative is that these ecomorphs are the best adapted to conditions on small islands, unrelated to their great degree of sexual dimorphism.

2. Size dimorphism increases after colonization of solitary islands. Colonizers of empty islands tend to have relatively high levels of size dimorphism, but subsequently evolve even higher levels (Poe et al., 2007). For example, in the Greater Antilles, solitary anole species all have as their sister taxa either

trunk-crown or trunk-ground anoles; comparison to estimates of ancestral size dimorphism indicates increased size dimorphism in these solitary species.[409]

3. Size dimorphism decreases during adaptive radiation with increased species number. Jamaica, the island with the fewest anole species, has the highest median size dimorphism, whereas the two most species-rich islands, Cuba and Hispaniola, have the lowest dimorphism. This trend has several causes. First, among the ecomorphs common to all four islands, size dimorphism within each ecomorph is inversely related to species number on an island (analysis of covariance, heterogeneity of slopes non-significant, island species number effect, $F_{1,11} = 3.97$, $p = 0.036$, one-tailed). Second, the ecomorphs found only on the larger, and more species-rich, islands—grass-bush and trunk—have relatively low dimorphism. Third, most Greater Antillean unique anoles, which occur only on the two largest islands (with one exception), also tend to have intermediate-to-low dimorphism.[410]

The relationship between sexual shape dimorphism and number of species has only been examined in one comparison: the species in the Jamaican radiation have a higher mean shape dimorphism than the anoles of Puerto Rico (Butler et al., 2007). Whether, as would be predicted, Lesser Antillean anoles have even greater dimorphism, and Hispaniolan and Cuban anoles even less dimorphism, remains to be tested.

These trends support the hypothesis that sexual dimorphism evolves adaptively in response to the presence or absence of other species, presumably as a result of resource competition. Moreover, they indicate that the degree of dimorphism decreases during adaptive radiation, both because species within microhabitats evolve decreased dimorphism and because the microhabitats occupied only in species-rich radiations tend to be filled by species with low dimorphism.

THE RELATIVE IMPORTANCE OF SEXUAL DIMORPHISM VERSUS
INTERSPECIFIC DIFFERENTIATION IN ADAPTIVE RADIATION

A second question about sexual dimorphism concerns how substantial a role it plays in adaptive radiation. Most research has implicitly assumed that sexual dimorphism is a minor contributor to the ecomorphological diversity within an adaptively radiating clade. In theory, however, there is no reason that much of the niche differentiation that occurs within a clade could not be manifested as differences between the sexes within species (Fig. 15.11). No study to date has examined the role that sexual dimorphism plays in adaptive radiation.

---

409. This analysis was limited to species endemic to solitary islands and did not consider populations of species also found on islands with other species.

410. Data from Schwartz and Henderson (1991) and Butler et al. (2000). The Cuban aquatic anole, A. *vermiculatus* and its sister taxon, the rock-wall anole, A. *bartschi,* are conspicuous exceptions to the generalization that unique anoles have low dimorphism.

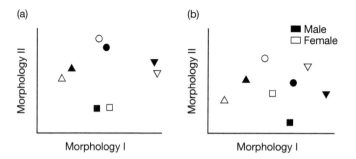

FIGURE 15.11

The role of sexual dimorphism in adaptive radiation. Sexual dimorphism could be a minor (a) or a major (b) component of morphological differentiation. Symbols represent different species, shaded symbols are males and open symbols are females.

Butler et al. (2007) examined the positions of both sexes of Puerto Rican and Jamaican anoles in multivariate morphological and ecological space. They found that the lion's share of the variation was accounted for by consistent differences among the ecomorph classes. Nonetheless, a substantial additional portion of the variation was explained by sexual differences within species, as well as a small amount due to variation that occurred between sexes in some ecomorphs and not others.[411] Moreover, because of sexual dimorphism, morphological and ecological space were much more fully occupied than if no sexual differences had existed—the morphospace volume occupied by both sexes on these two islands is 59% greater than that occupied just by females and 88% greater than that occupied by males. Similarly, both sexes occupy 33% more multivariate ecological space than females alone and 47% more than males.

These data indicate that sexual size and shape dimorphism play an important role in anole adaptive radiation. In islands with few species, much of the ecomorphological variation among anoles is partitioned between the sexes. As radiation proceeds, dimorphism decreases as species' niches become compressed by the presence of competitors, but it still accounts for an important part of the ecological and morphological variation.

Clearly, work is needed on patterns of shape dimorphism on islands both larger and smaller than the two studied to date. In addition, experimental studies on the evolutionary dynamics of sexual dimorphism could prove quite interesting. One would predict, for example, that the addition of a second species to a site previously occupied by only one species would lead to selection for the sexes to become more similar in the original species. Alternatively, patterns of selection might differ among the sexes, with the sex more similar to the introduced species being affected more greatly.[412] Anoles could prove to be a model system for the study of the evolution of sexual dimorphism, as well as of its role in adaptive radiation.

411. The ecomorph-by-dimorphism interaction term.
412. Alternatively, the same questions could be investigated by looking at the effect of introduced species on the sexual dimorphism of native species.

The importance of sexual dimorphism in anole adaptive radiation has one additional implication. As discussed in several previous chapters, the ecomorphs differ in degree of sexual size and shape dimorphism, as well as in social structure and social behavioral traits (e.g., display rate). These differences indicate that the ecomorph phenomenon represents more than just morphological adaptations to moving on different sized structures. Rather, occupation of different structural microhabitats has led to divergent adaptation not just in limb length and toepad size, but also in social structure, display and foraging behavior, size and shape dimorphism, and other characteristics (Chapters 3, 8 and 9). One possibility is that these disparate evolutionary changes are in response to independent aspects of structural microhabitat; that is, limb and toepads may evolve in response to selection for efficient locomotion in the different structural microhabitats, dimorphism may evolve in response to differences among microhabitats that affect the strength of sexual selection, foraging mode may evolve in response to effects of structural microhabitat on prey availability, and so on.

Alternatively, however, these features may be causally linked, representing an evolutionary syndrome of features related to structural microhabitat. For example, the short legs of twig anoles, necessary for locomotion on narrow surfaces (Chapter 13), may make rapid movements to capture prey and escape predators impossible. As a result, twig anoles may need to be more cryptic than other anoles, and thus may display less. In addition, they may need to forage more widely for less active prey, both because their slow speed precludes them from catching more active prey and because their microhabitat limits the area they can scan for active prey. This active lifestyle may lead to increased home range size and a lessened ability to defend territories, thus possibly decreasing the strength of intrasexual selection among males, but increasing the opportunity for female mate choice. At the other extreme, the broad surfaces that trunk-ground anoles use select for long legs: the great sprint speed these legs impart allow these lizards to display frequently in exposed places. Moreover, the large area they can survey for prey allows them to remain stationary, at the same time keeping an eye out for intruders, which can be quickly repelled, thus increasingly the ability of males to exclude others from their territories and possibly limiting opportunities for female choice. In this way, locomotor behavior and morphology, foraging behavior, social structure, and sexual dimorphism all may be integrated aspects of evolutionary adaptation to different structural microhabitats.

IS THE TERM "ADAPTIVE RADIATION" MEANINGFUL? A COMPARATIVE TEST TO INVESTIGATE WHETHER A CLADE CONSTITUTES AN ADAPTIVE RADIATION

I'll conclude the chapter by asking a simple question: do anoles constitute an adaptive radiation? Certainly, *Anolis* is speciose and ecologically diverse, and much of this diversification appears to have been adaptive. But this could probably be said about many clades of organisms. Given enough time, almost all clades will diversify, and a substantial proportion of the ensuing diversity is likely to be adaptive. Does that mean that most clades constitute adaptive radiations? Certainly, many workers who specialize on a particular group refer to their study subject as an adaptive radiation—isn't that more exciting than studying an "ordinary" group?[413]

But this approach renders the term meaningless. If adaptive radiation is the normal, expected outcome of evolutionary diversification, then why have the term at all? Designating a clade as an adaptive radiation would add no extra information. Although arguing about whether a clade is an adaptive radiation or not might seem an insignificant debate over terminology, the issue actually is significant. Evolutionary biologists often are interested in trying to explain why a particular clade is so diverse. Before this question can be investigated, however, we need to know which are the exceptional clades upon which to focus—the diversity of clades that represent the usual expected outcome of evolutionary diversification requires no special explanation. For this reason, reserving the term "adaptive radiation" for those clades which are exceptionally diverse is important.[414]

But how do we recognize those clades that are exceptional? The first question is, what is the metric to compare clades? Many studies have compared the species richness of different clades (e.g., Barraclough et al., 1999; Owens et al., 1999; Ricklefs et al., 2007). Although investigating what causes some clades to be species rich and others to be species poor is interesting and important, it is not the same as asking whether a clade constitutes an adaptive radiation. The reason is simple: clades can be ecologically and morphologically extremely diverse, despite containing few species (consider Darwin's finches, with only 14 species [Grant, 1986; Grant and Grant, 2008]), or they can be species rich, but ecologically and morphologically homogeneous (e.g., plethodontid salamanders [Kozak et al., 2006]). Thus, species richness and ecological and phenotypic disparity are distinct aspects of evolutionary diversification, both of which are considered in this chapter. To examine adaptive radiation, however, we need to focus on phenotypic disparity, which quantifies the extent to which members of a clade have evolved adaptations to using different parts of the environment.

Borrowing a page from community ecology, Miles and I developed a null model[415] to test the hypothesis that a clade has exceptionally great ecomorphological disparity (Losos

413. Of course, some contrarians pride themselves on studying species-rich groups with exceptionally little adaptive variation, which have been given the name "nonadaptive radiations" (in fact, a whole lexicon of types of radiations has been proposed, including "developmental," "architectural," etc. [Erwin, 1992; Givnish, 1997]).

414. Put another way, if *Anolis* is not exceptional, why should we pay particular attention to it, as compared to any other group of lizards? Why should I write this book, and why should you read it? The unusual breadth and integration of research on anoles is certainly an alternative reason, but I think much of the interest in anoles in the general scientific community is based on the idea that anoles are, indeed, special, in the extent of their evolutionary diversification.

415. An ecological null model is "a pattern-generating model that is based on randomization of ecological data or random sampling from a known or imagined distribution. The null model is designed with respect to some ecological or evolutionary process of interest. Certain elements of the data are held constant, and others are allowed to vary stochastically to create new assemblage patterns. The randomization is designed to produce a pattern that would be expected in the absence of a particular ecological mechanism" (Gotelli and Graves, 1996, pp. 3–4). Null models became famous in the context of debates over whether communities exhibited

and Miles, 2002). To employ this null model, one needs to: 1) establish the set of clades included in the comparison;[416] 2) quantify the disparity in putatively adaptive traits (i.e., traits for which an adaptive basis for interspecific variation has been established, such as limb length and lamella number);[417] and 3) determine whether some clades have exceptionally great (or little) disparity compared to what would be expected by chance if clades did not differ in their evolutionary propensities.

Miles and I implemented this approach to ask whether any of the subclades of iguanid lizards are exceptionally disparate in the sort of ecomorphological characters studied in anoles and other lizards. These clades form an appropriate pool to compare because, with one exception, all are similar in basic aspects of natural history such as diet, foraging and territorial behavior, body size and general morphology.[418] Also, the clades all appear to be of approximately the same age (Macey et al., 1997; Schulte et al., 1998; Wiens et al., 2006), so comparisons are not confounded by differences in the amount of time they have had to accumulate differences. To establish a null model of expected disparity, we randomized species among clades (standardizing species' values to account for interclade differences) and then compared the observed values of clade disparity to those generated by the null model.

The results of this analysis are clearcut. The clade to which anoles belong, the Polychrotinae,[419] has the greatest disparity (Fig. 15.12), which is significantly greater than

---

nonrandom patterns of species co-occurrence or phenotypic similarity. They were used to ask questions such as "Do particular species co-occur in communities less often than expected by chance?" and "Are coexisting species less similar in body size than would be expected by chance?". The history of these debates is summarized in Gotelli and Graves (1996); the exchange between Diamond, Gilpin, Simberloff, and Connor in the Strong et al. (1984) volume on community structure is instructive regarding both the science and the sociology of the debate.

416. Adaptive radiation must be viewed as a comparative concept: a clade is judged as an adaptive radiation compared to some universe of other clades comparable in some respects, such as evolutionary age. If not, then all life itself is an adaptive radiation, and all other clades pale in comparison. Or, to make a more narrow comparison, if, as many argue, placental mammals—the clade that includes whales, bats, elephants, and shrews—constitute an adaptive radiation, then any smaller and more restricted clade of mammals would by comparison likely not be considered an adaptive radiation. Thus, for this reason, adaptive radiation is a matter of scale; a clade can only be meaningfully judged in relation to a set of comparable clades.

Some will contend that the only appropriate means to test an evolutionary hypothesis is through sister group comparisons. Based on the logic that sister taxa are of the same age and should be similar in many respects due to their common ancestry, such comparisons are the appropriate and preferred comparison for many questions in evolutionary biology (Cracraft, 1981; Brooks and McClennan, 1991, 2002). However, heretical as it may be to some, sister group comparisons are not always appropriate. In this case, the sister-group approach might judge an unexceptional clade to be exceptional if its sister is even less diverse or, conversely, might fail to identify an exceptional clade if its sister is even more exceptional—compared to placentals, for examples, marsupials, diverse as they are, would not be considered an adaptive radiation (Losos and Miles, 2002). For this reason, the appropriate comparison is between a focal clade and a universe of other clades as similar as possible in age, natural history, geography and other attributes.

417. Disparity can be quantified in a number of ways (Foote, 1997; Erwin, 2007); perhaps the simplest is to calculate the mean pairwise distance between all species in a multivariate space defined by the characters under study: the greater the mean distance, the greater the phenotypic differences among species.

418. The exception are iguanas (the Iguaninae) because they are distinct from all other iguanids (ironically enough) in being herbivorous, with concomitant differences in body size, foraging and territorial behavior, physiology and many other aspects of their biology. For this reason, they were excluded from the analysis. This point is discussed at greater length in Losos and Miles (2002), as are more details about the method.

419. Note that the monophyly of the Polychrotinae has come into question, as discussed in Chapter 6.

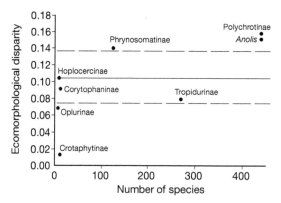

FIGURE 15.12

Ecomorphological disparity of subclades of the Iguanidae. Clades falling above the top dashed line have unusually great ecomorphological disparity, whereas clades falling below the lower line have unusually low disparity. *Anolis*, though younger than the seven subclades, still has exceptionally great disparity, as does the subclade (Polychrotinae) to which it belongs. This analysis also indicates that the relationship between disparity and species richness of clades is positive, but not very strong. Modified with permission from Losos and Miles (2002).

would be expected by chance. Moreover, the disparity value for the anole clade[420] within the Polychrotinae is significantly greater than expected by chance, even though anoles have been diverging for a shorter period of time than the clades that comprise the null pool (Fig. 15.12). The bottom line is that, at least in comparison to a set of similar clades, *Anolis* exhibits exceptional ecomorphological disparity and thus merits designation as an adaptive radiation.

Of course, a theme of this book is that anoles comprise not one radiation, but at least six (four on the Greater Antilles and two, and possibly more, on the mainland). Given that much of the disparity of *Anolis* recurs on each island, each island radiation likely exhibits exceptional disparity, but Losos and Miles (2002) did not sample widely enough to test this proposition. A study directed at this question would require collecting data on appropriate comparison clades (the ones in Losos and Miles [2002] being too old) and would require modifying the test to account for the non-monophyly of most of the anole radiations. Anole phylogeny makes clear that evolutionary diversification has occurred entirely independently only on Jamaica; by contrast, a moderate amount of inter-island reticulation exists among clades on the other three islands of the Greater Antilles (Chapter 6). This pattern of relationship means that the fauna of none of the three islands is the result of a single initial colonizing species. Nonetheless, the number of inter-island connections is small and most of them occurred early in anole history (Chapter 6). Moreover, the observation that sister clades on different islands are almost always ecomorphologically different indicates that the evolutionary diversification that has produced today's anole faunas occurred in situ; the diversity that exists on each island

420. Represented by a variety of Greater Antillean species plus *Anolis* Phenacosaurus *heterodermus* from the mainland.

today is not the product of species that had already evolved their differences on other islands coming into coexistence by multiple colonization events (i.e., ecological sorting [Chapter 7]).

Thus, although only Jamaica exactly meets the postulated first step of adaptive radiation, the history of the other islands agrees with it in spirit, even if several of the clades present on an island did not initially diverge there. An appropriately designed null model could examine whether the ecomorphological diversity on these islands is greater than expected for a radiation comprised of multiple clades; my feeling is that such a null model would be strongly rejected in all cases, supporting the existence of multiple adaptive radiations in the Greater Antilles and on the mainland.

FUTURE DIRECTIONS

In this chapter, I have attempted to take a synthetic approach to understand the progression of anole adaptive radiation. As has been plainly evident, the speculation-to-empiricism ratio in this chapter has been much higher than in previous chapters, and throughout the chapter I have highlighted what remains to be learned. For this reason, I will not summarize future directions in this and the next two—also synthetic— chapters.

16

THE FIVE FAUNAS RECONSIDERED

The *Anolis* evolutionary pageant exhibits a fundamental duality. On one hand, the Greater Antillean ecomorphs are renowned for convergence of entire communities, with the same set of ecomorphs evolving repeatedly. On the other hand, only one of the other four anole faunas—the anoles of the small islands of the Greater Antilles—contains many types of ecomorphs. The story of three of the other anole faunas—the mainland, the Lesser Antilles, and the unique anoles of the Greater Antilles—is primarily one of non-convergence, both internally and with the ecomorph radiations.

The simplest explanation for this contrast is that the environments in the Greater Antilles select for the same set of phenotypes, whereas the environments in the other localities select for different phenotypes. By environments, I mean abiotic factors such as temperature and humidity, as well as the structures which anoles use, the food they eat, and the other species with which they interact as predators, prey, and competitors.

This idea can be cast in the framework of an adaptive landscape in which the x- and y-axes represent different aspects of the phenotype and the height of the z-axis represents the extent to which multivariate phenotypes are favored by selection (reviewed in Fear and Price, 1998; Schluter, 2000; Arnold et al., 2001). In this light, the simple hypothesis above would suggest that adaptive peaks are in the same place in the Greater Antilles, and in different places in the other areas (Fig. 16.1).[421]

---

421. Keep in mind the abstract nature of figures like 16.1. Although the adaptive landscape for a single population in a static environment is mathematically defined and analytically tractable, the extension to consideration of the landscape for multiple co-occurring species in an evolving clade should be viewed as a heuristic analogy. Technically, the adaptive landscape specifically refers to how a population will evolve in a

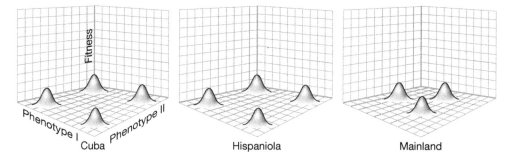

FIGURE 16.1

Similar adaptive landscapes on the islands of the Greater Antilles (represented here by Cuba and Hispaniola) may be responsible for the evolution of the same set of ecomorphs on each island, whereas a different landscape on the mainland could account for the different patterns of morphological evolution found there.

This hypothesis makes a major assumption, that evolution is completely predictable; that is, movement through phenotype space is unhindered such that species will always evolve to the highest available peak (assuming, in addition, that once a peak is occupied by one species, it cannot then be occupied evolutionarily by another species). In addition, as an explanation of ecomorph convergence, the hypothesis also assumes a unitary match between the environmental factors that impose selection and the possible phenotypic responses—that is, that only one phenotypic solution exists for problems posed by the environment.[422]

These assumptions need not be true, and if they are not, then the simple hypothesis above may be incorrect: convergent evolutionary radiations may not necessarily imply similarity in environments, and lack of convergence in radiations may occur even in very similar environments.

---

particular selective setting, and multiple peaks on that landscape indicate different regions of phenotypic space favored by selection. However, the existence of multiple peaks for a single population does not necessarily imply that multiple, sympatric species in that environment would evolve to the different peaks. Rather, the landscapes for each of the species would not necessarily be the same, because each species would occur with a different complement of co-occurring species (i.e., each species is part of the environment for other species). Consequently, to envision whether the same set of phenotypes would be favored on different islands, we would need to look at the landscapes for each species separately. However, these landscapes might not be static, but rather might change each time a new species joins the community. In addition, as a species evolves, then so might the adaptive landscapes for all co-occurring species. None of these issues is incorporated into the mathematical underpinning of the adaptive landscape. Consequently, application of this concept to an evolving adaptive radiation should be considered a metaphor, albeit an extremely useful one.

422. This view of selection—the environment creating problems to which populations must adapt—has been criticized because organisms interact with their environment and these interactions shape the way in which natural selection operates (Lewontin, 2000). Certainly this is true in some ways for anoles. For example, by selecting which part of the environment they use, anoles determine the biophysical environment which they experience. Nonetheless, much of the discussion of anole evolution concerns the external environment—vegetation structures, regimes of temperature and humidity—which do set demands to which organisms must adapt if they are to use the environment successfully. In this sense, I feel this metaphor is a useful way to understand anole evolution.

In this chapter, I will examine the hypothesis that convergence among the Greater Antillean ecomorphs and non-convergence with the other anole faunas stems directly from similarities and differences in the adaptive landscapes they occupy. Along the way, I will consider a variety of alternative and interacting explanations for these patterns. At the outset, I want to make clear that expectations should be kept low. We have almost no hard data on any of these ideas. Consequently, this chapter is meant to be forward looking: my hope is to lay out ideas that may profitably be explored in the future, rather than to provide definitive tests of alternative hypotheses. Nonetheless, I will not refrain from providing my own intuition about which factors are most likely to be of primary importance in guiding anole evolution.

CONVERGENCE AND THE ADAPTIVE LANDSCAPE

Probably the single most notable fact about anole evolution is the convergence of entire communities that has occurred across the four islands of the Greater Antilles. Adaptive radiation on each of these islands presumably followed the scenario detailed in previous chapters, with resource-competition-driven character displacement being of paramount importance. The question is: why have these separate radiations produced extremely similar evolutionary outcomes?

The most parsimonious explanation for the repeated evolution of the ecomorphs is that the selective environment—the adaptive landscape—is the same on all four islands of the Greater Antilles. This is not an easy hypothesis to test. If we could test it, however, we might find one of three outcomes. The hypothesis would be supported if we found that adaptive landscapes are generally the same in the Greater Antilles, but that these landscapes differ from those elsewhere. At the other extreme, we might find either that landscapes everywhere are all the same or that they are all different. Either of these findings would suggest that factors other than the environment have played a role in shaping the anole radiations.

A third possibility, which could occur regardless of whether landscapes in different areas are similar, is that we might find unoccupied adaptive peaks. These vacancies could occur for two reasons. First, they might represent ecological opportunities that, for whatever reason, have not been exploited by anoles. Conversely, they might represent alternative adaptive responses to particular ecological conditions. That is, more than one way of adapting to a given situation might exist. For example, when faced with prey that contains a toxic substance, predators may evolve resistance or simply avoid eating the part of the body that contains the toxin (cf. Farrell et al. [1991] and Berenbaum and Zangerl [1992] on diverse responses in herbivores to plant defenses). In a similar vein, in the presence of predators, potential prey may respond by evolving greater crypticity, ability to flee, or ability to defend themselves (e.g., Losos et al., 2002).

Regardless of the explanation, the presence of unoccupied adaptive peaks would suggest that the external environment may not be solely responsible for determining patterns of convergence and divergence: factors internal to a population also might play a role in determining which peaks are occupied and which are not.

TESTING THE HYPOTHESIS THAT CONVERGENCE RESULTS FROM SIMILARITY IN THE ADAPTIVE LANDSCAPE

In theory, the topography of the adaptive landscape could be discovered in two ways.

MEASUREMENTS OF NATURAL SELECTION COMBINED WITH EXPERIMENTAL APPROACHES

The first method would be to measure selection on existing species. The expectation would be that selection would maintain ecomorphs in more or less their current state, either through stabilizing selection or through selection that might be directional for one generation, but for which temporal changes in selection ended up with no net change over time (Grant and Grant, 2002). Such a test would confirm the existence of selection favoring the phenotypes of the ecomorphs in an environment occupied by those ecomorphs. It would not, however, be able to assess the form of selection in areas of phenotypic space not currently occupied.

This problem theoretically could be solved by conducting experimental introductions of phenotypically different species to islands on which they did not occur, to measure selection in portions of morphological space not naturally occupied on that island. As suggested in Chapter 14, perhaps introducing only males in experimental enclosures (á la Pacala and Roughgarden, 1982; Rummel and Roughgarden, 1985; Malhotra and Thorpe, 1991) containing the native anole fauna and following their fate through their lifespan might be a way to get around the obvious ethical difficulties with such an approach.[423]

Studies such as these would characterize the selective pressures operating on anole communities today, in the presence of the ecomorphs. They presumably would show that the ecomorph phenotypes that occur today are maintained by selection. What they would show about phenotypes not naturally present on an island is harder to predict. If, for example, one established a population of grass-bush anoles or a rock-wall specialist like *A. bartschi* on Jamaica, would selection favor those phenotypes? This test would have to be conducted in two stages. If all individuals perished, then selection gradients could not be calculated because they involve comparing survivors to non-survivors; nonetheless, this result would strongly indicate that the particular phenotype occurs in an adaptive valley. If there were some survivors, then we could determine how selection would

---

423. Care would have to be taken to choose species that could not interbreed with native species. Even then, one might worry about the possibility of introducing diseases or parasites to which the native species were not adapted.

Unfortunately, another option, examining localities where species already have been introduced, would not work in this case. The reason is that no cases of introduction of species with ecomorphologies not already occurring on an island have been reported in the Greater Antilles (i.e., there have been no introductions of unique anole species or of the absent ecomorphs on Puerto Rico or Jamaica).

operate on the population's phenotype. Would stabilizing selection maintain their phenotypes or would strong directional selection prevent the phenotype from persisting for long, perhaps by transforming the population into one of the ecomorphs? Ideally, enough different phenotypes could be introduced to cover a broad swath of anole ecomorphospace, though they probably couldn't all be introduced at the same time and place.

Such a study would be incomplete, however, because it would only examine the adaptive landscape in the presence of the ecomorphs. If species interact, then the selective optimum for one species might change depending on what other species are present—character displacement is an example of the different position of adaptive peaks in the presence of competitors. What we are really interested in asking is whether the environments on different islands have driven adaptive radiation in the same direction. To ask this question, we would need to estimate the adaptive landscape in the presence of different numbers and combinations of other species. Perhaps the place to start would be to estimate the landscape for a single species by itself. By placing different phenotypes in an enclosure with no other species, we might be able to estimate the phenotype favored on a Greater Antillean island in the absence of other anole species. Perhaps by then placing different combinations of pairs of species, we could envision the adaptive landscape at the two-species stage. This would be easier if the optimum phenotype at the one-species stage corresponded with the phenotype of an extant species. By examining enough combinations of species numbers and phenotypes, we might be able to get a sense of what the adaptive landscape looks like, and how it changes through the course of a radiation.

Of course, even if such an approach were possible, difficulties would abound. First, we would have to assume that somewhere among the anole phenotypes existing today are species similar to the ancestral anoles that existed in the early stages of radiation. If not, we might fail to estimate a crucial part of the anole landscape.[424] Second, to conduct these experiments thoroughly, they ideally would be carried out over a number of years because selection can vacillate from one year to the next (Grant and Grant, 2002). Third, the experiments should probably be conducted in a wide variety of different localities because environmental conditions vary among and within islands. Finally, fourth, it is a leap of faith to assume that the environments today mirror those encountered by anoles during their evolution, even aside from the vast alterations caused by humans in recent years. Probably for these reasons, as well as the tremendous amount of work that would be required, no study of this sort has ever been conducted on any

---

424. Some studies have hybridized different forms to create phenotypes not extant today (e.g., Schluter, 1994; Lexer et al., 2003). Unfortunately, most anole species are unlikely to reproduce with other species with very dissimilar phenotypes either because they have been separated evolutionarily for many millions of years and thus are unlikely to be interfertile, or because they coexist with closely related dissimilar forms and have evolved pre-mating reproductive isolating mechanisms. Nonetheless, I am not aware of any study that has tried to hybridize different species either naturally or through in vitro means.

organism. Nonetheless, much could be learned—these reservations notwithstanding—and anoles might be a good group on which to attempt such a study.

PREDICTING THE ADAPTIVE LANDSCAPE FROM KNOWLEDGE OF THE ENVIRONMENT AND THE FORM-FUNCTION RELATIONSHIP

A complementary approach to inferring the adaptive landscape based on measurements of selection would be to derive it from first principles concerning the ways anoles interact with the environment. That is, start with the resources available in the environment and then, based on an understanding of how morphology relates to functional performance and in turn to resource use, predict the phenotypes that would be favored in that environment. In other words, invert the approach that has been taken to date; rather than starting with the species and its morphology and asking why those particular traits are adaptive in the environment in which they occur (Chapter 13), we need to focus on the environment and ask whether we can predict which traits would be favored in that environment.

This approach is exemplified by work on the evolution of beak size in Darwin's finches, which proceeded in several steps (Schluter and Grant, 1984; summarized in Schluter [2000]).[425] The authors proceeded as follows:

1. They quantified the availability of seeds of different sizes on a number of islands.

2. They determined the maximum seed size that could be cracked by a finch with a given beak size.

3. They determined the minimum seed size taken by finches with a given beak size (presumably, the minimum size was related to the efficiency with which small seeds could be manipulated and ingested, but this was not directly examined).

4. For each beak size, they calculated the total density of seeds on an island between the minimum and maximum values.

5. For each beak size, they converted seed density to predicted finch density by means of an empirically derived equation describing the relationship between seed density, finch body mass (which is related to beak size), and population density.

6. For each island, they plotted the relationship between beak size and predicted finch density, with the assumption that the beak sizes with the highest densities represented adaptive peaks.

Based on this analysis, Schluter and Grant (1984) found that most islands had multiple adaptive peaks (Fig. 16.2). Moreover, a reasonably close match was observed between

425. Case (1979) took a somewhat similar approach, minus the functional component, to understand body size evolution in *Cnemidophorus* lizards (see Chapter 17).

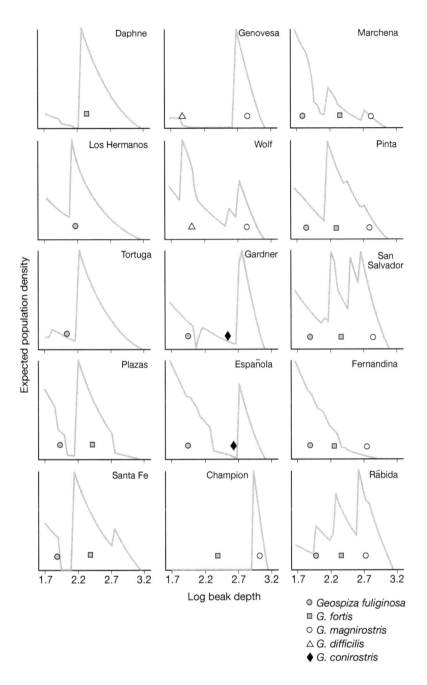

FIGURE 16.2

The adaptive landscape for beak depth in ground finches on the Galápagos. Based on the distribution of seeds on an island and the empirical relationship between beak size and population density, the population size of finches could be predicted as a function of beak depth. Most islands have multiple adaptive peaks, and the morphology of finch species lies close to these peaks on most islands. Modified with permission from Schluter and Grant (1984).

predicted and observed beak size on most islands, and these results were robust to incorporation of information on the beak sizes of sympatric species on an island.

Variation in limb length in anoles is the best candidate as an analog to beak size in finches. Would this approach work for anoles? The relationship for anoles between limb length and surface diameter is not as straightforward as the beak size–seed size function in finches. Two observations seem particularly relevant. First, the extent to which sprint speed is affected by surface diameter is a function of limb length: long-legged species are greatly affected, whereas short-legged species hardly notice differences in surface diameter; second, the more sensitive a species is to perch diameter, the narrower its breadth of habitat use and the more it avoids surfaces on which its sprint speed is greatly submaximal (Irschick and Losos, 1999; Chapter 13).

From these data, we can see how derivation of a performance-based adaptive landscape might begin. Clearly, long-legged species should be affected by the availability of broad surfaces. By contrast, short-legged species might be expected to occur everywhere. Three questions would have to be addressed to make progress:

1. What is the relationship between habitat availability and population size? As with the finch example, we can imagine measuring the availability of suitable vegetation (i.e., surface diameters at which a species could run at 50% or 80% [or some other arbitrary cut-off] of maximal speed). Then, we would need to establish the empirical relationship between vegetation availability and population size. Schluter and Grant (1984) simply summed all seeds within the acceptable range; we might want to develop a more precise equation that weighted different-sized supports by how much they affected sprint performance and how frequently they were used.[426] A more sophisticated approach might consider not just how sprint performance changes on different surfaces, but also how prey capture and predator risk vary as well. These would be a function not only of the lizard's performance, but also of the abundance of prey and predators on different surfaces.

2. How does the presence of other species affect habitat use and, as a result, population size? We know that anole species shift their habitat use in the presence of other species (Chapter 11). Presumably this results either from interspecific aggression or resource depletion, or both (or intra-guild predation when the species differ in size; see Chapter 11). These habitat shifts would have to be incorporated into the adaptive landscape model to predict how adaptive peaks would shift in the presence of other species.

426. A comparable approach was tried in the Darwin's finch study, but did not qualitatively change the results (D. Schluter, pers. comm.).

3. How should the distribution of surfaces at different heights be included? Limb length and sprinting capability are not obviously related to perch height in any mechanistic way, yet long-legged species generally occur relatively low to the ground. Most likely, perch height is related to toepad structure (Chapter 13). Two possible approaches would be either to limit measures of habitat availability to the height ranges occupied by different species, or to extend the analysis to a multivariate adaptive landscape and consider toepad structure along with hindlimb length. This would require further examination of the functional and ecological consequences of variation in toepad structure, which is not as well understood as the consequences of limb length variation (Chapter 13).

Obviously, this proposed work is very conjectural, with many loose ends and much more data needed. Certainly, we would want to include other characteristics beside limb length, not only toepad structure, but tail length, head dimensions and other traits, whether in one big multivariate analysis, or in separate univariate landscapes. Needless to say, this would require considerable effort. Whether we could actually build an anole adaptive landscape from first principles, and thus test the extent to which the environment drives convergence across the Greater Antilles, but not elsewhere, is unclear, but I think it would be worth a try.

In theory, both of these approaches—the development of selective and functional landscapes—are practical, but they may not occur any time soon. In the meantime, we have no actual data supporting the proposition that convergence of the ecomorph radiations is the result of similarity in underlying adaptive landscapes. In the absence of such data, I now turn to consider the evidence, also quite meager, that other factors might have shaped the anole radiations.

MORPHOLOGY-PERFORMANCE RELATIONSHIPS

Selection does not act directly on phenotypes, but rather on the functional capabilities produced by phenotypes (Arnold, 1983; Garland and Losos, 1994). For example, selection presumably didn't favor long legs in cheetahs because they are aesthetically pleasing, but because they allow the cats to run very fast. As discussed in Chapter 13, no straightforward relationship may exist between morphology and functional capabilities. Rather, radically different phenotypes may confer the same functional capabilities (Simpson, 1953; Bock and Miller, 1959; Losos and Miles, 1994).

The upshot of many-to-one mapping of morphology onto performance capabilities is that the adaptive landscape is determined by two relationships: the mapping of selection onto performance, and of performance onto phenotype (Fig. 16.3). If a one-to-one relationship exists between phenotype and performance, then selection will favor only a single phenotype for each selective peak in the performance landscape. However, if the

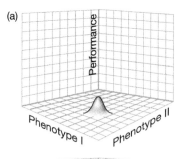

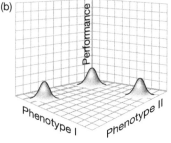

FIGURE 16.3

Phenotype, performance, fitness and the adaptive landscape. If the relationships between phenotype and performance and between performance and fitness are unimodal, then only a single peak may exist in the adaptive landscape for a population. Conversely, if multiple combinations of phenotypic characters can produce the same performance, then the adaptive landscape will necessarily contain multiple peaks, even if the performance-fitness relationship is unimodal.

relationship between phenotype and performance is many-to-one, then multiple phenotypic optima may exist for each selective peak in the performance landscape. The result is that two clades radiating independently in similar landscapes might nonetheless produce different phenotypes (Alfaro et al., 2005; Stayton, 2006; Collar and Wainwright, 2006; Wainwright, 2007; Young et al., 2007).

Could the many-to-one phenomenon explain differences between the anole faunas? For example, might mainland and Greater Antillean unique anoles be functionally convergent with the ecomorphs, even though they are phenotypically disparate? For the most part, the possibility of many-to-one functional relationships has been little studied, although some preliminary studies hint that they might exist (Chapter 13). However, if that were the case, we would expect to see species that parallel the ecomorphs in ecology and behavior, but not in morphology. This explanation might pertain to some species, but wouldn't apply to the divergent habitat use of many Greater Antillean unique anoles, nor to the behavioral differences between mainland and West Indian anoles (discussed below).

EVOLUTIONARY CONSTRAINTS

In the preceding discussion, evolution is dictated solely by external conditions: the environmental setting determines the adaptive landscape, and species necessarily evolve to occupy the highest peaks. This scenario assumes that a species can evolve with equal ease in any direction. However, for a variety of reasons (e.g., the genetic covariances among traits, the way in which development proceeds), evolutionary change may be constrained such that a species may more easily evolve in some directions than in others, and some phenotypes may not be attainable at all (Arthur and Farrow, 1999; Gould, 2002; Schwenk and Wagner, 2003, 2004; Brakefield, 2006).

The existence of such constraints might make convergence either more or less likely. On one hand, two clades radiating in similar environments might evolve in different ways if their genetic and developmental systems were different such that evolution was constrained to progress in different directions (Fig. 16.4a). Alternatively, if the clades share the same genetic and developmental systems, they might be biased to evolve in similar ways, even in environments that are not identical (Fig. 16.4b).[427]

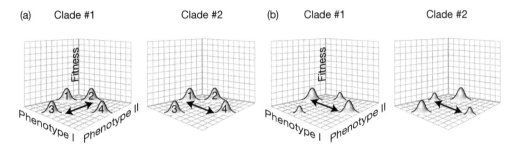

FIGURE 16.4

The effect of constraints on the direction of evolution in an adaptive landscape. In (a) two clades exhibit evolutionary constraints that bias them to evolve in different directions: the arrows indicate the direction in which each clade evolves most readily. Such biases could arise because of genetic linkages (termed covariances) among traits or because the way in which development proceeds, making evolutionary change in the developmental system easier in some ways than in others (these two explanations may represent the same phenomenon, because developmental systems are under genetic control). The result of such biases is that the two clades may radiate in different ways from the same initial starting point in the same adaptive landscape: in the panel on the left, species with phenotypes #2 and #3 would be more likely to evolve, whereas in the panel on the right, phenotypes #1 and #4 likely would evolve. Conversely, in (b), clades with the same biases may radiate the same way, even though occurring in different adaptive landscapes.

427. Perhaps Gould had the anole ecomorphs in mind when he wrote in his usual inimitable style (2002, p. 1174):

> . . . the markedly inhomogeneous occupation of morphospace—surely one of the cardinal, most theoretically, and most viscerally fascinating aspects of life's history on earth—must be explained largely by the limits and channels of historical constraint, and not by the traditional mapping of organisms upon the clumped and nonrandom distribution of adaptive peaks in our current ecological landscapes. In other words, the inhomogeneous occupation of morphospace largely records the influence of structural rules and regularities emerging "from the inside" of inherited genetic and developmental systems of organisms, and does not only (or even primarily) reflect the action of functional principles realized by the mechanisms of natural selection imposed "from the outside."

Actually, Gould (2002) probably wasn't thinking about the anole adaptive landscape because he focused on evolutionary change occurring deeper in phylogenetic history. In fact, although Gould certainly knew *Anolis* from his field work in the Bahamas, to him they were "just a fleeting shadow running across a snail-studded ground" (1997, p. 16). There is no evidence that the anole ecomorph story entered into his thinking at all, even though he occupied an office in the Museum of Comparative Zoology only 24 m from Ernest Williams' for many years (actually, the distance was only 15 m as the anole hops, but a locked door [under which an anole could pass] required a circuitous sidestep into another hall. Perhaps it was this extra 9-m detour that prevented Gould from fully appreciating the many-splendored lessons of *Anolis*).

Two commonly discussed forms of constraints involve genetic correlations among traits, promoting evolution along the "genetic lines of least resistance" (Schluter, 1996, 2000; Blows and Hoffman, 2005), and developmental pathways, which also would bias the variation available within a population (Maynard Smith et al., 1985; Gould, 2002). If genetic and developmental systems are stable through time, then such constraints could have long lasting effects on evolutionary diversification; this, however, is a big if (Shaw et al., 1995; Schluter, 2000). Currently, few data are available to evaluate the role of constraints in shaping anole evolution. No studies have examined the genetic variance-covariance structure of any *Anolis* species,[428] and little information on anole development is available; in fact, the first embryological staging series for an anole species has just been published (Sanger et al., 2008b).

Although few direct data are available, the hypothesis that evolutionary constraints have played a large role in directing anole evolution seems unlikely. The traits that characterize the different ecomorph types—such as limb lengths, toepad dimensions and body size—are all continuous, quantitative characters. In general, substantial additive genetic variation is usually present for such morphological characters (Mousseau and Roff, 1987; Falconer and Mackay, 1996). Even though genes of large effect that account for substantial amounts of variation among species and populations are increasingly being discovered for all sorts of quantitative characters of this sort (Abzhanov et al., 2004, 2006; Shapiro et al., 2004; Colosimo et al., 2005), including limb length (Storm et al., 1994), these traits generally conform to the properties of heritability and response to selection as predicted by quantitative genetics theory (reviewed in Roff, 2007). For this reason, these traits should readily respond to selective pressures; lack of suitable genetic variation for other phenotypes is unlikely to explain the repeated evolution of ecomorphs.

In theory, genetic correlations among traits may favor the evolution of some multivariate phenotypes and preclude the evolution of others. However, such correlations would have to have persisted for tens of millions of years to have been the primary cause for the repeated evolution of ecomorphs across the Greater Antilles (Revell et al., 2007a). Although no relevant data are presently available to test genetic constraint hypotheses for *Anolis*, the ability to investigate such questions will be facilitated both by the availability of the *A. carolinensis* genome and by ongoing anole breeding projects, and I expect that before too long we will have a better understanding of the genetic architecture underlying ecomorphologically important traits.

Another reason that evolutionary constraint is unlikely to be responsible for the repeated evolution of the ecomorphs on the Greater Antilles is purely empirical: ample evidence exists that, in fact, evolution has produced a plethora of species that do not correspond to any ecomorph. Examples include many Lesser Antillean species, the unique species of the Greater Antilles and, most of the mainland fauna. These species

428. However, phenotypic variance-covariance matrices have been compared among populations of *A. cristatellus* (Revell et al., 2007a).

are interspersed throughout the anole phylogeny, which indicates that the ability to evolve out of the ecomorph mold is not a special condition of a particular clade. This empirical record would seem to contradict the hypothesis that developmental or genetic biases are responsible for the repeated evolution of the ecomorphs. Nonetheless, more data on anole developmental and genetic systems would be extremely useful to examine these ideas directly.

HISTORICAL CONTINGENCIES

Gould (1989, 2002) was the strongest proponent of the view that the outcome of evolution is historically contingent, which he defined as "an unpredictable sequence of antecedent states, where any major change in any step of the sequence would have altered the final result. This final result is therefore dependent, or contingent, upon everything that came before—the unerasable and determining signature of history" (Gould, 1989, p. 283).[429]

This perspective considers the predictability of evolution: can we foresee the course of evolution from an initial starting point? Gould's answer is "no": unpredictable events will happen along the way, and without foreknowledge of what those events will be, the evolutionary outcome is indeterminate. This view accords with Gould's (1989) famous analogy of "re-winding the evolutionary tape": if one could turn back the clock and start over again, from the same ancestral form living in the same place, evolution would be unlikely to take the same course.

In the context of the adaptive landscape and anole evolution, we may look at the question slightly differently and ask: does the history of a clade affect how it diversifies? Or, conversely: is the landscape deterministic such that any clade evolving on the same adaptive landscape will converge upon the same evolutionary outcome, regardless of its history?

Just what aspects of history are we talking about? Two types seem to be the most likely to affect the eventual evolutionary outcome:

1. The starting point of a radiation (Gould's "happenstance of a realized beginning" [2002, p.1160]): the biology of the ancestral species—its phenotype, natural history, even the amount and type of genetic variation—can affect subsequent evolutionary change (Travisano et al., 1995; Price et al., 2000). Ancestral forms will have their own evolutionary predispositions, resulting from genetic constitution, developmental systems, behavior patterns and a variety of other, interrelated factors that will make evolutionary change more likely in some directions than in others, particularly if these constraints are maintained through the course of a clade's history (Arnold, 1994; Donoghue, 2005). To exaggerate, had the ancestral anole

---

429. For a review of Gould's ideas on contingency and the concept itself, see Beatty (2006, 2008).

been limbless or possessed wings, the course of subsequent evolutionary diversification would have been very different.

2. Chance events: the occurrence and order in which mutations occur might play an important role in directing evolutionary change (Mani and Clarke, 1990; Wichman et al., 1999; Ortlund et al., 2007; but see Weinreich et al. [2006]). Similarly, random events—lightning or a falling tree killing a particular individual, an ill-timed volcanic eruption, or any other matter of happenstance—could push evolutionary change in one direction or another.

Recognition of the importance of historical contingencies does not mean that natural selection and adaptation do not occur. Rather, this perspective emphasizes that even in the presence of natural selection, evolutionary outcomes are not necessarily predictable. An important consideration in this light is the shape of the adaptive landscape. Consider a population evolving in a landscape with a single adaptive peak. Regardless of any of the possible contingencies just discussed, natural selection will tend to drive that population up that peak, or as close to the peak as possible given the variation that can be produced by genetic and developmental systems of the population (Fig. 16.5a).

By contrast, consider a more rugged adaptive landscape in which there are several high peaks, and in which no way exists to move from one peak to another without traversing an adaptive valley (Fig. 16.5b). On this landscape, historical contingencies may matter a great deal. Even if the peaks are the same height—i.e., they are equally favored by selection, none superior to the others—the actual peak that a population ascends may be affected by where the population begins—selection generally favoring movement up the nearest peak—and the pattern of constraint affecting the directions in which the population can most easily move on the landscape. Furthermore, for the same reasons, a population may end up on a suboptimal peak; once on such a peak, selection may have trouble moving the population to a higher peak because it would require first evolving in the direction of lower fitness into an adaptive valley, something selection by itself generally will not do (Fig. 16.5c).[430]

Historical contingency can thus prevent convergence: species evolving on the same adaptive landscape may evolve in different directions. However, contingency is a two-edged sword: species experiencing the same contingent events (e.g., the same ancestral phenotype) might converge, even on adaptive landscapes that are quite different (Fig. 16.6).

The possibility of contingency applies not only to species, but to entire communities. Community ecologists have long known that alternative stable equilibria may exist for the structure of a community (e.g., Scheffer et al., 2001; Chase, 2003a,b; Persson et al., 2007). In other words, given a set of resources in a particular environmental setting,

430. The topic of evolutionary transitions from one peak to another is actually much more complicated than this (Lande, 1986; Arnold et al., 2001), but I present this simple version for heuristic purposes.

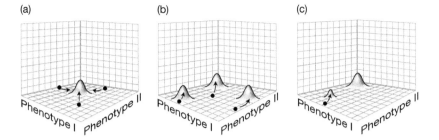

FIGURE 16.5

Historical contingency and the adaptive landscape. In (a) a species is likely to end up on or near the same peak regardless of constraints and where it starts. By contrast, in (b) initial starting conditions, as well as constraints (Fig. 16.4), may determine which peak is occupied because species are most likely to ascend the nearest peak unless constraints push them toward a different peak. This phenomenon can lead to species ending up on a suboptimal peak (c).

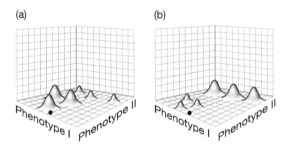

FIGURE 16.6

Initial starting conditions can cause species diversification to converge even in very different adaptive landscapes. In both panels, phenotypes in the lower left quadrant are most likely to evolve due to the clade's initial starting condition, even though the adaptive landscapes are quite different.

multiple ways may exist for a set of species to divide these resources, and each of these community configurations may be stable and resistant to replacement by other possible communities. These multiple ways of existence would correspond to alternative adaptive peaks mentioned at the outset of this chapter. Why one community structure may occur rather than another may be a result of the same historical vagaries—starting conditions, availability of particular mutations, random chance—that affect individual species (Fukami et al., 2007).

How important was historical contingency in anole evolution? For example, would the evolutionary trajectory of an anole radiation have differed depending on whether the ancestor was a twig anole, a crown giant, a grass-bush anole, or something else? This question is difficult to address for two reasons. First, as discussed in the previous chapter, inferring the ancestral phenotype of each of the anole radiations is problematic. For this reason, determining whether the radiations were initiated from similar phenotypic starting points is not possible. Second, we have little idea of how rugged the adaptive landscape is—that is, how many adaptive peaks there are and how likely a population is to become stuck on a sub-optimal peak. The more rugged the landscape, the more important a species' starting point will be and the more likely that radiations initiated by phenotypically different ancestors would explore different portions of morphological space. For the same reasons, chance events are more likely to have lasting consequences when the landscape is rugged.

We can address this second point to a limited extent. If the adaptive landscape is rugged, then transitions between ecomorph types that are most closely situated in the adaptive landscape might be more likely. The inability to infer ancestral states complicates such an analysis; however, we can ask whether certain pairs of ecomorphs tend to be closely related, which would suggest that transitions from one type to the other occur more readily than other possible transitions. Although no formal analysis has been conducted, examination of the phylogeny of Greater Antillean anoles indicates no obvious patterns of this sort (Chapter 7). However, one intriguing bit of evidence supports the possibility that transitions may occur more readily between adjacent peaks: in all three cases in which one ecomorph type arose from within another ecomorph type (grass-bush/trunk-ground in Cuba and Puerto Rico and trunk-crown/crown-giant in Jamaica [Chapter 7]), the two ecomorphs are ecologically and, to some extent, morphologically proximate.

EVOLUTIONARY DIVERSIFICATION AND THE *ANOLIS* ADAPTIVE LANDSCAPE

The preceding discussion makes clear that definitive conclusions about causes of the differences among the anole faunas will be hard to come by. Nonetheless, in the remainder of the chapter, I will discuss what we can and cannot say about the differences among the faunas. I take as my starting point the premise that similarity in adaptive landscapes across the Greater Antilles has driven convergence of the ecomorphs. I begin by examining patterns of occurrence of the different ecomorph types and evolution on species-poor islands in the West Indies to see if any general conclusions can be made about the anole adaptive landscape in the West Indies. I then explore non-convergence in the Lesser Antilles, among the Greater Antillean unique anoles, and on the mainland and discuss why evolution may have gone in different directions in these areas.

PATTERNS OF ECOMORPH OCCURRENCE: THE CASE OF THE MISSING ECOMORPHS

Not all ecomorphs are present at all locations in the Greater Antilles for two reasons: failure of some ecomorph types to evolve on some islands and failure of ecomorphs present on an island to occur in some localities. Consideration of both of these phenomena suggests that we can make some conclusions about the shape of the anole adaptive landscape.

The ecomorph radiations are not perfectly convergent; rather, trunk anoles are absent from Puerto Rico and trunk and grass-bush anoles from Jamaica. At first glance, the requisite structural habitat for trunk anoles, large tree trunks, appears to occur in abundance on these islands. Hispaniolan trunk anoles are voracious consumers of ants (Chapter 8; the diet of the Cuban trunk anole, *A. loysianus*, is unknown [Rodríguez Schettino, 1999]), and ants also seem common on these islands. Similarly, the prerequisites for grass-bush anoles would seem to be present in Jamaica.[431] Thus, the absence of these ecomorphs is not obviously attributable to environmental deficiencies on these islands.

The concept of "empty niches" has fallen into disfavor in recent years. Lewontin (1978, 1985) summarized the argument against them: one can imagine almost any combination of traits that could exist, such as flying mollusks, so speaking of their absence is pointless; niches don't exist independent of the organisms that occupy them. On the other hand, Lewontin (2000) also makes clear that his critique is directed toward designation of a niche in the absence of any species that has ever filled it. Convergent evolution has long been considered evidence for a predictable environment-organism interaction which suggests that the environment repeatedly elicits similar evolutionary outcomes (see discussion in Schoener, 1989; Harmon et al., 2005). Thus, it does not seem too much of a stretch to consider the niche for a trunk or grass-bush anole existing prior to its evolution.

Why, then, are some ecomorphs absent on Puerto Rico and Jamaica? One possibility is that their niches don't actually occur there. As just argued, this seems implausible—grass and tree-trunks abound on both islands[432]—but a more detailed analysis would be useful. In the case of trunk anoles, an alternative ecological possibility is that the trunk ecomorph niche has been usurped by the small trunk-crown anoles, *A. stratulus* (Puerto Rico) and *A. opalinus* (Jamaica), which, though good trunk-crown anoles in terms of morphology and ecology (Chapter 3), do nonetheless often occur on tree trunks. Perhaps this is an example of alternative phenotypes capable of utilizing the same set of resources? By contrast, for some unknown reason, the small trunk-crown anoles of Cuba and Hispaniola are generally restricted to montane localities, thus leaving the "trunk anole niche" open for trunk anoles over most of these islands. This explanation, however, would not account for the missing grass-bush anole of Jamaica; even though Jamaican anoles are less differentiated morphologically than the ecomorphs on other islands, none of the Jamaican species seems to greatly utilize typical grass-bush habitats.

Explanations based on non-adaptive factors should also be explored, though none are particularly compelling a priori. Perhaps genetic or developmental constraints exist in

---

431. Indeed, as anyone who has walked around tourist areas and been accosted by local peddlers can attest, grass is readily available just about anywhere in Jamaica.

432. Keeping in mind, of course, that the vegetation of these islands has been greatly altered by humans over the last several hundred years. Most of Jamaica was probably forested prior to human arrival (Eyre, 1996). Although open, grassy habitats previously may have been less common in Jamaica than they are today, many grass-bush species (e.g., most Cuban species, Puerto Rican *A. krugi*) occur in forested habitats.

the *grahami* Series (Jamaica) and *cristatellus* Series (Puerto Rico) preventing the production of appropriate phenotypes? These clades have diversified over otherwise much the same ecomorphological space as anoles on the other Greater Antillean islands, so neither this possibility, nor the ancestral starting condition for the radiations, seems likely to have had an impact. Finally, the Jamaican radiation is substantially younger than the other three Greater Antillean radiations, which raises the possibility that not enough time has been available to evolve more than four ecomorph types, although 24 million years[433] would seem long enough (Chapter 6).[434]

In sum, the evolutionary absence of these ecomorphs is a mystery for which we have no good explanation at the present time. However, these are not the only cases of missing Greater Antillean ecomorphs; even when an ecomorph is present on an island, it is often not found everywhere (Chapter 11). Trunk-ground and trunk-crown anoles are generally present in most localities in the Greater Antilles, but other ecomorph types can be more patchy in distribution.

As with the absence of ecomorphs from an entire island, the explanation for these local lacunae relies either on ecology or contingency. Ecologically, the explanations are effectively the same: appropriate habitat is unavailable either because it doesn't exist or is usurped by other taxa. However, the contingency explanation is a little different. Many of the absences seem to relate to thermal and hydric physiology. For some reason, on some islands ecomorph clades exhibit greater physiological versatility—either within or between species—than on other islands. For example, twig anoles occur commonly in the lowlands on Jamaica and Cuba, but not in Hispaniola or Puerto Rico. Assuming that ecological physiology accounts for these distributional patterns, research could be directed toward investigating why some clades are able to evolve greater versatility than others.

In contrast to the Greater Antilles themselves, ecomorph absences on landbridge islands near the Greater Antilles are more readily explainable. Prior to the rise in sea levels, land-bridge islands presumably harbored the full complement of ecomorphs present on the larger landmass to which they were connected (either a Greater Antillean

433. Even the 7 or 13 mya dates for initial within-island divergence suggested by earlier studies seem adequate (Hedges and Burnell, 1990; Jackman et al., 2002).

434. In addition to missing ecomorphs, the island radiations differ in other ways as well. For example, the Jamaican ecomorphs seem less differentiated than those on other islands (Beuttell and Losos, 1999). A quantitative analysis confirmed the imperfection of ecomorph convergence: although most morphological variation among Greater Antillean ecomorph species is explained by ecomorph type, some variation is accounted for by island effects (Langerhans et al., 2006). For example, Cuban ecomorph species tend to have the shallowest heads and Hispaniolan anoles the deepest heads. Differences in the environment across the islands could account for these effects. However, historical/phylogenetic effects—such as constraints or differences in ancestral phenotypes that have persisted to the present—could also be responsible because anoles on each island generally are more closely related to each other than to species on other islands; statistical analysis was unable to separate island and phylogenetic effects.

One particularly interesting phylogenetic effect was evident in the analysis: Cuban trunk-crown anoles have shorter limbs than other trunk-crown anoles, and are also the only trunk-crown anoles that have twig anoles, the shortest-legged of the ecomorphs, as their sister taxa (Langerhans et al., 2006). Possibly, the short-leggedness of the Cuban trunk-crown anoles is related to their being a member of a particularly short-legged clade, thus making them susceptible to whatever short-legged evolutionary biases that clade may possess.

island or the Great Bahama Bank). Consequently, their diminished fauna today is primarily the result of extinction and is related to island area: the smaller the island, the fewer the species. These extinctions have not been random. Rather, trunk-ground species are almost universally present, and if a second species occurs, it is almost always a trunk-crown species. The identity of the third and fourth ecomorph is consistent within a region, but varies across regions (Chapter 4).

The consistency of these patterns strongly argues that the environment determines patterns of ecomorph occurrence and that it does so in substantially the same way throughout the Greater Antilles. These islands might be a good place to develop or test models about the adaptive landscape. One particular question of interest would be whether the environment is unsuitable for ecomorphs that are absent, or whether those ecomorphs are excluded by the presence of other ecomorphs better adapted to environmental conditions. Why, for example, is the twig anole *A. angusticeps* often absent from small islands in the Great Bahamas Bank when appropriate habitat—an abundance of narrow vegetation—occurs on most of these islands? One possibility is that many islands lack some other attribute necessary for these twig anoles, such as the appropriate prey species, but another is that for some reason, other ecomorphs can exclude *A. angusticeps* from these islands, but not from larger ones.

EVOLUTIONARY DIVERSIFICATION ON SPECIES-POOR ISLANDS

If the adaptive landscape changes with the addition of new species, we would not expect the four ecomorphs found on Jamaica to also occur on more ecomorph-rich islands (ditto for Puerto Rico's five ecomorphs on Hispaniola and Cuba). The fact that they do suggests that the adaptive landscape is relatively static and that the positions of the adaptive peaks are relatively independent of each other.

We can test this hypothesis by examining patterns of evolutionary diversification on islands with relatively few ecomorphs. Assuming that these islands are environmentally similar to the Greater Antilles (a big assumption), if the adaptive landscape is static, we would expect to find typical ecomorph species.

To examine this idea, I focus only on oceanic islands because landbridge islands probably had a larger fauna in the recent past. Small islands in the Greater Antilles have been colonized primarily by trunk-ground and trunk-crown anoles (although the ancestral form of *A. acutus* on St. Croix is indeterminate [Chapter 4]). For the most part, these species are still recognizable as members of their ancestral ecomorph type; those species that have diverged generally occur in morphological space in positions intermediate between trunk-ground and trunk-crown anoles (Losos et al., 1994; Losos and de Queiroz, 1997; Chapter 15).

Evolutionary diversification in the Lesser Antilles has produced somewhat greater ecomorphological diversity than that seen on 1- or 2-species islands in the Greater Antilles. Although many species appear to be trunk-crown anoles, a few are as large as crown-giants, and the rest lie in intermediate positions in morphological space, again generally between trunk-crown and trunk-ground anoles (Chapter 4).

The faunas of these small islands could be interpreted in two ways with regard to the idea that the adaptive landscape changes as a function of the number of species present. The occurrence of ecomorph species on these islands might suggest that the same adaptive landscapes exist there as on larger islands, and thus that landscapes do not change depending on the number of species present. Exceptions would be explained as islands that are environmentally different. Alternatively, the glass-half-empty viewpoint would emphasize those species that do not fit neatly into any of the ecomorph categories. Ultimately, direct measurement of the adaptive landscape is needed to assess the extent to which environmental differences among islands drive these patterns.

Nonetheless, two observations are clear. First, when communities—anywhere in the West Indies, including the Greater Antilles—contain 1–2 ecomorphs, those ecomorphs are almost always trunk-ground and/or trunk-crown anoles. Moreover, on species-poor islands, species that do not belong to any ecomorph category are often most phenotypically similar to these two ecomorphs. Second, islands with 1–2 species almost never contain species resembling trunk, grass-bush, or twig species,[435] and nothing like these types has evolved on those small islands on which substantial evolutionary divergence has occurred. Notably, two of these types—grass-bush and trunk—are the ones that are missing from some Greater Antillean islands.

I draw three conclusions from these observations: first, adaptive landscapes throughout the West Indies are similar in that the highest peaks generally correspond to trunk-ground and trunk-crown anoles, or something like them. Second, the twig, grass-bush, and trunk ecomorph peaks seem to be lower, and thus are filled later in the course of faunal development. A corollary of this statement is that the absence of these forms from many islands results not because their niches do not occur on the islands, but simply because not enough species are found there, due to impediments on colonization and speciation. Third, it follows that genetic and developmental constraints and historical contingencies are of secondary importance in shaping patterns of ecomorphological evolution in West Indian anoles.

These are bold statements, perhaps easier to make because they will not be easy to test. Nonetheless, I believe that some of the ideas outlined in this chapter provide the means, at least in theory, to go about testing them. Obviously, the two-species islands of the Lesser Antilles are the biggest challenge, given that many of the species on these islands cannot be assigned to an ecomorph category (see Chapters 4 and 15).

---

435. Note that in contrast to the situation with landbridge islands—in which the absence of some ecomorph types may be the result of lack of appropriate habitat on small islands—oceanic islands (e.g., the Cayman Islands, St. Croix, the Lesser Antilles) are generally fairly large and contain well developed habitats that seem comparable to habitats which maintain the full complement of ecomorphs on the Greater Antilles.

The two-species islands in the Lesser Antilles are notable in a second respect. Although species from solitary Lesser Antillean islands are quite similar regardless of location, the species composition of two-species islands differs greatly between the north and the south in three ways:

- Although sympatric species almost always differ substantially in body size by approximately the same amount (differences slightly greater in the north), the species are larger in the south (Schoener, 1970b; Roughgarden, 1995).
- Sympatric species in the north differ in perch height, with the larger species found high in the tree and the smaller species near the ground; species on the same island in the south both occur at approximately the same, intermediate height (Roughgarden et al., 1983; Buckley and Roughgarden, 2005b).
- Species on the same island in the south differ in body temperature and segregate by habitat type, whereas species in the north attain similar body temperatures and do not partition habitat types (Roughgarden et al., 1981, 1983; Buckley and Roughgarden, 2005b).

Environmental variables could explain some of these differences. The southern islands are warmer, being closer to the equator, and they also have greater insect abundance (Buckley and Roughgarden, 2005a); both of these factors might promote higher growth rates and hence larger size (e.g., Roughgarden and Fuentes, 1977). In addition, the greater amount of high elevation—hence cooler—habitat in the more mountainous southern islands might promote the evolution of habitat segregation, whereas the more limited range of habitats available in the northern islands might have led to within-habitat niche partitioning (Roughgarden et al., 2003; Buckley and Roughgarden, 2005b).

On the other hand, in this case historical contingencies may play a role as well. The different evolutionary paths taken in the Lesser Antilles could indicate the existence of alternative adaptive peaks and alternative possible community structures. Perhaps either configuration of species is equally likely on these islands and the vagaries of history are responsible for the different outcomes. In this light, the different evolutionary endpoints might be the result of different initial starting conditions. The two areas were colonized by distantly related anole clades, the south by a member of the basal Dactyloa clade from South America, and the north by a member of the *cristatellus* Series (Fig. 5.6). These clades differ in a number of respects: *cristatellus* Series anoles are small-to-medium in size and usually heliothermic; by contrast, Dactyloa anoles often are quite large. Unfortunately, the ecology of few mainland Dactyloa clade anoles is well known, so generalizing about the ecology of this clade is difficult; however, many Dactyloa species occur in deep forest and probably are not heliothermic (e.g., Vitt et al., 2003a). Moreover, the phylogeny of Dactyloa is not well understood. Given these difficulties, inferring the ancestral condition for the two Lesser Antillean clades is impractical, but the possibility remains that the clades were initiated from different starting points, and that these differences affected how they subsequently evolved and which adaptive peaks they ultimately occupied.

Anolis wattsi, a small species from the northern Lesser Antilles that is usually found near the ground, has been introduced to St. Lucia in the southern half of the island chain (Fig. 16.7; Corke, 1987), and also to Trinidad, which was previously inoculated by humans with several southern Lesser Antillean anoles (White and Hailey, 2006). Follow-up studies on the outcome of these introductions might provide some insights about whether environment or contingency is responsible for the different evolutionary pathways taken by anoles in the two halves of the Lesser Antilles: successful invasion of *A. wattsi* would support the contingency hypothesis by suggesting that the evolutionary absence of species that use low microhabitats in the southern Lesser Antilles is not the result of environmental inhospitality.

FIGURE 16.7

Hybrid Lesser Antillean community on St. Lucia. Thanks to human introductions, two southern Lesser Antillean species, the native *A. luciae* (a) and *A. extremus* introduced from Barbados (b), now coexist with the small northern Lesser Antillean species, *A. wattsi* (c). How these species interact ecologically and evolutionarily may provide insights on why anole communities in the northern and southern Lesser Antilles are structured differently.

The unique anoles of the Greater Antilles are interesting in two respects: first, many of them are greatly divergent from the ecomorphs, in contrast to the pattern seen in the Lesser Antilles and the small islands of the Greater Antilles. This divergence occurs both in morphology—e.g., Chamaeleolis, Chamaelinorops, *A. vermiculatus*, *A. fowleri*, *A. eugenegrahami*, *A. bartschi*—and in microhabitats occupied—e.g., streams, leaf litter, rock walls, cave entrances (see descriptions in Appendix 4.1). Second, these forms are utterly non-convergent; none of these "unique" anoles has a morphological counterpart, nor an ecological one, on another island.[436]

The second anomaly about the unique anoles is that the Hispaniolan species and the single Jamaican species are found only in the mountains and generally have relatively small geographic ranges. By contrast, most of the Cuban unique species can be found at low elevations and some have quite broad geographic distributions.

What's going on with these species? Explanations based on environmental differences between islands have already been discussed in Chapter 4 and been found wanting—for the most part, the microhabitats occupied by these species occur across all of the Greater Antilles. But what other explanations are there? One salient observation is that these species are found almost exclusively on the two islands that have both the most species and the greatest number of ecomorphs, Cuba and Hispaniola. Perhaps these anoles have evolved to occupy minor adaptive peaks, ones that only are filled once the ecomorph peaks are already occupied?

If this were the case, we might expect unique anoles to have evolved relatively recently and from an ecomorph ancestor. However, this is not the case. Most unique anoles are on branches that go back deep into the phylogeny, and none has evolved from within a clade composed of another ecomorph type (Fig. 7.1). Of course, the ecomorphs themselves mostly evolved early in anole phylogeny, and the inability to infer ancestral states prevents a clear examination of the history of the unique anoles. Still, the phylogeny provides no support for the idea that unique anoles are late stages added after ecomorph radiation has been completed. Moreover, this hypothesis would not account for the non-convergence of these unique ecomorphological types across islands.

The deep ancestry of the unique anoles also precludes comparisons to sister taxa to see if particular species are similar to their close relatives. For the most part, the sister taxa of unique anoles are large and diverse clades.[437] One exception is Chamaeleolis, which is in the same clade as the Hispaniolan and Puerto Rican crown-giants. One

---

436. The closest appear to be the stream anoles of Cuba and Hispaniola, *A. vermiculatus* and *A. eugenegrahami*. However, not only are they greatly different in morphology, but they also appear to interact with the environment in different ways (Leal et al., 2002). Comparison of species often found on rock surfaces—such as the little-known *A. monticola* Series in Haiti and *A. lucius* and *A. bartschi* in Cuba—might also prove interesting.

437. In other cases, the phylogeny is too uncertain to unambiguously identify sister taxon relationships deep in the tree (Chapter 5).

scenario is that the ancestral Chamaeleolis initially was a crown-giant that emigrated from Puerto Rico or Hispaniola, but finding that niche already occupied in Cuba by the *equestris* Series,[438] it diverged to use different parts of the available habitat and food resource spectrum. This might be an example of a historical contingency; the Chamaeleolis way of life might most easily evolve from a species that was already very large, so sympatry of two crown-giant clades might be particularly likely to have channeled evolutionary diversification in this direction. This, however, is rampant speculation, particularly given that Chamaeleolis and the crown giants do not appear to be sister taxa (even though they are in the same clade), which makes tenuous even the original premise that the ancestral Chamaeleolis was a crown-giant.

Speculating about why particular ways of life evolve in one place but not another is always interesting. If it weren't for the existence of the ecomorphs, unique anole species wouldn't be so enigmatic. Rather, the Greater Antilles would be just another case of a species-area relationship, in which larger islands have not only more species, but also a greater diversity of functional types of species. But anole evolution in the Greater Antilles is dominated by convergent evolution, and it is in this light that evolution of the unique anoles is fascinating. Unfortunately, at this point I think we have few good leads to follow.

THE ANOLES OF THE MAINLAND

Mainland anoles are comparable to those of the Greater Antilles in the extent of their morphological and ecological diversity (Chapter 4). Nonetheless, most mainland anoles do not belong to any of the ecomorph categories. Quantitative analyses have found only a few cases in which a species qualifies as an ecomorph on both ecological and morphological grounds (Irschick et al., 1997; Velasco and Herrel, 2007): *A. auratus* is a grass-bush anole and *A. frenatus* and *A. biporcatus* may be crown-giants (Fig. 4.9). Qualitatively, a few other species seem to fit the ecomorph bill: both *A. pentaprion*[439] and the species in the Phenacosaurus clade appear to be twig anoles (Fig. 4.9), and probably some other arboreal species pass muster as trunk-crown or crown-giant anoles. On the other hand, some mainland species are morphologically similar to one ecomorph class, but ecologically similar to another (e.g., *A. ortonii* [Irschick et al., 1997]), and many mainland anoles are dissimilar to all ecomorphs in morphology, ecology, or both (Chapter 4).

Despite the lack of ecomorphs, mainland anoles for the most part use the same parts of the environment as the West Indian species—basically, all parts of the vegetation from near the ground to the canopy. Even some of the unusual microhabitats of the Greater Antillean unique anoles have their parallels in the mainland, including leaf litter (e.g., *A. humilis*, *A. nitens* [Fig. 4.11; Talbot, 1977; Vitt et al., 2001]), rock wall (*A. taylori* [Fitch and

---

438. Figure 5.6 suggests that the *equestris* Series originated slightly before the Chamaeleolis clade.
439. And probably its close relatives, *A. vociferans* and *A. fungosus* (Myers, 1971).

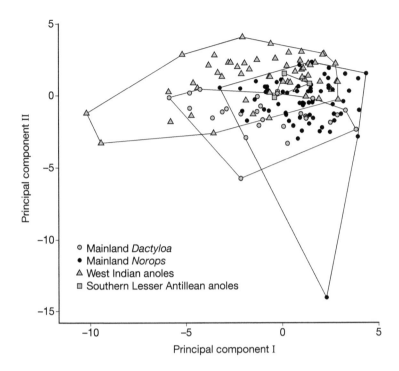

FIGURE 16.8

Relative position of mainland and West Indian anoles in morphological space. Data from a principal components analysis on size-adjusted morphometric variables. Modified with permission from Pinto et al. (2008).

Henderson, 1976), and aquatic anoles (e.g., *A. barkeri, A. oxylophus* [Vitt et al., 1995; Birt et al., 2001]).

Although they use the same suite of habitats, mainland and West Indian anoles have adapted to them in different ways, as outlined in Chapter 13. This lack of convergence extends to the entirety of the radiations in the two areas, which overlap only partially in morphological space (Fig. 16.8; Irschick et al., 1997; Velasco and Herrel, 2007; Pinto et al., 2008).[440] Interpreting this difference in position is difficult because the multivariate analyses are not entirely in agreement either within (Velasco and Herrel, 2007) or between studies; nonetheless, one common pattern is that mainland anoles often have more poorly developed toepads relative to Greater Antillean species (in agreement with Macrini et al. [2003]).

While considering explanations for differences between mainland and Greater Antillean anoles, the phylogenetic interrelationships of these two groups should be kept in mind (Chapter 5). The West Indies were colonized twice from Central or South America

440. Keep in mind, however, that these studies have included only a relatively small portion of mainland diversity.

and examination of Figure 16.8 indicates that both West Indian clades have radiated through parts of morphological space not explored by their mainland ancestors (Dactyloa). Similarly, the mainland Norops clade arose from within this West Indian clade, and members of that clade have radiated in part in an area of morphological space in which West Indian anoles are absent; moreover, to a large extent, this part of the mainland Norops radiation has involved returning to space occupied by mainland Dactyloa.

One explanation for this pattern of shifts in position in morphological space accompanying island-mainland transitions is that in each case the colonizing species experienced a radical reorganization of its genetic or developmental system that allowed evolutionary exploration of new morphological frontiers; in other words, preexisting constraints were broken, and new ones developed (e.g., Mayr's [1963] "genetic revolutions"). Given the arguments made against the importance of constraints in *Anolis* earlier in the chapter, this hypothesis seems unlikely.

A second possibility is that mainland and West Indian anoles have experienced similar radiations in terms of their functional capabilities, but that different morphological means of producing identical functions have evolved in the two areas. As discussed in Chapter 13, few data are available to evaluate the possibility of many-to-one relationships between morphology and performance in anoles. However, a second point is probably more significant in this context: mainland and West Indian species behave differently (Chapter 8). Consequently, selection in these two areas is likely to favor the different functional capabilities that are appropriate to these behaviors, rendering the many-to-one hypothesis insufficient as an explanation for mainland-island differences.

The other main class of explanation relies on environmental differences between the mainland and the West Indies. Central and South America differ from the West Indies in many ways: topography, climate, geology, to name just a few. The most important differences, however, are probably biotic: the mainland hosts not only many more species in total, but also many more types of species (e.g., salamanders, mammalian carnivores), as well as larger and more complicated food webs.

One or all of these differences could have played a role in sculpting differences in the anole faunas of these areas, but two factors that seem particularly relevant to anoles are the vegetation structure and the abundance of predators. Given that much of the ecomorphological work on anoles has focused on how differences in morphology have evolved to exploit different parts of a tree, vegetation structure would seem to be an important determinant of anole evolution. However, even within a Greater Antillean island, great variety exists in vegetation, from xeric scrub through dry forest to rainforest and cloud forest, yet the same basic ecomorph types occur widely throughout each island. Although certainly some differences in the structure of habitats occur between mainland and West Indian islands, it is not obvious that these differences matter to anoles. That is, anoles use the same variety of structures—e.g., tree trunks, twigs, leaves—in both areas. Even if the mainland in general had taller or broader trees or more lianas, how this would drive anole evolution in significantly different directions is not obvious.

Nonetheless, these thoughts represent just my intuition, and detailed study of how vegetation structure affects anole behavior, ecology, and morphology (e.g., Johnson et al., 2006), both within and between regions, would be instructive.

The difference in predator diversity in the two regions, by contrast, could be of major significance. Consider, for example, the vertebrate predator fauna of La Selva in the Atlantic lowland rainforest of Costa Rica, which includes more than 100 species of snakes, raptors, and members of the Carnivora (Greene, 1988). Although many of these species do not eat anoles, many other types of predators do—e.g., monkeys, peccaries, frogs, a variety of birds, spiders, and army ants. By contrast, the West Indies are a fairly benign place in which to be an anole. Birds and snakes are a threat, of course, but their diversity is less than on the mainland, and many other kinds of potential predators are not represented at all. At the El Verde Field Station in Puerto Rico, for example, anoles are eaten by only 14 species of birds, two species of snakes,[441] and one introduced mammal, as well as several species of frogs and invertebrates (Reagan et al., 1996). A conservative estimate is that at least twice as many species prey on anoles at La Selva (H. Greene, pers. comm.).

Greater predator species richness does not necessarily translate into greater predator abundance and higher rates of predation; each predatory species may be less abundant, or may include anoles as a smaller part of their diet. Nonetheless, the higher mortality rates of mainland anoles are plausibly a result of greater rates of predation (Chapter 8). A similar relationship between predator richness and mortality occurs among Bahamian islands (Schoener and Schoener, 1982b).[442]

More significant than sheer numbers of predators, however, is the diversity of predatory tactics, which is vastly greater on the mainland. The limited number of predatory species in the West Indies means that anoles only have to cope with a few types of predation. By contrast, mainland anoles have to deal with predators of all shapes and sizes, differing in means of locomotion, sensory system, foraging mode, and activity time.

When I first considered the role of predator differences in shaping the anole faunas, I focused on escape performance. I figured that a mainland anole living in the exact same habitat as a West Indian species needed to be faster and stronger to get away from all of these predators. This selection for greater maximal performance in theory could lead to differences in ecomorphological relationships and morphological diversity.

However, in retrospect, this perspective was pretty naïve. Consider an anole in Costa Rica, say *A. limifrons* (Fig. 16.9). Life must be pretty scary for this little lizard. The forest is full of eyes, in the canopy, on the ground, in the trees. And those eyes belong to predators that can attack in many different ways. Although some approaching predators can

441. And probably a third species (Wiley, 2003).

442. Though island size is a confounding factor in this case. More generally, note that even in the absence of increased mortality rates, predators can have a great effect on the ecology—and presumably the evolution—of species and communities by leading to changes in behavior, habitat use, physiology, and even morphology (the latter by inducing phenotypically plastic morphological changes [Lima, 1998; Ripple and Beschta, 2004; Schmidt and Van Buskirk, 2004; Hoverman et al., 2005]).

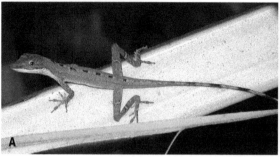

FIGURE 16.9

Predator and prey in the Costa Rican rainforest. *Anolis limifrons* (a) must contend with many different predators with diverse foraging styles, including the sit-and-wait foraging eyelash viper (*Bothriechis schlegeli*) (b). Lizard photo courtesy of J.D. Willson; snake photo courtesy of Harry Greene).

be seen a long way off, others materialize seemingly out of nowhere, either by stealth or quickness. Many of these predators have excellent vision, and some are consummate sit-and-wait foragers, perching in trees and scanning, looking for motion (e.g., Orians, 1969). No matter how fast an anole is, it may have little chance of escaping once a predator notices it.

Consequently, the best way for a mainland anole to avoid being eaten may be to avoid being seen.[443] This hypothesis predicts that a mainland and a West Indian anole using the same structural microhabitat might behave very differently. The mainland anole might be much less active, and might confine its activities to less exposed areas. Such differences might be accentuated by other considerations: fewer intraspecific competitors for food or mates might allow the mainland anole to be more selective about where and when it displayed and chased food. These differences would likely lead to very different selective pressures between mainland and West Indian species occupying the same structural microhabitat; for example, selection for high sprint speed might be less important than selection for crypticity in many mainland anoles.

A corollary to this hypothesis is that the most important factor affecting morphological differentiation among mainland anoles may not be differences in perch height or diameter, as in the West Indies, but distance to cover: some species may spend most of their time in relatively safe microhabitats, whereas others may be out in the open much more often. If this is the case, then we might expect mainland species differing in microhabitat use to experience different selective pressures for functional capabilities and morphology (cf. Pulliam and Mills, 1977; Lima and Valone, 1991).

443. The risks to a mainland anole of conspicuous behavior are well illustrated by Fleishman's (1991) observation of a Panamanian grass anole, *A. auratus*, that began displaying to another anole and was immediately captured by a vine snake, *Oxybelis aeneus*.

The predation hypothesis is consistent with the life history differences that exist between mainland and West Indian species (Chapter 8). In addition, the limited behavioral data also are in agreement: compared to West Indian anoles, mainland species seem warier, forage less, and rely more on crypsis and immobility to avoid predators (Losos et al., 1991; see Chapter 8).

The hypothesis that these life history and behavioral defenses are a result of differing predation pressures is plausible, but can they explain differences in morphological variation? The hypothesis makes three predictions: first, that mainland anoles interact with the environment in a fundamentally different way than do West Indian anoles; second, that differences in predation pressures are the cause; and third, that as a result, mainland anole evolutionary diversification has occurred in very different ways than in the West Indies.

Although differences in the relationship between habitat use and morphology have been reported (Table 13.3), we need much more detailed information on how mainland anoles interact with their environment. Is it correct that two species—one on the mainland, the other West Indian—using essentially the same microhabitat (e.g., tree trunks near the ground) nonetheless behave in very different ways? Assuming that these differences exist, the next question is whether differences in predation regime are the cause. This is a difficult prediction to test, but comparative analyses of habitat use and behavior between areas differing in predator faunas could be instructive;[444] examination of localities in which some predators have been introduced or extirpated by humans could add a quasi-experimental perspective.[445] In addition, experimental additions or removals could examine the extent of potential behavioral plasticity inherent within species, although evolved differences may be much greater in magnitude.

Testing the macroevolutionary sequelae of this hypothesis will be more difficult. A fairly large body of theoretical literature predicts that the presence of predators can spur diversification in different directions than would occur in their absence, but empirical data are relatively few (reviewed in Vamosi, 2005; Langerhans, 2006). For example, Zimmerman (1970) attributed some of the unusual behaviors and lifestyles of Hawaiian insects to lack of predators and noted that introduced predators have wiped out some of the species. Conversely, Doucette et al. (2004), working on Icelandic sticklebacks, suggested that the presence of predators may lead prey species to partition refuges sites, promoting subsequent morphological divergence (see also Rundle et al., 2003). Similarly, the evolution of different anti-predator strategies (e.g., fight versus flight) might lead to divergence in a variety of different behavioral, physiological and anatomical traits (e.g., Losos et al., 2002).

444. For example, Lister and Aguayo (1992) report that Mexican *A. nebulosus* males are much more active and display considerably more on an offshore island that lacks most predators than in a nearby population on the Mexican mainland (see Chapter 8 for examples of predator-induced shifts in habitat use).

445. For example, in the West Indies, introduction of mongooses resulted in the extinction of a number of ground snakes (Tolson and Henderson, 2006).

In addition, predator-prey coevolutionary dynamics may also have a large effect on patterns of prey diversification. For example, the development of greater predatory ability (faster speed, better shell-crushing ability) in predators may be parried by the evolution of counter-adaptations among prey (faster speed, thicker shells [Bakker, 1983; Vermeij, 1987]). Studies of the fossil record have shown how this escalation can lead to evolutionary diversification of prey in ways that do not occur in the absence of the predators (Vermeij, 1987). Nonetheless, for mainland anoles, the effect of predators probably results more from their increased presence, relative to the West Indies, than from predator-prey coevolution; the selective pressure probably comes not from functional improvements in mainland predators, but rather from an increase in the number and types of predation threats.

The approach that will need to be taken to study the effect of predation on anole diversification will need to be the same as for the study of evolutionary adaptation in general (Chapter 13): careful examination of the functional demands caused by the environment (in this case, predators), analysis of the behavior and ecology of the anoles in the context of these demands, and evaluation of functional and behavioral consequences of phenotypic differences that have evolved in the presence of different predator faunas. This approach can be coupled with studies of selection to examine how selective pressures vary in areas differing in predator communities; experimental approaches would certainly be possible with at least some types of predators.

Although I have focused on the role of predators, the greater species richness of mainland localities could affect anoles in other ways. An obvious alternative candidate is interspecific competition resulting from the greater diversity of insectivores on the mainland. The increased prey size and reduced foraging rate of mainland species was interpreted as a result of reduced intraspecific competition because of lower population densities that result from increased predation (Chapter 8). Alternatively, however, reduced population sizes could result from increased interspecific competition from non-anoles (although the observed higher growth rates, larger prey and greater feeding rates wouldn't be predicted results of increased competition; see Chapters 8 and 11). Moreover, independent of population size effects, the presence of more non-anole competitors may have forced anoles to shift to capturing different types of prey or foraging in different ways. The competition and predation hypotheses are not mutually exclusive; investigations of the effects of competitors should be conducted with the same approaches taken to studying predation.

One broader issue remains concerning mainland anole evolution. Clearly, the mainland radiation has not followed the path of the West Indian ecomorphs. But does a different ecomorph syndrome exist on the mainland? We know that convergence is rampant in the West Indies; is it equally prevalent on the mainland, but in the form of a different set of ecomorphs?

Currently, I have no answer to this question. No data are available to evaluate whether mainland communities are composed of similar sets of habitat specialists. Moreover, given uncertainties concerning phylogenetic relationships among mainland anoles, even if community similarity exists across the mainland, we wouldn't know whether ecologically similar species in different localities were the result of convergent evolution or close relationship. To date, the existence of two clades of twig anoles and three of aquatic anoles are the only clearcut cases of convergence in the mainland (Chapter 7).

Obviously, I have many more questions than answers. Moreover, many of the questions are posed in very vague terms, without clearly defined approaches to answer them. I can understand how those who like clearly defined hypotheses and research programs would be unhappy with the research agenda laid out in this chapter. In my defense, all I will say is that the general issues discussed here are not specific to anoles. Rather, many of the most exciting and challenging questions in evolutionary biology revolve around the processes generating large scale patterns of macroevolution. Methods for their study are still very much in their infancy, and I propose that *Anolis* may be an excellent group in which to develop and fine-tune them.

17

ARE ANOLES SPECIAL, AND IF SO, WHY?

What's so great about anoles? Why have I written a whole book about them—and spent more than 20 years studying them—and why have you read the book? Of course, they're attractive and engaging little creatures, with great variety and entertaining behavior. But if that were their only claim to fame, this book would be of limited interested.

Quite the contrary, anoles are receiving ever-increasing attention: more and more papers, by more and more research groups, on increasingly diverse topics; even the anole genome is being sequenced. What, if anything, makes them so special?

I suggest that the interest in anoles stems from three factors:

1. The exceptional extent to which the adaptive radiation of anoles has been studied.
2. The great diversity and disparity exhibited by anole evolution.
3. The replicated adaptive radiations in the Greater Antilles.

In this last, concluding chapter, I will consider whether anoles really are so special and if so, why. I'll then conclude the book by looking forward to consider what the future holds for the lizards themselves.

ANOLIS AS A MODEL TAXON FOR STUDIES
OF BIOLOGICAL DIVERSITY

In the Prologue, I suggested that *Anolis* is nearly unrivalled in the depth and breadth of knowledge about its biological diversity, spanning fields as disparate as phylogenetics, ecology, physiology, behavior and evolution, and including both laboratory and field studies and experimental and observational approaches. After having read through the book, you can decide for yourself whether our knowledge of anoles is broader and more integrated than that of other diverse groups of organisms.

Why has so much work been conducted on anoles? The answer is simple. For many types of studies, anoles—particularly Greater Antillean species—are ideal subjects. They are often abundant and easy to observe, they can be manipulated in the field to answer behavioral and ecological questions, they can be brought into the lab for a wide variety of different studies, and they can be marked and followed over reasonably short generation times. Plus, many species co-occur, facilitating studies of interspecific interactions. Finally, the patterns of convergence add statistical replication to evolutionary analyses. The only glaring shortcoming in our knowledge of anoles is our lack of understanding of the genetic basis underlying phenotypic variation, and that is likely to change radically in the near future.

For these reasons, anoles have been useful subjects to develop new approaches and to test important and general questions in a wide variety of fields. Moreover, the ability to integrate knowledge concerning so many different aspects of their biology has made them an ideal group for synthetic studies of biodiversity and evolution, an attribute that will only grow in the future as we learn more about them.

Anoles are particularly useful for macroevolutionary studies for two additional reasons. Grant (1986), following Lack (1947), suggested that Darwin's finches are at just the right stage of evolutionary diversification to combine studies of pattern and process; that is, they are diverse enough to illustrate interesting patterns of adaptive radiation, yet they are similar enough that process-based studies in behavioral, ecological and microevolutionary time can provide meaningful insight about how and why adaptive diversification occurred. I would argue that the same can be said about anoles; indeed, that has been the primary theme of this book.

We can contrast cases like Darwin's finches and anoles with case studies at either end of the spectrum. On one hand, studies of closely related species in the process of diverging and speciating provide wonderful insights into these processes. Studies on sticklebacks, walking sticks, and columbines (e.g., Rundle et al., 2000; Nosil et al., 2004; Colosimo et al., 2005; Whittall et al., 2006)—to name just three—are at the cutting edge of evolutionary biology, applying modern methods and approaches to advance our knowledge of the evolutionary process. Nonetheless, groups such as these are not adaptive radiations; they simply don't display enough ecological and phenotypic diversity. Studies on these groups certainly are informative concerning microevolutionary processes, and the groups themselves may be nascent adaptive radiations,[446] but adaptive radiations they are not, and the extent to which we can scale up from studies of groups such as these to macroevolutionary levels is not clear.[447]

At the other end of the spectrum, some of the most famous examples of adaptive radiation—such as beetles, placental mammals and angiosperms—represent old and extremely diverse groups. Although their disparity is the hallmark of adaptive radiation, these clades are so diverse in so many ways that it is hard to imagine how process-based studies could be informative about the origin of these differences (Grant, 1986). Consider placental mammals, and more specifically the subclade Afrotheria: what sorts of studies could help us understand why this clade differentiated to produce golden moles, aardvarks, elephants, and other taxa? In other words, the macroevolutionary pattern is present, but it is not clear how we can devise studies to understand the processes that drove evolutionary diversification in these old and disparate groups.

Anoles exhibit a second advantage for the study of adaptive diversification, one not shared by Darwin's finches and some other groups: the ability to conduct manipulative experiments in nature, over both ecological and evolutionary timescales.[448] For the last quarter century, ecologists have emphasized the importance of manipulative experiments for hypothesis testing; in recent years, evolutionary biologists are increasingly taking the same approach, though experimental studies in natural settings are still rare (Reznick, 2005). Studies in laboratory microcosms have demonstrated the utility of experimental methods to the study of adaptive radiation (Rainey et al., 2000; MacLean and Bell, 2002; Kassen et al., 2004; Meyer and Kassen, 2007); now is the time to extend this approach to the field.

This is where anoles have their greatest advantage as a macroevolutionary study system. Experimental work on anoles is feasible at all time scales: behavioral, ecological, and microevolutionary. Moreover, quasi-experiments established by anole introductions and natural experiments created by nature via replicated evolution all provide powerful means for hypothesis testing. By synthesizing these experimental approaches with observational studies on extant taxa and phylogenetic studies of evolutionary history, *Anolis* is an excellent system for the yin and yang of hypothesis generation and testing, as well as for the mutual illumination of historical and present-day studies discussed in Chapter 1.

It is for these reasons that *Anolis* has been—and continues to be—an excellent group for a wide variety of studies, and particularly for synthetic, broad-scale integrative work.

446. Or members of larger clades that do constitute adaptive radiations.

447. In the most authoritative treatment of adaptive radiation in half a century (and maybe ever), Schluter (2000) relied heavily on *Anolis* and Darwin's finches as examples, but he and I differ slighty in emphasis: whereas I focus on adaptively disparate groups, he emphasizes the ability to study processes in recently diverging clades (see pp. 8–9 of his book).

448. Such studies cannot be conducted on Darwin's finches because research in the Galápagos is stringently regulated and limited.

A major goal of this book has been to not only make this point, but to illustrate that abundant opportunity still exists to jump on the *Anolis* bandwagon—all are welcome, and the more, the merrier!

ANOLIS ADAPTIVE RADIATION

But enough cheerleading—let's get down to the nitty-gritty: is the evolutionary diversification of *Anolis* exceptional and, if so, why have these lizards evolved such diversity and disparity?

ARE ANOLES SPECIAL?

To decide if *Anolis* is exceptional, we need to delineate an appropriate pool of comparison clades. In Chapter 15, I presented one approach, arguing that the appropriate comparison is to a sample of clades that share similarities in biology, natural history, and age. Based on this approach, I found that both *Anolis* and the Polychrotinae (the larger clade to which *Anolis* belongs) exhibit significantly great ecomorphological disparity (Fig. 15.12).

A second, more traditional, approach is to compare *Anolis* to its sister group.[449] As discussed in Chapter 6, uncertainty currently exists about the sister taxon of *Anolis*. Nonetheless, all of the candidates that have been mentioned in the literature are clades that contain few species and little ecological and morphological variety. It seems safe to conclude that, in comparison to its sister group, *Anolis* is exceptionally species rich and ecomorphologically diverse.

A third approach would be to compare *Anolis* to other clades which diversified in the same biogeographic region. If we consider first the West Indies, no reptile clade comes even remotely close to rivaling *Anolis* in species richness or ecomorphological diversity. Expanding to all vertebrates, the only comparable group is eleutherodactyline frogs, with about 150 species and extensive, though little studied, ecomorphological diversity (Hedges, 1989; Hedges et al., 2008). Even if we expand the scope to consider the Neotropics, anoles, eleuths, and perhaps dendrobatid frogs (Grant et al., 2006) seem to be exceptional, certainly among amphibians and reptiles. Comparisons in this case are more difficult because there are so many more groups on the mainland, but few other candidates exhibit comparable diversity and disparity. Of course, one could argue that the comparison is unfair; anoles and eleuths are exceptionally old clades (Chapter 6; Heinicke et al., 2007; Hedges et al., 2008), so the appropriate comparison should be to Neotropical clades of comparable age. In the absence of detailed and dated phylogenies for other groups, this point cannot be resolved, but few contenders exist among other amphibian and reptile groups, nor all that many among mammals, birds, or fish, either.[450]

449. For reasons discussed in Footnote 416, I prefer the first approach.
450. Poeciliid fish (Meffe and Snelson, 1989; Hrbek et al., 2007) and hummingbirds (McGuire et al., 2007b) are possible examples.

In summary, by whatever criterion one wants to use, *Anolis* stands out as an exceptionally diverse and ecomorphologically disparate clade.

WHAT IS RESPONSIBLE FOR THE EXCEPTIONAL DIVERSIFICATION OF ANOLES?

ECOLOGICAL OPPORTUNITY

Ecologists and evolutionary biologists often identify ecological opportunity as an important stimulus to adaptive radiation (Simpson, 1953; Schluter, 1988a,b, 2000). Remote islands are particularly good candidates because their depauperate faunas mean that colonizing species may find a surfeit of resources and few competitors. Indeed, many of the most famous examples of adaptive radiation occur on distant oceanic islands, such as Hawaii and the Galápagos, and in their aquatic counterparts, inland lakes such as the African Rift Lakes and Lake Baikal.[451]

Groups radiating on such islands often exhibit substantially greater ecomorphological disparity than their close relatives in mainland settings (Carlquist, 1974). This evolutionary ebullience is usually credited to niche expansion in the absence of other competing taxa. The result is that species in the radiating clade diverge, occupying a wide array of different niches that are usually utilized by other clades in mainland settings (reviewed in Schluter [2000]; for a recent example, see Chiba [2004]). As outlined in Chapter 11, Greater Antillean *Anolis* fulfill this scenario very well.

Nonetheless, ecological opportunity cannot be the whole story, because not all clades radiate under such conditions. In the Galápagos, for example, Darwin's finches are the only birds to have diversified to any extent; similarly, some plant, insect and mollusk groups have radiated extensively in this archipelago, but many others have not (Jackson, 1994). In Hawaii and any other isolated island or island group, the story is the same (e.g., Zimmerman, 1970; Carlquist, 1974). Greater Antillean anoles again fit the picture: in the West Indies, few other taxa (including only one other reptile clade, *Sphaerodactylus* geckoes) have radiated to any substantial extent, even though most have been present in the West Indies as long as anoles (Crother and Guyer, 1996; see Thorpe et al., 2008).[452]

DIVERSITY OF A CLADE'S CLOSE RELATIVES

Why, then, do some clades radiate and not others? One predictor may be the diversity of a clade's relatives elsewhere (Carlquist, 1974). Consider, for example, Hawaiian honeycreepers and Darwin's finches. Both of these clades have radiated extensively,[453] and their sister taxa on the mainland also exhibit substantial—though not as great—ecomorphological diversity (Burns et al., 2002; Lovette et al., 2002). By contrast, two

451. Lakes surrounded by terrestrial habitats are, for freshwater denizens, the evolutionary equivalent of islands surrounded by water.

452. Length of residence in an area is an important consideration because the radiation of an early colonist may preclude diversification by later arrivals (Carlquist, 1974; for an interesting counterexample, see the discussion of the tropheine cichlids in Lake Tanganyika in Salzburger et al. [2005]).

453. They have radiated so much that their ecomorphological disparity is almost as great as that seen within all passerine birds (Burns et al., 2002; Lovette et al., 2002).

clades that have not radiated to any substantial extent despite having been present on these islands just as long, Hawaiian thrushes and Galápagos mockingbirds, belong to clades that also show little disparity on the mainland (Lovette et al., 2002; Arbogast et al., 2006; Grant and Grant, 2008). A corollary of this pattern is that some clades seem to diversify repeatedly on different islands, whereas others diversify rarely. For example, some clades of African cichlids radiate in many different lakes, whereas other clades never exhibit much diversification (Seehausen, 2006).

However, it is probably premature to consider this to be a general rule of adaptive radiation because some clades that radiate on islands are not diverse elsewhere in their range, such as *Tetragnatha* spiders and aglycyderid weevils (Gillespie et al., 1994; Paulay, 1994) and cichlid fish in most African rivers (Joyce et al., 2005), and no overall assessment of the generality of this phenomenon has been conducted. Clearly, whether the clades that adaptively radiate on islands can be predicted by the diversity of their relatives elsewhere would make for an interesting study. Nonetheless, to the extent that this rule does hold, anoles would seem to be a good example, given that they have diversified greatly both in the West Indies and on the mainland.

ECOLOGICAL OPPORTUNITY AND THE MAINLAND RADIATION

All in all, Greater Antillean *Anolis* would seem to be a classic example of island adaptive radiations resulting from ecological opportunity. Nonetheless, this conclusion leads to a question: if ecological opportunity prompted the anole radiation in the West Indies, how do we account for the comparable ecomorphological variety on the mainland (Chapter 16)? Has ecological opportunity played a role there, as well?

It is easy to imagine anoles arriving on a proto-West Indian island brimming with empty niches, but the mainland is a different story. Today the mainland is full of animals of all sorts that vie with anoles for arboreal insects (Chapter 11). In the absence of fossils and detailed phylogenetic analyses, we don't know what other taxa were present on the mainland 40 or more million years ago, and thus whether anoles initially diversified in the presence of other arboreal insectivores. Nonetheless, we might expect that mainland communities were diverse and species rich in the distant past, even if we don't know what kind of species were present. And if that is the case, then the evolutionary success of mainland anoles suggests that ecological opportunity may not be a prerequisite for anole adaptive radiation.

On the other hand, few data support such a supposition, and we shouldn't discount the possibility that ecological opportunity was abundant in the early days of mainland anole diversification. For example, few mammalian insectivores[454] are known from the Neotropics in the Eocene and Oligocene (MacFadden, 2006). Although the fossil record of bird diversity is scant, molecular studies suggest that modern Neotropical clades, at least, were not diverse in the Eocene or much of the Oligocene. In particular, Amazonian

---

454. Or, for that matter, any potential mammalian predators of anoles.

forest canopy and scrub habitats today are dominated by North American clades, which began to diversify in Amazonia only 12 million years ago. Perhaps the most likely scenario is that these clades displaced suboscine passerines, but even those avian clades have only been diversifying in South America for the last 32 million years (Ricklefs, 2002). Although many lizard clades have probably been present in the neotropics for a long period of time, few of these clades contain arboreal insectivores (Chapter 11); similarly, being primarily nocturnal, frogs probably do not compete with anoles to a great extent (Chapter 11). Thus, it is conceivable that mainland anole diversification, at least in its early stage, occurred in a relatively empty ecological theater. This possibility applies particularly to the older Dactyloa clade; by contrast, the more diverse Norops clade colonized the mainland more recently (Fig. 6.1), when birds and mammals were more diverse and Dactyloa also was already present (although possibly restricted to southern Central America and South America, as it is today).

WHY HAVE ANOLES RADIATED WHERE OTHER TAXA HAVE NOT?

Regardless of the role that ecological opportunity has played in anole diversification, we still must ask why anoles have diversified to so much greater an extent than other taxa with which they coexist. Even if ecological opportunity was the stimulus to diversification, many other clades had the same opportunity but failed to take evolutionary advantage of it.

In Chapter 15, I put forth my hypothesis: the evolution of toepads provided anoles with the evolutionary flexibility to adapt to many different aspects of arboreal existence, allowing species to specialize to use twigs, grass blades, the canopy, and other parts of the environment. In this regard, the evolution of toepads in anoles would be a classic example of a key innovation allowing a clade to utilize the environment in a different way and thus leading to adaptive diversification within this new adaptive zone, just as the evolution of wings prompted the adaptive radiation of birds into a variety of niches unavailable to their theropod ancestors.

One way of distinguishing the power of the toepad versus ecological opportunity would be to see how anoles do when introduced to other parts of the world (Chapter 11). The success of anoles in Bermuda (Wingate, 1965), Micronesia (Rodda et al., 1991), and islands near Japan (Hasegawa et al., 1988; Okochi et al., 2006) indicates that anoles can infiltrate other ecosystems; however, these are all islands, where ecological opportunity may have been great. The real test will be if and when anoles are introduced to continental settings in the Old World, where ecological opportunity may be limited.[455] Will the

455. I have mentioned the utility of studying introduced populations repeatedly in the last few chapters, so I want to reemphasize that I in no way condone such introductions. Nonetheless, given the extent of global commerce and the ease with which anoles stow away, it is probably inevitable that *A. sagrei*, *A. carolinensis* or some other species will eventually arrive in many far-off destinations. Of course, in some places, such as Madagascar, toepadded, arboreal and diurnal insectivorous lizards already exist and have radiated widely, as I will discuss shortly. Even if toepads are a key innovation, they may be of little use to invading anoles in such places because their potential niches may already have been preempted.

possession of toepads be sufficient to allow anoles to become established and diversify in such settings?

As important as toepads may have been, they are not the whole story. Toepads may have allowed anoles to diverge into different structural microhabitats, but anoles also show repeated divergence and convergence in their occupation of thermal microhabitats. Repeatedly within ecomorph clades, species have differentiated in the thermal microhabitats they occupy, with concomitant adaptation in thermal physiology. Indeed, the rate of evolution in thermal biology is even higher than in ecomorphology (Hertz et al., in review). The lability in thermal biology is particularly notable because thermal biology is evolutionarily conservative among most lizard clades (Bogert, 1949; Huey, 1982; Hertz et al., 1983; Andrews, 1998; but see Castilla et al. [1999]). Why anoles exhibit so much greater evolutionary flexibility in thermal physiology than other types of lizards is unknown.

Another factor that may be important in adaptive radiation is "evolvability," simply the ability to evolve readily into diverse forms (Schluter, 2000). Perhaps this seems self-evident, but taxa that are limited in their ability to evolve will change more slowly or not at all; populations that can readily adjust will be able to adapt to local circumstances (Lovette et al, 2002; Arbogast et al., 2006). Evolvability is an attribute of a population; consequently, data on genetics and response to selection is the best way to measure it. For the time being, we don't have a good measure of anole evolvability; however, interspecific comparisons indicate that anoles are evolutionarily labile, displaying great variety in both morphology and thermal physiology compared to other clades (e.g., Warheit et al., 1999). To the extent that anoles are more evolvable than other taxa, a variety of different factors could be responsible.

- *Modularity.* Phenotypically and genetically, aspects of the anole phenotype may be structured independently (i.e., they are compartmentalized or modular), allowing aspects of the phenotype to evolve independently of each other. This idea has been discussed in phenotypic (Liem, 1974; Vermeij, 1974) and quantitative genetic (Cheverud, 1996; Wagner and Altenberg, 1996) terms for many years; recently the parallel idea has been developed at the genomic level (Kirschner and Gerhart, 1998; Rutherford and Lindquist, 1998). How this idea might apply to anoles is not clear. Interspecific morphometric variation in toepad characteristics, limb dimensions, body size, and sexual size dimorphism are uncorrelated (Harmon et al., 2005), and none of these characteristics is likely to covary with thermal physiology, so in this sense anole adaptive responses may occur along several independent pathways. Whether analogous compartmentalization exists in anole genomes is unknown, though such questions will be increasingly amenable to study in the near future.

- *Broad Niche Use.* Although specialized to use particular parts of the environment, anoles are nonetheless highly flexible in their habitat use and behavior: any species can be found almost anywhere in the environment, at least occasionally (Chapter 3).[456] A similar phenomenon is seen in cichlid fish which, despite specializations of the jaw for particular trophic niches, can eat a broad range of different types of food (Galis and Metz, 1998; Kornfield and Smith, 2000). As a result, given the opportunity to expand their habitat use by the absence of competitors or predators, or forced to shift habitat use by their presence, anoles can do so and subsequently adapt to the new conditions in which they occur (Chapters 11–13).

· *Phenotypic Plasticity.* The potential evolutionary significance of phenotypic plasticity has attracted increasing interest in recent years (e.g., West-Eberhard, 1989, 2003; Schlichting and Pigliucci, 1998; DeWitt and Scheiner, 2004; Ghalambor et al., 2007). Adaptive phenotypic plasticity has been discovered in two anole species: individuals of *A. carolinensis* and *A. sagrei* that grow up using broad surfaces develop relatively longer hindlimbs than those that grow on narrow surfaces (Chapter 12). Presumably, such plasticity could allow a population of lizards to persist in a habitat in which it would otherwise perish; given enough time, advantageous genetic variation would appear and spread through the population, leading to genetic adaptation and elaboration of the traits.[457] Whether hindlimb plasticity, much less plasticity in other traits, occurs to a greater extent in anoles than in other taxa is unknown.

· *High Rate of Speciation.* An alternative perspective is that anoles speciate at a rate greater than that of other clades, and the resulting abundance of species sets the stage for evolutionary divergence in adaptive phenotypic traits. In Chapter 15, I suggested that the reliance of anoles on visual signals for communication increases the likelihood that populations in different environments will diverge and become reproductively isolated. A high rate of speciation could promote adaptive diversification in two ways. First, the incidence of ecologically similar species becoming secondarily sympatric and undergoing character displacement is likely to be a function of the number of species in a region. Second, to the extent that gene flow constrains evolutionary divergence (Mayr, 1963; Moore et al., 2007), then an increased likelihood that populations will become reproductively isolated should increase the rate of evolutionary divergence (Futuyma, 1987).

Whether, in fact, any of these possibilities explains the extensive evolutionary diversity of anoles relative to other taxa is unknown. For one thing, we don't even know

456. This refers more to structural than thermal microhabitat. Crown-giants occasionally are seen on the ground, and trunk-ground and grass-bush anoles every now and then climb high into a tree. However, deep forest anoles aren't often found in the middle of a sunny field, nor open habitat anoles in deep forest.

457. Note that mutations are random with respect to their selective value. Particularly beneficial mutations do not arise in response to particular environmental exigencies. For this reason, the potential for phenotypic plasticity to facilitate subsequent evolutionary adaptation is in no way Lamarckian, as sometimes is supposed.

whether these factors differ between anoles and other taxa. Whether anoles exhibit particularly great compartmentalization, niche breadth, or plasticity compared to other
lizard clades or other Neotropical taxa is unknown and would make for an interesting
study, as would investigation of the extent to which greater species diversity promotes
phenotypic differentiation.

In sum, anoles display many of the characteristics exhibited by other adaptive radiations. At least in the Greater Antilles, and possibly on the mainland, they took advantage
of ecological opportunity to diversify widely. The possession of toepads allowed them to
diversify throughout the arboreal realm, which was underutilized by other taxa. In addition, anoles exhibit a variety of other characteristics that may explain their great evolutionarily lability. In many of these regards, anoles appear exceptional relative to most
other lizard clades and most other neotropical taxa, but share similarities with other
clades that have radiated adaptively.

REPLICATE ADAPTIVE RADIATIONS

What is particularly exceptional about *Anolis* is the fact that independent radiations on
four separate islands have produced communities composed of the same set of habitat
specialists. The idea that communities in similar environments—such as deserts or
Mediterranean habitats—should exhibit similar structure and composition has a long
pedigree (Orians and Paine, 1983; Blondel et al., 1984; Pianka, 1986; Wiens, 1989;
Losos, 1996c; Kelt et al., 1996). If these habitats occur in far-off lands, they usually will
be occupied by distantly related taxa, and thus similarity in community structure likely
would be convergent (Schluter, 1986). Note, however, that communities can converge in
overall structure (e.g., species richness, pattern of spacing in ecological or morphological space) while their constituent species may differ greatly (Ricklefs and Travis, 1980;
Schluter, 1990). Communities that are composed of species exhibiting the same set of
convergently evolved phenotypes—termed "species-for-species" matching—are quite
rare, and it is this phenomenon that is *Anolis*'s number one claim to fame.[458]

458. The null model debate of the late 1970s and early 1980s (Chapter 11 and Footnote 415), acrimonious as
it was, had one salutary effect: it made clear that before making a claim that a community is structured by
deterministic processes, one must first assess the possibility that the community patterns could have resulted
from random processes.

In this vein, it would be nice to conduct a null model analysis of the Greater Antillean anole radiations to ask
if the apparent species-for-species matching is greater than would be expected by chance (cf. Schluter [2000]).
The observations are that the same four ecomorphs occur on all four islands, the same five ecomorphs occur on
three islands, and the same six on two islands; and that phylogenetic analysis indicates that in almost all cases,
the presence of the same ecomorph on multiple islands is the result of convergence (Chapter 7).

This species-for-species matching is impressive, but imperfect, given the absence of several ecomorphs
from two islands. Moreover, the unique anoles—one in Jamaica, eight in Hispaniola, 12 in Cuba—are not
matched. The question then becomes: given these non-matched components, is the extent of species-for-species
matching among the ecomorphs greater than would be expected to occur by chance? Put another way, if
evolutionary diversification occurred randomly (i.e., morphological change occurred in random directions as
species diversified), producing the same number of species on each island as are observed today with the same

In laboratory experiments, replicated microbial systems will diversify to produce identical communities composed of the same set of 2–3 habitat specialists (Rainey and Travisano, 1998; Meyer and Kassen, 2007). By contrast, among communities of organisms in nature, very few examples of species-for-species matching exist. Evidence from mainland settings is almost non-existent; communities in different mainland areas, even in similar environments, tend to be composed of dissimilar species;[459] this is true even when higher level properties of these communities, such as species richness or niche packing, do show evidence of convergence (see reviews in Orians and Paine [1983]; Wiens [1989]; Melville et al. [2006]).

Replicate adaptive radiations, when they do occur, are almost always found on islands or in lakes. Young, post-glacial lakes in the northern hemisphere provide the most extensive example of replicated adaptive radiation (see reviews in Schluter [2000] and Snorasson and Skúlason [2004]). In such lakes, which have only been colonized since the end of the last Ice Age and which generally have low diversity, fish repeatedly diversify into two ecomorphs that utilize pelagic and benthic habitats. Examples of this divergence are known from Alaska, Canada, Iceland, Ireland, Scandinavia, Scotland and elsewhere; in some clades, the same pattern of divergence has occurred independently in multiple

---

phylogenetic relationships, how likely would it be to generate a pattern in which there is as much species-for-species matching as there is today among the Greater Antilles? A more elaborate null model might also include the caveat that not only would there have to be as much species-for-species matching, but the species or clades that converge across islands would have to be those that are among the most abundant and geographically widespread on the island (i.e., the convergence wouldn't include a clade with an extremely restricted range on one island, because none of the ecomorph clades on any of the islands has such a distribution).

Even the simpler analysis would be complicated in many ways. For example, the existence of clades of similar species, all members of the same ecomorph category, means that the match across islands would sometimes be between species and sometimes between clades. Moreover, given the stasis in ecological morphology evident in recent times (as evidenced by these clades of morphologically similar species), simulations would use a non-Brownian Motion model of character evolution to incorporate this pattern of evolution into the null model.

I have not conducted such an analysis. Nonetheless, I think it unlikely that the convergence of the ecomorphs across four islands is likely under a random model. Given the vast swath of morphological space occupied by anoles, even just by Greater Antillean anoles, it seems unlikely that a radiation producing six species (i.e., Jamaica) would manage to produce four ecomorphs that also have evolved on all three other islands. Similarly unlikely would be a radiation of 10 species (Puerto Rico) producing four types shared by three other islands, and a fifth type shared by two others. That Hispaniola and Cuba could produce the same six set of ecomorphs by chance seems less implausible; if these were the only two islands, I would be less convinced, but the congruence of the four islands seems to me to be highly unlikely to have arisen by coincidence.

459. Molecular systematic studies sometimes reveal that morphologically dissimilar species in a local area are not, as previously thought, each related to morphologically similar species elsewhere, but, rather, are closely related to each other and thus represent an in situ radiation (e.g., Australian corvids [Sibley and Ahlquist, 1990; Barker et al., 2002]; Malagasy songbirds [Yamagishi et al., 2001]). In some cases, these findings indicate the existence of multiple cases of convergence across regions, such as in Malagasy and Asian ranid frogs (Bossuyt and Milinkovitch, 2000), *Myotis* bats (Ruedi and Mayer, 2001; Stadelmann et al., 2007), and African and Laurasian mammals (Madsen et al., 2001). However, such cases usually fall short of constituting replicate adaptive radiations because most species in each region are probably not convergent with species in the other region. This lack of widespread convergence is certainly true for the placental mammal faunas of different regions; more complete analyses of ranid frogs and *Myotis* are needed to evaluate the extent to which those radiations are matched across regions. As discussed in the previous footnote, quantitative statistical methods are needed to investigate whether in any of these cases, radiations in different regions are more similar than would be expected by chance.

lakes (e.g., Taylor and McPhail, 2000; Østbye et al., 2006; Landry et al., 2007). This pattern of evolution into pelagic and benthic ecomorphs has occurred in a wide variety of fish, including sticklebacks, charr, salmon, trout, and whitefish. Patterns of morphological divergence usually are similar, with the pelagic planktivores tending to be smaller, more slender and possessing a greater number of gill rakers than the benthic carnivores.

In some cases, evolutionary divergence in these lakes has proceeded beyond the two-species stage. As with Greater Antillean anoles, ecomorph occurrence is nested among post-glacial lakes, with the ecomorphs present in two-species lakes always present in lakes with a greater number of species. In all cases, lakes with three or four species include at least one benthic and one pelagic ecomorph; additional species either subdivide the benthic niche according to depth or are piscivorous.

The most famous case of replicated adaptive radiation in lake fish is the cichlids of the East African Great Lakes (reviewed in Fryer and Iles, 1972; Stiassny and Meyer, 1999; Kornfield and Smith, 2000; Kocher, 2004; Salzburger and Meyer, 2004; Salzburger et al., 2005; Seehausen, 2006; Genner et al., 2007). Approximately 2000 species occur in these lakes, but what is particularly remarkable is the extraordinary radiations that have occurred in Lake Tanganyika (9–12 million years old, 250 species), Lake Malawi (2–5 million years old, 1000 species) and Lake Victoria (less than—possibly much less than—200,000 years old, 500–1000 species). These lakes have experienced independent evolutionary radiations and have each produced a dazzling array of ecomorphological diversity, including plankton grazers, algae scrapers, sand filterers, egg predators, piscivores, sit-and-wait and rapid pursuit predators, species that pluck insect larvae from crevices, fish scale eaters that rasp scales off the sides of other fish (with species with curved heads and jaws specialized to eat from either the left or the right side of the prey), molluscivores, and piscivores (Fryer and Iles, 1972). Moreover, a number of these habitat specialists have evolved convergently in two or all three of these lakes (Fig. 17.1; Fryer and Iles, 1972).

There can be no doubt that the extent of adaptive radiation of African lake cichlids is extraordinary, particularly given the young age of the Lake Victoria radiation. Further, a picture is a worth a thousand words, and illustrations such as Figure 17.1 convincingly suggest that adaptive convergence has occurred among fish in the different lakes. Nonetheless, in many respects, our understanding of replicated adaptive radiation in cichlids lags well behind that of anoles. In particular, two sorts of data are still lacking.

First, although cases of ecomorphological convergence between the lakes certainly exist, we have no idea how common this convergence is: no quantitative analyses have examined the entire faunas of the lakes (although Joyce et al. [2005] is a nice start in this direction). Are these faunas ecomorphologically matched, or do only a few instances of convergence exist, embedded in a larger sea of non-convergence between the lakes? That is, is the situation in the African lakes more like that of the anoles of the Greater Antilles, in which a few unique forms exist, but to a large extent, species-for-species matches occur across islands; or are the lakes more similar to the comparison of placental and

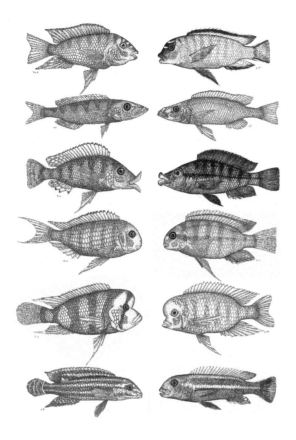

FIGURE **17.1**

Convergent evolution in cichlid fishes in the African Great Lakes. Fish in the left column are from Lake Tanganyika and fish on the right are from Lake Malawi. Phylogenetic analyses indicate that independent evolutionary radiations have occurred in these lakes, and thus that these forms are convergent (Kocher et al., 1993). Reprinted with permission from Albertson and Kocher (2006).

marsupial mammals, in which some convergent examples exist, but the faunas are overall not all that similar?[460] Fryer and Iles' (1972) monograph suggests that the lake situation may be more like the latter; although a number of cases of convergence exist, the lakes differ in their degree of divergence and specialization, and many ecomorphological types in each lake apparently have no counterpart in the others.[461]

Second, although visually compelling, documentation of cichlid convergence would be more convincing if it were supplemented by quantitative morphometric analysis indicating that forms truly are convergent (e.g., Rüber and Adams, 2001; Joyce et al.,

460. The marsupial-placental example is a favorite of textbook writers (including me!), but as an example of replicated adaptive radiation, the case falls short. First, Australian marsupials are generally not compared to the fauna of any particular place, but rather to placentals in general. Second, although stunning examples of convergence exist (thylacine-wolf, dasyurid-cat, phalanger-flying squirrel), these are cherry-picked case studies with no overall quantitative assessment. Certainly, there are no marsupial equivalents of cetaceans, bats, and many other placentals, nor any placental equivalent to kangaroos (for a nice introduction to marsupial diversity and parallels, or lack thereof, to placentals, see Springer et al. [1997]). I make these points not to cast aspersion on the wonderful utility of the marsupial-placental comparison as an example of convergent evolution, but simply to say that this example is not a case study of replicated adaptive radiation. See also Leigh et al. (2007), which provides a fascinating discussion of convergence of other mammalian faunas.

461. Fryer and Iles (1972, p. 517) provided a table listing 16 types of "morphologically and/or ecologically equivalent species" found in all three lakes, but point out that in some of these cases, species filling the same ecological niche are not morphologically similar. Thus, the extent of species-for-species matching of ecological equivalents across these lakes is unclear.

2005), and by functional, ecological, and behavioral data investigating the adaptive basis for this convergence.[462]

The number of examples of replicated adaptive radiation on islands is quite small. Probably the best case of replicated adaptive radiation in a terrestrial setting, other than *Anolis*, is the land snails of the genus *Mandarina* in the Bonin Islands near Japan (Chiba, 2004). Ecologically, four types of microhabitat specialists exist: arboreal, semi-arboreal, sheltered ground, and exposed ground. Sympatric species differ in microhabitat use and members of the same microhabitat specialist class do not coexist. Morphologically, the snails cluster into four groups corresponding to their microhabitat use. Phylogenetic analysis indicated that these different ecomorphs have evolved independently multiple times among the islands, except possibly the exposed ground ecomorph, which may be ancestral to the others (Fig. 17.2).

The spiny leg clade of Hawaiian long-jawed spiders (*Tetragnatha*) is another example (Gillespie, 2004). These spiders come in four microhabitat specialist types: species morphologically adapted to leaf litter, moss, twigs, and bark. Communities contain 2–4

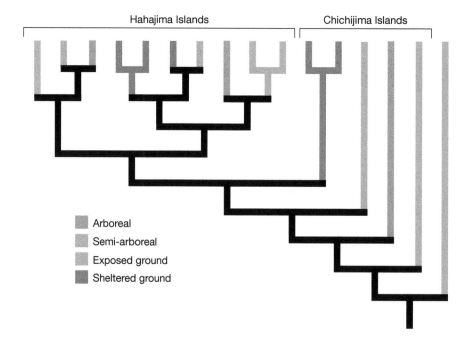

FIGURE 17.2

Replicated adaptive radiation of *Mandarina* snails in islands near Japan. Multiple islands with distinct snail species occur in both the Hahajimas and Chichijimas. Species occupying different microhabitats are morphologically differentiated. Modified from Chiba (2004) with permission.

462. Indeed, although the adaptive basis for ecomorphological differentiation is well studied in pelagic-benthic species pairs (reviewed in Schluter, 2000), for most other cases of replicated adaptive radiation, it has not received much detailed investigation along the lines discussed in Chapter 13.

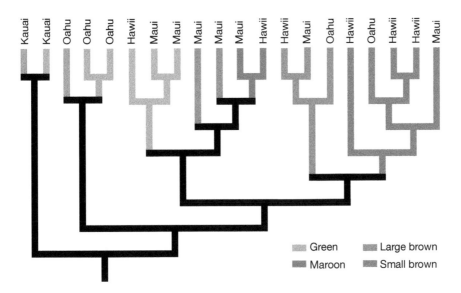

FIGURE 17.3

Evolutionary diversification of *Tetragnatha* spiders in the Hawaiian Islands. The occurrence of similar ecomorphs on different islands results in part from convergent evolution and in part from colonization. Modified with permission from Gillespie (2004).

"ecomorphs," but no site has more than one member of an ecomorph class. At least three of the ecomorphs are found on every island, but this similarity is only in part a result of convergent evolution; although some ecomorphs have evolved independently, and thus are more closely related to different ecomorphs on the same island, other ecomorphs have evolved only once or twice and have dispersed from one island to another (Fig. 17.3). Overall, a parsimony reconstruction of ecomorph evolution suggests the occurrence of six transitions from one ecomorph to another and eight instances of dispersal of an ecomorph from one island to another.

From this review, we can draw a number of conclusions about replicated adaptive radiations: in particular, they are quite rare, and limited almost exclusively to closely related taxa with poor dispersal ability that occur on islands or lakes in the same region. I will examine each of these points in turn.

THE RARITY OF REPLICATED ADAPTIVE RADIATIONS

Given the amount of attention paid to adaptive radiation in recent years, as well as the fact that the idea of community convergence has been discussed for more than three decades, the paucity of well documented cases can't be a result of no one looking for them. Certainly, as more and more taxa are studied, additional unexpected cases will come to light, particularly in non-morphological characters, for which divergence within radiations and convergence among them may be harder to detect. Nonetheless, it seems unlikely that replicated adaptive radiation is a widespread phenomenon that simply has not yet been noticed.

SPECIES-FOR-SPECIES MATCHING IS LIMITED TO CLOSELY RELATED TAXA

Many cases of evolutionary convergence of communities have been investigated for distantly related taxa, but evidence for species-for-species matching is rarely found (Wiens, 1989; Schluter, 1990; Price et al., 2000). Species-for-species matching has almost exclusively been detected among relatively closely related species, such as cichlid fish or anoles. The only exception to this generality is the possibility that benthic and pelagic fish are matched in postglacial lakes in different regions, even though they occur in distantly related fish families. However, such matching has not been demonstrated.[463]

Consider, for example, the adaptive radiation of day geckos (*Phelsuma*) on Indian Ocean islands (Fig. 17.4). Despite their nocturnal, gekkonid heritage, day geckoes—diurnal, as their name implies—show many similarities to anoles (see references in

FIGURE 17.4

Day geckos (*Phelsuma*). (a) *P. astriata*, Seychelles; (b) *P. ornata*, Mauritius. Photo courtesy of Luke Harmon.

463. Although divergence into benthic and pelagic ecomorphs has occurred in many different fish families, I am unaware of any study that has quantitatively compared the morphologies of different species pairs to investigate whether the pelagic ecomorphs of different types of fish are more similar to each other than any pelagic ecomorph is to its benthic counterpart, as the replicated adaptive radiation hypothesis would suggest. An alternative possibility is that even though evolutionary divergence has occurred in the same manner in each lake, this differentiation has not been great enough to override preexisting differences among clades (Stayton, 2006; Revell et al., 2007b).

Harmon et al., 2007, 2008). They are relatively small, arboreal, insectivorous, sit-and-wait foraging lizards which have large toepads and are usually green. Further like anoles, they are highly territorial and communicate through head movements.[464] *Phelsuma* has experienced independent radiations in the Mascarene, Seychelles, and Comoros islands, all of which are embedded phylogenetically in the much larger radiation on Madagascar, the presumed ancestral home of these lizards (Austin et al., 2004; Rocha et al., 2007; Harmon et al., 2008). Within each radiation, species have diversified morphologically and ecologically; as many as five species can occur sympatrically, and ecomorphological relationships similar to those in anoles have been detected (Harmon, 2005; Harmon et al., 2008). Moreover, sympatric species partition the habitat and shift their habitat use in the presence of other species (Harmon et al., 2007).

In other words, if ever there were two distantly related clades that seemed likely to have produced replicated adaptive radiations, *Phelsuma* and *Anolis*—separated evolutionarily by approximately 175 million years since their last common ancestor (Wiens et al., 2006)—are the ones. Yet, their radiations aren't mirror images. Compared to anoles, *Phelsuma* exhibits relatively little variation in limb or tail length, toepad size or habitat use. No twig day geckos exist, nor grass-bush species. There are—or were[465]—giant day geckos as large as the largest anole, but they tended to use rocks frequently and the largest species apparently was nocturnal (Vinson and Vinson, 1969). Microhabitat partitioning among sympatric day geckos sometimes occurs by tree type (palm versus non-palm), a phenomenon unknown in anoles (Thorpe and Crawford, 1979; Harmon et al., 2007). All in all, despite their many similarities, *Anolis* and *Phelsuma* have not diversified in the same ways, although in broad terms their radiations exhibit many similarities.[466]

Why haven't anoles and day geckos traveled down exactly the same evolutionary paths? All of the potential explanations for non-convergence mentioned in Chapter 16 are possibilities. For example, Indian Ocean and West Indian island environments may be different. One obvious example is that Madagascar, the ancestral cradle of *Phelsuma*,

---

464. But they move their heads side to side, rather than vertically up and down like iguanid lizards (Marcellini, 1977; Delheusy and Bels, 1994; Murphy and Myers, 1996).

465. The largest day gecko, *P. gigas*, which reached 190 mm SVL, went extinct on Rodrigues Island in the 19th century (Vinson and Vinson, 1969). The largest living species, *P. guentheri*, reaches a respectable 160 mm SVL (Austin et al., 2004), larger than most crown-giant anoles.

466. We might also wonder whether the *Phelsuma* radiations in the different island groups in the Indian Ocean have produced matched outcomes. This question has not yet been explicitly analyzed: a preliminary morphometric analysis suggests some cases of cross-island convergence, but also some species on one island—particularly in the Mascarenes and Madagascar—are unlike any species found on other islands (Harmon et al., in press).

A question more suited for *Animal Planet* concerns what would happen if *Anolis* and *Phelsuma* ever came together. Would the species interact? If so, who would win? This is more than a thought experiment, as both anoles and day geckos have been introduced to Hawaii (McKeown, 1996), and the Madagascar giant day gecko, *P. madagascariensis*, has not only been introduced to the Florida Keys (Krysko et al., 2003), but has been observed eating an *A. carolinensis* (J. Kolbe, pers. comm.). Anecdotal reports from Oahu claim that the day geckos are kicking the anoles' butts (i.e., supplanting them from areas previously colonized), but I am unaware of any scientific study of this battle of the arboreal green lizard radiations.

is also home to another large radiation of arboreal, diurnal, and insectivorous lizards. The presence of chameleons—specialized to use narrow, arboreal surfaces[467]—may have constrained the ecological diversification of day geckos.[468]

Alternatively, the differences in anole and day gecko radiations may reflect the different evolutionary potentialities of geckos and iguanid lizards. Geckos, for example, tend to have more laterally oriented limbs than iguanid lizards, which may place limits on the way geckos can adapt to different microhabitats. Moreover, gecko toepads have setal hairs that are elaborated to a much greater extent than the relatively simple setae of anoles, but anole setal densities are higher (Ruibal and Ernst, 1965; Williams and Peterson, 1982). Although a preliminary study found no difference in clinging ability between anoles and geckos (Irschick et al., 1996), further study would be useful because anecdotal evidence suggests that geckos are better clingers (e.g., many geckos will readily run across a ceiling upside down, something that anoles rarely do). If day geckos do, indeed, have greater clinging ability than anoles, then they may not have needed to diverge in limb length as much as anoles to adapt to using different microhabitats.[469] These, as well as a myriad of other differences, may have steered anole and day gecko evolution down different evolutionary paths, even if the adaptive landscapes in the two areas were extremely similar.

The *Anolis–Phelsuma* example is probably representative of most similar situations. As discussed in Chapter 16, similar clades diversifying in what appears to be similar environmental situations may realize very different evolutionary trajectories for two primary reasons. First, they are unlikely to occupy identical adaptive landscapes. For the most part, distantly related clades that are ecologically similar are unlikely to radiate in the same geographic area. As a result, such clades are not likely to experience the same patterns of selection because environments in different areas are unlikely to be the same; if nothing else, interactions with different sets of other clades are likely to produce different evolutionary outcomes. Conversely, when distantly related clades diversify in the same geographic area, they are likely to radiate in different ways to prevent competitive exclusion (Malagasy chameleons and day geckos possibly being an example).[470]

Second, distantly related clades tend to differ in so many ways that it is unlikely that entire evolutionary radiations will unfold in the same way. The differences between *Phelsuma* and *Anolis* would constitute different initial starting points for radiation, but also probably reflect different genetic and development constraints (see Chapter 16).

467. Although, paradoxically, one clade, *Brookesia*, is primarily terrestrial, despite possessing the modifications of the hands and feet for grasping narrow surfaces.

468. Chameleon species also occur naturally alongside *Phelsuma* in the Comoros and on some islands in the Seychelles, but are not found naturally in the Mascarene Islands.

469. In this regard, I should add that day geckos have no claws! Whether this clawlessness is a testament to the efficacy of gecko toepads or a constraint on habitat use, or both, is unknown.

470. In theory, one could imagine an archipelago in which Clade A radiates in half the islands and Clade B in the other half so that the two clades do not coexist, but I am unaware of any such cases.

Certainly, cases of convergence among distantly related species are common (Conway Morris, 2003), but it may be too much to expect that entire radiations of distant relatives will evolve in lockstep. Only closely related clades are likely both to start with similar initial phenotypes and to have developmental and genetic systems that bias evolutionary diversification to occur in similar ways.

REPLICATED ADAPTIVE RADIATIONS LIMITED TO ISLANDS AND LAKES

The reason that replicated adaptive radiations are limited to islands and lakes is an extension of the reason they they only occur among closely related clades. Radiations on different continents usually, though not always, will be accomplished by distantly related clades which are likely to diversify in different ways (Pianka, 1986; Cadle and Greene, 1993; Losos, 1994a). Moreover, clades radiating on different continents are unlikely to experience identical selective pressures. Not only will the different biota lead to divergent adaptive landscapes due to variation in regimes of predation, competition, disease, and so on, but the number of simultaneously radiating clades that co-occur in continental settings will be greater. That is, the depauperate faunas on islands allow a single clade to radiate by itself into wide open ecological space. By contrast, when such space occurs in continental settings (perhaps due to appearance of a new resource or extinction of a previously dominant group), many clades may radiate simultaneously, limiting the opportunities available to any one clade.

POOR DISPERSAL ABILITY

Few cases of replicated adaptive radiation are known in flying organisms.[471] The reason is obvious. Evolutionary replication is most likely when it occurs on separate islands or lakes in the same region, so that the environments are likely to be as similar as possible. However, if species in the radiating clade are able to move back and forth between evolutionary arenas, then independent radiations will not occur. The faunas in the different areas may end up being matched perfectly, but that will result because the matching species are closely related, rather than convergent. This phenomenon is seen to some extent in the Hawaiian *Tetragnatha* discussed above. By contrast, for non-flying animals such as lizards or frogs, dispersal between islands probably occurs much less frequently (Chapter 6), setting the stage for replicated adaptive radiation.

In summary, replicated adaptive radiations are very rare, and *Anolis* is perhaps the most extensive and best documented example. Why replicated adaptive radiation has occurred in these lizards seems straightforward. Earlier in the chapter I discussed why

---

471. The only potential example of which I'm aware is the convergence of *Myotis* bats in different regions of the northern hemisphere discussed in Footnote 459.

Anolis has radiated to such a great extent; here I've shown that the reason for evolutionary replication is that Greater Antillean *Anolis* has all the necessary ingredients: radiation of closely related, relatively poorly-dispersing species on isolated islands with low diversity in the same general region.

Still, we might ask why replicated adaptive radiation is so uncommon, particularly given that it is seen so readily in laboratory experiments with microbial systems. One possibility, of course, is that the environment—so easy to control in the laboratory—is rarely so similar in different localities in nature. In other words, the lack of replicated adaptive radiation reflects a lack of replicated adaptive landscapes. The other possibility is that adaptive radiation doesn't occur all that often, and rarely occurs multiple times in closely related clades—with sufficiently similar phenotypes, ecology and evolutionary potentiality—in sufficiently similar environments. If we accept the view that the acquisition of different developmental and genetic systems and other constraining factors prevent all but closely related taxa from diversifying in the same way, then it may simply be that closely related taxa rarely get the opportunity to radiate multiple times in highly similar environments, and *Anolis* on Greater Antillean islands may be one of those few exceptions.

PARALLELISM, GENETIC CONSTRAINT, AND ANOLE ADAPTIVE RADIATION

One reason that closely related clades may diversify in the same way is that they share similar developmental and genetic systems. Hence, when species from such clades are subjected to the same selective conditions, they may adapt in genetically and developmentally similar ways (Haldane, 1932; Gould, 2002; Hoekstra, 2006). Recent studies have provided many examples in a wide range of organisms and traits in which parallel phenotypic change in multiple populations or closely related species is caused by similar genetic changes (e.g., Sucena et al., 2003; Colosimo et al., 2005; Derome and Bernatchez, 2006; Derome et al., 2006; Hoekstra et al., 2006; Protas et al., 2006; Shapiro et al., 2006; Whittall et al., 2006).[472] Whether convergence of the anole ecomorphs similarly has been accomplished by the same genetic means remains to be seen; the combination of the *A. carolinensis* genome and the status of the vertebrate limb and craniofacial region as model systems in developmental biology (e.g., Niswander, 2002; Tickle, 2002; Abzhanov et al., 2004, 2006; Stopper and Wagner, 2005) suggests that we may soon have an answer to this question.

In the previous chapter, I argued that genetic constraints are unlikely to have played a role in shaping the convergence of the anole ecomorphs. Nonetheless, if this convergence

---

472. Of course, this is not always the case; some times convergent phenotypic evolution is accomplished by different genetic changes, even in closely related species (e.g., Hoekstra and Nachman, 2003; Hoekstra et al., 2006; Wittkopp et al., 2004).

has been accomplished by the same genetic changes, then we may have to look more carefully at the possibility that not just adaptation alone, but the interplay between adaptation and constraint, has been responsible for the replicated adaptive radiation of Greater Antillean anoles (Gould, 2002).

However, even if convergence in *Anolis* has occurred by way of identical genetic changes, it does not necessarily follow that limited genetic options—i.e., constraints—have played an important role in shaping the anole radiations. Rather, even if they were completely unconstrained in terms of the direction in which they could evolve, species with similar genetic architecture might be expected to adapt to similar selective conditions by means of the same genetic changes (Gould, 2002).

ANOLE FUTURES: BIODIVERSITY, CONSERVATION, AND THE FATE OF *ANOLIS*

It seems appropriate to end this book by discussing anole biological diversity and the extent to which it is likely to be imperiled in the years to come. On the positive side, anole biodiversity may be substantially greater than we presently realize. New species are being discovered at a high rate, mostly in Central and South America, but also in Cuba, primarily in the mountains in the east (e.g., Fong and Garrido, 2000; McCranie et al., 2000; Garrido and Hedges, 2001; Köhler et al., 2001, 2007; Köhler and Sunyer, 2008; Navarro et al., 2001; Pacheco and Garrido, 2004; Hulebak et al., 2007; Poe and Ibañez, 2007; Poe and Yañez-Miranda, 2007; Ugueto et al., 2007). Most of these are genuinely new, previously unknown taxa, although in some cases the new species result from breaking of one species into several.[473] Given the regularity with which these new forms are being discovered, who knows how many anole species there are? Moreover, as discussed in Chapter 14, molecular data raise the possibility that many widespread species may actually be complexes of parapatric species. Anole diversity is probably substantially underestimated.

On the negative side, anoles experience the same pressures that confront much of the world's fauna and flora: habitat destruction, global climate change, invasive species, and overexploitation (Wilcove et al., 1998; Gibbon et al., 2000). Some of these, however, are much graver threats than others.

HABITAT DESTRUCTION

As is often the case (Wilcove et al., 1998; Gibbon et al., 2000), habitat destruction is probably the biggest threat. The most extreme case is Haiti, where less than 1% of the land has forest cover (Hedges and Woods, 1993) and several species—most notably the aquatic anole, *A. eugenegrahami*—are in grave jeopardy. More generally, approximately

---

473. This taxonomic "splitting" perhaps has been excessive in a few cases.

90% of most West Indian habitats have been degraded; to a large extent, much of the change in West Indian habitats has involved a shift from closed forest to open forest and agricultural lands (Mittermeier et al., 1999). The disappearance of *A. roosevelti*, last seen more than 75 years ago, may be a result of the extensive habitat destruction that occurred on the islands near Puerto Rico early in the last century (Mayer, 1989). Similarly, much of the original forest—both rainforest and dry forest—in Central America is gone or severely degraded and deforestation rates in some areas are among the highest in the world (Janzen, 1988; Mittermeier et al., 1999). One species from Mexico, *A. naufragus*, is known only from one locality, which was almost totally deforested subsequent to its discovery (Campbell et al., 1989). Other than *A. roosevelti* and *A. naufragus*, no species are currently suspected to have gone extinct, but this will change in the years to come.

One ironic twist resulting from this habitat degradation is that the most common anoles today probably were much less plentiful before the arrival of humans. In Cuba, for example, the most abundant species are *A. sagrei* and the green anoles, *A. porcatus* and *A. allisoni*, species which occur in open, sunny habitats and which are common in and around human habitations. In contrast, within intact forests throughout much of the island, *A. sagrei* is much less abundant and the green anoles less commonly seen (although they may be more abundant in the sun-drenched canopy). In prehistoric times, when Cuba was mostly forested, these species must have been much less plentiful and more patchily distributed than they are today. Similarly, *A. sericeus*, a Central American species often found in edge habitats, is probably more common today than it was in the past (Henderson and Fitch, 1975). Conversely, many forest-dwelling species, particularly those that require pristine forest, probably were much more abundant in times past.[474] Such species, particularly those with small geographic ranges today, face an uncertain future in many places.

GLOBAL CLIMATE CHANGE

Global warming poses many threats to species and ecosystems. The most direct is from increased temperature and changes in precipitation, to which populations could respond in three ways: by adapting, by shifting their range, or by going extinct (Parmesan, 2006). Given the evolutionary lability of anole thermal and hydric physiology (Chapters 10 and 12), we might expect that anoles—more than many other taxa—may be able to adapt

---

474. These recent shifts caution against evolutionary interpretations based on current distributions and abundance. On the other hand, the major conclusions of this book concerning ecomorph ecology and evolution are not affected by the realization that much forested habitat has been converted to more open habitats because the ecomorphs usually occur in all but the most degraded habitats, albeit sometimes represented by different species in closed and open forest. Thus, general conclusions from work conducted today about ecomorph ecology and evolution probably apply to the conditions that existed prior to the arrival of humans, even if the relative mix of open and closed habitats has changed. Research conducted in the most degraded habitats (e.g., agricultural fields), where usually only 1–2 anole species occur, usually at low densities (e.g., Glor et al., 2001a), probably has little applicability to prehistoric times, but relatively little work is conducted in such areas.

to changing temperatures and precipitation regimes. On the other hand, these changes may occur too rapidly and anole species may be forced to shift their ranges if they are to avoid extinction.

Broad scale predictive analyses using interpolated climate data and remote sensing approaches (Chapter 10) have not yet been performed for anoles, but one such study for Mexican butterflies, birds, and mammals predicted relatively few extinctions, but widespread range shifts and changes in the composition of local communities (Peterson et al., 2002). Montane populations may be particularly vulnerable because their geographic ranges are often small and the potential to shift to higher elevations as temperature increases may be limited; at the extreme, populations shifting upward may run out of mountain (Parmesan, 2006). Just that has apparently happened in the cloud forests of Costa Rica, where many frog species have disappeared (Pounds et al., 1999, 2006). Even in lowland areas, relatively cool-adapted, closed forest species may be imperiled as temperatures increase and the habitat becomes more suitable for more warm-adapted, open habitat species (Tewksbury et al., 2008).

The only relevant data on anoles comes from the Monteverde Cloud Forest Preserve at 1,540 m elevation in Costa Rica, where two formerly abundant montane species, *A. tropidolepis* and *A. altae*, disappeared in the mid-1990s, while *A. intermedius*, a species also found at lower elevations and thus presumably better adapted to warmer conditions, has not experienced a change in population size (Fig. 17.5; Pounds et al., 1999, 2006).

Climate change can also affect populations in many indirect ways, by altering the composition of communities and by changing the functioning of ecosystems (Parmesan, 2006). For example, the disappearance of montane frogs may not be due to changes in temperature and moisture levels per se, but rather to the resulting spread of pathogenic chytrid fungus facilitated by these changes (Pounds et al., 2006). One possible example involving anoles relates to the substantial decline in leaf-litter anoles at the La Selva Biological Station in Costa Rica, which may be related to reduced litter accumulation due to changing patterns of rainfall (Whitfield et al., 2007).

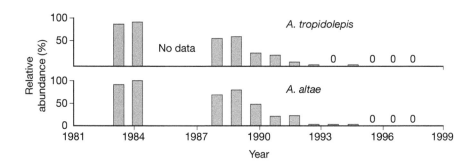

FIGURE 17.5

Decline in Costa Rican montane anole populations. No data were collected 1984–1987. Modified with permission from Pounds et al. (1999).

Invasive species have had calamitous impacts on native species and ecosystems (Wilcove et al., 1998; Mooney and Hobbs, 2000). Documented effects on anoles, however, have been relatively minor. Despite the many introductions of anoles from one place to another throughout the West Indies and elsewhere, few examples exist of introduced species negatively affecting the natives. Rather, in most cases, introduced anoles have had relatively little success when invading in the presence of ecologically similar species (Chapter 11), and many invaders are restricted to human environs and severely disturbed habitats (e.g., Fitch et al., 1989; Henderson and Powell, 2001; Greene et al., 2002; White and Hailey, 2006; Powell and Henderson, 2008b).

Probably the best known counterexample is the effect of *A. sagrei* on *A. carolinensis* in Florida. Concomitant with the expansion of *A. sagrei* throughout Florida, *A. carolinensis* has apparently become much scarcer. There can be no doubt that *A. sagrei* has a negative effect on *A. carolinensis* (or its close relatives elsewhere [Schoener, 1975; Losos and Spiller, 1999; Campbell, 2000]), but the conversion of much of Florida into parking lots, roadways, and other prime *A. sagrei* habitat probably has something to do with *A. carolinensis's* decline as well. More generally, though, the survival of *A. carolinensis* is probably not threatened. Rather, after colonizing Florida several million years ago, *A. carolinensis* probably experienced ecological release in the absence of other anoles. Now that *A. sagrei* is present, *A. carolinensis* seems to have retreated to its ancestral, trunk-crown niche, reestablishing the pattern of niche partitioning and sympatric coexistence that initially evolved in Cuba between the *carolinensis* and *sagrei* clades and which is evident today throughout Cuba, the Bahamas, and Little Cayman (Chapter 11; Losos, 1996c).

Aside from this case, few examples of negative effects of an introduced anole on other anole species have been reported. In several cases, an introduced species has caused habitat shifts in other species, either native (e.g., Losos et al., 1993a) or introduced (e.g., Wingate, 1965; Salzburg, 1984). Evidence of population declines resulting from the introduced species is also scant and limited to urban settings. For example, in parts of Santo Domingo, the introduced Cuban green anole, *A. porcatus*, seems to have had a negative effect on the Hispaniolan green anole, *A. chlorocyanus* (Powell et al., 1990; Powell and Henderson, 2008b; see also Fitch et al. [1989] for a similar example).

Effects of other introduced species on anoles have also been rarely documented. The only clearly detrimental impact is the introduction of the brown tree snake to Guam, which has eliminated *A. carolinensis*, also introduced, from natural habitats (Fritts and Rodda, 1998). Mongooses have been widely introduced throughout the West Indies and have ravaged populations of many species of mammals, birds, and reptiles (Seaman and Randall, 1962; Case and Bolger, 1991; Powell and Henderson, 2005). Although anoles are often a major component of mongoose diets (Waide and Reagan, 1983; Vilella, 1998; Wilson and Vogel, 1999), I am unaware of any reports of substantial population level effects, although they probably occur in some places.

Anoles are not widely used by local people for any purpose. As far as I know, anoles are not eaten by people anywhere—for good reason, as I imagine they'd be pretty crunchy. On the other hand, anoles are commercially collected, primarily for export for the pet trade. I am unaware of global data on the magnitude of the trade, but it can be substantial. For example, from 1998–2002, more than 250,000 *A. carolinensis* and more than 100,000 *A. sagrei* were legally exported from the United States; in the same period, as many as 30,000 anoles of various species may have been imported into the U.S. (M. Schlaepfer, pers. comm.).[475] Figures for imports into other countries are unavailable, but may be large because there are many reptile hobbyists in Europe. The United States is the only country likely to have much domestic trade in anoles, and these numbers, too, are great because many *A. carolinensis* and *A. sagrei* are captured and sold within the United States, not only for the pet trade, but also to laboratories, educational supply companies, and zoos.[476] Data on the magnitude of this trade is scarce, but more than 250,000 anoles were collected in Florida in a four-year period in the early 1990s (Enge, 2005);[477] in Louisiana, nearly a million *A. carolinensis* a year were collected in the mid-1990s, but that number has declined to around 350,000 per year in 2006, apparently as a result of declining demand, rather than shortage of anoles (J. Boundy, pers. comm.).

These are not insignificant numbers, and the pet trade can certainly threaten species, particularly if they have small geographic ranges and are easily collected (Stuart et al., 2006). Nonetheless, most of the anole species being collected are very abundant and the trade in most other species is probably much smaller. Occasionally there are claims on the internet or elsewhere that collecting is threatening particular anole species, usually those found on small islands. Although this is certainly possible, no data are available to substantiate such claims.

WHITHER *ANOLIS*?

What will the future hold for *Anolis*? Certainly, species will be lost. Indeed, who knows how many species—unknown and unlamented—have disappeared in Central and South America as a result of loss of their habitat before they could be discovered? No doubt, more species will perish as their environment is destroyed. Moreover, habitat fragmentation will hinder the ability of species to shift their geographic and elevational ranges as climate changes. Invasive species and collecting for the pet trade may have some effect as well. Without question, anole biodiversity will take a hit.

---

475. Data from the Lemis data base of the United States Fish and Wildlife Service. Importation numbers may be overestimates because exports are sometimes mistakenly recorded as imports (Schlaepfer et al., 2005).

476. Where they are often fed to other animals!

477. This number may be a substantial underestimate because dealers were not required to report the number of the introduced *A. sagrei* and as a result, most did not do so.

On the other hand, the survival of the clade as a whole is not jeopardized, and anoles will fare much better than many other taxa. Quite a few anole species do well in human-disrupted habitats (Henderson and Powell, 2001; Powell and Henderson, 2008b) and, with their great behavioral and evolutionary flexibility, anoles are better prepared than most species to adjust to changing conditions in both the short- and long-term.

A theme of this book has been the marriage of observation and experiment, of historical inference and present-day investigation. It is regrettable that humans have messed up the world in so many ways, and that our fellow fauna and flora have paid so heavy a price, and will continue to do so. Nonetheless, these disruptions set the evolutionary stage for the sort of research that could scarcely be imagined, much less intentionally be put into practice.

Several of the hallmarks of anole evolution are that they they adapt quickly to new environmental conditions; they respond behaviorally, ecologically, and evolutionary to selective pressures resulting from the presence of other species; and they diversify evolutionarily in response to ecological opportunity and the absence of other, similar species. In this book, I have laid out the evidence to support these claims and have suggested small scale ways to test them.

But we humans are creating the opportunity to test these ideas on a much more massive scale. Can anoles really adapt rapidly to environmental change? We're changing the environment in a myriad of ways, and we will see just how rapidly they can evolve, whether some types of change are more easily accommodated than others, and whether some types of species are more evolutionarily adept than others. Does the presence of other species spark evolutionary adjustment? We're adding and subtracting species all over the place. Does adaptive radiation result when anoles colonize new areas with open environmental space? Let's see what they'll do in Hawaii, Taiwan, Guam, and the many other previously anole-free places they'll eventually occupy.

Don't get me wrong. I'd much rather appreciate and study anoles in pristine habitats in a world spared the ravages of mankind. But this is the world in which we live. History is in the past, and usually we are hard pressed to study the processes underlying it, but anoles may be an exception. Environmental disruptions have recreated all aspects of the factors thought to have been important in the genesis of their incredibly rich biological diversity. Even as we strive to minimize further environmental damage, it is our rare opportunity to study in the present the same phenomena and processes that were so generous to *Anolis* in the past.

Of course, such studies are just an adjunct to ongoing studies of natural populations in less disrupted habitats. We have learned much from such studies over the course of the past four decades, knowledge that has been valuable not only for understanding anole biology, but also for addressing broader questions in ecology, evolutionary biology, and other disciplines. As this book has made clear, however, we have much more yet to learn. Indeed, the more we learn, and the more we develop new methods and new ideas, the more we realize what we have yet to discover. Most of the general statements about anole biology made in this book are based on data from relatively few species, usually less than 10% of the nearly 400 described anole species. For many interesting and important topics, we have data only from a handful of species. More detailed study on many species—directed, where possible, toward addressing questions of broad and general interest—is needed to fully comprehend the patterns and underlying processes involved in the genesis of anole biological diversity.

Anoles are an evolutionary marvel. They, along with eleutherodactyline frogs, are the dominant vertebrate element of West Indian ecosystems. In the mainland neotropics, they are nearly unrivalled in terms of their species diversity. They are excellent—nearly perfect—subjects for scientific studies of biological diversity. More generally, they are simply delightful creatures to observe and study. Reverend Lockwood (1876, p.16) had it right more than a century and a quarter ago when he said that *Anolis* "is everything that is commendable: clean, inoffensive, pretty and wonderfully entertaining; provoking harmless mirth, and stirring up in the thinker the profoundest depths of his philosophy."

AFTERWORD

AN ANOLE BESTIARY

In this section, I present a list of all West Indian anole species and of all mainland species mentioned in the text. In addition, Figure A.1 presents the complete phylogeny from Nicholson et al. [2005] that served as the basis for several figures in this book and was used for all original statistical analyses presented here.

WEST INDIAN SPECIES

This list is based primarily on Caribherp (http://evo.bio.psu.edu/caribherp/lists/wi-list .htm), last modified December 6, 2007 (at the time of writing). I have not included several island populations that are normally considered as subspecies of *A. marmoratus* or *A. sagrei* (e.g., *A. m. kahouannensis* from the island of Kahouanne offshore from Guadeloupe and *A. s. luteosignifer* from Cayman Brac) and for which no recent phylogenetic analysis has presented a compelling argument for elevation to species status. The two species from Isla Providencia and San Andrés in the southwestern Caribbean are included. Islands in the Lesser Antilles are only distinguished into northern and southern groups because some species occur on multiple islands. Ecomorph designations are based on Beuttel and Losos (1999); species not included in that study are assigned to ecomorph based on natural history information in the literature and examination of specimens. Ecomorph designations are not applied to Lesser Antillean species, although

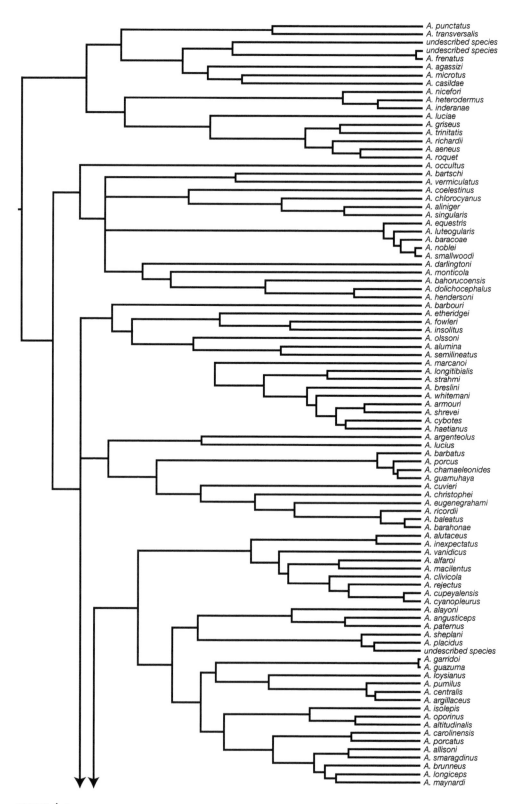

FIGURE A.1

Phylogeny of anoles used for figures and analyses in this book from Nicholson et al. (2005). Branch lengths were made proportional to time using the program r8s (Sanderson, 2003).

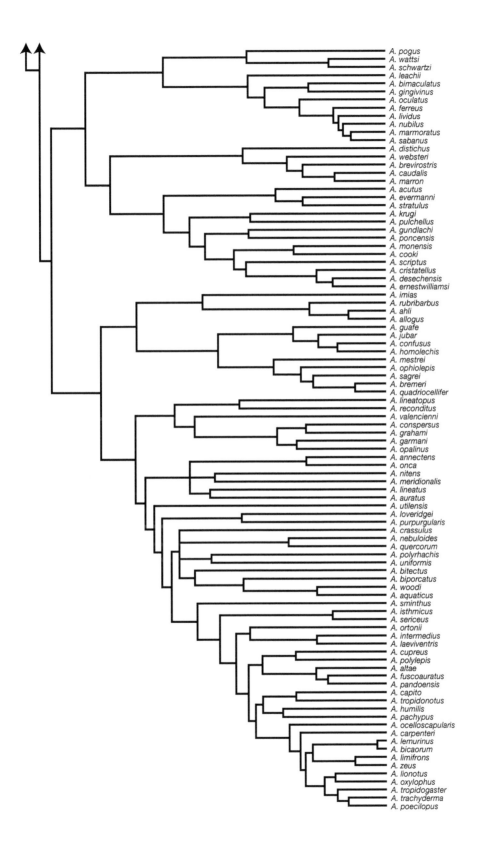

A. pogus
A. wattsi
A. schwartzi
A. leachii
A. bimaculatus
A. gingivinus
A. oculatus
A. ferreus
A. lividus
A. nubilus
A. marmoratus
A. sabanus
A. distichus
A. websteri
A. brevirostris
A. caudalis
A. marron
A. acutus
A. evermanni
A. stratulus
A. krugi
A. pulchellus
A. gundlachi
A. poncensis
A. monensis
A. cooki
A. scriptus
A. cristatellus
A. desechensis
A. ernestwilliamsi
A. imias
A. rubribarbus
A. ahli
A. allogus
A. guafe
A. jubar
A. confusus
A. homolechis
A. mestrei
A. ophiolepis
A. sagrei
A. bremeri
A. quadriocellifer
A. lineatopus
A. reconditus
A. valencienni
A. conspersus
A. grahami
A. garmani
A. opalinus
A. annectens
A. onca
A. nitens
A. meridionalis
A. lineatus
A. auratus
A. utilensis
A. loveridgei
A. purpurgularis
A. crassulus
A. nebuloides
A. quercorum
A. polyrhachis
A. uniformis
A. bitectus
A. biporcatus
A. woodi
A. aquaticus
A. sminthus
A. isthmicus
A. sericeus
A. ortonii
A. intermedius
A. laeviventris
A. cupreus
A. polylepis
A. altae
A. fuscoauratus
A. pandoensis
A. capito
A. tropidonotus
A. humilis
A. pachypus
A. ocelloscapularis
A. carpenteri
A. lemurinus
A. bicaorum
A. limifrons
A. zeus
A. lionotus
A. oxylophus
A. tropidogaster
A. trachyderma
A. poecilopus

some species qualify as members of particular ecomorph classes (see Chapter 4 and Losos and de Queiroz [1997]). This column is also left blank for unique anoles.

Series/Clade names correspond to those in Fig. 5.6 and follow Savage and Guyer [1989] and Brandley and de Queiroz [2004]. The following Series were recognized so that all taxa would be monophyletic: the *insolitus* Series, comprised of *A. insolitus*, *A. etheridgei*, and *A. fowleri*; the *bartschi* Series composed of *A. bartschi* and *A. vermiculatus*; and the *alutaceus*, *hendersoni*, and *semilineatus* Series (each raised from Species Group status). The clades Chamaeleolis and Chamaelinorops might also be considered series. The two members of the mainland Norops radiation that have recolonized the West Indies are listed simply as Norops because phylogenetic relationships within this clade are not well established (Chapter 5).

SPECIES	ISLAND	ECOMORPH	SERIES/CLADE
Anolis acutus	St. Croix		*cristatellus*
Anolis aeneus	Southern Lesser Antilles		*roquet*
Anolis Chamaeleolis *agueroi*	Cuba		Chamaeleolis
Anolis ahli	Cuba	Trunk-Ground	*sagrei*
Anolis alayoni	Cuba	Twig	*angusticeps*
Anolis alfaroi	Cuba	Grass-Bush	*alutaceus*
Anolis aliniger	Hispaniola	Trunk-Crown	*chlorocyanus*
Anolis allisoni	Cuba	Trunk-Crown	*carolinensis*
Anolis allogus	Cuba	Trunk-Ground	*sagrei*
Anolis altavelensis	Hispaniola	Trunk	*distichus*
Anolis altitudinalis	Cuba	Trunk-Crown	*carolinensis*
Anolis alumina	Hispaniola	Grass-Bush	*semilineatus*
Anolis alutaceus	Cuba	Grass-Bush	*alutaceus*
Anolis anfiloquioi	Cuba	Grass-Bush	*alutaceus*
Anolis angusticeps	Cuba, Bahamas	Twig	*angusticeps*
Anolis argenteolus	Cuba		*lucius*
Anolis argillaceus	Cuba		*angusticeps*
Anolis armouri	Hispaniola	Trunk-Ground	*cybotes*
Anolis bahorucoensis	Hispaniola	Grass-Bush	*hendersoni*
Anolis baleatus	Hispaniola	Crown-Giant	*ricordii*
Anolis baracoae	Cuba	Crown-Giant	*equestris*
Anolis barahonae	Hispaniola	Crown-Giant	*ricordii*
Anolis Chamaeleolis *barbatus*	Cuba		Chamaeleolis
Anolis Chamaelinorops *barbouri*	Hispaniola		Chamaelinorops

SPECIES	ISLAND	ECOMORPH	SERIES/CLADE
Anolis bartschi	Cuba		bartschi
Anolis bimaculatus	Northern Lesser Antilles		bimaculatus
Anolis birama	Cuba	Trunk-Ground	sagrei
Anolis bremeri	Cuba	Trunk-Ground	sagrei
Anolis breslini	Hispaniola	Trunk-Ground	cybotes
Anolis brevirostris	Hispaniola	Trunk	distichus
Anolis brunneus	Bahamas	Trunk-Crown	carolinensis
Anolis caudalis	Hispaniola	Trunk	distichus
Anolis centralis	Cuba		angusticeps
Anolis Chamaeleolis chamaeleonides	Cuba		Chamaeleolis
Anolis chlorocyanus	Hispaniola	Trunk-Crown	chlorocyanus
Anolis christophei	Hispaniola		christophei
Anolis clivicola	Cuba	Grass-Bush	alutaceus
Anolis coelestinus	Hispaniola	Trunk-Crown	chlorocyanus
Anolis concolor	San Andrés		Norops
Anolis confusus	Cuba	Trunk-Ground	sagrei
Anolis conspersus	Grand Cayman	Trunk-Crown	grahami
Anolis cooki	Puerto Rico	Trunk-Ground	cristatellus
Anolis cristatellus	Puerto Rico	Trunk-Ground	cristatellus
Anolis cupeyalensis	Cuba	Grass-Bush	alutaceus
Anolis cuvieri	Puerto Rico	Crown-Giant	ricordii
Anolis cyanopleurus	Cuba	Grass-Bush	alutaceus
Anolis cybotes	Hispaniola	Trunk-Ground	cybotes
Anolis darlingtoni	Hispaniola	Twig	darlingtoni
Anolis delafuentei	Cuba	Trunk-Ground	sagrei
Anolis desechensis	Desecheo	Trunk-Ground	cristatellus
Anolis distichus	Hispaniola, Bahamas	Trunk	distichus
Anolis dolichocephalus	Hispaniola	Grass-Bush	hendersoni
Anolis equestris	Cuba	Crown-Giant	equestris
Anolis ernestwilliamsi	Carrot Rock	Trunk-Ground	cristatellus
Anolis etheridgei	Hispaniola		insolitus
Anolis eugenegrahami	Hispaniola		eugenegrahami
Anolis evermanni	Puerto Rico	Trunk-Crown	cristatellus
Anolis extremus	Southern Lesser Antilles		roquet

(Continued on following page)

SPECIES	ISLAND	ECOMORPH	SERIES/CLADE
Anolis fairchildi	Bahamas	Trunk-Crown	*carolinensis*
Anolis ferreus	Northern Lesser Antilles		*bimaculatus*
Anolis fowleri	Hispaniola		*insolitus*
Anolis fugitivus	Cuba	Grass-Bush	*alutaceus*
Anolis garmani	Jamaica	Crown-Giant	*grahami*
Anolis garridoi	Cuba	Twig	*angusticeps*
Anolis gingivinus	Northern Lesser Antilles		*bimaculatus*
Anolis grahami	Jamaica	Trunk-Crown	*grahami*
Anolis griseus	Southern Lesser Antilles		*roquet*
Anolis guafe	Cuba	Trunk-Ground	*sagrei*
Anolis Chamaeleolis *guamuhaya*	Cuba		Chamaeleolis
Anolis guazuma	Cuba	Twig	*angusticeps*
Anolis gundlachi	Puerto Rico	Trunk-Ground	*cristatellus*
Anolis haetianus	Hispaniola	Trunk-Ground	*cybotes*
Anolis hendersoni	Hispaniola	Grass-Bush	*hendersoni*
Anolis homolechis	Cuba	Trunk-Ground	*sagrei*
Anolis imias	Cuba	Trunk-Ground	*sagrei*
Anolis incredulus	Cuba	Trunk-Crown	*carolinensis*
Anolis inexpectatus	Cuba	Grass-Bush	*alutaceus*
Anolis insolitus	Hispaniola	Twig	*insolitus*
Anolis isolepis	Cuba	Trunk-Crown	*carolinensis*
Anolis juangundlachi	Cuba	Grass-Bush	*alutaceus*
Anolis jubar	Cuba	Trunk-Ground	*sagrei*
Anolis koopmani	Hispaniola	Grass-Bush	*monticola*
Anolis krugi	Puerto Rico	Grass-Bush	*cristatellus*
Anolis leachii	Northern Lesser Antilles		*bimaculatus*
Anolis lineatopus	Jamaica	Trunk-Ground	*grahami*
Anolis litoralis	Cuba		*angusticeps*
Anolis lividus	Northern Lesser Antilles		*bimaculatus*
Anolis longiceps	Navassa	Trunk-Crown	*carolinensis*
Anolis longitibialis	Hispaniola	Trunk-Ground	*cybotes*
Anolis loysianus	Cuba	Trunk	*angusticeps*
Anolis luciae	Southern Lesser Antilles		*roquet*
Anolis lucius	Cuba		*lucius*
Anolis luteogularis	Cuba	Crown-Giant	*equestris*
Anolis macilentus	Cuba	Grass-Bush	*alutaceus*

SPECIES	ISLAND	ECOMORPH	SERIES/CLADE
Anolis marcanoi	Hispaniola	Trunk-Ground	*cybotes*
Anolis marmoratus	Northern Lesser Antilles		*bimaculatus*
Anolis marron	Hispaniola	Trunk	*distichus*
Anolis maynardi	Little Cayman	Trunk-Crown	*carolinensis*
Anolis mestrei	Cuba	Trunk-Ground	*sagrei*
Anolis monensis	Mona	Trunk-Ground	*cristatellus*
Anolis monticola	Hispaniola		*monticola*
Anolis noblei	Cuba	Crown-Giant	*equestris*
Anolis nubilis	Northern Lesser Antilles		*bimaculatus*
Anolis occultus	Puerto Rico	Twig	*occultus*
Anolis oculatus	Northern Lesser Antilles		*bimaculatus*
Anolis olssoni	Hispaniola	Grass-Bush	*semilineatus*
Anolis opalinus	Jamaica	Trunk-Crown	*grahami*
Anolis ophiolepis	Cuba	Grass-Bush	*sagrei*
Anolis oporinus	Cuba	Trunk-Crown	*carolinensis*
Anolis paternus	Cuba	Twig	*angusticeps*
Anolis pigmaequestris	Cuba	Crown-Giant	*equestris*
Anolis pinchoti	Providencia		Norops
Anolis placidus	Hispaniola	Twig	*angusticeps*
Anolis pogus	Northern Lesser Antilles		*bimaculatus*
Anolis poncensis	Puerto Rico	Grass-Bush	*cristatellus*
Anolis porcatus	Cuba	Trunk-Crown	*carolinensis*
Anolis Chamaeleolis *porcus*	Cuba		Chamaeleolis
Anolis pulchellus	Puerto Rico	Grass-Bush	*cristatellus*
Anolis pumilus	Cuba		*angusticeps*
Anolis quadriocellifer	Cuba	Trunk-Ground	*sagrei*
Anolis reconditus	Jamaica		*grahami*
Anolis rejectus	Cuba	Grass-Bush	*alutaceus*
Anolis richardii	Southern Lesser Antilles		*roquet*
Anolis ricordii	Hispaniola	Crown-Giant	*ricordii*
Anolis rimarum	Hispaniola		*monticola*
Anolis roosevelti	Puerto Rico Bank	Crown-Giant	*ricordii*
Anolis roquet	Southern Lesser Antilles		*roquet*
Anolis rubribarbus	Cuba	Trunk-Ground	*sagrei*
Anolis ruibali	Cuba		*angusticeps*
Anolis rupinae	Hispaniola		*monticola*

(*Continued on following page*)

SPECIES	ISLAND	ECOMORPH	SERIES/CLADE
Anolis sabanus	Northern Lesser Antilles		*bimaculatus*
Anolis sagrei	Cuba, Bahamas, Other islands	Trunk-Ground	*sagrei*
Anolis schwartzi	Northern Lesser Antilles		*bimaculatus*
Anolis scriptus	Inagua	Trunk-Ground	*cristatellus*
Anolis semilineatus	Hispaniola	Grass-Bush	*semilineatus*
Anolis sheplani	Hispaniola	Twig	*angusticeps*
Anolis shrevei	Hispaniola	Trunk-Ground	*cybotes*
Anolis singularis	Hispaniola	Trunk-Crown	*chlorocyanus*
Anolis smallwoodi	Cuba	Crown-Giant	*equestris*
Anolis smaragdinus	Bahamas	Trunk-Crown	*carolinensis*
Anolis spectrum	Cuba	Grass-Bush	*alutaceus*
Anolis strahmi	Hispaniola	Trunk-Ground	*cybotes*
Anolis stratulus	Puerto Rico	Trunk-Crown	*cristatellus*
Anolis terraealtae	Northern Lesser Antilles		*bimaculatus*
Anolis terueli	Cuba		*angusticeps*
Anolis toldo	Cuba	Trunk-Crown	*carolinensis*
Anolis trinitatis	Southern Lesser Antilles		*roquet*
Anolis valencienni	Jamaica	Twig	*grahami*
Anolis vanidicus	Cuba	Grass-Bush	*alutaceus*
Anolis vermiculatus	Cuba		*bartschi*
Anolis vescus	Cuba	Grass-Bush	*alutaceus*
Anolis wattsi	Northern Lesser Antilles		*bimaculatus*
Anolis websteri	Hispaniola	Trunk	*distichus*
Anolis whitemani	Hispaniola	Trunk-Ground	*cybotes*

MAINLAND SPECIES

The many mainland species described since the last published list of anole species (Savage and Guyer, 1989) preclude an accurate listing of all species. For this reason, I only list those mainland species mentioned in the text. I mention notable ecomorphological information in "Notes." Some species are assigned to an ecomorph class following Irschick et al. (1997), information in the literature (particularly Savage [2002]), or personal observations. I indicate only whether species belong to the Norops or Dactyloa clades because the lower level systematics of mainland anoles is in flux (see Chapter 5). I include in this list several species that occur on islands in the Pacific Ocean off the coast of northern South America.

SPECIES	LOCALITY	CLADE	NOTES
Anolis agassizi	Malpelo Island	Dactyloa	Rocky surfaces, large
Anolis altae	Costa Rica	Norops	Low to ground
Anolis aquaticus	Costa Rica and Panama	Norops	Aquatic anole
Anolis auratus	Widespread in Central America to northern South America	Norops	Grass-Bush anole
Anolis barkeri	Mexico	Norops	Aquatic anole
Anolis biporcatus	Widespread in Central America to northern South America	Norops	Crown-Giant[478]
Anolis capito	Widespread in Central America	Norops	Near ground, relatively large
Anolis cupreus	Widespread in Central America	Norops	Low to ground
Anolis frenatus	Costa Rica to Colombia	Dactyloa	Crown-Giant
Anolis fungosus	Costa Rica and Panama	Norops	Twig anole
Anolis fuscoauratus	Amazonia	Norops	Low to ground
Anolis gadovi	Mexico	Norops	
Anolis gorgonae	Gorgona Island	Dactyloa	Arboreal
Anolis humilis	Costa Rica and Panama	Norops	Ground litter inhabitant
Anolis insignis	Costa Rica and Panama	Dactyloa	Crown anole, large
Anolis intermedius	Costa Rica and Panama	Norops	Low to ground
Anolis limifrons	Widespread in Central America	Norops	Near the ground; often narrow diameter vegetation
Anolis macrolepis	South America	Norops	Aquatic anole
Anolis naufragus	Mexico	Norops	
Anolis nebulosus	Mexico	Norops	Ground to high in trees

(Continued on following page)

478. Mean perch height for *A. biporcatus* in Irschick et al. [1997], based on observations of five individuals, was lower than that of West Indian crown-giants. However, my unpublished observations in Panama and Costa Rica and those of others (e.g., Leenders [2001]) indicate that it often goes high into the canopy, much like crown-giants.

SPECIES	LOCALITY	CLADE	NOTES
Anolis nitens	Amazonia	Norops	Leaf litter
Anolis onca	Venezuela	Norops	Ground-dwelling, sandy areas
Anolis ortonii	Amazonia	Norops	Low to ground
Anolis oxylophus	Widespread in Central America	Norops	Aquatic anole
Anolis pentaprion	Widespread in Central America to Colombia	Norops	Twig anole
Anolis polylepis	Costa Rica and Panama	Norops	Moderately arboreal
Anolis proboscis	Ecuador	Dactyloa	
Anolis sericeus	Widespread in Central America	Norops	Moderately arboreal
Anolis taylori	Mexico	Norops	Rocky surfaces
Anolis transversalis	Amazonia	Dactyloa	Arboreal
Anolis tropidolepis	Costa Rica	Norops	Montane, low to ground
Anolis vociferans	Costa Rica	Norops	Twig anole
Phenacosaurus clade	South America		Twig anoles, some quite large

REFERENCES

Abzhanov, A., W.P. Kuo, C. Hartmann, B.R. Grant, P.R. Grant, and C.J. Tabin. 2006. The calmodulin pathway and evolution of elongated beak morphology in Darwin's finches. *Nature* 442:563–567.

Abzhanov, A., M. Protas, B.R. Grant, P.R. Grant, and C.J. Tabin. 2004. *BMP4* and morphological variation of beaks in Darwin's finches. *Science* 305:1462–1465.

Ackerly, D.D., D.W. Schwilk, and C.O. Webb. 2006. Niche evolution and adaptive radiation: testing the order of trait divergence. *Ecology* 87:S50–S61.

Alfaro, M.E., D.I. Bolnick, and P.C. Wainwright. 2005. Evolutionary consequences of many-to-one mapping of jaw morphology to mechanics in labrid fishes. *American Naturalist* 165:e140–e154.

Anderson, R.A., and W.H. Karasov. 1981. Contrasts in energy intake and expenditure in sit-and-wait and widely foraging lizards. *Oecologia* 49:67–72.

Andersson, M. 1994. *Sexual Selection*. Princeton University Press: Princeton, NJ.

Andersson, M., and L.W. Simmons. 2006. Sexual selection and mate choice. *Trends in Ecology and Evolution* 21:296–302.

Andrews, R.M. 1971. Structural habitat and time budget of a tropical *Anolis* lizard. *Ecology* 52:262–270.

Andrews, R.M. 1976. Growth rate in island and mainland anoline lizards. *Copeia* 1976:477–482.

Andrews, R.M. 1979. Evolution of life histories: A comparison of *Anolis* lizards from matched island and mainland habitats. *Breviora* 454:1–51.

Andrews, R.M. 1985a. Mate choice by females of the lizard, *Anolis carolinensis*. *Journal of Herpetology* 19:284–289.

Andrews, R.M. 1985b. Oviposition frequency of *Anolis carolinensis*. *Copeia* 1985:259–262.

Andrews, R.M. 1988. Demographic correlates of variable egg survival for a tropical lizard. *Oecologia* 76:376–382.

Andrews, R.M. 1991. Population stability of a tropical lizard. *Ecology* 72:1204–1217.

Andrews, R.M. 1998. Geographic variation in field body temperature of *Sceloporus* lizards. *Journal of Thermal Biology* 23:329–334.

Andrews, R.M., and T. Asato. 1977. Energy utilization of a tropical lizard. *Comparative Biochemistry and Physiology* 58A:57–62.

Andrews, R.M., and A.S. Rand. 1974. Reproductive effort in anoline lizards. *Ecology* 55:1317–1327.

Andrews, R.M., and A.S. Rand. 1983. Limited dispersal of juvenile *Anolis limifrons*. *Copeia* 1983:429–434.

Andrews, R.M., and J.D. Nichols. 1990. Temporal and spatial variation in survival rates of the tropical lizard *Anolis limifrons*. *Oikos* 57:215–221.

Andrews, R.M., and O.J. Sexton. 1981. Water relations of the eggs of *Anolis auratus* and *Anolis limifrons*. *Ecology* 62:556–562.

Andrews, R.M., and J.A. Stamps. 1994. Temporal variation in sexual size dimorphism of *Anolis limifrons* in Panama. *Copeia* 1994:613–622.

Angilleta, M.J. Jr., A.F. Bennett, H. Guderley, C.A. Navas, F. Seebacher, and R.S. Wilson. 2006. Coadaptation: A unifying principle in evolutionary thermal biology. *Physiological and Biochemical Zoology* 79:282–294.

Anker, A., S.T. Ahyong, P.Y. Noel, and A.R. Palmer. 2006. Morphological phylogeny of alpheid shrimps: Parallel preadaptation and the origin of a key morphological innovation, the snapping claw. *Evolution* 60:2507–2528.

Arbogast, B.S., S.V. Drovetski, R.L. Curry, P.T. Boag, G. Seutin, P.R. Grant, B.R. Grant, and D.J. Anderson. 2006. The origin and diversification of Galapagos mockingbirds. *Evolution* 60:370–382.

Arbogast, B.S., S.V. Edwards, J. Wakeley, P. Beerli, and J.B. Slowinski. 2002. Estimating divergence times from molecular data on phylogenetic and population genetic timescales. *Annual Review of Ecology and Systematics* 33:707–740.

Arim, M., and P.A. Marquet. 2004. Intraguild predation: a widespread interaction related to species biology. *Ecology Letters* 7:557–564.

Arnold, D.L. 1980. Geographic variation in *Anolis brevirostris* (Sauria:Iguanidae) in Hispaniola. *Breviora* 461:1–31.

Arnold, E.N. 1984. Evolutionary aspects of tail shedding in lizards and their relatives. *Journal of Natural History* 18:127–169.

Arnold, E.N. 1988. Caudal autotomy as a defense. Pp. 235–273 in C. Gans and R.B. Huey, Eds., *Biology of the Reptilia, Volume 16, Ecology B: Defense and Life History*. Alan R. Liss: New York, NY.

Arnold, E.N. 1994. Do ecological analogues assemble their common features in the same order? An investigation of regularities in evolution, using sand-dwelling lizards as examples. *Philosophical Transactions of the Royal Society of London B* 344:277–290.

Arnold, M.L. 1997. *Natural Hybridization and Evolution*. Oxford University Press: Oxford, UK.

Arnold, S.J. 1983. Morphology, performance, and fitness. *American Zoologist* 23:347–361.

Arnold, S.J., M.E. Pfrender, and A.G. Jones. 2001. The adaptive landscape as a conceptual bridge between micro- and macroevolution. *Genetica* 112/113:9–32.

Arthur, W., and M. Farrow. 1999. The pattern of variation in centipede segment number as an expression of developmental constraint in evolution. *Journal of Theoretical Biology* 200:183–191.

Austen, N.L. 1867. The crested anolis. *Land and Water* 4(79):9.

Austin, J.J., E.N. Arnold, and C.G. Jones. 2004. Reconstructing an island radiation using ancient and recent DNA: the extinct and living day geckoes (*Phelsuma*) of the Mascarene Islands. *Molecular Phylogeny and Evolution* 31:109–122.

Autumn, K. 2006. How gecko toes stick. *American Scientist* 94:124–132.

Autumn, K. 2007. Gecko adhesion: Structure, function, and applications. *MRS Bulletin* 32:473–478.

Autumn, K., A. Dittmore, D. Santos, M. Spenko, and M. Cutkosky. 2006. Frictional adhesion: A new angle on gecko attachment. *Journal of Experimental Biology* 209:3569–3579.

Autumn, K., Y.A. Liang, S.T. Hsieh, W. Zeach, W.P. Chan, T.W. Kenny, R. Fearing, and R.J. Full. 2000. Adhesive force of a single gecko foot-hair. *Nature* 405:681–685.

Autumn, K., and J.B. Losos. 1997. Notes on jumping ability and thermal biology of the enigmatic anole *Chamaelinorops barbouri*. *Journal of Herpetology* 31:442–444.

Autumn, K., and A.M. Peattie. 2002. Mechanisms of adhesion in geckos. *Integrative and Comparative Biology* 42:1081–1090.

Autumn, K., M. Sitti, Y.A. Liang, A.M. Peattie, W.R. Hansen, S. Sponberg, T.W. Kenny, R. Fearing, J.N. Israelachvili, and R.J. Full. 2002. Evidence for van der Waals adhesion in gecko setae. *Proceedings of the National Academy of Sciences of the United States of America* 99:12252–12256.

Bakken, G.S. 1992. Measurement and application of operative and standard operative temperatures in ecology. *American Zoologist* 32:194–216.

Bakken, G.S., and D.M. Gates. 1975. Heat transfer analysis of animals: Some implications for field ecology, physiology and evolution. Pp. 255–290 in D.M. Gates and R.B. Schmerl, Eds., *Perspectives of Biophysical Ecology*. Springer-Verlag: New York, NY.

Bakker, R.T. 1983. The deer flees, the wolf pursues: Incongruities in predator-prey evolution. Pp. 350–382 in D.J. Futuyma and M. Slatkin, Eds., *Coevolution*. Sinauer Associates: Sunderland, MA.

Ballinger, R.E. 1973. Experimental evidence of the tail as a balancing organ in the lizard, *Anolis carolinensis*. *Herpetologica* 29:65–66.

Ballinger, R.E., K.R. Marion, and O.J. Sexton. 1970. Thermal ecology of the lizard, *Anolis limifrons* with comparative notes on three additional Panamanian anoles. *Ecology* 51:246–254.

Barbour, T. 1930. The anoles. I. The forms known to occur on the Neotropical islands. *Bulletin of the Museum of Comparative Zoology* 70:105–144.

Barker, K.F, G.F. Barrowclough, and J.G. Groth. 2002. A phylogenetic analysis for passerine birds: taxonomic and biogeographic implications of an analysis of nuclear DNA sequence data. *Proceedings of the Royal Society B* 269:295–308.

Barraclough, T.G., J.E. Hogan, and A.P. Vogler. 1999. Testing whether ecological factors promote cladogenesis in a group of tiger beetles (Coleoptera: Cicindelidae). *Proceedings of the Royal Society of London B* 266:1061–1067.

Barraclough, T.G., and A.P. Vogler. 2000. Detecting the geographical pattern of speciation from species-level phylogenies. *American Naturalist* 155:419–434.

Bartlett, R.D., and P. Bartlett. 2003. *Reptiles and Amphibians of the Amazon: An Ecotourist's Guide*. University Press of Florida: Gainesville, FL.

Baum, D.A., and M.J. Donoghue. 1995. Choosing among alternative "phylogenetic" species concepts. *Systematic Botany* 20:560–573.

Baum, D.A., and A. Larson. 1991. Adaptation reviewed: A phylogenetic methodology for studying character macroevolution. *Systematic Zoology* 40:1–18.

Baxter, L.R., Jr. 2003. Basal ganglia systems in ritualistic social displays: Reptiles and humans; function and illness. *Physiology and Behavior* 79:451–460.

Beatty, J. 2006. Replaying life's tape. *Journal of Philosophy* 103:336–362.

Beatty, J. 2008. Chance variation and evolutionary contingency: Darwin, Simpson *the Simpsons*, and Gould. Pp. 189–210 in M. Ruse, Ed., *The Oxford Handbook of Philosophy of Biology*. Oxford University Press: Oxford, UK.

Bellairs, A. 1969. *The Life of Reptiles*. Weidenfeld and Nicholson: London, UK.

Bels, V.L. 1990. The mechanism of dewlap extension in *Anolis carolinensis* (Reptilia: Iguanidae) with histological analysis of the hyoid apparatus. *Journal of Morphology* 206: 225–244.

Bels, V.L., J.P. Theys, M.B. Bennett, and L. Legrand. 1992. Biomechanical analysis of jumping in *Anolis carolinensis* (Reptilia: Iguanidae). *Copeia* 1992:492–505.

Bennett, A.F. 1980. The thermal dependence of lizard behaviour. *Animal Behaviour* 28: 752–762.

Bennett, A.F., T.T. Gleeson, and G.C. Gorman. 1981. Anaerobic metabolism in a lizard (*Anolis bonairensis*) under natural conditions. *Physiological Zoology* 54:237–241.

Bennett, A.F., and R.B. Huey. 1990. Studying the evolution of physiological performance. Pp. 251–284 in D. Futuyma and J. Antonovics, Eds., *Oxford Surveys in Evolutionary Biology, Volume 7*. Oxford University Press: Oxford, UK.

Berenbaum, M.R., and A.R. Zangerl. 1992. Genetics of physiological and behavioral resistance to host furanocoumarins in the parsnip webworm. *Evolution* 46:1373–1384.

Bergmann, P.J., and D.J. Irschick. 2005. Effects of temperature on maximum clinging ability in a diurnal gecko: Evidence for a passive clinging mechanism. *Journal of Experimental Zoology* 303A:785–791.

Berovides Álvarez, V., and A. Sampedro Marin. 1980. Competición en especies de lagartos iguánidos de Cuba. *Ciencias Biológicas* 5:115–122.

Beuttell, K., and J.B. Losos. 1999. Ecological morphology of Caribbean anoles. *Herpetological Monographs* 13:1–28.

Bickford, D., D.J. Lohman, N.S. Sodhi, P.K.L. Ng, R. Meier, K. Winker, K.K. Ingram, and I. Das. 2007. Cryptic species as a window on diversity and conservation. *Trends in Ecology and Evolution* 22:148–155.

Biewener, A.A. 2003. *Animal Locomotion*. Oxford University Press: Oxford, UK.

Birkhead, T.R., and T. Pizzari. 2002. Postcopulatory sexual selection. *Nature Reviews Genetics* 3:262–273.

Birt, R.A., R. Powell, and B.D. Greene. 2001. Natural history of *Anolis barkeri*, a semi-aquatic lizard from southern México. *Journal of Herpetology* 35:161–166.

Bjørklund, M. 1997. Are 'comparative methods' always necessary? *Oikos* 80:607–612.

Blake, J. 1983. A chromosomal C-banding in *Anolis grahami*. Pp. 621–625 in A.G.J. Rhodin and K. Miyata, Eds., *Advances in Herpetology and Evolutionary Biology: Essays in Honor of Ernest E. Williams*. Museum of Comparative Zoology, Harvard University: Cambridge, MA.

Bloch, N., and D.J. Irschick. 2004. Toe-clipping dramatically reduces clinging performance in a pad-bearing lizard (*Anolis carolinensis*). *Journal of Herpetology* 37:293–298.

Bloch, N., and D.J. Irschick. 2006. An analysis of inter-population divergence in visual display behavior of the green anole lizard (*Anolis carolinensis*). *Ethology* 112:370–378.

Blondel, J., F. Vuilleumier, L.F. Marcus, and E. Terouanne. 1984. Is there ecomorphological convergence among Mediterranean bird communities of Chile, California, and France? *Evolutionary Biology* 18:141–213.

Blouin-Demers, G., and P. Nadeau. 2005. The cost-benefit model of thermoregulation does not predict lizard thermoregulatory behavior. *Ecology* 86:560–566.

Blows, M.W., and A.A. Hoffman. 2005. A reassessment of genetic limits to evolutionary change. *Ecology* 86:1371–1384.

Bock, W.J., and W.D. Miller. 1959. The scansorial foot of the woodpeckers, with comments on the evolution of perching and climbing feet in birds. *American Museum Novitates* 1931:1–45.

Bogert, C.M. 1949. Thermoregulation in reptiles, a factor in evolution. Evolution 3:195–211.

Bolnick, D.I. 2004. Can intraspecific competition drive disruptive selection? An experimental test in natural populations of sticklebacks. *Evolution* 58:608–618.

Bolnick, D.I., R. Svanbäck, MS. Araújo, and L. Persson. 2007. Comparative support for the niche variation hypothesis that more generalized populations also are more heterogeneous. *Proceedings of the National Academy of Sciences of the United States of America* 104:10075–10079.

Boncoraglio, G., and N. Saino. 2007. Habitat structure and the evolution of bird song: A meta-analysis of the evidence for the acoustic adaptation hypothesis. *Functional Ecology* 21:134–142.

Bonser, R.H. 1999. Branching out in locomotion: The mechanics of perch use in birds and primates. *Journal of Experimental Biology* 202:1459–1463.

Borges-Landáez and Shine. 2003. Influence of toe-clipping on running speed in *Eulamprus quoyii*, an Australian scincid lizard. *Journal of Herpetology* 37:592–595.

Bossuyt, F., and M.C. Milinkovitch. 2000. Convergent adaptive radiations in Madagascan and Asian ranid frogs reveal covariation between larval and adult traits. *Proceedings of the National Academy of Sciences of the United States of America* 97:6585–6590.

Boughman, J.W. 2002. How sensory drive can promote speciation. *Trends in Ecology and Evolution* 17:571–577.

Boumans, L., D.R. Vieites, F. Glaw, and M. Vences. 2007. Geographical patterns of deep mitochondrial differentiation in widespread Malagasy reptiles. *Molecular Phylogeny and Evolution* 45:822–839.

Brakefield, P.M. 2006. Evo-devo and constraints on selection. *Trends in Ecology and Evolution* 21:362–368.

Brandley, M.C., and K. de Queiroz. 2004. Phylogeny, ecomorphological evolution, and historical biogeography of the *Anolis cristatellus* series. *Herpetological Monographs* 18:90–126.

Brattstrom, B.H. 1978. Learning studies in lizards. Pp. 173–182 in N. Greenberg and P.D. MacLean, Eds., *Behavior and Neurology of Lizards*. National Institute of Mental Health: Rockville, MD.

Breuil, M. 2002. Histoire Naturelle des Amphibiens et Reptiles Terrestres de l'Archipel Guadeloupeen. Guadeloupe, Saint-Martin, Saint-Barthelemy. *Patrimoines Naturels* 54:1–339.

Britton, T., C.L. Anderson, D. Jacquet, S. Lundqvist, and K. Bremer. 2007. Estimating divergence times in large phylogenetic trees. *Systematic Biology* 56:741–752.

Brodie, E.D., III, A.J. Moore, and F.J. Janzen. 1995. Visualizing and quantifying natural selection. *Trends in Ecology and Evolution* 10:313–318.

Bromham, L., and D. Penny. 2003. The modern molecular clock. *Nature Reviews Genetics* 4:216–224.

Brooks, D.R., and D.A. McLennan. 1991. *Phylogeny, Ecology, and Behavior: A Research Program in Comparative Biology.* University of Chicago Press: Chicago, IL.

Brooks, D.R., and D.A. McLennan. 2002. *The Nature of Diversity: An Evolutionary Voyage of Discovery.* University of Chicago Press: Chicago, IL.

Brower, A.V.Z. 1994. Rapid morphological radiation and convergence among races of the butterfly *Heliconius erato* inferred from patterns of mitochondrial DNA evolution. *Proceedings of the National Academy of Sciences of the United States of America* 91:6491–6495.

Brown, J.H., and M.V. Lomolino. 1998. *Biogeography,* 2nd Ed. Sinauer Associates: Sunderland, MA.

Brown, J.L., S. Vargo, E.F. Connor, and M.S. Nuckols. 1997. Causes of vertical stratification in the density of *Cameraria hamadryadella. Ecological Entomology* 22:16–25.

Brown, P.R., and A.C. Echternacht. 1991. Interspecific behavioral interaction of adult male *Anolis* sagrei and gray-throated *Anolis carolinensis* (Sauria: Iguanidae): a preliminary field study. Pp. 21–30 in J.B. Losos and G.C. Mayer, Eds., *Anolis Newsletter IV.* Division of Amphibians and Reptiles, National Museum of Natural History, Smithsonian Institution: Washington, DC.

Brown, W.L., and E.O. Wilson. 1956. Character displacement. *Systematic Zoology* 5:49–64.

Browne, J. 1995. *Charles Darwin: Voyaging.* Princeton University Press: Princeton, NJ.

Browne, J. 2002. *Charles Darwin: The Power of Place.* Princeton University Press: Princeton, NJ.

Buckley, C.R., M. Jackson, M. Youssef, D.J. Irschick, and S.C. Adolph. 2007. Testing the persistence of phenotypic plasticity after incubation in the western fence lizard, *Sceloporus occidentalis. Evolutionary Ecology Research* 9:169–183.

Buckley, L.B., and J. Roughgarden. 2005a. Effect of species interactions on landscape abundance patterns. *Journal of Animal Ecology* 74:1182–1194.

Buckley, L.B., and J. Roughgarden. 2005b. Lizard habitat partitioning on islands: The interaction of local and landscape scales. *Journal of Biogeography* 32:2113–2121.

Buckley, L.B., and J. Roughgarden. 2006. Climate, competition, and the coexistence of island lizards. *Functional Ecology* 20:315–322.

Buckley, L.B., and W. Jetz. 2007. Insularity and the determinants of lizard population density. *Ecology Letters* 10:481–489.

Buden, D.W. 1974. Prey remains of barn owls in the southern Bahamas. *Wilson Bulletin* 86:336–343.

Bullock, D.J., H.M. Jury, and P.G.H. Evans. 1993. Foraging ecology in the lizard *Anolis oculatus* (Iguanidae) from Dominica, West Indies. *Journal of Zoology* 230:19–30.

Burghardt, G. 1964. Effects of prey size and movement on the feeding behavior of the lizards *Anolis carolinensis* and *Eumeces fasciatus. Copeia* 1964:576–578.

Burghardt, G.M. 1977. Learning processes in reptiles. Pp. 555–681 in C. Gans and D.W. Tinkle, Eds., *Biology of the Reptilia, Vol. 7: Ecology and Behaviour A*. Academic Press: London, UK.

Burnell, K.L., and S.B. Hedges. 1990. Relationships of West Indian *Anolis* (Sauria: Iguanidae): an approach using slow-evolving protein loci. *Caribbean Journal of Science* 26:7–30.

Burns, J.K., C.A. Cunningham, R.A. Dupuis, M.N. Trask, J.S. Tulloch, R. Powell, J.S. Parmerlee, Jr., K.L. Kopecky, and M.L. Jolley. 1992. Lizards of the Cayos Siete Hermanos, Dominican Republic, Hispaniola. *Bulletin of the Chicago Herpetological Society* 27:225–232.

Burns, K.J., S.J. Hackett, and N.K. Klein. 2002. Phylogenetic relationships and morphological diversity in Darwin's finches and their relatives. *Evolution* 56:1240–1252.

Buskirk, E.R., K.L. Andersen, and J. Brozek. 1956. Unilateral activity and bone and muscle development in the forearm. *Research Quarterly* 27:127–131.

Buskirk, R.E. 1985. Zoogeographic patterns and tectonic history of Jamaica and the northern Caribbean. *Journal of Biogeography* 12:445–461.

Bustard, H.R. 1968. The ecology of the Australian gecko *Heteronotia binoei* in northern New South Wales. *Journal of Zoology* 156:483–497.

Buth, D.G., G.C. Gorman, and C.S. Lieb. 1980. Genetic divergence between *Anolis carolinensis* and its Cuban progenitor, *Anolis porcatus*. *Journal of Herpetology* 14:279–284.

Butler, M.A. 2005. Foraging mode of the chameleon, *Bradypodion pumilum*: A challenge to the sit-and-wait versus active forager paradigm? *Biological Journal of the Linnean Society* 84:797–808.

Butler, M.A. 2007. *Vive le différence*! Sexual dimorphism and adaptive patterns in lizards of the genus *Anolis*. *Integrative and Comparative Biology* 47:272–284.

Butler, M.A., and J.B. Losos. 1997. Testing for unequal amounts of evolution in a continuous character on different branches of a phylogenetic tree using linear and squared-change parsimony: An example using Lesser Antillean *Anolis* lizards. *Evolution* 51:1623–1635.

Butler, M.A., and J.B. Losos. 2002. Multivariate sexual dimorphism, sexual selection, and adaptation in Greater Antillean *Anolis* lizards. *Ecological Monographs* 72:541–559.

Butler, M.A., S.A. Sawyer, and J.B. Losos. 2007. Sexual dimorphism and adaptive radiation in *Anolis* lizards. *Nature* 447:202–205.

Butler, M.A., T.W. Schoener, and J.B. Losos. 2000. The relationship between sexual size dimorphism and habitat use in Greater Antillean *Anolis* lizards. *Evolution* 54:259–272.

Cadle, J.E., and H.W. Greene. 1993. Phylogenetic patterns, biogeography, and the ecological structure of neotropical snake assemblages. Pp. 281–293 in R.E. Ricklefs and D. Schluter, Eds., *Species Diversity in Ecological Communities: Historical and Geographical Perspectives*. University Chicago Press: Chicago, IL.

Calder, W.A., III. 1984. *Size, Function, and Life History*. Harvard University Press: Cambridge, MA.

Caldwell, J.P., and L.J. Vitt. 1999. Dietary asymmetry in leaf litter frogs and lizards in a transitional northern Amazonian rain forest. *Oikos* 84:383–397.

Calsbeek, R. 2008. An ecological twist on the morphology-performance-fitness axis. *Evolutionary Ecology Research* 10:197–212.

Calsbeek, R., and C. Bonneaud. 2008. Postcopulatory fertilization bias as a form of cryptic sexual selection. *Evolution* 62:1137–1148.

Calsbeek, R., C. Bonneaud, S. Prabhu, N. Manoukis, and T.B. Smith. 2007a. Multiple paternity and sperm storage lead to increased genetic diversity in *Anolis* lizards. *Evolutionary Ecology Research* 9:495–503.

Calsbeek, R., C. Bonneaud, and T.B. Smith. 2008. Differential fitness effects of immunocompetence and neighbourhood density in alternative female lizard morphs. *Journal of Animal Ecology* 77:103–109.

Calsbeek, R., and D.J. Irschick. 2007. The quick and the dead: correlational selection on morphology, performance, and habitat use in island lizards. *Evolution* 61:2493–2503.

Calsbeek, R., J.H. Knouft, and T.B. Smith. 2006. Variation in scale numbers is consistent with ecologically based natural selection acting within and between lizard species. *Evolutionary Ecology* 20:377–394.

Calsbeek, R., and T.B. Smith. 2003. Ocean currents mediate evolution in island lizards. *Nature* 426:552–555.

Calsbeek, R., and T.B. Smith. 2007. Probing the adaptive landscape using experimental islands: Density-dependent natural selection on lizard body size. *Evolution* 61:1052–1061.

Calsbeek, R., and T.B. Smith. 2008. Experimentally replicated disruptive selection on performance traits in a Caribbean lizard. *Evolution* 62:478–484.

Calsbeek, R., T.B. Smith, and C. Bardeleben. 2007b. Intraspecific variation in *Anolis sagrei* mirrors the adaptive radiation of Greater Antillean anoles. *Biological Journal of the Linnean Society* 90:189–199.

Campbell, J.A., D.M. Hillis, and W.W. Lamar. 1989. A new lizard of the genus *Norops* (Sauria: Iguanidae) from the cloud forest of Hidalgo, Mexico. *Herpetologica* 45:232–241.

Campbell, T., and C. Bleazy. 2000. Natural history notes: *Anolis carolinensis* (green anole). Nectivory and flower pollination. *Herpetological Review* 31:239.

Campbell, T.S. 2000. *Analyses of the Effects of an Exotic Lizard* (Anolis sagrei) *on a Native Lizard* (Anolis carolinensis) *in Florida, Using Islands as Experimental Units*. Ph.D. Dissertation, University of Tennessee, Knoxville, TN.

Campbell, T.S., and A.C. Echternacht. 2003. Introduced species as moving targets: Changes in body sizes of introduced lizards following experimental introductions and historical invasions. *Biological Invasions* 5:193–212.

Cannatella, D.C., and K. de Queiroz. 1989. Phylogenetic systematics of the anoles: is a new taxonomy warranted? *Systematic Zoology* 38:57–68.

Carlquist, S. 1974. *Island Biology*. Columbia University Press: New York, NY.

Caro, T. 2005. *Antipredator Defenses in Birds and Mammals*. University of Chicago Press: Chicago, IL.

Carothers, J.H. 1984. Sexual selection and sexual dimorphism in some herbivorous lizards. *American Naturalist* 124:244–254.

Carpenter, C.C. 1962. Patterns of behavior in two Oklahoma lizards. *American Midland Naturalist* 67:132–152.

Carroll, S.B., J.K. Grenier, and S.D. Weatherbee. 2005. *From DNA to Diversity: Molecular Genetics and the Evolution of Animal Design*, 2nd Ed. Blackwell Scientific: Malden, MA.

Carroll, S.P., S.P. Klassen, and H. Dingle. 1998. Rapidly evolving adaptations to host ecology and nutrition in the soapberry bug. *Evolutionary Ecology* 12:955–968.

Carson, H.L., and D.A. Clague. 1995. Geology and biogeography of the Hawaiian Islands. Pp. 14–29 in W.L. Wagner and V.A. Funk, Eds., *Hawaiian Biogeography*. Smithsonian Institution Press: Washington, DC.

Carstens, B.C., and C.L. Richards. 2007. Integrating coalescent and ecological niche modeling in comparative phylogeography. *Evolution* 61:1439–1454.

Cartmill, M. 1985. Climbing. Pp. 73–88 in M. Hildebrand, D.M. Bramble, K.F. Liem, and D.B. Wake, Eds., *Functional Vertebrate Morphology*. Belknap Press: Cambridge, MA.

Carvalho, P., J.A.F. Diniz-Filho, and L.M. Bini. 2006. Factors influencing changes in trait correlations across species after using phylogenetic independent contrasts. *Evolutionary Ecology* 20:591–602.

Case, S.M. 1990. Dewlap and other variation in the lizards *Anolis distichus* and *A. brevirostris* (Reptilia: Iguanidae). *Biological Journal of the Linnean Society* 40:373–393.

Case, S.M., and E.E. Williams. 1984. Study of a contact zone in the *Anolis distichus* complex in the Central Dominican Republic. *Herpetologica* 40:118–137.

Case, S.M., and E.E. Williams. 1987. The cybotoid anoles and *Chamaelinorops* lizards (Reptilia: Iguanidae): Evidence of mosaic evolution. *Zoological Journal of the Linnean Society* 91:325–341.

Case, T.J. 1978. A general explanation for insular body size trends in terrestrial vertebrates. *Ecology* 59:1–18.

Case, T.J. 1979. Character displacement and coevolution in some *Cnemidophorus* lizards. *Fortschritte der Zoologie* 25:235–282.

Case, T.J. 1983. Sympatry and size similarity in *Cnemidophorus*. Pp. 297–325 in R.B. Huey, E.R. Pianka and T.W. Schoener, Eds., *Lizard Ecology: Studies of a Model Organism*. Harvard University Press: Cambridge, MA.

Case, T.J. 1990. Patterns of coexistence in sexual and asexual *Cnemidophorus* lizards. *Oecologia* 83:220–227.

Case, T.J., and D.T. Bolger. 1991. The role of interspecific competition in the biogeography of island lizards. *Trends in Ecology and Evolution* 6:135–139.

Case, T.J., and R. Sidell. 1983. Pattern and chance in the structure of model and natural communities. *Evolution* 37:832–849.

Cast, E.E., M.E. Gifford, K.R. Schneider, A.J. Hardwick, J.S. Parmerlee, Jr., and R. Powell. 2000. Natural history of an anoline lizard community in the Sierra Baoruco, Dominican Republic. *Caribbean Journal of Science* 36:258–266.

Castilla, A.M., R. Van Damme, and D. Bauwens. 1999. Field body temperatures, mechanisms of thermoregulation and evolution of thermal characteristics in lacertid lizards. *Natura Croatica* 8:253–274.

Castro-Herrera, F. 1988. *Niche Structure of an Anole Community in a Tropical Rain Forest within the Choco Region of Colombia*. Ph.D. Dissertation, North Texas State University, Denton, TX.

Censky, E.J., K. Hodge, and J. Dudley. 1998. Over-water dispersal of lizards due to hurricanes. *Nature* 395:556.

Chandler, C.R., and P.J. Tolson. 1990. Habitat use by a boid snake, *Epicrates monensis*, and its anoline prey, *Anolis cristatellus*. *Journal of Herpetology* 24:151–157.

Charlesworth, B., R. Lande, and M. Slatkin. 1982. A neo-Darwinian commentary on macroevolution. *Evolution* 36:474–498.

Chase, J.M. 2003a. Community assembly: When should history matter? *Oecologia* 136: 489–498.

Chase, J.M. 2003b. Experimental evidence for alternative stable equilibria in a benthic pond food web. *Ecology Letters* 6:733–741.

Chase, J.M. 2007. Drought mediates the importance of stochastic community assembly. *Proceedings of the National Academy of Sciences of the United States of America* 104:17430–17434.

Chase, J.M., P.A. Abrams, J.P. Grover, S. Diehl, P. Chesson, R.D. Holt, S.A. Richards, R.M. Nisbet, and T.J. Case. 2002. The interaction between predation and competition: A review and synthesis. *Ecology Letters* 5:302–315.

Cheverud, J.M. 1996. Developmental integration and the evolution of pleiotropy. *American Zoologist* 36:44–50.

Chiba, S. 2004. Ecological and morphological patterns in communities of land snails of the genus *Mandarina* from the Bonin Islands. *Journal of Evolutionary Biology* 17:131–143.

Christian, K.A., and B.W. Weavers. 1996. Thermoregulation of monitor lizards in Australia: An evaluation of methods in thermal biology. *Ecological Monographs* 66:139–157.

Christian, K.A., and G.S. Bedford. 1995. Seasonal changes in thermoregulation by the frill-neck lizard, *Chlamydosaurus kingii*, in tropical Australia. *Ecology* 76:124–132.

Cisper, G.L., C. Huntington, D.D. Smith, R. Powell, J.S. Parmerlee, Jr., and A. Lathrop. 1995. Four new Coccidi (Apicomplexa: Eimeriidae) from anoles (Lacertilia: Polychrotidae) in the Dominican Republic. *Journal of Parasitology* 81:252–255.

Clark, D.L., and J.C. Gillingham. 1990. Sleep-site fidelity in two Puerto Rican lizards. *Animal Behaviour* 39:1138–1148.

Clark, D.L., J.M. Macedonia, and G.G. Rosenthal. 1997. Testing video playback to lizards in the field. *Copeia* 1997:421–424.

Clark, D.R. Jr. 1971. Branding as a marking technique for amphibians and reptiles. *Copeia* 1971:148–151.

Clark, D.R., and J.C. Kroll. 1974. Thermal ecology of anoline lizards: temperate versus tropical strategies. *Southwestern Naturalist* 19:9–19.

Cleland, C.E. 2002. Methodological and epistemic differences between historical science and experimental science. *Philosophy of Science* 69:474–496.

Coddington, J.A. 1988. Cladistic tests of adaptational hypotheses. *Cladistics* 4:3–22.

Coddington, J.A. 1990. Bridges between evolutionary pattern and process. *Cladistics* 6:379–386.

Coddington, J.A. 1994. The roles of homology and convergence in studies of adaptation. Pp. 53–78 in R. Vane-Wright and P. Eggleton, Eds. *Phylogenetics and Ecology.* Academic Press: London, UK.

Collar, D.C., and P.C. Wainwright. 2006. Discordance between morphological and mechanical diversity in the feeding mechanism of centrarchid fishes. *Evolution* 60:2575–2584.

Collette, B.B. 1961. Correlations between ecology and morphology in anoline lizards from Havana, Cuba and southern Florida. *Bulletin of the Museum of Comparative Zoology* 125:137–162.

Collins, J.P. 1971. Ecological observations on a little known South American anole: *Tropidodactylus onca. Breviora* 370:1–6.

Colosimo, P.F., K.E. Hosemann, S. Balabhadra, G. Villarreal, Jr., M. Dickson, J. Grimwood, J. Schmutz, R.M. Myers, D. Schluter, and D.M. Kingsley. 2005. Widespread parallel evolution in sticklebacks by repeated fixation of ectodysplasin alleles. *Science* 307:1928–1933.

Connell, J.H. 1980. Diversity and the coevolution of competitors, or the ghost of competition past. *Oikos* 35:131–138.

Connell, J.H. 1983. On the prevalence and relative importance of interspecific competition: Evidence from field experiments. *American Naturalist* 122:661–696.

Conner, J., and D. Crews. 1980. Sperm transfer and storage in the lizard, *Anolis carolinensis*. *Journal of Morphology* 163:331–348.

Conrad, J.L., O. Rieppel, and L. Grande. 2007. An Eocene iguanian (Squamata: Reptilia) from Wyoming, U.S.A. *Journal of Paleontology* 81:1375–1383.

Conway Morris, S. 1998. *The Crucible of Creation: The Burgess Shale and the Rise of Animals.* Oxford University Press: Oxford, UK.

Conway Morris, S. 2003. *Life's Solution: Inevitable Humans in a Lonely Universe.* Cambridge University Press: Cambridge, UK.

Cooper, W.E. Jr. 2005a. Ecomorphological variation in foraging behaviour by Puerto Rican *Anolis* lizards. *Journal of Zoology* 265:133–139.

Cooper, W.E. Jr. 2005b. The foraging mode controversy: Both continuous variation and clustering of foraging movement occurs. *Journal of Zoology* 267:179–190.

Cooper, W.E. Jr. 2006. Risk factors affecting escape behaviour by Puerto Rican *Anolis* lizards. *Canadian Journal of Zoology* 84:495–504.

Cooper, W.E. Jr. 2007. Foraging modes as suites of coadapted movement traits. *Journal of Zoology* 272:45–56.

Cooper, W.E. Jr., and N. Greenberg. 1992. Reptilian coloration and behavior. Pp. 298–422 in C. Gans and D. Crews, Eds. *Biology of the Reptilia, Volume 18, Physiology E: Hormones, Brain, and Behavior.* University of Chicago Press: Chicago, IL.

Corey, D.T. 1988. Comments on a wolf spider feeding on a green anole lizard. *Journal of Arachnology* 16:319–392.

Corke, D. 1987. Reptile conservation on the Maria Islands (St. Lucia, West Indies). *Biological Conservation* 40: 263–279.

Corn, M.J. 1971. Upper thermal limits and thermal preferenda for three sympatric species of *Anolis. Journal of Herpetology* 5:17–21.

Corn, M.J. 1981. *Ecological Separation of* Anolis *Lizards in a Costa Rican Rain Forest.* Ph.D. Dissertation, University of Florida: Gainesville, FL.

Cowles, R.B., and C.M. Bogert. 1944. A preliminary study of the thermal requirements of desert reptiles. *Bulletin of the American Museum of Natural History* 83:265–296.

Cox, C.B., and P.D. Moore, 2000. *Biogeography: An Ecological and Evolutionary Approach,* 6th Ed. Blackwell Publishing: Oxford, UK.

Coy Otero, A., and N. Lorenzo Hernandez. 1982. Lista de los helmintos parásitos de los vertebrados silvestres cubanos. *Poeyana* 235:1–57.

Coyne, J.A., and H.A. Orr. 2004. *Speciation.* Sinauer Associates: Sunderland, MA.

Cracraft, J. 1981. Pattern and process in paleobiology: The role of cladistic analysis in systematic paleontology. *Paleobiology* 7:456–468.

Cracraft, J. 1990. The origin of evolutionary novelties: Pattern and process at different hierarchical levels. Pp. 21–44 in M.H. Nitecki, Ed., *Evolutionary Innovations.* University of Chicago Press: Chicago, IL.

Creer, D.A., K. de Queiroz, T.R. Jackman, J.B. Losos, and A. Larson. 2001. Systematics of the *Anolis roquet* series of the Southern Lesser Antilles. *Journal of Herpetology* 35:428–441.

Crews, D. 1973. Coition-induced inhibition of sexual receptivity in female lizards (*Anolis carolinensis*). *Physiology and Behavior* 11:463–468.

Crews, D. 1975. Psychobiology of reptilian reproduction. *Science* 189:1059–1065.

Crews, D., and M.C. Moore. 2005. Historical contributions of research on reptiles to behavioral neuroendocrinology. *Hormones and Behavior* 48:384–394.

Crother, B.I., and C. Guyer. 1996. Caribbean historical biogeography: Was the dispersal-vicariance debate eliminated by an extraterrestrial bolide? *Herpetologica* 52:440–465.

Crowley, S.R., and R.D. Pietruszka. 1983. Aggressiveness and vocalization in the leopard lizards (*Gambelia wislizenii*): The influence of temperature. *Animal Behaviour* 31:1055–1060.

Cruz, A. 1976. Food and foraging ecology of the American kestrel in Jamaica. *Condor* 78:409–423.

Cullen, D.J., and R. Powell. 1994. A comparison of food habits of a montane and a lowland population of *Anolis distichus* (Lacertilia: Polychrotidae) from the Dominican Republic. *Bulletin of the Maryland Herpetological Society* 30:62–66.

Currin, S., and G.J. Alexander. 1999. How to make measurements in thermoregulatory studies: The heating debate continues. *African Journal of Herpetology* 48:33–40.

Dalrymple, G.H. 1980. Comments on the density and diet of a giant anole *Anolis equestris*. *Journal of Herpetology* 14:412–415.

Darwin, C. 1859. *On the Origin of Species by Means of Natural Selection, or the Preservation of Favoured Races in the Struggle for Life*. John Murray: London, UK.

Darwin. C. 1871. *The Descent of Man, and Selection in Relation to Sex*. John Murray: London, UK.

Daudin, F.M. 1802. *Histoire Naturelle, Générale et particulière des Reptiles, Volume 4*. F. Dufart: Paris, France.

Dayan, T., and D. Simberloff. 2005. Ecological and community-wide character displacement: The next generation. *Ecology Letters* 8:875–894.

Dayton, P.K., and E. Sala. 2001. Natural history: The sense of wonder, creativity and progress in ecology. *Scientia Marina* 65:199–206.

D'Cruze, N.C. 2005. Natural history observations of sympatric *Norops* (Beta *Anolis*) in a subtropical mainland community. *Herpetological Bulletin* 91:10–18.

de Queiroz, A. 2002. Contingent predictability in evolution: Key traits and diversification. *Systematic Biology* 51:917–929.

de Queiroz, K. 2005. Ernst Mayr and the modern concept of species. *Proceedings of the National Academy of Sciences of the United States of America* 102:6600–6607.

de Queiroz, K. 2007. Species concepts and species delimitation. *Systematic Biology* 56:879–886.

de Queiroz, K., and P.D. Cantino. 2001. Phylogenetic nomenclature and the PhyloCode. *Bulletin of Zoological Nomenclature* 58:254–271.

de Queiroz, K., L.-R. Chu, and J.B. Losos. 1998. A second *Anolis* lizard in Dominican amber and the systematics and ecological morphology of Dominican amber anoles. *American Museum Novitates* 3249:1–23.

de Queiroz, K., and J. Gauthier. 1992. Phylogenetic taxonomy. *Annual Review of Ecology and Systematics* 23:449–480.

de Queiroz, K., and D.A. Good. 1997. Phenetic clustering in biology: A critique. *Quarterly Review of Biology* 72:3–30.

Debrot, A.O., J.A. De Freitas, A. Brouwer, and M. Van Marwijk Kooy. 2001. The Curaçao barn owl: Status and diet, 1987–1989. *Caribbean Journal of Science* 37:185–193.

Deckel, A.W. 1995. Laterality of aggressive responses in *Anolis*. Journal of Experimental Zoology 272:194–200.

Deckel, A.W. 1998. Hemispheric control of territorial aggression in *Anolis carolinensis*: effects of mild stress. *Brain, Behavior and Evolution* 51:33–39.

Decourcy, K.R., and T.A. Jenssen. 1994. Structure and use of male territorial headbob signals by the lizard *Anolis carolinensis*. *Animal Behaviour* 47:251–262.

Delheusy, V., and V. Bels. 1994. Comportement agonistique du gecko géant diurne *Phelsuma madagascariensis grandis*. Amphibia-Reptilia 15:63–79.

DeMarco, V.G. 1985. Maximum prey size of an insectivorous lizard, *Sceloporus undulatus garmani*. Copeia 1985:1077–1080.

Derome, N., and L. Bernatchez. 2006. The transcriptomics of ecological convergence between 2 limnetic coregonine fishes (Salmonidae). *Molecular Biology and Evolution* 23:2370–2378.

Derome, N., P. Duchesne, and L. Bernatchez. 2006. Parallelism in gene transcription among sympatric lake whitefish (*Coregonus clupeaformis* Mitchill) ecotypes. *Molecular Ecology* 15:1239–1249.

DeWitt, T.J., and S.M. Scheiner. 2004. *Phenotypic Plasticity: Functional and Conceptual Approaches*. Oxford University Press: New York, NY.

Dial, R., and J. Roughgarden. 1995. Experimental removal of insectivores from rain forest canopy: Direct and indirect effects. *Ecology* 76:1821–1834.

Dial, R., and J. Roughgarden. 1996. Natural history observations of *Anolisomyia rufianalis* (Diptera: Sarcophagidae) infesting *Anolis* lizards in a rain forest canopy. *Environmental Entomology* 25:1325–1328.

Dial, R., and J. Roughgarden. 2004. Physical transport, heterogeneity, and interactions involving canopy anoles. Pp. 270–296 in M. Lowman and B. Rinker, Eds. *Forest Canopies*, 2nd Ed. Academic Press: New York, NY.

Diamond, J. 1986. Overview: laboratory experiments, field experiments, and natural experiments. Pp. 3–22 in J. Diamond and T.J. Case, Eds., *Community Ecology*. Harper & Row: New York, NY.

Diamond, J.M., and T. J. Case. 1986. *Community Ecology*. Harper and Row: New York, NY.

Díaz, L.M., A.R. Estrada, and L.V. Moreno. 1996. A new species of *Anolis* (Sauria: Iguanidae) from the Sierra de Trinidad, Sancti Spíritus, Cuba. *Caribbean Journal of Science* 32:54–58.

Díaz, L.M., N. Navarro, and O.H. Garrido. 1998. Nueva especie de *Chamaeleolis* (Sauria: Iguanidae) de la Meseta de Cabo Cruz, Granma, Cuba. *Avicennia* 8/9:27–34.

Díaz-Uriarte, R., and T. Garland, Jr. 1996. Testing hypotheses of correlated evolution using phylogenetically independent contrasts: sensitivity to deviations from Brownian motion. *Systematic Biology* 45:27–47.

Dmi'el, R., G. Perry, and J. Lazell. 1997. Evaporative water loss in nine insular populations of the *Anolis cristatellus* group in the British Virgin Islands. *Biotropica* 29:111–116.

Dobson, A.P., S.W. Pacala, J.D. Roughgarden, E.R. Carper, and E.A. Harris. 1992. The parasites of *Anolis* lizards in the northern Lesser Antilles I. Patterns of distribution and abundance. *Oecologia* 91:110–117.

Dobzhansky, T. 1937. *Genetics and the Origin of Species*. Columbia University Press: New York, NY.

Dodd, C.K. Jr. 1993. The effects of toeclipping on sprint performance of the lizard *Cnemidophorus sexlineatus*. *Journal of Herpetology* 27:209–213.

Doebeli, M., and U. Dieckmann. 2000. Evolutionary branching and sympatric speciation caused by different types of ecological interactions. *American Naturalist* 156:S77–S101.

Doiron, S., L. Bernatchez, and P.U. Blier. 2002. A comparative mitogenomic analysis of the potential adaptive value of Arctic charr mtDNA introgression in brook charr populations (*Salvelinus fontinalis* Mitchill). *Molecular Biology and Evolution* 19:1902–1909.

Dolman, G., and C. Moritz. 2006. A multilocus perspective on refugial isolation and divergence in rainforest skinks (*Carlia*). *Evolution* 60:573–582.

Donoghue, M.J. 2005. Key innovations, convergence, and success: Macroevolutionary lessons from plant phylogeny. *Paleobiology* 31(supplement):77–93.

Donoghue, M.J., and D.D. Ackerly. 1996. Phylogenetic uncertainties and sensitivity analyses in comparative biology. *Philosophical Transactions of the Royal Society* 351:1241–1249.

Donoghue, M.J., and J.A. Gauthier. 2004. Implementing the PhyloCode. *Trends in Ecology and Evolution* 19:281–282.

Donoghue, P.J., and M.J. Benton. 2007. Rocks and clocks: Calibrating the Tree of Life using fossils and molecules. *Trends in Ecology and Evolution* 22:424–431.

Dorit, R.L. 1990. The correlates of high diversity in Lake Victoria haplochromine cichlids: a neontological perspective. Pp. 322–353 in R.M. Ross and W.D. Allmon, Eds., *Causes of Evolution: a Paleontological Perspective*. University of Chicago Press: Chicago, IL.

Doucette, L.I., S. Skúlason, and S.S. Snorrason. 2004. Risk of predation as a promoting factor of species divergence in threespine sticklebacks (*Gasterosteus aculeatus* L.). *Biological Journal of the Linnean Society* 82:189–203.

Drosophila 12 Genomes Consortium. 2007. Evolution of genes and genomes on the *Drosophila* phylogeny. *Nature* 450:203–218.

Duellman, W.E. 1978. The biology of an equatorial herpetofauna in Amazonian Ecuador. *Miscellaneous Publications of the Museum of Natural History, University of Kansas* 65:1–352.

Duellman, W.E. 1987. Lizards in an Amazonian rain forest community: resource utilization and abundance. *National Geographic Research* 3:489–500.

Duellman, W.E. 2005. *Cuso Amazónico: The Lives of Amphibians and Reptiles in an Amazonian Rainforest*. Cornell University Press: Ithaca, NY.

Dunham, A.E., D.B. Miles, and D.N. Reznick. 1988. Life history patterns in squamate reptiles. Pp. 441–522 in C. Gans and R.B. Huey Eds., *Biology of the Reptilia, Volume 16, Ecology B. Defense and Life History*. Alan R. Liss, Inc.: New York, NY.

Dunn, E.R. 1944. The lizard genus *Phenacosaurus*. *Caldasia* 3:57–62.

Dzialowski, E.M. 2005. Use of operative temperature and standard operative temperature models in thermal biology. *Journal of Thermal Biology* 30:317–334.

Eales, J., R.S. Thorpe, and A. Malhotra. 2008. Weak founder signal in a recent introduction of Caribbean *Anolis*. *Molecular Ecology* 17:1416–1426.

Eaton, J.M., K.G. Howard, and R. Powell. 2001. Geographic Distribution: *Anolis carolinensis* (Green anole). Anguilla. *Herpetological Review* 32:118.

Eaton, J.M., S.C. Larimer, K.G. Howard, R. Powell, and J.S. Parmerlee, Jr. 2002. Population densities and ecological release of the solitary lizard *Anolis gingivinus* in Anguilla, West Indies. *Caribbean Journal of Science* 38:27–36.

Eberhard, W.G. 1996. *Sexual Selection by Cryptic Female Choice*. Princeton University Press: Princeton, NJ.

Echelle, A.F., A.A. Echelle, and H.S. Fitch. 1978. Inter- and intraspecific allometry in a display organ: The dewlap of *Anolis* (Iguanidae) species. *Copeia* 1978:245–250.

Echternacht, A.C., and G.P. Gerber. 2000. Natural history notes. *Anolis conspersus*. Nectivory. *Herpetological Review* 31:173.

Edwards, J.G. 1954. A new approach to infraspecific categories. *Systematic Zoology* 3:1–20.

Eldredge, N., and J. Cracraft. 1980. *Phylogenetic Patterns and the Evolutionary Process: Method and Theory in Comparative Biology*. Columbia University Press: New York, NY.

Elstrott, J., and D.J. Irschick. 2004. Evolutionary correlations among morphology, habitat use and clinging performance in Caribbean *Anolis* lizards. *Biological Journal of the Linnean Society* 83:389–398.

Endler, J.A. 1977. *Geographic Variation, Speciation, and Clines*. Princeton University Press: Princeton, NJ.

Endler, J.A. 1980. Natural selection on color patterns in *Poecilia reticulata*. *Evolution* 34:76–91.

Endler, J.A. 1986. *Natural Selection in the Wild*. Princeton University Press: Princeton, NJ.

Endler, J.A. 1992. Signals, signal conditions, and the direction of evolution. *American Naturalist* 139:S125–153.

Endler, J.A. 1993. The color of light in forests and its implications. *Ecological Monographs* 63:1–27.

Enge, K.M. 2005. Commercial harvest of amphibians and reptiles in Florida for the pet trade. Pp. 198–214 in W.E. Meshaka, Jr., and K.J. Babbitt, Eds., *Amphibians and Reptiles: Status and Conservation in Florida*. Krieger Publishers: Malabar, FL.

Erwin, D.H. 1992. A preliminary classification of evolutionary radiations. *Historical Biology* 6:133–147.

Erwin, D.H. 2007. Disparity: Morphological pattern and developmental context. *Paleontology* 50:57–73.

Estes, R., and E.E. Williams. 1984. Ontogenetic variation in the molariform teeth of lizards. *Journal of Vertebrate Paleontology* 4:96–107.

Estrada, A.R., and A. Silva Rodriguez. 1984. Análisis de la ecomorfología de 23 especies de lagartos Cubanos del género *Anolis*. *Ciencias Biológicas* 12:91–104.

Estrada, A.R., and J. Novo Rodríguez. 1986a. Subnicho estructural de *Anolis bartschi* (Sauria: Iguanidae) en la Sierra de los Órganos, Pinar del Río, Cuba. *Poeyana* 316:1–10.

Estrada, A.R., and J. Novo Rodriquez. 1986b. Nuevos datos sobre las puestas comunales de *Anolis bartschi* (Sauria: Iguanidae) en la sierra de los Organos, Pinas del Río, Cuba. *Ciencias Biológicas* 15:135–136.

Estrada, A.R., and S.B. Hedges. 1995. A new species of *Anolis* (Sauria:Iguanidae) from eastern Cuba. *Caribbean Journal of Science* 31:65–72.

Etheridge, R.E. 1959. *The Relationships of the Anoles (Reptilia: Sauria: Iguanidae): An Interpretation Based on Skeletal Morphology.* Ph.D. Dissertation, University of Michigan: Ann Arbor, MI.

Etheridge, R.E. 1964. Late Pleistocene lizards from Barbuda, British West Indies. *Bulletin of the Florida State Museum, Biological Sciences* 9:43–75.

Etheridge, R.E. 1965. Fossil lizards from the Dominican Republic. *Quarterly Journal of the Florida Academy of Sciences* 28:83–105.

Etheridge, R.E. 1967. Lizard caudal vertebrae. *Copeia* 1967:693–721.

Etheridge, R.E., and K. de Queiroz. 1988. A phylogeny of Iguanidae. pp. 283–367 in R. Estes and G. Pregill, Eds., *Phylogenetic Relationships of the Lizard Families.* Stanford University Press: Stanford, CA.

Eyre, L.A. 1996. The tropical rainforests of Jamaica. *Jamaica Journal* 26(1):26–37.

Eyre-Walker, A. 2006. The genomic rate of adaptive evolution. *Trends in Ecology and Evolution* 21:569–575.

Falconer, D.S., and T.F.C. Mackay. 1996. *Introduction to Quantitative Genetics,* 4th Ed. Longman: Essex, UK.

Farrell, B.D., D.E. Dussourd, and C. Mitter. 1991. Escalation of plant defense: Do latex and resin canals spur plant diversification? *American Naturalist* 138:881–900.

Fauth, J.E., J. Bernardo, M. Camara, W.J. Resetarits, Jr., J. van Buskirk, and S.A. McCollum. 1996. Simplifying the jargon of community ecology: A conceptual approach. *American Naturalist* 147:282–286.

Fear, K.K., and T. Price. 1998. The adaptive surface in ecology. *Oikos* 82:440–448.

Feder, M.E., and T. Mitchell-Olds. 2003. Evolutionary and ecological functional genomics. *Nature Reviews Genetics* 4:651–657.

Felsenstein, J. 1985. Phylogenies and the comparative method. *American Naturalist* 125:1–15.

Felsenstein, J. 1988. Phylogenies and quantitative characters. *Annual Review of Ecology and Systematics* 19:445–472.

Felsenstein, J. 2004. *Inferring Phylogenies.* Sinauer Associates: Sunderland, MA.

Fisher, M., and A. Muth. 1989. A technique for permanently marking lizards. *Herpetological Review* 20:45–46.

Fitch, H.S. 1972. Ecology of *Anolis tropidolepis* in Costa Rican cloud forest. *Herpetologica* 28:10–21.

Fitch, H.S. 1973a. A field study of Costa Rican lizards. *University of Kansas Science Bulletin* 50:39–126.

Fitch, H.S. 1973b. Observations on the population ecology of the Central American iguanid lizard *Anolis cupreus. Caribbean Journal of Science* 13:215–229.

Fitch, H.S. 1975. Sympatry and interrelationships in Costa Rican anoles. *Occasional Papers of the Museum of Natural History, the University of Kansas, Lawrence, Kansas* 40:1–60.

Fitch, H.S. 1976. Sexual size differences in the mainland anoles. *Occasional Papers of the Museum of Natural History, the University of Kansas* 50:1–21.

Fitch, H.S. 1981. Sexual size differences in reptiles. *Miscellaneous Publications of the Museum of Natural History, University of Kansas* 70:1–72.

Fitch, H.S., and D.M. Hillis. 1984. The *Anolis* dewlap: Interspecific variability and morphological associations with habitat. *Copeia* 1984:315–323.

Fitch, H.S., and R.W. Henderson. 1976. A field study of the rock anoles (Reptilia, Lacertilia, Iguanidae) of Southern Mexico. *Journal of Herpetology* 10:303–311.

Fitch, H.S., and R.W. Henderson. 1987. Ecological and ethological parameters in *Anolis bahorucoensis*, a species having rudimentary development of the dewlap. *Amphibia-Reptilia* 8:69–80.

Fitch, H.S., R.W. Henderson, and H. Guarisco. 1989. Aspects of the ecology of an introduced anole: *Anolis cristatellus* in the Dominican Republic. *Amphibia-Reptilia* 10:307–320.

Fite, K.V., and B.C. Lister. 1981. Bifoveal vision in *Anolis* lizards. *Brain, Behavior and Evolution* 19:144–154.

Fitting, H. 1926. *Die Ökologische Morphologie der Pflanzen*. Gustav Fischer: Jena, Germany.

Fitzpatrick, B.M., and M. Turelli. 2006. The geography of mammalian speciation: Mixed signals from phylogenies and range maps. *Evolution* 60:601–615.

Fleishman, L.J. 1985. Cryptic movement in the vine snake *Oxybelis aeneus*. *Copeia* 1985: 242–245.

Fleishman, L.J. 1988a. Sensory and environmental influences on display form in *Anolis auratus*, a grass anole from Panama. *Behavioral Ecology and Sociobiology* 22:309–316.

Fleishman, L.J. 1988b. The social behavior of *Anolis auratus*, a grass anole from Panama. *Journal of Herpetology* 22:13–23.

Fleishman, L.J. 1988c. Sensory influences on physical design of a visual display. *Animal Behaviour* 36:1420–1424.

Fleishman, L.J. 1991. Design features of the displays of anoline lizards. Pp. 33–48 in J.B. Losos and G.C. Mayer, Eds., *Anolis Newsletter IV*. National Museum of Natural History, Smithsonian Institution: Washington, DC.

Fleishman, L.J. 1992. The influence of sensory system and the environment on motion patterns in the visual displays of anoline lizards and other vertebrates. *American Naturalist* 139:S36–S61.

Fleishman, L.J. 2000. Signal function, signal efficiency and the evolution of anoline lizard dewlap color. Pp. 209–236 in Y. Espmark, T. Amundsen, and G. Rosenqvist, eds., *Animal Signals: Signalling and Signal Design in Animal Communication*. Tapir Academic Press: Trondheim, Norway.

Fleishman, L.J., M. Bowman, D. Saunders, W.E. Miller, M.J. Rury, and E.R. Loew. 1997. The visual ecology of Puerto Rican anoline lizards: habitat light and spectral sensitivity. *Journal of Comparative Physiology A* 181:446–460.

Fleishman, L.J., E.R. Loew, and M. Leal. 1993. Ultraviolet vision in lizards. *Nature* 365:397.

Fleishman, L.J., W.J. McClintock, R.B. D'Eath, D.H. Brainard, and J.A. Endler. 1998. Colour perception and the use of video playback experiments in animal behaviour. *Animal Behaviour* 56:1035–1040.

Fleishman, L.J., and M. Persons. 2001. The influence of stimulus and background colour on signal visibility in the lizard *Anolis cristatellus*. *Journal of Experimental Biology* 204: 1559–1575.

Fleming, T.H., and R.S. Hooker. 1975. *Anolis cupreus*: The response of a lizard to tropical seasonality. *Ecology* 56:1243–1261.

Flores, G., J.H. Lenzycki, and J. Palumbo, Jr. 1994. An ecological study of the endemic Hispaniolan anoline, *Chamaelinorops barbouri* (Lacertilia: Iguanidae). *Breviora* 499:1–23.

Floyd, H.G., and T.A. Jenssen. 1983. Food habits of the Jamaican lizard, *Anolis opalinus*: Resource partitioning and seasonal effects examined. Copeia 1983:319–331.

Fong, A., and O.H. Garrido. 2000. Nueva especie de *Anolis* (Sauria: Iguanidae) de la región norte de Cuba oriental. *Revista de Biologia Tropical* 48:665–670.

Font, E., and L.C. Rome. 1990. Functional morphology of dewlap extension in the lizard *Anolis equestris* (Iguanidae). *Journal of Morphology* 206:245–258.

Fontenot, B.E., M.E. Gifford, and R. Powell. 2003. Seasonal variation in dietary preferences of a Hispaniolan anole, *Anolis longitibialis*. *Herpetological Bulletin* 86:2–4.

Foote, M. 1993. Discordance and concordance between morphological and taxonomic diversity. *Paleobiology* 19:185–204.

Foote, M. 1997. The evolution of morphological diversity. *Annual Review of Ecology and Systematics* 28:129–152.

Foote, M. 1999. Morphological diversity in the evolutionary radiation of Paleozoic and post-Paleozoic crinoids. *Paleobiology Memoir* 1:1–115.

Forsgaard, K. 1983. The axial skeleton of *Chamaelinorops*. Pp. 284–295 in A.G.J. Rhodin and K. Miyata, Eds., *Advances in Herpetology and Evolutionary Biology: Essays in Honor of Ernest E. Williams*. Museum of Comparative Zoology, Harvard University: Cambridge, MA.

Fortey, R., 2000. *Trilobites: Eyewitness to Evolution*. Harper-Collins: London, UK.

Fox, W. 1948. Effect of temperature on development of scutellation in the garter snake, *Thamnophis elegans atratus*. Copeia 1948:252–262.

Fox, W. 1963. Special tubules for sperm storage in female lizards. *Nature* 198:500–501.

Fox, W., C. Gordon, and M.H. Fox. 1961. Morphological effects of low temperatures during the embryonic development of the garter snake, *Thamnophis elegans*. *Zoologica* 46:57–71.

Frankie, G.W., H.G. Baker, and P.A. Opler. 1974. Comparative phenological studies of trees in tropical wet and dry forests in the lowlands of Costa Rica. *Journal of Ecology* 62:881–919.

Franz, R., and D. Cordier. 1986. *Herpetofaunas of the National Parks of Haiti*. Report prepared for USAID/Haiti.

Franz, R., and D.F. Gicca. 1982. Observations on the Haitian snake *Antillophis parvifrons alleni*. *Journal of Herpetology* 16:419–421.

Fritts, T.H., and G.H. Rodda. 1998. The role of introduced species in the degradation of island ecosystems: A case history of Guam. *Annual Review of Ecology and Systematics* 29:113–140.

Frost, D.R., and D.M. Hillis. 1990. Species in concept and practice: Herpetological applications. *Herpetologica* 46:87–104.

Frost, D.R., and R. Etheridge. 1989. A phylogenetic analysis and taxonomy of iguanian lizards (Reptilia: Squamata). *University of Kansas Museum of Natural History Miscellaneous Publications* 81:1–65.

Frost, D.R., R. Etheridge, D. Janies, and T.A. Titus. 2001. Total evidence, sequence alignment, evolution of polychrotid lizards, and a reclassification of the *Iguania* (Squamata: Iguania). *American Museum Novitates* 3343:1–38.

Frumhoff, P.C., and H.K. Reeve. 1994. Using phylogenies to test hypotheses of adaptation: A critique of some current proposals. *Evolution* 48:172–180.

Fryer, G., and T.D. Iles. 1972. *The Cichlid Fishes of the Great Lakes of Africa: Their Biology and Evolution*. Oliver and Boyd: Edinburgh, UK.

Fukami, T., H.J.E. Beaumont, X.-X. Zhang and P.B. Rainey. 2007. Immigration history controls diversification in experimental adaptive radiation. *Nature* 446:436–439.

Fuller, R.C., C.F. Baer, and J. Travis. 2005. How and when selection experiments might actually be useful. *Integrative and Comparative Biology* 45:391–404.

Futuyma, D.J. 1987. On the role of species in anagenesis. *American Naturalist* 130:465–473.

Futuyma, D.J. 2005. Progress on the origin of species. *PLoS Biology* 3:197–199.

Futuyma, D.J., and G. Moreno. 1988. The evolution of ecological specialization. *Annual Review of Ecology and Systematics* 19:207–234.

Galis, F. 2001. Key innovations and radiations. Pp. 581–605 in G.P. Wagner, Ed., *The Character Concept in Evolutionary Biology*. Academic Press: San Diego, CA.

Galis, F., and J.A.J. Metz. 1998. Why are there so many cichlid species? *Trends in Ecology and Evolution* 13:1–2.

Gans, C. 1974. *Biomechanics: An Approach to Vertebrate Biology*. University of Michigan Press: Ann Arbor, MI.

Garcea, R., and G. Gorman. 1968. A difference in male territorial display behavior in two sibling species of *Anolis*. *Copeia* 1968:419–420.

García-Paris, M., D.A. Good, G. Parra-Olea, and D.B. Wake. 2000. Biodiversity of Costa Rican salamanders: Implications of high levels of genetic differentiation and phylogeographic structure for species formation. *Proceedings of the National Academy of Sciences of the United States of America* 97:1640–1647.

Garland, T. Jr., A.F. Bennett, and E.L. Rezende. 2005. Phylogenetic approaches in comparative physiology. *Journal of Experimental Biology* 208:3015–3035.

Garland, T. Jr., and J.B. Losos. 1994. Ecological morphology of locomotor performance in squamate reptiles. Pp. 240–302 in P.C. Wainwright and S.M. Reilly, Eds., *Ecological Morphology: Integrative Organismal Biology*. University of Chicago Press: Chicago, IL.

Garland, T. Jr., P.E. Midford, and A.R. Ives. 1999. An introduction to phylogenetically based statistical methods, with a new method for confidence intervals on ancestral values. *American Zoologist* 39:374–388.

Garrido, O.H. 1975. Nuevos reptiles del archipiélago cubano. *Poeyana* 141:1–58.

Garrido, O.H., and S.B. Hedges. 1992. Three new grass anoles from Cuba (Squamata: Iguanidae). *Caribbean Journal of Science* 28:21–29.

Garrido, O.H., and S.B. Hedges. 2001. A new anole from the northern slope of the Sierra Maestra in eastern Cuba (Squamata: Iguanidae). *Journal of Herpetology* 35:378–383.

Gassett, J.W., T.H. Folk, K.J. Alexy, K.V. Miller, B.R. Chapman, F.L. Boyd, and D.I. Hall. 2000. Food habits of cattle egrets on St. Croix, U.S. Virgin Islands. *Wilson Bulletin* 112:268–271.

Gavrilets, S. 2000a. Rapid evolution of reproductive barriers driven by sexual conflict. *Nature* 403:886–889.

Gavrilets, S. 2000b. Waiting time to parapatric speciation. *Proceedings of the Royal Society of London B* 267:2483–2492.

Gavrilets, S. 2004. *Fitness Landscapes and the Origins of Species*. Princeton University Press: Princeton. NJ.

Genner, M.J., O. Seehausen, D.H. Lunt, D.A. Joyce, P.W. Shaw, G.R. Carvalho, and G.F. Turner. 2007. Age of cichlids: New dates for ancient lake fish radiations. *Molecular Biology and Evolution* 24:1269–1282.

Gerber, G.P. 1999. A review of intraguild predation and cannibalism in *Anolis*. Pp. 28–39 in J.B. Losos and M. Leal, Eds. *Anolis Newsletter V*. Washington University: Saint Louis. MO.

Gerber, G.P., and A.C. Echternacht. 2000. Evidence for asymmetrical intraguild predation between native and introduced *Anolis* lizards. *Oecologia* 124:599–607.

Gerhardt, R.P. 1994. The food habits of sympatric *Ciccaba* owls in northern Guatemala. *Journal of Field Ornithology* 65:258–264.

Ghalambor, C.K., J.K. McKay, S.P. Carroll, and D.N. Reznick. 2007. Adaptive versus non-adaptive phenotypic plasticity and the potential for contemporary adaptation in new environments. *Functional Ecology* 21:394–407.

Giannasi, N., R.S. Thorpe, and A. Malhotra. 2000. A phylogenetic analysis of body size evolution in the *Anolis roquet* group (Sauria: Iguanidae): Character displacement or size assortment? *Molecular Ecology* 9:193–202.

Gibbon, J.W., D.E. Scott, T.J. Ryan, K.A. Buhlmann, T.D. Tuberville, B.S. Metts, J.L. Greene, T. Mills, Y. Leiden, S. Poppy, and C.T. Winne. 2000. The global decline of reptiles, déjá vu amphibians. *Bioscience* 50:653–666.

Gibbons, J.W., and K.M. Andrews. 2004. PIT tagging: Simple technology at its best. *Bioscience* 54:447–454.

Gibbs, H.L., S.J. Corey, G. Blouin-Demers, K.A. Prior, and P.J. Weatherhead. 2006. Hybridization between mtDNA-defined phylogeographic lineages of black rat snakes (*Pantherophis sp.*). *Molecular Ecology* 15:3755–3767.

Gillespie, R.G. 2004. Community assembly through adaptive radiation in Hawaiian spiders. *Science* 303:356–359.

Gillespie, R.G., H.B. Croom, and S.R. Palumbi. 1994. Multiple origins of a spider radiation in Hawaii. *Proceedings of the National Academy of Sciences of the United States of America* 91:2290–2294.

Gittenberger, E. 1991. What about non-adaptive radiation? *Biological Journal of the Linnean Society* 43:263–272.

Gittleman, J.L. 1981. The phylogeny of parental care in fishes. *Animal Behaviour* 29:936–941.

Gittleman, J.L., and H.-K. Luh. 1994. Phylogeny, evolutionary models, and comparative methods: a simulation study. Pp. 103–122 in P. Eggleton and D. Vane-Wright, Eds., *Pattern and Process: Phylogenetic Approaches to Ecological Problems*. Academic Press: London, UK.

Givnish, T.J. 1997. Adaptive radiation and molecular systematics: issues and approaches. Pp. 1–54 in T.J. Givnish and K.J. Sytsma, Eds., *Molecular Evolution and Adaptive Radiation*. Cambridge University Press: Cambridge, UK.

Glor, R.E. 2003. Rediscovering the diversity of Dominican anoles. Pp. 141–152 in R.W. Henderson and R. Powell, Eds., *Islands and the Sea: Essays on Herpetological Exploration in the West Indies*. Society for the Study of Amphibians and Reptiles: Ithaca, NY.

Glor, R.E., A.S. Flecker, M.F. Benard, and A.G. Power. 2001a. Lizard diversity and agricultural disturbance in a Caribbean forest landscape. *Biodiversity and Conservation* 10:711–723.

Glor, R.E., M.E. Gifford, A. Larson, J.B. Losos, L. Rodríguez Schettino, A.R. Chamizo Lara, and T.R. Jackman. 2004. Partial island submergence and speciation in an adaptive radiation: A multilocus analysis of the Cuban green anoles. *Proceedings of the Royal Society of London B* 271:2257–2265.

Glor, R.E., J.J. Kolbe, R. Powell, A. Larson, and J.B. Losos. 2003. Phylogenetic analysis of ecological and morphological diversification in Hispaniolan trunk-ground anoles (*Anolis cybotes* group). *Evolution* 57:2383–2397.

Glor, R.E., J.B. Losos, and A. Larson. 2005. Out of Cuba: Overwater dispersal and speciation among lizards in the *Anolis carolinensis* subgroup. *Molecular Ecology* 14:2419–2432.

Glor, R.E., L.J. Vitt, and A. Larson. 2001b. A molecular phylogenetic analysis of diversification in Amazonian *Anolis* lizards. *Molecular Ecology* 10:2661–2668.

Glossip, D., and J.B. Losos. 1997. Ecological correlates of number of subdigital lamellae in anoles. *Herpetologica* 53:192–199.

Gnanamuthu, C.P. 1930. The mechanism of the throat-fan in a ground lizard, *Sitana ponticeriana* Cuv. *Records of the Indian Museum* 32:149–159.

Goldberg, S.R., C.R. Bursey, and H. Cheam. 1997. Helminths of 12 species of *Anolis* lizards (*Polychrotidae*) from the Lesser Antilles, West Indies. *Journal of the Helminthological Society of Washington* 64:248–257.

Goldwasser, L., and J. Roughgarden. 1993. Construction and analysis of a large Caribbean food web. *Ecology* 74:1216–1233.

Goodman, D. 1971. Differential selection of immobile prey among terrestrial and riparian lizards. *American Midland Naturalist* 86:217–219.

Gorman, G.C. 1968. The relationships of *Anolis* of the *roquet* species group (Sauria: Iguanidae)—III. Comparative study of display behavior. *Breviora* 284: 1–31.

Gorman, G.C. 1973. The chromosomes of the Reptilia, a cytotaxonomic interpretation. Pp. 349–424 in A.B. Chiarelli and E. Capanna, Eds., *Cytotaxonomy and Vertebrate Evolution*. Academic Press: London, UK.

Gorman, G.C. 1980. *Anolis occultus*, a small cryptic canopy lizard: Are there pair bonds? *Caribbean Journal of Science* 15:29–31.

Gorman, G.C., and L. Atkins. 1968. New karyotypic data for 16 species of *Anolis* (Sauria: Iguanidae) from Cuba, Jamaica, and the Cayman Islands. *Herpetologica* 24:13–21.

Gorman, G.C., and L. Atkins. 1969. The zoogeography of Lesser Antillean *Anolis* lizards—an analysis based upon chromosomes and lactic dehydrogenases. *Bulletin of the Museum of Comparative Zoology* 138:53–80.

Gorman, G.C., D.G. Buth, M. Soulé, and S.Y. Yang. 1980. The relationship of the *Anolis cristatellus* species group: Electrophoretic analysis. *Journal of Herpetology* 14:269–278.

Gorman, G.C., D. Buth, M. Soulé, and S.Y. Yang. 1983. The relationships of the Puerto Rican *Anolis*: Electrophoretic and karyotypic studies. Pp. 626–642 in A.G.J. Rhodin and K. Miyata, Eds., Museum of Comparative Zoology, Harvard University: Cambridge, MA.

Gorman, G.C., and R. Harwood. 1977. Notes on population density, vagility, and activity patterns of the Puerto Rican grass lizard, *Anolis pulchellus* (Reptilia, Lacertilia, Iguanidae). *Journal of Herpetology* 11:363–368.

Gorman, G.C., and S. Hillman. 1977. Physiological basis for climatic niche partitioning in two species of Puerto Rican *Anolis* (Reptilia, Lacertilia, Iguanidae). Journal of Herpetology 11:337–340.

Gorman, G.C., and Y.J. Kim. 1975. Genetic variation and genetic distance among populations of *Anolis* lizards on two Lesser Antillean island banks. *Systematic Zoology* 24:369–373.

Gorman, G.C., and Y.J. Kim. 1976. *Anolis* lizards of the eastern Caribbean: A case study in evolution. II. Genetic relationships and genetic variation of the *bimaculatus* group. *Systematic Zoology* 25:62–77.

Gorman, G.C., and P. Licht. 1974. Seasonality in ovarian cycles among tropical *Anolis* lizards. *Copeia* 55:360–369.

Gorman, G.C., C.S. Lieb, and R.H. Harwood. 1984. The relationships of *Anolis gadovi*: Albumin immunological evidence. *Caribbean Journal of Science* 20:145–152.

Gorman, G.C., and B. Stamm. 1975. The *Anolis* lizards of Mona, Redonda, and La Blanquilla: chromosomes, relationships, and natural history notes. *Journal of Herpetology* 9:197–205.

Gorman, G.C., A.C. Wilson, and M. Nakanishi. 1971. A biochemical approach towards the study of reptilian phylogeny: Evolution of serum albumin and lactic dehydrogenase. *Systematic Zoology* 20:167–185.

Gorman, G.C., and S.Y. Yang. 1975. A low level of backcrossing between the hybridizing *Anolis* lizards of Trinidad. *Herpetologica* 31:196–198.

Gotelli, N.J., and G.R. Graves. 1996. *Null Models in Ecology*. Smithsonian Institution Press: Washington, DC.

Gould, S.J. 1984. Toward the vindication of punctuational change. Pp. 9–34 in W.A. Berggren and J.A. Van Couvering, Eds., *Catastrophes and Earth History: The New Uniformitarianism*. Princeton University Press: Princeton, NJ.

Gould, S.J. 1989. *Wonderful Life: The Burgess Shale and the Nature of History*. W.W. Norton: New York, NY.

Gould, S.J. 1997. The paradox of the visibly irrelevant. *Natural History* 106(11):12–18, 60–66.

Gould, S.J. 2002. *The Structure of Evolutionary Theory*. Harvard University Press: Cambridge, MA.

Gould, S.J., N.L. Gilinsky, and R.Z. German. 1987. Asymmetry of lineages and the direction of evolutionary time. *Science* 236:1437–1441.

Graham, M.H. 2003. Confronting multicollinearity in ecological multiple regression. *Ecology* 84:2809–2815.

Grant, B.W., and A.E. Dunham. 1988. Thermally imposed constraints on the activity of the desert lizard *Sceloporus merriami*. Ecology 69:167–176.

Grant, P.R. 1986. *Ecology and Evolution of Darwin's Finches*. Princeton University Press: Princeton, NJ.

Grant, P.R., and I. Abbott. 1980. Interspecific competition, island biogeography and null hypotheses. *Evolution* 34:332–341.

Grant, P.R., and B.R. Grant. 1992. Hybridization of bird species. *Science* 256:193–197.

Grant, P.R., and B.R. Grant. 1996. Speciation and hybridization in island birds. *Philosophical Transactions of the Royal Society of London* 351:765–772.

Grant, P.R., and B.R. Grant. 2002. Unpredictable evolution in a 30-year study of Darwin's finches. *Science* 296:707–711.

Grant, P.R., and B.R. Grant. 2006a. Evolution of character displacement in Darwin's finches. *Science* 313:224–226.

Grant, P.R., and B.R. Grant. 2006b. Species before speciation is complete. *Annals of the Missouri Botanical Garden* 93:94–102.

Grant, P.R., and B.R. Grant. 2008. *How and Why Species Multiply: The Radiation of Darwin's Finches.* Princeton University Press: Princeton, NJ.

Grant, T., D.R. Frost, J.P. Caldwell, R. Gagliardo, C.F.B. Haddad, P.J.R. Kok, D.B. Means, B.P. Noonan, W.E. Schargel, and W.C. Wheeler. 2006. Phylogenetic systematics of dart-poison frogs and their relatives (Amphibia: Athesphatanura: Dendrobatidae). *Bulletin of the American Museum of Natural History* 299:1–262.

Grazulis, T.P. 2001. *The Tornado: Nature's Ultimate Windstorm.* University of Oklahoma Press: Norman, OK.

Greenberg, B., and G.K. Noble. 1944. Social behavior of the American chameleon, *Anolis carolinensis* Voight. *Physiological Zoology* 17:392–439.

Greenberg, N. 2002. Ethological aspects of stress in a model lizard, *Anolis carolinensis.* *Integrative and Comparative Biology* 42:526–540.

Greenberg, N. 2003. Sociality, stress, and the corpus striatum of the green *Anolis* lizard. *Physiology and Behavior* 79:429–440.

Greenberg, N., and D. Crews. 1990. Endocrine and behavioral responses to aggression and social dominance in the green anole lizard, *Anolis carolinensis.* *General and Comparative Endocrinology* 77:246–255.

Greenberg, N., and L. Hake. 1990. Hatching and neonatal behavior of the lizard, *Anolis carolinensis.* *Journal of Herpetology* 24:402–405.

Greene, B.T., D.T. Yorks, J.S. Parmerlee, Jr., R. Powell, and R.W. Henderson. 2002. Discovery of *Anolis sagrei* in Grenada with comments on its potential impact on native anoles. *Caribbean Journal of Science* 38:270–272.

Greene, H.W. 1986. Diet and arboreality in the emerald monitor, *Varanus prasinus*, with comments on the study of adaptation. *Fieldiana Zoology New Series* 31:1–12.

Greene, H.W. 1988. Species richness in tropical predators. Pp. 259–280 in F. Almeda and C.M. Pringle, Eds., *Tropical Rainforests: Diversity and Conservation.* California Academy of Sciences: San Francisco, CA.

Greene, H.W. 1994. Systematics and natural history, foundations for understanding and conserving biodiversity. *American Zoologist* 34:48–56.

Greene, H.W. 2005. Organisms in nature as a central focus for biology. *Trends in Ecology and Evolution* 20:23–27.

Greene, H. W., and F. M. Jaksić. 1983. Food niche relationships among sympatric predators: Effects of level of prey identification. *Oikos* 40:151–154.

Greene, H.W., and J.B. Losos. 1988. Systematics, natural history, and conservation. *Bioscience* 38:458–462.

Grosholz, E.D. 1992. Interactions of intraspecific, interspecific, and apparent competition with host-pathogen population dynamics. *Ecology* 73:507–514.

Gross, M.R. 1985. Disruptive selection for alternative life histories in salmon. *Nature* 313:47–48.

Gübitz, T., R.S. Thorpe, and A. Malhotra. 2005. The dynamics of genetic and morphological variation on volcanic islands. *Proceedings of the Royal Society of London B* 272:751–757.

Guisan, A., and N.E. Zimmerman. 2000. Predictive habitat distribution models in ecology. *Ecological Modelling* 135:147–186.

Guyer, C. 1988a. Food supplementation in a tropical mainland anole, *Norops humilis*: Demographic effects. *Ecology* 69:350–361.

Guyer, C. 1988b. Food supplementation in a tropical mainland anole, *Norops humilis*: Effects on individuals. *Ecology* 69:362–369.

Guyer, C. and M.A. Donnelly. 2005. *Amphibians and Reptiles of La Selva, Costa Rica, and the Caribbean slope: A Comprehensive Guide.* University of California Press: Berkeley, CA.

Guyer, C., and J.M. Savage. 1986. Cladistic relationships among anoles (Sauria: Iguanidae). *Systematic Zoology* 35:509–531.

Guyer, C., and J.M. Savage. 1992. Anole systematics revisited. *Systematic Biology* 41:89–110.

Haefner, J.W. 1988. Niche shifts in Greater Antillean *Anolis* communities: Effects of niche metric and biological resolution on null model tests. *Oecologia* 77:107–117.

Haldane, J.B.S. 1932. *The Causes of Evolution.* Harper and Brothers: London, UK.

Han, D., K. Zhou, and A.M. Bauer. 2004. Phylogenetic relationships among gekkotan lizards inferred from C-*mos* nuclear DNA sequences and a new classification of the Gekkota. *Biological Journal of the Linnean Society* 83:353–368.

Hardy, C.R. 2006. Reconstructing ancestral ecologies: Challenges and possible solutions. *Diversity and Distributions* 12:7–19.

Hardy, J.D. Jr. 1982. Biogeography of Tobago, West Indies, with special reference to amphibians and reptiles: A review. *Bulletin of the Maryland Herpetological Society* 18:37–142.

Harmon, L.J. 2005. *Competition and Community Structure in Day Geckos* (Phelsuma) *in the Indian Ocean.* Ph.D. Dissertation, Washington University: St. Louis, MO.

Harmon, L.J., and R. Gibson. 2006. Multivariate phenotypic evolution among island and mainland populations of the ornate day gecko, *Phelsuma ornata*. *Evolution* 60:2622–2632.

Harmon, L.J., L.L. Harmon, and C.G. Jones. 2007. Competition and community structure in diurnal arboreal geckos (genus *Phelsuma*) in the Indian Ocean. *Oikos* 116:1863–1878.

Harmon, L.J., J.J. Kolbe, J.M. Cheverud, and J.B. Losos. 2005. Convergence and the multidimensional niche. *Evolution* 59:409–421.

Harmon, L.J., and J.B. Losos. 2005. The effect of intraspecific sample size on Type I and Type II error rates in comparative studies. *Evolution* 59:2705–2710.

Harmon, L.J., J.B. Losos, J. Davies, R. Gillespie, J.L. Gittleman, W.B. Jennings, K. Kozak, A. Larson, M.A. McPeek, F. Moreno-Roarck, T. Near, A. Purvis, R.E. Ricklefs, D. Schluter, J.A. Schulte, II, O. Seehausen, B. Sidlauskas, O. Torres-Carvajal, J. Weir, and A.Ø. Mooers. In review. Constraints and the scaling of evolutionary rates.

Harmon, L.J., J. Melville, A. Larson, and J.B. Losos. 2008. The role of geography and ecological opportunity in the diversification of day geckos (*Phelsuma*). *Systematic Biology* 57:562–573.

Harmon, L.J., J.A. Schulte II, A. Larson, and J.B. Losos. 2003. Tempo and mode of evolutionary radiation in iguanian lizards. *Science* 301:961–964.

Harris, B.R., D.T. Yorks, C.A. Bohnert, J.S. Parmerlee, Jr., and R. Powell. 2004. Population densities and structural habitats in lowland populations of *Anolis* lizards on Grenada. *Caribbean Journal of Science* 40:31–40.

Harrison, R.G. 1998. Linking evolutionary pattern and process: The relevance of species concepts for the study of speciation. Pp. 19–31 in N. Greenberg and P.D. MacLean, Eds., *Endless Forms: Species and Speciation.* Oxford University Press: Oxford, UK.

Harvey, P.H., and M.D. Pagel. 1991. *The Comparative Method in Evolutionary Biology*. Oxford University Press: Oxford, UK.

Harvey, P.H., and A. Purvis. 1991. Comparative methods for explaining adaptations. *Nature* 351:619–624.

Hasegawa, M., T. Kusano, and K. Miyashita. 1988. Range expansion of *Anolis c. carolinensis* on Chichi-Jima, the Bonin Islands, Japan. *Japanese Journal of Herpetology* 12:115–118.

Hass, C.A., and S.B. Hedges. 1991. Albumin evolution in West Indian frogs of the genus *Eleutherodactylus* (Leptodactylidae): Caribbean biogeography and a calibration of the albumin immunological clock. *Journal of Zoology* 225:413–426.

Hass, C.A., S.B. Hedges, and L.R. Maxson. 1993. Molecular insights into the relationships and biogeography of West Indian anoline lizards. *Biochemical Systematics and Ecology* 21:97–114.

Hatcher, M.J., J.T.A. Dick, and A.M. Dunn. 2006. How parasites affect interactions between competitors and predators. *Ecology Letters* 9:1253–1271.

Heaney, L.R. 2007. Is a new paradigm emerging for oceanic island biogeography? *Journal of Biogeography* 34:753–757.

Heard, S.B., and D.L. Hauser. 1995. Key evolutionary innovations and their ecological mechanisms. *Historical Biology* 10:151–173.

Heatwole, H. 1968. Relationship of escape behavior and camouflage in anoline lizards. *Copiea* 1968:109–113.

Heatwole, H. 1977. Habitat selection in reptiles. Pp. 137–155 in C. Gans and D.W. Tinkle, Eds., *Biology of the Reptilia, Vol. 7*. Academic Press: New York, NY.

Heatwole, H., T.-H. Lin, E. Villalón, A. Muniz and A. Matta. 1969. Some aspects of the thermal ecology of Puerto Rican anoline lizards. *Journal of Herpetology* 3:65–77.

Heatwole, H., E. Ortiz, A.M. Diaz-Collazo, and A.R. Jiménez-Vélez. 1962. Aquatic tendencies in the Puerto Rican pasture-lizard *Anolis pulchellus*. *Herpetologica* 17:272–274.

Heatwole, H., and F. Torres. 1963. Escape of *Anolis cristatellus* by submerging in water. *Herpetologica* 19:223–224.

Hecht, M.K. 1951. Fossil lizards of the West Indian genus *Aristelliger* (Gekkonidae). *American Museum Novitates* 1538:1–34.

Hecht, M.K. 1952. Natural selection in the lizard genus *Aristelliger*. *Evolution* 6:112–124.

Hector, A.,B. Schmid, C. Beierkuhnlein, M.C. Caldeira, M. Diemer, P.G. Dimitrakopoulos, J.A. Finn, H. Freitas, P.S. Giller, J. Good, R. Harris, P. Högberg, K. Huss-Danell, J. Joshi, A. Jumpponen, C. Körner, P.W. Leadley, M. Loreau, A. Minns, C.P.H. Mulder, G. O'Donovan, S.J. Otway, J.S. Pereira, A. Prinz, D.J. Read, M. Scherer-Lorenzen, E.-D. Schulze, A.-S. Siamantziouras, E.M. Spehn, A.C. Terry, A.Y. Troumbis, F.I. Woodward, S. Yachi, and J.H. Lawton. 1999. Plant diversity and productivity experiments in European grasslands. *Science* 286:1123–11276.

Hedges, S.B. 1989. Evolution and biogeography of West Indian frogs of the genus *Eleutherodactylus*: Slow-evolving loci and the major groups. In C.A. Woods, Ed., *Biogeography of the West Indies: Past, Present, and Future*. E.J. Brill: Leiden, Netherlands.

Hedges, S.B. 2001. Biogeography of the West Indies: An overview. Pp. 15–33 in C.A. Woods and F.E. Sergile, Eds., *Biogeography of the West Indies: Patterns and Perspectives*. CRC Press: Boca Raton, FL.

Hedges, S.B., and K.L. Burnell. 1990. The Jamaican radiation of *Anolis* (Sauria: Iguanidae): An analysis of relationships and biogeography using sequential electrophoresis. *Caribbean Journal of Science* 26:31–44.

Hedges, S.B., W.E. Duellman, and M.P. Heinicke. 2008. New World direct-developing frogs (Anura: Terrarana): Molecular phylogeny, classification, biogeography, and conservation. *Zootaxa* 1737:1–182.

Hedges, S.B., and C.A. Woods. 1993. Caribbean hot spot. *Nature* 364:375.

Heinicke, M.P., W.E. Duellman, and S.B. Hedges. 2007. Major Caribbean and Central American frog faunas originated by ancient oceanic dispersal. *Proceedings of the National Academy of Sciences of the United States of America* 104:10092–10097.

Henderson, R.W., and B.I. Crother. 1989. Biogeographic patterns of predation in West Indian colubrid snakes. Pp. 479–518 in C.A. Wood, Ed., *Biogeography of the West Indies: Past, Present, and Future.* Sandhill Crane Press: Gainesville, FL.

Henderson, R.W., and H.S. Fitch. 1975. A comparative study of the structural and climatic habitats of *Anolis sericeus* (Reptilia: Iguanidae) and its syntopic congeners at four localities in southern Mexico. *Herpetologica* 31:459–471.

Henderson, R.W., and M.A. Nickerson. 1976. Observations on the behavioral ecology of three species of *Imantodes* (Reptilia, Serpentes, Colubridae). *Journal of Herpetology* 10:205–210.

Henderson, R.W., T.A. Noeske-Hallin, J.A. Ottenweiler, and A. Schwartz. 1987. On the diet of the boa *Epicrates striatus* on Hispaniola, with notes on *E. fordi* and *E. gracilis. Amphibia-Reptilia* 8:251–258.

Henderson, R.W., and R. Powell. 2001. Responses by the West Indian herpetofauna to human-influenced resources. *Caribbean Journal of Science* 37:40–50.

Henderson, R.W., and R. Powell. 2005. Geographic distribution. *Anolis sagrei* (brown anole). *Herpetological Review* 36:467.

Henderson, R.W., and R.A. Sajdak. 1996. Diets of West Indian racers (Colubridae: *Alsophis*): Composition and biogeographic implications. Pp. 327–338 in R. Powell and R.W. Henderson, Eds., *Contributions to West Indian Herpetology: A Tribute to Albert Schwartz.* Society for the Study of Amphibians and Reptiles: Ithaca, NY.

Henderson, R.W., A. Schwartz, and T.A. Noeske-Hallin. 1987. Food habits of three colubrid tree snakes (genus *Uromacer*) on Hispaniola. *Herpetologica* 43:241–248.

Hendry, A.P. 2001. Adaptive divergence and the evolution of reproductive isolation in the wild: An empirical demonstration using introduced sockeye salmon. *Genetica* 112–113:515–534.

Hendry, A.P., and M.T. Kinnison. 2001. An introduction to microevolution: Rate, pattern, process. *Genetica* 112–113:1–8.

Hennig, W. 1966. *Phylogenetic systematics.* University of Illinois Press: Urbana, IL.

Herrel, A., R. Joachim, B. Vanhooydonck, and D.J. Irschick. 2006. Ecological consequences of ontogenetic changes in head shape and bite performance in the Jamaican lizard *Anolis lineatopus. Biological Journal of the Linnean Society* 89:443–454.

Herrel, A., L.D. McBrayer, and P.M. Larson. 2007. Functional basis for sexual differences in bite force in the lizard *Anolis* carolinensis. *Biological Journal of the Linnean Society* 91:111–119.

Herrel, A., J.C. O'Reilly, and A.M. Richmond. 2002. Evolution of bite performance in turtles. *Journal of Evolutionary Biology* 15:1083–1094.

Herrel, A., R. Van Damme, B. Vanhooydonck, and F. de Bree. 2001. The implications of bite performance for diet in two species of lacertid lizard. *Canadian Journal of Zoology* 79:662–670.

Herrel, A., B. Vanhooydonck, R. Joachim, and D.J. Irschick. 2004. Frugivory in polychrotid lizards: Effects of body size. *Oecologia* 140:160–168.

Herrel, A., B. Vanhooydonck, J. Porck and D.J. Irschick. 2008. Anatomical basis of differences in locomotor behavior in *Anolis* lizards: a comparison between two ecomorphs. *Bulletin of the Museum of Comparative Zoology* 159:213–238.

Hertz, P.E. 1980a. Comparative physiological ecology of the sibling species *Anolis cybotes* and *A. marcanoi. Journal of Herpetology* 14:92–95.

Hertz, P.E. 1980b. Responses to dehydration in *Anolis* lizards sampled along altitudinal transects. *Copeia* 1980:440–446.

Hertz, P.E. 1981. Adaptation to altitude in two West Indian anoles (Reptilia: Iguanidae): Field thermal biology and physiological ecology. *Journal of Zoology* 195:25–37.

Hertz, P.E. 1983. Eurythermy and niche breadth in West Indian *Anolis* lizards: A reappraisal. Pp. 472–483 in A.G.J. Rhodin and K. Miyata, Eds., *Advances in Herpetology and Evolutionary Biology: Essays in Honor of Ernest E. Williams*. Museum of Comparative Zoology, Harvard University: Cambridge, MA.

Hertz, P.E. 1992a. Evaluating thermal resource partitioning in sympatric lizards *Anolis cooki* and *A. cristatellus*: A field test using null hypotheses. *Oecologia* 90:127–136.

Hertz, P.E. 1992b. Temperature regulation in Puerto Rican *Anolis* lizards: A field test using null hypotheses. *Ecology* 73:1405–1417.

Hertz, P.E., A. Arce-Hernandez, J. Ramirez-Vazquez, W. Tirado-Rivera, and L. Vazquez-Vives. 1979. Geographical variation of heat sensitivity and water loss rates in the tropical lizard, *Anolis gundlachi. Comparative Biochemistry and Physiology* 62A:947–953.

Hertz, P.E., L.J. Fleishman, and C. Armsby. 1994. The influence of light intensity and temperature on microhabitat selection in two *Anolis* lizards. *Functional Ecology* 8:720–729.

Hertz, P.E., and R.B. Huey. 1981. Compensation for altitudinal changes in the thermal environment by some *Anolis* lizards on Hispaniola. *Ecology* 62:515–521.

Hertz, P.E., R.B. Huey, and E. Nevo. 1982. Fight versus flight: Body temperature influences defensive responses of lizards. *Animal Behaviour* 30:676–679.

Hertz, P.E., R.B. Huey, and E. Nevo. 1983. Homage to Santa Anita: Thermal sensitivity of sprint speed in agamid lizards. *Evolution* 37:1075–1084.

Hertz, P.E., R.B. Huey, and R.D. Stevenson. 1993. Evaluating temperature regulation by field-active ectotherms: The fallacy of the inappropriate question. *American Naturalist* 142:796–818.

Hertz, P.E., R.B. Huey, and R.D. Stevenson. 1999. Temperature regulation in free-ranging ectotherms: What are the appropriate questions? *African Journal of Herpetology* 48:41–48.

Hertz, P.E., R.B. Huey, and T. Garland, Jr. 1988. Time budgets, thermoregulation, and maximal locomotory performance: Are reptiles Olympians or boy scouts? American Zoologist 28:927–938.

Hews, D.K., and R.A. Worthington. 2001. Fighting from the right side of the brain: Left visual field preference during aggression in free-ranging male tree lizards (*Urosaurus ornatus*). *Brain, Behavior and Evolution* 58:356–361.

Hicks, R., and T.A. Jenssen. 1973. New studies on a montane lizard of Jamaica, *Anolis reconditus*. *Breviora* 404:1–23.

Hicks, R.A., and R.L. Trivers. 1983. The social behavior of *Anolis valencienni*. Pp. 570–595 in A.G.J. Rhodin and K. Miyata, Eds., *Advances in Herpetology and Evolutionary Biology: Essays in Honor of Ernest E. Williams*. Museum of Comparative Zoology, Harvard University: Cambridge, MA.

Higham, T.E., M.S. Davenport, and B.C. Jayne. 2001. Maneuvering in an arboreal habitat: The effects of turning angle on the locomotion of three sympatric ecomorphs of *Anolis* lizards. *Journal of Experimental Biology* 204:4141–4155.

Hilburn, D.J., and R.L. Dow. 1990. Mediterranean fruit fly, *Ceratitis capitata*, eradicated from Bermuda. *Florida Entomologist* 73:342–343.

Hildebrand, M. 1985. Walking and running. Pp. 38–57 in M. Hildebrand, D.M. Bramble, K.F. Liem, and D.B. Wake, Eds., *Functional Vertebrate Morphology*. Belknap Press: Cambridge, MA.

Hiller, U. 1975. Comparative studies on the functional morphology of two gekkonid lizards. *Journal of the Bombay Natural History Society* 73:278–282.

Hillis, D.M., C. Moritz, and B.K. Mable. 1996. Eds., *Molecular Systematics*. Sinauer Associates: Sunderland, MA.

Hillman, S.S., and G.C. Gorman. 1977. Water loss, desiccation tolerance, and survival under desiccating conditions in 11 species of Caribbean *Anolis*: Evolutionary and ecological implications. *Oecologia* 29:105–116.

Hillman, S., G.C. Gorman, and R. Thomas. 1979. Water loss in *Anolis* lizards: Evidence for acclimation and intraspecific differences along a habitat gradient. *Comparative Biochemistry and Physiology* 62A:491–494.

Hite, J.L., C.A. Rodríguez Gómez, S.C. Larimer, A.M. Díaz-Lameiro, and R. Powell. 2008. Anoles of St. Vincent (Squamata: Polychrotidae): Population Densities and Structural Habitat Use. *Caribbean Journal of Science* 44:102–115.

Hoekstra, H.E. 2006. Genetics, development and evolution of adaptive pigmentation in vertebrates. *Heredity* 97:222–234.

Hoekstra, H.E., R.J. Hirschmann, R.A. Bundey, P.A. Insel, and J.P. Crossland. 2006. A single amino acid mutation contributes to adaptive beach mouse color pattern. *Science* 313:101–104.

Hoekstra, H.E., J.M. Hoekstra, D. Berrigan, S.N. Vignieri, A. Hoang, C.E. Hill, P. Beerli, and J.G. Kingsolver. 2001. Strength and tempo of directional selection in the wild. *Proceedings of the National Academy of Sciences of the United States of America* 98:9157–9160.

Hoekstra, H.E., and M.W. Nachman. 2003. Different genes underlie adaptive melanism in different populations of rock pocket mice. *Molecular Ecology* 12:1185–1194.

Holland, B., and W.R. Rice. 1998. Chase-away and sexual selection: Antagonistic seduction versus resistance. *Evolution* 52:1–7.

Holmes, M.M., and J. Wade. 2004. Seasonal plasticity in the copulatory neuromuscular system of green anole lizards: A role for testosterone in muscle but not motoneuron morphology. *Journal of Neurobiology* 60:1–11.

Holt, R.D. 1977. Predation, apparent competition and the structure of prey communities. *Theoretical Population Biology* 12:197–229.

Holt, R.D. 1984. Spatial heterogeneity, indirect interactions, and the coexistence of prey species. *American Naturalist* 124:377–406.

Holt, R.D., and J.H. Lawton. 1994. The ecological consequences of shared natural enemies. *Annual Review of Ecology and Systematics* 25:495–520.

Hosken, D., and R. Snook. 2005. How important is sexual conflict? *American Naturalist* 165:S1–S4.

Hotton, N., III. 1955. A survey of the adaptive relationships of dentition to diet in the North American Iguanidae. *American Midland Naturalist* 53:88–114.

Hover, E.L., and T.A. Jenssen. 1976. Descriptive analysis and social correlates of agonistic displays of *Anolis limifrons* (Sauria, Iguanidae). *Behaviour* 58:173–191.

Hoverman, J.T., J.R. Auld, and R.A. Relyea. 2005. Putting prey back together again: Integrating predator–induced behavior, morphology, and life history. *Oecologia* 144:481–491.

Howard, A.K., J.D. Forester, J.M. Ruder, J.S. Parmerlee, Jr., and R. Powell. 1999. Natural history of a terrestrial Hispaniolan anole: *Anolis barbouri*. *Journal of Herpetology* 33:702–706.

Hrbek, T., J. Seckinger and A. Meyer. 2007. A phylogenetic and biogeographic perspective on the evolution of poeciliid fishes. *Molecular Phylogenetics and Evolution* 43:986–998.

Hudson, S. 1996. Natural toe loss in southeastern Australian skinks: Implications for markings lizards by toe-clipping. *Journal of Herpetology* 30:106–110.

Huelsenbeck, J.P., R. Nielsen, and J.P. Bollback. 2003. Stochastic mapping of morphological characters. *Systematic Biology* 52:131–158.

Huelsenbeck, J.P., B. Rannala, and J.P. Masly. 2000. Accommodating phylogenetic uncertainty in evolutionary studies. *Science* 288:2349–2350.

Huey, R.B. 1974. Behavioral thermoregulation in lizards: Importance of associated costs. *Science* 184:1001–1003.

Huey, R.B. 1982. Temperature, physiology, and the ecology of reptiles. Pp. 25–91 in C. Gans and F.H. Pough, Eds., *Biology of the Reptilia, Vol. 12. Physiology (C)*. Academic Press: London, UK.

Huey, R.B. 1983. Natural variation in body temperature and physiological performance in a lizard (*Anolis cristatellus*). Pp. 484–490 in A.G.J. Rhodin and K. Miyata, Eds., *Advances in Herpetology and Evolutionary Biology: Essays in Honor of Ernest E. Williams*. Museum of Comparative Zoology, Harvard University: Cambridge, MA.

Huey, R.B. 1991. Physiological consequences of habitat selection. *American Naturalist* 137:S91–S115.

Huey, R.B., and A.F. Bennett. 1987. Phylogenetic studies of coadaptation: Preferred temperatures versus optimal performance temperatures of lizards. *Evolution* 41:1098–1115.

Huey, R.B., A.E. Dunham, K.L. Overall, and R.A. Newman. 1990. Variation in locomotor performance in demographically known populations of the lizard *Sceloporus merriami*. *Physiological Zoology* 63:845–872.

Huey, R.B., G.W. Gilchrist, M.L. Carlson, D. Berrigan, and L. Serra. 2000. Rapid evolution of a geographic cline in size in an introduced fly. *Science* 287:308–309.

Huey, R.B., and P.E. Hertz. 1984. Is a jack-of-all-temperatures a master of none? *Evolution* 38:441–444.

Huey, R.B., P.E. Hertz, and B. Sinervo. 2003. Behavioral drive versus behavioral inertia in evolution: A null model approach. *American Naturalist* 161:357–366.

Huey, R.E., W. Schneider, G.L. Erie, and R.D. Stevenson. 1981. A field-portable racetrack and timer for measuring acceleration and speed of small cursorial animals *Experientia* 37:1356–1357.

Huey, R.B., and M. Slatkin. 1976. Cost and benefits of lizard thermoregulation. *Quarterly Review of Biology* 51:363–384.

Huey, R.B., and R.D. Stevenson. 1979. Integrating thermal physiology and ecology of ectotherms: A discussion of approaches. *American Zoologist* 19:357–366.

Huey, R.B., and T.P. Webster. 1975. Thermal biology of a solitary lizard: *Anolis marmoratus* of Guadeloupe, Lesser Antilles. *Ecology* 56:445–452.

Huey, R.B., and T.P. Webster. 1976. Thermal biology of *Anolis* lizards in a complex fauna: The *cristatellus* group on Puerto Rico. *Ecology* 57:985–994.

Hug, L.A., and A.J. Roger. 2007. The impact of fossils and taxon sampling on ancient molecular dating analyses. *Molecular Biology and Evolution* 24:1889–1897.

Hugall, A.F., and M.S.Y. Lee. 2004. Molecular claims of Gondwanan age for Australian agamid lizards are untenable. *Molecular Biology and Evolution* 21:2102–2110.

Hugall, A., C. Moritz, A. Moussalli, and J. Stanisic. 2002. Reconciling paleodistribution models and comparative phylogeography in the Wet Tropics rainforest land snail *Gnarosophia bellendenkerensis* (Brazier 1875). *Proceedings of the National Academy of Sciences of the United States of America* 99:6112–6117.

Hughes, C. 1998. Integrating molecular techniques with field methods in studies of social behavior: A revolution results. *Ecology* 79:383–399.

Hulebak, E., S. Poe, R. Ibáñez, and E.E. Williams 2007. A striking new species of *Anolis* lizard (Squamata, Iguania) from Panama. *Phyllomedusa* 6:5–10.

Hunsaker, D., II, and P. Breese. 1967. Herpetofauna of the Hawaiian Islands. *Pacific Science* 21:423–428.

Hunter, J.P. 1998. Key innovations and the ecology of macroevolution. *Trends in Ecology and Evolution* 13:31–36.

Husak, J.F. 2006. Does survival depend on how fast you *can* run or how fast you *do* run? *Functional Ecology* 20:1080–1086.

Husak, J.F., S.F. Fox, M.B. Lovern, and R.A. Van den Bussche. 2006a. Faster lizards sire more offspring: Sexual selection on whole-animal performance. *Evolution* 60:2122–2130.

Husak, J.F., J.M. Macedonia, S.F. Fox, and R.C. Sauceda. 2006b. Predation cost of conspicuous male coloration in collared lizards (*Crotaphytus collaris*): An experimental test using clay-covered model lizards. *Ethology* 112:572–580.

Huyghe, K., A. Herrel, B. Vanhooydonck, J.J. Meyers, and D.J. Irschick. 2007. Microhabitat use, diet, and performance data on the Hispaniolan twig anole, *Anolis sheplani*: Pushing the boundaries of morphospace. *Zoology* 110:2–8.

Huyghe, K., B. Vanhooydonck, H. Scheers, M. Molina-Borja, and R. Van Damme. 2005. Morphology, performance and fighting capacity in male lizards, *Gallotia galloti*. *Functional Ecology* 19:800–807.

Inger, R.F. 1983. Morphological and ecological variation in the flying lizards (genus *Draco*). *Fieldiana Zoology New Series* 18:1–35.

Irschick, D.J. 2000. Effects of behaviour and ontogeny on the locomotor performance of a West Indian lizard, *Anolis lineatopus*. *Functional Ecology* 14:438–444.

Irschick, D.J. 2003. Measuring performance in nature: implications for studies of fitness within populations. *Integrative and Comparative Biology* 43:396–407.

Irschick, D.J., C.C. Austin, K. Petren, R.N. Fisher, J.B. Losos, and O. Ellers. 1996. A comparative analysis of clinging ability among pad-bearing lizards. *Biological Journal of the Linnean Society* 59:21–35.

Irschick, D.J., and T. Garland, Jr. 2001. Integrating function and ecology in studies of adaptation: Investigations of locomotory capacity as a model system. *Annual Review of Ecology and Systematics* 32:367–396.

Irschick, D.J., G. Gentry, A. Herrel, and B. Vanhooydonck. 2006a. Effects of sarcophagid fly infestations on green anole lizards (*Anolis carolinensis*): An analysis across seasons and age/sex classes. *Journal of Herpetology* 40:107–112.

Irschick, D.J., A. Herrel, and B. Vanhooydonck. 2006b. Whole-organism studies of adhesion in pad-bearing lizards: Creative evolutionary solutions to functional problems. *Journal of Comparative Physiology A* 192:1169–1177.

Irschick, D.J., and B.C. Jayne. 1999. Comparative three-dimensional kinematics of the hindlimb for high-speed bipedal and quadrupedal locomotion of lizards. *Journal of Experimental Biology* 202:1047–1065.

Irschick, D.J., and J.B. Losos. 1996. Morphology, ecology, and behavior of the twig anole, *Anolis angusticeps*. Pp. 291–301 in R. Powell and R.W. Henderson, Eds., *Contributions to West Indian Herpetology: A Tribute to Albert Schwartz*. Society for the Study of Amphibians and Reptiles: Ithaca, NY.

Irschick, D.J., and J.B. Losos. 1998. A comparative analysis of the ecological significance of maximal locomotor performance in Caribbean *Anolis* lizards. *Evolution* 52:219–226.

Irschick, D.J., and J.B. Losos. 1999. Do lizards avoid habitats in which performance is submaximal? The relationship between sprinting capabilities and structural habitat use in Caribbean anoles. *American Naturalist* 154:293–305.

Irschick, D.J., T.E. Macrini, S. Koruba, and J. Forman. 2000. Ontogenetic differences in morphology, habitat use, behavior, and sprinting capacity in two West Indian *Anolis* lizard species. *Journal of Herpetology* 34:444–451.

Irschick, D.J., L.J. Vitt, P. Zani, and J. B. Losos. 1997. A comparison of evolutionary radiations in mainland and West Indian *Anolis* lizards. *Ecology* 78:2191–2203.

Irwin, D.E. 2002. Phylogeographic breaks without geographic barriers to gene flow. *Evolution* 56:2383–2394.

Iturralde-Vinent, M.A. 2001. Geology of the amber-bearing deposits of the Greater Antilles. *Caribbean Journal of Science* 17:141–167.

Iturralde-Vinent, M.A. 2006. Meso-Cenozoic Caribbean paleogeography: Implications for the historical biogeography of the region. *International Geology Review* 48:791–827.

Iturralde-Vinent, M.A., and R.D.E. MacPhee. 1999. Paleogeography of the Caribbean region: Implications for Cenozoic biogeography. *Bulletin of the American Museum of Natural History* 238:1–95.

Jackman, T.R., D.J. Irschick, K. de Queiroz, J.B. Losos, and A. Larson. 2002. Molecular phylogenetic perspective on evolution of lizards of the *Anolis grahami* Series. *Journal of Experimental Zoology: Molecular and Developmental Evolution* 294:1–16.

Jackman, T.R., A. Larson, K. de Queiroz, and J.B. Losos. 1999. Phylogenetic relationships and tempo of early diversification in *Anolis* lizards. *Systematic Biology* 48:254–285.

Jackman, T.R., J.B. Losos, A. Larson, and K. de Queiroz. 1997. Phylogenetic studies of convergent adaptive radiation in Caribbean *Anolis* lizards. Pp. 535–557 in T.J. Givnish and K.J. Sytsma, Eds., *Molecular Evolution and Adaptive Radiation*. Cambridge University Press: Cambridge, UK.

Jackson, M.H. 1994. *Galápagos: A Natural History*, 2nd Ed. University of Calgary Press: Alberta, Canada.

James, K.H. 2006. Arguments for and against the Pacific origin of the Caribbean Plate: Discussion, finding for an inter-American origin. *Geologica Acta* 4:279–302.

James, R.S., C.A. Navas, and A. Herrel. 2007. How important are skeletal muscle mechanics in setting limits on jumping performance? *Journal of Experimental Biology* 210:923–933.

Janzen, D. 1998. Tropical dry forests: The most endangered major tropical ecosystem. Pp. 130–137 in E.O. Wilson, Ed., *Biodiversity*. National Academy Press: Washington, DC.

Jayne, B.C., and A.F. Bennett. 1990. Selection on locomotor performance capacity in a natural population of garter snakes. *Evolution* 44:1204–1229.

Jayne, B.C., and D.J. Irschick. 2000. A field study of incline use and preferred speeds for the locomotion of lizards. *Ecology* 81:2969–2983.

Jenssen, T.A. 1970a. Female response to filmed displays of *Anolis nebulosus* (Sauria, Iguanidae). *Animal Behaviour* 18:640–647.

Jenssen, T.A. 1970b. The ethoecology of *Anolis nebulosus* (Sauria, Iguanidae). *Journal of Herpetology* 4:1–38.

Jenssen, T.A. 1973. Shift in the structural habitat of *Anolis opalinus* due to congeneric competition. Ecology 54:863–869.

Jenssen, T.A. 1977. Evolution of anoline lizard display behavior. *American Zoologist* 17:203–215.

Jenssen, T.A. 1978. Display diversity in anoline lizards and problems of interpretation. Pp. 269–285 in N. Greenberg and P.D. MacLean, Eds., *Behavior and Neurology of Lizards*. National Institute of Mental Health: Rockville, MD.

Jenssen, T.A. 1979a. Display behavior of male *Anolis opalinus* (Sauria, Iguanidae): A case of weak display stereotypy. *Animal Behaviour* 27:173–184.

Jenssen, T.A. 1979b. Display modifiers of *Anolis opalinus* (Lacertilia: Iguanidae). *Herpetologica* 35:21–30.

Jenssen, T.A. 1996. A test of assortative mating between sibling lizard species, *Anolis websteri* and *A. caudalis*, in Haiti. Pp. 303–316 in R. Powell and R.W. Henderson, Eds., *Contributions to West Indian Herpetology: A Tribute to Albert Schwartz*. Society for the Study of Amphibians and Reptiles: Ithaca, NY.

Jenssen, T.A., K.R. DeCourcy, and J.D. Congdon. 2005. Assessments in contests of male lizards (*Anolis carolinensis*): How should smaller males respond when size matters? *Animal Behaviour* 69:1325–1336.

Jenssen, T.A., and P.C. Feely. 1991. Social behavior of the male anoline lizard *Chamaelinorops barbouri*, with a comparison to *Anolis*. *Journal of Herpetology* 25:454–461.

Jenssen, T.A., and N.L. Gladson. 1984. A comparative display analysis of the *Anolis brevirostris* complex in Haiti. *Journal of Herpetology* 18:217–230.

Jenssen, T.A., N. Greenberg, and K.A. Hovde. 1995. Behavioral profile of free-ranging male lizards, *Anolis carolinensis*, across breeding and post-breeding seasons. *Herpetological Monographs* 8:41–62.

Jenssen, T.A., K.A. Hovde, and K.G. Taney. 1998. Size-related habitat use by nonbreeding *Anolis carolinensis* lizards. *Copeia* 1998:774–779.

Jenssen, T.A., M.B. Lovern, and J.D. Congdon. 2001. Field-testing the protandry-based mating system for the lizard, *Anolis carolinensis*: Does the model organism have the right model? *Behavioral Ecology and Sociobiology* 50:162–172.

Jenssen, T.A., D.L. Marcellini, C.A. Pague, and L.A. Jenssen. 1984. Competitive interference between two Puerto Rican lizards, *Anolis cooki* and *Anolis cristatellus*. *Copeia* 1984:853–861.

Jenssen, T.A., and S.C. Nunez. 1994. Male and female reproductive cycles of the Jamaican lizard, *Anolis opalinus*. *Copeia* 1994:767–780.

Jenssen, T.A., and S.C. Nunez. 1998. Spatial and breeding relationships of the lizard, *Anolis carolinensis*: Evidence of intrasexual selection. *Behaviour* 135:981–1003.

Jenssen, T.A., K.S. Orrell, and M.B. Lovern. 2000. Sexual dimorphisms in aggressive signal structure and use by a polygynous lizard, *Anolis carolinensis*. *Copeia* 2000:140–149.

Johns, G.C., and J.C. Avise. 1998. A comparative summary of genetic distances in the vertebrates from the mitochondrial cytochrome *b* gene. *Molecular Biology and Evolution* 15: 1481–1490.

Johnson, M. 2007. *Behavioral Ecology of Caribbean* Anolis *Lizards: A Comparative Approach*. Ph.D. Dissertation, Washington University: Saint Louis, MO.

Johnson, M.A., M. Leal, L. Rodríguez Schettino, A. Chamizo Lara, L.J. Revell, and J.B. Losos. 2008. A phylogenetic perspective on foraging mode evolution and habitat use in West Indian *Anolis* lizards. *Animal Behaviour* 75:555–563.

Johnson, M.A., R. Kirby, S. Wang, and J.B. Losos. 2006. What drives variation in habitat use by *Anolis* lizards: Habitat availability or selectivity? *Canadian Journal of Zoology* 84:877–886.

Jones, J.K. Jr. 1989. Distribution and systematics of bats in the Lesser Antilles. Pp. 645–660 in C.A. Woods, Ed. *Biogeography of the West Indies: Past, Present, and Future*. Sandhill Crane Press: Gainesville, FL.

Jones, R.E., L.J. Guillette, Jr., C.H. Summers, R.R. Tokarz, and D. Crews. 1983. The relationship among ovarian condition, steroid hormones, and estrous behavior in *Anolis carolinensis*. *Journal of Experimental Zoology* 227:145–154.

Jones, R.E., K.H. Lopez, T.A. Maldonado, T.R. Summers, C.H. Summers, C.R. Propper, and J.D. Woodling. 1997. Unilateral ovariectomy influences hypothalamic monoamine asymmetries in a lizard (*Anolis*) that exhibits alternation of ovulation. *General and Comparative Endocrinology* 108:306–315.

Joyce, D.A., D.H. Lunt, R. Bills, G.F. Turner, C. Katongo, N. Duftner, C. Sturmbauer, and O. Seehausen. 2005. An extant cichlid fish radiation emerged in an extinct Pleistocene lake. *Nature* 435:90–95.

Kaiser, H., D.M.Green, and M. Schmid. 1994. Systematics and biogeography of eastern Caribbean frogs (Leptodactylidae: *Eleutherodactylus*), with a description of a new species from Dominica. *Canadian Journal of Zoology* 72:2217–2237.

Karr, J.R., and F.C. James. 1975. Eco-morphological configurations and convergent evolution in species and communities. Pp. 258–291 in M.L. Cody and J.M. Diamond, Eds., *Ecology and Evolution of Communities*. Belknap Press: Cambridge, MA.

Kassen, R.,M. Llewellyn, and P.B. Rainey. 2004. Ecological constraints on diversification in a model adaptive radiation. *Nature* 431:984–988.

Kästle, W. 1998. Studies on the ecology and behaviour of *Sitana sivalensis* spec. nov. *Veröffentlichungen aus dem Fuhlrott-Museum* 4:121–206.

Kattan, G.H. 1984. Sleeping perch selection in the lizard *Anolis ventrimaculatus*. *Biotropica* 16:328–329.

Kattan, G.H., and H.B. Lillywhite. 1989. Humidity acclimation and skin permeability in the lizard *Anolis carolinensis*. *Physiological Zoology* 262:593–606.

Kearney, M., and W.P. Porter. 2004. Mapping the fundamental niche: Physiology, climate, and the distribution of a nocturnal lizard. *Ecology* 85:3119–3131.

Kelt, D.A., J.H. Brown, E.J. Heske, P.A. Marquet, S.R. Morton, J.W. Reid, K.A. Rogovin, and G. Shenbrot. 1996. Community structure of desert small mammals: Comparisons across four continents. *Ecology* 77:746–761.

Kiester, A.R. 1979. Conspecifics as cues: A mechanism for habitat selection in the Panamanian grass anole (*Anolis auratus*). *Behavioral Ecology and Sociobiology* 5:323–330.

Kiester, A.R., G.C. Gorman, and D.C. Arroyo. 1975. Habitat selection behavior of three species of *Anolis* lizards. *Ecology* 56: 220–225.

Kingsolver, J.G., H.E. Hoekstra, J.M. Hoekstra, D. Berrigan, S.N. Vignieri, C.E. Hill, A. Hoang, P. Gilbert, and P. Beerli. 2001. The strength of phenotypic selection in natural populations. *American Naturalist* 157:245–261.

Kireeva, G.D. 1958. Some ecological morphology of *Schwagerina* of the Bafumutskoi Basin and Donetz Basin. Problem of micropaleontology. *Academy of Sciences USSR*, Report 2:9–41.

Kirschner, J., and M. Gerhart. 1998. *Cells, Embryos, and Evolution: Toward a Cellular and Developmental Understanding of Phenotypic Variation and Evolutionary Adaptability*. Blackwell Publishing: Oxford, UK.

Kitazoe, Y., H. Kishino, P.J. Waddell, N. Nakajima, T. Okabayashi, T. Watabe, and Y. Okuhara. 2007. Robust time estimation reconciles views of the antiquity of placental mammals. *PLoS One* 2:e384.

Klecka, W.R. 1980. *Discriminant Analysis*. Sage University Paper: Beverly Hills, CA.

Knouft, J.H., J.B. Losos, R.E. Glor, and J.J. Kolbe. 2006. Phylogenetic analysis of the evolution of the niche in lizards of the *Anolis sagrei* group. *Ecology* 87:S29–S38.

Knowles, L.L., and C.L. Richards. 2005. Importance of genetic drift during Pleistocene divergence as revealed by analyses of genomic variation. *Molecular Ecology* 14:4023–4032.

Knox, A.K., J.B. Losos, and C.J. Schneider. 2001. Adaptive radiation versus intraspecific differentiation: Morphological variation in Caribbean *Anolis* lizards. *Journal of Evolutionary Biology* 14:904–909.

Kocher, T.D. 2004. Adaptive evolution and explosive speciation: The cichlid fish model. *Nature Reviews Genetics* 5:288–298.

Köhler, G. 2003. *Reptiles of Central America*. Herpeton: Offenbach, Germany.

Köhler, G. 2005. *Incubation of Reptile Eggs*. Krieger Publishing Company: Malabar, FL.

Köhler, G., J.R. McCranie, and L.D. Wilson. 2001. A new species of anole from western Honduras (Squamata: Polychrotidae). *Herpetologica* 57:247–255.

Köhler, G., M. Ponce, J. Sunyer, and A. Batista. 2007. Four new species of anoles (genus *Anolis*) from the Serranía de Tabasará, West-Central Panama (Squamata: Polychrotidae). *Herpetologica* 63:375–391.

Köhler, G., and J. Sunyer. 2008. Two new species of anoles formerly referred to as *Anolis limifrons* (Squamata: Polychrotidae). *Herpetologica* 64:92–108.

Kolbe, J.J., P.L. Colbert, and B.E. Smith. 2008a. Niche relationships and interspecific interactions in Antiguan lizard communities. *Copeia* 2008:261–272.

Kolbe, J.J., R.E. Glor, L. Rodríguez Schettino, A. Chamizo Lara, A. Larson, and J. B. Losos. 2004. Genetic variation increases during biological invasion by a Cuban lizard. *Nature* 431:177–181.

Kolbe, J.J., R.E. Glor, L. Rodríguez Schettino, A. Chamizo Lara, A. Larson, and J.B. Losos. 2007a. Multiple sources, admixture, and genetic variation in introduced *Anolis* lizard populations. *Conservation Biology* 21:1612–1625.

Kolbe, J.J., A. Larson, J.B. Losos, and K. de Queiroz. 2008. Admixture determines genetic diversity and population differentiation in the biological invasion of a lizard. *Biology Letters* 4:434–437.

Kolbe, J.J., A. Larson, and J.B. Losos. 2007b. Differential admixture shapes morphological variation among invasive populations of the lizard, *Anolis sagrei*. *Molecular Ecology* 16:1579–1591.

Kolbe, J.J., and J.B. Losos. 2005. Hind-limb length plasticity in *Anolis carolinensis*. *Journal of Herpetology* 39:674–678.

Komorowski, J.-C., G. Boudon, M. Semet, F. Beauducel, C. Anténor-Habazac, S. Bazin and G. Hammouya. 2005. Guadeloupe. Pp. 65–102 in J.M. Lindsay, R. Robertson, J. Shepherd, and S. Ali, Eds., *Volcanic Hazard Atlas of the Lesser Antilles*. Seismic Research Unit, University of the West Indies: Trinidad and Tobago.

Kornfield, I., and P.F. Smith. 2000. African cichlid fishes: Model systems for evolutionary biology. *Annual Review of Ecology and Systematics* 31:163–196.

Kozak, K.H., R.A. Blaine, and A. Larson. 2006. Gene lineages and eastern North American palaeodrainage basins: Phylogeography and speciation in salamanders of the *Eurycea bislineata* species complex. *Molecular Ecology* 15:191–207.

Kozak, K.H., and J.J. Wiens. 2006. Does niche conservatism promote speciation? A case study in North American salamanders. *Evolution* 60:2604–2621.

Kruuk, L.E.B., J. Merilá, and B.C. Sheldon. 2001. Phenotypic selection on a heritable size trait revisited. *American Naturalist* 158:557–571.

Krysko, K.L., A.N. Hooper, and C.M. Sheehy, III. 2003. The Madagascar giant day gecko, *Phelsuma madagascariensis grandis* Gray 1870 (Sauria: Gekkonidae): A new established species in Florida. *Florida Scientist* 66:222–225.

Lack, D. 1947. *Darwin's Finches*. Cambridge University Press: Cambridge, UK.

Lailvaux, S.P., A. Herrel, B. Vanhooydonck, J.J. Meyers and D.J. Irschick. 2004. Performance capacity, fighting tactics, and the evolution of life-stage morphs in the green anole lizard (*Anolis carolinensis*). *Proceedings of the Royal Society of London B* 271:2501–2508.

Lailvaux, S.P., and D.J. Irschick. 2006. No evidence for female association with high-performance males in the green anole lizard, *Anolis carolinensis*. *Ethology* 112:707–715.

Lailvaux, S.P., and D.J. Irschick. 2007a. Effects of temperature and sex on jump performance and biomechanics in the lizard *Anolis carolinensis*. *Functional Ecology* 21:534–543.

Lailvaux, S.P., and D.J. Irschick. 2007b. The evolution of performance-based male fighting ability in Caribbean *Anolis* lizards. *American Naturalist* 170:573–586.

Lande, R. 1981. Models of speciation by sexual selection on polygenic traits. *Proceedings of the National Academy of Sciences of the United States of America* 78:3721–3725.

Lande, R. 1986. The dynamics of peak shifts and the pattern of morphological evolution. *Paleobiology* 12:343–354.

Lande, R., and S.J. Arnold. 1983. The measurement of selection on correlated characters. *Evolution* 37:1210–1226.

Landry, L., W.F. Vincent, and L. Bernatchez. 2007. Parallel evolution of lake whitefish dwarf ecotypes in association with limnological features of their adaptive landscape. *Journal of Evolutionary Biology* 20:971–984.

Langerhans, R.B. 2006. Evolutionary consequences of predation: Avoidance, escape, reproduction, and diversification. Pp. 177–220 in A.M.T. Elewa, Ed., *Predation in Organisms: A Distinct Phenomenon.* Springer Verlag: Heidelberg, Germany.

Langerhans, R.B., J.H. Knouft, and J.B. Losos. 2006. Shared and unique features of diversification in Greater Antillean *Anolis* ecomorphs. *Evolution* 60:362–369.

Lappin, A.K., and J.F. Husak. 2005. Weapon performance, not size, determines mating success and potential reproductive output in the collared lizard (*Crotaphytus collaris*). *American Naturalist* 166:426–436.

Larimer, S.C., R. Powell, and J.S. Parmerlee, Jr. 2006. Effects of structural habitat on the escape behavior of the lizard, *Anolis gingivinus. Amphibia-Reptilia* 27:569–574.

Larson, A., and J.B. Losos. 1996. Phylogenetic systematics of adaptation. Pp. 187–220 in M.R. Rose and G.V. Lauder, Eds., *Adaptation.* Academic Press: San Diego, CA.

Laska, A.L. 1970. The structural niche of *Anolis scriptus* on Inagua. *Breviora* 349:1–6.

Lauder, G.V. 1981. Form and function: Structural analysis in evolutionary morphology. *Paleobiology* 7:430–442.

Lauder, G.V., and K.F. Liem. 1989. The role of historical factors in the evolution of complex organismal functions. In D.B. Wake and G. Roth, Eds., *Complex Organismal Functions: Integration and Evolution.* Pp. 63–78. John Wiley: New York, NY.

Lazell, J.D. Jr. 1964. The anoles (Sauria: Iguanidae) of the Guadeloupéen archipelago. *Bulletin of the Museum of Comparative Zoology* 131:359–401.

Lazell, J.D. Jr. 1965. An *Anolis* (Sauria, Iguanidae) in amber. *Journal of Paleontology* 39:379–382.

Lazell, J.D. Jr. 1966. Studies on *Anolis reconditus* Underwood and Williams. *Bulletin of the Institute of Jamaica Science Series* 18:1–15.

Lazell, J.D. Jr. 1969. The genus *Phenacosaurus* (Sauria: Iguanidae). *Breviora* 325:1–24.

Lazell, J.D. Jr. 1972. The anoles (Sauria: Iguanidae) of the Lesser Antilles. *Bulletin of the Museum of Comparative Zoology* 143:1–115.

Lazell, J.D. Jr. 1983. Biogeography of the herpetofauna of the British Virgin Islands, with description of a new anole (Sauria: Iguanidae). Pp. 99–117 in A.G.J. Rhodin and K. Miyata, Eds., *Advances in Herpetology and Evolutionary Biology: Essays in Honor of Ernest E. Williams.* Museum of Comparative Zoology, Harvard University: Cambridge, MA.

Lazell, J.D. Jr. 1987. A new flying lizard from the Sangihe Archipelago, Indonesia. *Breviora* 488:1–9.

Lazell, J.D. Jr. 1992. New flying lizards and predictive biogeography of two Asian archipelagos. *Bulletin of the Museum of Comparative Zoology* 152:475–505.

Lazell, J.D. Jr. 1996. Careening Island and the Goat Islands: Evidence for the arid-insular invasion wave theory of dichopatric speciation in Jamaica. Pp. 195–205 in R. Powell and R.W. Henderson, Eds., *Contributions to West Indian Herpetology: A Tribute to Albert Schwartz*. Society for the Study of Amphibians and Reptiles: Ithaca, NY.

Lazell, J.D. Jr. 1999. Giants, dwarfs, and rock-knockoffs: Evolution of diversity in Antillean anoles. Pp. 55–56 in J.B. Losos and M. Leal, Eds., *Anolis Newsletter V*. Washington University: Saint Louis, MO, USA.

Lazell, J.D. Jr. 2005. *Island: Fact and Theory in Nature*. University of California Press: Berkeley, CA.

Lazell, J.D. Jr. and S. McKeown. 1998. Identity of the knight anole introduced to Oahu, Hawaiian Islands. *Bulletin of the Chicago Herpetological Society* 33:181.

Le Galliard, J.-F., J. Clobert, and R. Ferriére. 2004. Physical performance and Darwinian fitness in lizards. *Nature* 432:502–505.

Leal, M. 1999. Honest signalling during prey-predator interactions in the lizard *Anolis cristatellus*. *Animal Behaviour* 58:521–526.

Leal, M., and L.J. Fleishman. 2002. Evidence for habitat partitioning based on adaptation to environmental light in a pair of sympatric lizard species. *Proceedings of the Royal Society of London B* 269:351–359.

Leal, M., and L.J. Fleishman. 2004. Differences in visual signal design and detectability between allopatric populations of *Anolis* lizards. *American Naturalist* 163:26–39.

Leal, M., A.K. Knox, and J.B. Losos. 2002. Lack of convergence in aquatic *Anolis* lizards. *Evolution* 56:785–791.

Leal, M., and J.B. Losos. 2000. Behavior and ecology of the Cuban "Chipojo bobos" *Chamaeleolis barbatus* and *C. porcus*. *Journal of Herpetology* 34:318–322.

Leal, M., and J.A. Rodríguez-Robles. 1995. Antipredator responses of *Anolis cristatellus* (Sauria: Polychrotidae). *Copeia* 1995:155–162.

Leal, M., and J.A. Rodríguez-Robles. 1997a. Antipredator responses of the Puerto Rican giant anole, *Anolis cuviei* (Squamata: Polychrotidae). *Biotropica* 29:372–375.

Leal, M., and J.A. Rodríguez-Robles. 1997b. Signalling displays during predator-prey interactions in a Puerto Rican anole, *Anolis cristatellus*. *Animal Behaviour* 54:1147–1154.

Leal, M., J.A. Rodríguez-Robles, and J.B. Losos. 1998. An experimental study of interspecific interactions between two Puerto Rican *Anolis* lizards. *Oecologia* 117:273–278.

Leal, M., and R. Thomas. 1992. *Eleutherodactylus coqui* (Puerto Rican coquí). Prey. *Herpetological Review* 23:79–80.

Leenders, T. 2001. *A Guide to Amphibans and Reptiles of Costa Rica*. Zona Tropical, S.A.: Miami, FL.

Leigh, E.G. Jr., A. Hladik, C.M. Hladik, and A. Jolly. 2007. The biogeography of large islands, or how does the size of the ecological theater affect the evolutionary play? *Revue E'cole (Terre Vie)* 62:105–168.

Leroi, A.M., M.R. Rose, and G.V. Lauder. 1994. What does the comparative method reveal about adaptation? *American Naturalist* 143:381–402.

Lever, C.L. 1987. *Naturalized Birds of the World*. Longman Scientific and Technical: Harlow, UK.

Lewontin, R. 1978. Adaptation. *Scientific American* 239:212–229.

Lewontin, R. 1985. Adaptation. Pp. 65–84 in R. Levins and R. Lewontin, Eds., *The Dialectical Biologist*. Harvard University Press: Cambridge, MA.

Lewontin, R. 2000. *The Triple Helix: Gene, Organism, and Environment*. Harvard University Press: Cambridge, MA.

Lexer, C., M.E. Welch, J.L. Durphy, and L.H. Rieseberg. 2003. Natural selection for salt tolerance quantitative trait loci (QTLs) in wild sunflower hybrids: Implications for the origin of *Helianthus paradoxus*, a diploid hybrid species. *Molecular Ecology* 12:1225–1235.

Licht, P. 1967. Thermal adaptation in the enzymes of lizards in relation to preferred body temperatures. Pp. 131–145 in C.L. Prosser, Ed., *Molecular Mechanisms of Temperature Regulation*. American Association for the Advancement of Science: Washington, DC.

Licht, P. 1974. Response of *Anolis* lizards to food supplementation in nature. *Copeia* 1974: 215–221.

Licht, P., and G.C. Gorman. 1970. Reproductive and fat cycles in Caribbean *Anolis* lizards. *University of California Publications in Zoology* 95:1–52.

Lieberman, S.S. 1986. Ecology of the leaf litter herpetofauna of a neotropical rain forest: La Selva, Costa Rica. *Acta Zoologica Mexicana Nueva Serie* 15:1–72.

Liem, K.F. 1974. Evolutionary strategies and morphological innovations: Cichlid pharyngeal jaws. *Systematic Zoology* 22:425–441.

Lighty, R.G., I.G. Macintyre, and R. Stuckenrath. 1979. Holocene reef growth on the edge of the Florida Shelf. *Nature* 278:281–282.

Lima, S.L. 1998. Nonlethal effects in the ecology of predator-prey interaction. *Bioscience* 48:25–34.

Lima, S.L., and T.J. Valone. 1991. Predators and avian community organization: An experiment in a semi-desert grassland. *Oecologia* 105–112.

Liner, E.A. 1996. Natural History Notes. *Anolis carolinensis carolinensis* (green anole). Nectar feeding. *Herpetological Review* 27:78.

Lister, B.C. 1976a. The nature of niche expansion in West Indian *Anolis* lizards I. Ecological consequences of reduced competition. *Evolution* 30: 659–676.

Lister, B.C. 1976b. The nature of niche expansion in West Indian *Anolis* lizards II. Evolutionary components. *Evolution* 30:677–692.

Lister, B.C. 1981. Seasonal niche relationships of rain forest anoles. *Ecology* 62:1548–1560.

Lister, B.C., and A. Garcia Aguayo. 1992. Seasonality, predation, and the behavior of a tropical mainland anole. *Journal of Animal Ecology* 61:717–733.

Lockwood, S. 1876. The Florida chameleon. *American Naturalist* 10:4–16.

Loew, E.R., L.J. Fleishman, R.G. Foster, and I. Provencio. 2002. Visual pigments and oil droplets in diurnal lizards: A comparative study of Caribbean anoles. *Journal of Experimental Biology* 205:927–938.

Lomolino, M.V. 2000. Ecology's most general, yet protean pattern: The species-area relationship. *Journal of Biogeography* 27:17–26.

Losos, J.B. 1985a. An experimental demonstration of the species recognition role of *Anolis* dewlap color. *Copeia* 1985:905–910.

Losos, J.B. 1985b. Male aggressive behavior in a pair of sympatric sibling species. *Breviora* 484:1–30.

Losos, J.B. 1990a. A phylogenetic analysis of character displacement in Caribbean *Anolis* lizards. *Evolution* 44:558–569.

Losos, J.B. 1990b. Concordant evolution of locomotor behaviour, display rate, and morphology in *Anolis* lizards. *Animal Behaviour* 39:879–890.

Losos, J.B. 1990c. Ecomorphology, performance capability, and scaling of West Indian *Anolis* lizards: An evolutionary analysis. *Ecological Monographs* 60:69–388.

Losos, J.B. 1990d. The evolution of form and function: Morphology and locomotor performance in West Indian *Anolis* lizards. *Evolution* 44:1189–1203.

Losos, J.B. 1990e. Thermal sensitivity of sprinting and clinging performance in the Tokay gecko (*Gekko gecko*). *Asiatic Herpetological Research* 3:54–59.

Losos, J.B. 1992a. A critical comparison of the taxon-cycle and character-displacement models for size evolution of *Anolis* lizards in the Lesser Antilles. *Copeia* 1992:279–288.

Losos, J.B. 1992b. The evolution of convergent structure in Caribbean *Anolis* communities. *Systematic Biology* 41:403–420.

Losos, J.B. 1994a. An approach to the analysis of comparative data when a phylogeny is unavailable or incomplete. *Systematic Biology* 43:117–123.

Losos, J.B. 1994b. Historical contingency and lizard community ecology. Pp. 319–333 in L.J. Vitt and E.R. Pianka, Eds., *Lizard Ecology: Historical and Experimental Perspectives*. Princeton University Press: Princeton, NJ.

Losos, J.B. 1996a. Dynamics of range expansion by three introduced species of *Anolis* lizards on Bermuda. *Journal of Herpetology* 30:204–210.

Losos, J.B. 1996b. Ecological and evolutionary determinants of the species-area relation in Caribbean anoline lizards. *Philosophical Transactions of the Royal Society of London* 351:847–854.

Losos, J.B. 1996c. Phylogenetic perspectives on community ecology. *Ecology* 77:1344–1354.

Losos, J.B. 1999. Uncertainty in the reconstruction of ancestral character states and limitations on the use of phylogenetic comparative methods. *Animal Behaviour* 58:1319–1324.

Losos, J.B. 2001. Evolution: A lizard's tale. *Scientific American* 284(3): 64–69.

Losos, J.B. 2004. Adaptation and speciation in Greater Antillean anoles. Pp. 335–343 in U. Dieckmann, M. Doebeli, J.A.J. Metz, and D. Tautz, Eds. *Adaptive Speciation*. Cambridge University Press: Cambridge, UK.

Losos, J.B. 2007. Detective work in the West Indies: Integrating historical and experimental approaches to study island lizard evolution. *Bioscience* 57:585–597.

Losos, J.B., R.M. Andrews, O.J. Sexton, and A.L. Schuler. 1991. Behavior, ecology, and locomotor performance of the giant anole, *Anolis frenatus*. *Caribbean Journal of Science* 27:173–179.

Losos, J.B., M. Butler, and T.W. Schoener. 2003a. Sexual dimorphism in body size and shape in relation to habitat use among species of Caribbean *Anolis* lizards. Pp. 356–380 in S. F. Fox, J.K. McCoy and T.A. Baird, Eds., *Lizard Social Behavior*. Johns Hopkins Press: Baltimore, MD.

Losos, J.B., and L.-R. Chu. 1998. Examination of factors potentially affecting dewlap size in Caribbean anoles. *Copeia* 1998:430–438.

Losos, J.B., D.A. Creer, D. Glossip, R. Goellner, A. Hampton, G. Roberts, N. Haskell, P. Taylor, and J. Etling. 2000. Evolutionary implications of phenotypic plasticity in the hindlimb of the lizard *Anolis sagrei*. *Evolution* 54:301–305.

Losos, J.B., and K. de Queiroz. 1997. Evolutionary consequences of ecological release in Caribbean *Anolis* lizards. *Biological Journal of the Linnean Society* 61:459–483.

Losos, J.B., M.R. Gannon, W.J. Pfeiffer, and R.B. Waide. 1990. Notes on the ecology and behavior of the lagarto verde, *Anolis cuvieri*, in Puerto Rico. *Caribbean Journal of Science* 26:65–66.

Losos, J.B., and R.E. Glor. 2003. Phylogenetic comparative methods and the geography of speciation. *Trends in Ecology and Evolution* 18:220–227.

Losos, J.B., and D.J. Irschick. 1996. The effect of perch diameter on escape behaviour of *Anolis* lizards: Laboratory predictions and field tests. *Animal Behaviour* 51:593–602.

Losos, J.B., D.J. Irschick, and T.W. Schoener. 1994. Adaptation and constraint in the evolution of specialization of Bahamian *Anolis* lizards. *Evolution* 48:1786–1798.

Losos, J.B., T.R. Jackman, A. Larson, K. de Queiroz, and L. Rodríguez-Schettino. 1998. Contingency and determinism in replicated adaptive radiations of island lizards. *Science* 279:2115–2118.

Losos, J.B., M. Leal, R.E. Glor, K. de Queiroz, P.E. Hertz, L. Rodríguez Schettino, A. Chamizo Lara, T.R. Jackman, and A. Larson. 2003b. Niche lability in the evolution of a Caribbean lizard community. *Nature* 423:542–545.

Losos, J.B., J.C. Marks, and T.W. Schoener. 1993a. Habitat use and ecological interactions of an introduced and a native species of *Anolis* lizard on Grand Cayman, with a review of the outcomes of anole introductions. *Oecologia* 95:525–532.

Losos, J.B., P.L.N. Mouton, R. Bickel, I. Cornelius, and L. Ruddock. 2002. The effect of body armature on escape behaviour in cordylid lizards. *Animal Behaviour* 64:313–321.

Losos, J.B., and D.B. Miles. 1994. Adapation, constraint, and the comparative method: Phylogenetic issues and methods. Pp. 60–98 in P.C. Wainwright and S.M. Reilly, Eds., *Ecological Morphology: Integrative Organismal Biology*. University of Chicago Press: Chicago, IL.

Losos, J.B., and D.B. Miles. 2002. Testing the hypothesis that a clade has adaptively radiated: Iguanid lizard clades as a case study. *American Naturalist* 160:147–157.

Losos, J.B., and D. Schluter. 2000. Analysis of an evolutionary species–area relationship. *Nature* 408:847–850.

Losos, J.B., T.W. Schoener, R.B. Langerhans, and D.A. Spiller. 2006. Rapid temporal reversal in predator-driven natural selection. *Science* 314:1111.

Losos, J.B., T.W. Schoener, and D.A. Spiller. 2003c. Effect of immersion in seawater on egg survival in the lizard *Anolis sagrei*. *Oecologia* 137:360–362.

Losos, J.B., T.W. Schoener, and D.A. Spiller. 2004. Predator-induced behaviour shifts and natural selection in field-experimental lizard populations. *Nature* 432:505–508.

Losos, J.B., T.W. Schoener, K.I. Warheit, and D. Creer. 2001. Experimental studies of adaptive differentiation in Bahamian *Anolis* lizards. *Genetica* 112–113:399–416.

Losos, J.B., and B. Sinervo. 1989. The effect of morphology and perch diameter on sprint performance of *Anolis* lizards. *Journal of Experimental Biology* 145:23–30.

Losos, J.B., and D.A. Spiller. 1999. Differential colonization success and asymmetrical interactions between two lizard species. *Ecology* 80:252–258.

Losos, J.B., and D.A. Spiller. 2005. Natural history notes. *Anolis smaragdinus* (Bahamian green anole). Dispersal. *Herpetological Review* 36:315–316.

Losos, J.B., B.M. Walton, and A.F. Bennett. 1993b. Trade-offs between sprinting and clinging ability in Kenyan chameleons. *Functional Ecology* 7:281–286.

Losos, J.B., K.I. Warheit, and T.W. Schoener. 1997. Adaptive differentiation following experimental island colonization in *Anolis* lizards. *Nature* 387:70–73.

Lovern, M.B., M.M. Holmes, and J. Wade. 2004. The green anole (*Anolis carolinensis*): A reptilian model for laboratory studies of reproductive orphology and behavior. *ILAR Journal* 45:54–64.

Lovern, M.B., and T.A. Jenssen. 2003. Form emergence and fixation of head bobbing displays in the green anole lizard (*Anolis carolinensis*): A reptilian model of signal ontogeny. *Journal of Comparative Psychology* 117:133–141.

Lovern, M.B., T.A. Jenssen, K.S. Orrell, and T. Tuchak. 1999. Comparisons of temporal display structure across contexts and populations in male *Anolis carolinensis*: Signal stability or lability? *Herpetologica* 55:222–234.

Lovette, I.J., and E. Bermingham. 1999. Explosive speciation in the new world *Dendroica* warblers. *Proceedings of the Royal Society of London B* 266:1629–1636.

Lovette, I.J., E. Bermingham, and R.E. Ricklefs. 2002. Clade-specific morphological diversification and adaptive radiation in Hawaiian songbirds. *Proceedings of the Royal Society of London B* 269:37–42.

Luckan, L. 1917. Ecological morphology of *Abutilon theophrasti*. *Kansas University Science Bulletin* 10:219–228.

MacArthur, R.H. 1972. *Geographical Ecology: Patterns in the Distribution of Species*. Princeton University Press: Princeton, NJ.

MacDonald, D., and A.C. Echternacht. 1991. Red-throated and gray-throated *Anolis carolinensis*: do females know the difference? Pp. 92–100 in J.B. Losos and G.C. Mayer, Eds., *Anolis Newsletter IV*. National Museum of Natural History, Smithsonian Institution: Washington, DC.

Macedonia, J.M. 2001. Habitat light, colour variation, and ultraviolet reflectance in the Grand Cayman anole, *Anolis conspersus*. *Biological Journal of the Linnean Society* 73:299–320.

Macedonia, J.M., A.C. Echternacht, and J.W. Walguarnery. 2005. Color variation, habitat light, and background contrast in *Anolis carolinensis* along a geographical transect in Florida. *Journal of Herpetology* 37:467–478.

Macedonia, J.M., C.S. Evans, and J.B. Losos. 1994. Male *Anolis* lizards discriminate video-recorded conspecific and heterospecific displays. *Animal Behaviour* 47:1220–1223.

Macedonia, J.M., and J.A. Stamps. 1994. Species recognition in *Anolis grahami* (Sauria, Iguanidae): Evidence from responses to video playbacks of conspecific and heterospecific displays. *Ethology* 98:246–264.

Macey, J.R., A. Larson, N.B. Ananjeva, and T.J. Papenfuss. 1997. Evolutionary shifts in three major structural features of the mitochondrial genome among iguanian lizards. *Journal of Molecular Evolution* 44:660–674.

Macey, J.R., J.A. Schulte, N.B. Ananjeva, A. Larson, N. Rastegar-Pouyani, S.M. Shammakov, and T.J. Papenfuss. 1998a. Phylogenetic relationships among agamid lizards of the *Laudakia caucasia* species group: Testing hypotheses of biogeographic fragmentation and an area cladogram for the Iranian Plateau. *Molecular Phylogeny and Evolution* 10:118–131.

Macey, J.R., J.A. Schulte, A. Larson, Z. Fang, Y. Wang, B.S. Tuniyev and T.J. Papenfuss. 1998b. Phylogenetic relationships of toads of the *Bufo bufo* complex from the eastern escarpment of the Tibetan Plateau: A case of vicariance and dispersal. *Molecular Phylogeny and Evolution* 9:80–87.

MacFadden, B.J. 2005. Fossil horses: Evidence for evolution. *Science* 307:1728–1730.

MacFadden, B.J. 2006. Extinct mammalian biodiversity of the ancient New World tropics. *Trends in Ecology and Evolution* 21:157–165.

MacFadden, B.J., and R.C. Hulbert. 1988. Explosive speciation at the base of the adaptive radiation of Miocene grazing horses. *Nature* 336:466–468.

MacLean, R.C., and G. Bell. 2002. Experimental adaptive radiation in *Pseudomonas*. *American Naturalist* 160:569–581.

Macrini, T.E., and D.J. Irschick. 1998. An intraspecific analysis of trade-offs in sprinting performance in a West Indian lizard species (*Anolis lineatopus*). *Biological Journal of the Linnean Society* 63:579–591.

Macrini, T.E., D.J. Irschick, and J.B. Losos. 2003. Ecomorphological differences in toepad characteristics between mainland and island anoles. *Journal of Herpetology* 37:52–58.

Maddison, W.P. 1990. A method for testing the correlated evolution of two binary characters: are gains or losses concentrated on certain branches of a phylogenetic tree? *Evolution* 44: 539–557.

Maddison, W., and D. Maddison. 1992. *MacClade 3: Interactive Analysis of Phylogenetic and character Evolution.* Sinauer Associates: Sunderland, MA.

Madsen, O., M. Scally, C.J. Douady, D.J. Kao, R.W. DeBry, R. Adkins, H.M. Amrine, M.J. Stanhope, W.W. de Jong, and M.S. Springer. 2001. Parallel adaptive radiations in two major clades of placental mammals. *Nature* 409:610–614.

Malhotra, A., and R.S. Thorpe. 1991. Experimental detection of rapid evolutionary response in natural lizard populations. *Nature* 353:347–348.

Malhotra, A., and R.S. Thorpe. 1994. Parallels between island lizards suggests selection on mitochondrial DNA and morphology. *Proceedings of the Royal Society of London B* 257:37–42.

Malhotra, A., and R.S. Thorpe. 1997a. Microgeographic variation in scalation of *Anolis oculatus* (Dominica, West Indies): A multivariate analysis. *Herpetologica* 53:49–62.

Malhotra, A., and R.S. Thorpe. 1997b. Size and shape variation in a Lesser Antillean anole, *Anolis oculatus* (Sauria: Iguanidae) in relation to habitat. *Biological Journal of the Linnean Society* 60:53–72.

Malhotra, A., and R.S. Thorpe. 2000. The dynamics of natural selection and vicariance in the Dominican anole: Patterns of within-island molecular and morphological divergence. *Evolution* 54:245–258.

Manamendra-Arachchi, K., and S. Liyange. 1994. Conservation and distribution of the agamid lizards of Sri Lanka with illustrations of the extant species. *Journal of South Asian Natural History* 1:77–96.

Mani, G.S., and B.C.Clarke. 1990. Mutational order: A major stochastic process in evolution. *Proceedings of the Royal Society of London B* 240:29–37.

Mank, J.E. 2007. Mating preferences, sexual selection and patterns of cladogenesis in ray-finned fishes. *Journal of Evolutionary Biology* 20:597–602.

Manthey, U., and N. Schuster. 1996. *Agamid Lizards.* T.F.H. Publications, Inc.: Neptune, NJ.

Marcellini, D.L. 1977. Acoustic and visual display behavior of gekkonid lizards. *American Zoologist* 17:251–260.

Marcellini, D.L., T.A. Jenssen, and C.A. Pague. 1985. The distribution of *Anolis cooki*, with comments on its possible future extinction. *Herpetological Review* 16:99–102.

Marcus, L.F., M. Corti, A. Loy, G.J.P. Naylor, and D.E. Slice. 1996. *Advances in Morphometrics.* Plenum Publishers: New York, NY.

Marsh, R.L., and A.F. Bennett. 1985. Thermal dependence of isotonic contractile properties of skeletal muscle and sprint performance of the lizard *Dipsosaurus dorsalis*. *Journal of Comparative Physiology B* 155:541–551.

Marsh, R.L., and A.F. Bennett. 1986a. Dependence of sprint speed performance of the lizard *Sceloporus occidentalis*. *Journal of Experimental Biology* 126:79–87.

Marsh, R.L., and A.F. Bennett. 1986b. Thermal dependence of contractile properties of skeletal muscle from the lizard *Sceloporus occidentalis* with comments on methods of fitting and comparing force-velocity curves. *Journal of Experimental Biology* 126:63–77.

Marshall, D.C., C. Simon, and T.R. Buckley. 2006. Accurate branch length estimation in partitioned Bayesian analyses requires accommodation of among-partition rate variation and attention to branch length priors. *Systematic Biology* 55:993–1003.

Martin, J. 1992. *Masters of Disguise: A Natural History of Chameleons*. Checkmark Books: New York, NY.

Martínez-Meyer, E., and A.T. Peterson. 2006. Conservation of ecological niche characteristics in North American plant species over the Pleistocene-to-recent transition. *Journal of Biogeography* 33:1779–1789.

Martins, E.P. 1996. Conducting phylogenetic comparative studies when the phylogeny is not known. *Evolution* 50:12–22.

Martins, E.P. 1999. Estimation of ancestral states of continuous characters: A computer simulation study. *Systematic Biology* 48:642–650.

Martins, E.P., and T. Garland, Jr. 1991. Phylogenetic analyses of the correlated evolution of continuous characters: A simulation study. *Evolution* 45:534–557.

Martins, E.P., T.J. Ord, and S.W. Davenport. 2005. Combining motions into complex displays: Playbacks with a robotic lizard. *Behavioral Ecology and Sociobiology* 58: 351–360.

Martins, M., and M.E. Oliveira. 1998. Natural history of snakes in forests of the Manaus region, central Amazonia, Brazil. *Herpetological Natural History* 6:78–150.

Mattingly, W.B., and B.C. Jayne. 2004. Resource use in arboreal habitats: Structure affects locomotion of four ecomorphs of *Anolis* lizards. *Ecology* 85:1111–1124.

Mattingly, W.B., and B.C. Jayne. 2005. The choice of arboreal escape paths and its consequences for the locomotor behaviour of four species of *Anolis* lizards. *Animal Behaviour* 70:1239–1250.

Mayer, G.C. 1989. *Deterministic Patterns of Community Structure in West Indian reptiles and Amphibians*. Ph.D. Dissertation, Harvard University: Cambridge, MA.

Maynard Smith, J., R. Burian, S. Kaufman, P. Alberch, J. Campbell et al., B. Goodwin, R. Lande, D. Raup, and L. Wolpert. 1985. Developmental constraints and evolution. *Quarterly Review of Biology* 60:265–287.

Mayr, E. 1963. *Animal Species and Evolution*. Belknap Press: Cambridge, MA.

Mayr, E. 1969. *Principles of Systematic Zoology*. McGraw-Hill: New York, NY.

Mayr, E. 2004. *What Makes Biology Unique?* Cambridge University Press: Cambridge, UK.

McCoid, M.J. 1993. The "new" herpetofauna of Guam, Mariana Islands. *Herpetological Review* 24:16–17.

McCranie, J.R., G. Köhler, and L.D. Wilson. 2000. Two new species of anoles from northwestern Honduras related to *Norops laeviventris* (Wiegmann 1834). *Senckenbergiana Biologica* 80:213–223.

McFarlane, D.A., and K.L. Garrett. 1989. The prey of common barn-owls (*Tyto alba*) in dry limestone scrub forest of southern Jamaica. *Caribbean Journal of Science* 25:21–23.

McGuire, J.A., and A.C. Alcala. 2000. A taxonomic revision of the flying lizards (Iguana: Agamidae: *Draco*) of the Philippine Islands with a description of a new species. *Herpetological Monographs* 14:81–138.

McGuire, J.A., R.M. Brown, Mumpuni, A. Riyanto, and N. Andayani. 2007a. The flying lizards of the *Draco lineatus* group (Squamata: Iguania: Agamidae): A taxonomic revision with descriptions of two new species. *Herpetological Monographs* 21:179–212.

McGuire, J.A., C.C. Witt, D.L. Altshuler, and J.V. Remsen, Jr. 2007b. Phylogenetic systematics and biogeography of hummingbirds: Bayesian and maximum likelihood analyses of partitioned data and selection of an appropriate partitioning strategy. *Systematic Biology* 56:837–856.

McKeown, S. 1996. *A Field Guide to Reptiles and Amphibians in the Hawaiian Islands*. Diamond Head Publishing, Inc.: Los Osos, CA.

McKinnon, J.S., S. Mori, B.K. Blackman, L. David, D.M. Kingsley, L. Jamieson, J. Chou, and D. Schluter. 2004. Evidence for ecology's role in speciation. *Nature* 429:294–298.

McLaughlin, J.P., and J. Roughgarden. 1989. Avian predation on *Anolis* lizards in the northeastern Caribbean: An inter-island contrast. *Ecology* 70:617–628.

McMann, S. 2000. Effects of residence time on displays during territory establishment in a lizard. *Animal Behaviour* 59:513–522.

McMann, S. and A.V. Paterson. 2003. The relationship between location and displays in a territorial lizard. *Journal of Herpetology* 37:414–416.

Medvin, M.B. 1990. Sex differences in coloration and optical signalling in the lizard *Anolis carolinensis* (Reptilia, Lacertilia, Iguanidae). *Animal Behaviour* 39:192–193.

Meffe, G. K., and F. F. Snelson. 1989. *Ecology and Evolution of Livebearing Fishes (Poeciliidae)*. Prentice Hall: Englewood Cliffs, NJ.

Melville, J., L.J. Harmon, and J.B. Losos. 2006. Intercontinental community convergence of ecology and morphology in desert lizards. *Proceedings of the Royal Society of London B* 273:557–563.

Meshaka, W.E. Jr., B.P. Butterfield, and J.B. Hauge. 2004. *The Exotic Amphibians and Reptiles of Florida*. Krieger Publishing Co.: Malabar, FL.

Meshaka, W.E. Jr., and K.G. Rice. 2005. The knight anole: Ecology of a successful colonizing species in extreme southern mainland Florida. Pp. 225–230 in W.E. Meshaka, Jr., and K.J. Babbitt, Eds., *Amphibians and Reptiles: Status and Conservation in Florida*. Krieger Publishing Co.: Malabar, FL.

Mesquita, D.O., G.C. Costa, and G.R. Colli. 2006. Ecology of an Amazonian savanna lizard assemblage in Monte legre, Pará state, Brazil. *South American Journal of Herpetology* 1:61–71.

Meyer, J.R., and R. Kassen. 2007. The effects of competition and predation on diversification in a model adaptive radiation. *Nature* 446:432–435.

Miles, D.B. 2004. The race goes to the swift: Fitness consequences of variation in sprint performance in juvenile lizards. *Evolutionary Ecology Research* 6:63–75.

Miles, D.B., and A.E. Dunham. 1996. The paradox of the phylogeny: Character displacement of analyses of body size in island *Anolis*. *Evolution* 50:594–603.

Miller, A.H. 1949. Some ecologic and morphologic considerations in the evolution of higher taxonomic categories. Pp. 84–88 in E. Mayr and E. Schuz, Eds., *Ornithologie als Biologische Wissenschaft*. C. Winter: Heidelberg, Germany.

Milton, T.H., and T.A. Jenssen. 1979. Description and significance of vocalizations of *Anolis grahami* (Sauria: Iguanidae). *Copeia* 1979:481–489.

Mitchell, B.J. 1989. *Resources, Group Behavior, and Infant Development in White-Faced Capuchin Monkeys,* Cebus capucinus. Ph.D. Disseration, University of California: Berkeley, CA.

Mittermeier, R.A., N. Myers, P. Robles Gil, and C. Goettsch Mittermeier. 1999. *Hotspots: Earth's Biologically Richest and Most Endangered Terrestrial Ecoregions.* Cemex, S.A.: Mexico City, Mexico.

Miyata, K. 1983. Notes on *Phenacosaurus heterodermus* in the Sabana de Bogotá, Colombia. *Journal of Herpetology* 17:102–105.

Moermond, T.C. 1979a. Habitat constraints on the behavior, morphology, and community structure of *Anolis* lizards. *Ecology* 60:152–164.

Moermond, T.C. 1979b. The influence of habitat structure on *Anolis* foraging behavior. *Behaviour* 70:147–167.

Moermond, T.C. 1981. Prey-attack behavior of *Anolis* lizards. *Zeitschrift für Tierpsychologie* 56:128–136.

Moermond, T.C. 1983. Competition between *Anolis* and birds: A reassessment. Pp. 507–520 in A.G.J. Rhodin and K. Miyata, Eds., *Advances in Herpetology and Evolutionary Biology: Essays in Honor of Ernest E. Williams.* Museum of Comparative Zoology, Harvard University: Cambridge, MA.

Monks, S.P. 1881. A partial biography of the green lizard. *American Naturalist* 15:96–99.

Mooney, H.A., and R.J. Hobbs. 2000. *Invasive Species in a Changing World.* Island Press: Covelo, CA.

Moore, J.-S., J.L. Gow, E.B. Taylor, and A.P. Hendry. 2007. Quantifying the constraining influence of gene flow on adaptive divergence in the lake-stream threespine stickleback system. *Evolution* 61:2015–2026.

Mori, A., and T. Hikida. 1994. Field observations on the social behavior of the flying lizard, *Draco volans sumatranus,* in Borneo. *Copeia* 1994:124–130.

Morin, P. 1999. Productivity, intraguild predation, and population dynamics in experimental food webs. *Ecology* 80:752–760.

Mousseau, T.A., and D.A. Roff. 1987. Natural selection and the heritability of fitness components. *Heredity* 59:181–197.

Murphy, T.J., and A.A. Myers. 1996. The behavioral ecology of *Phelsuma astriata semicarinata* on Aride Island Nature Reserve, Seychelles. *Journal of Herpetology* 30:117–123.

Myers, C.W. 1971. Central American lizards related to *Anolis pentaprion:* Two new species from the Cordillera de Talamanca. *American Museum Novitates* 2471:1–40.

Myers, C.W. 1982. Blunt-headed vine snakes (*Imantodes*) in Panama, including a new species and other revisionary notes. *American Museum Novitates* 2738:1–50.

Naganuma, K.H., and J.D. Roughgarden. 1990. Optimal body size in Lesser Antillean *Anolis* lizards—a mechanistic approach. *Ecological Monographs* 60:239–256.

Navarro P., N., A. Fernandez V., and O.H. Garrido. 2001. Reconsideración taxonómica de *Anolis centralis litoralis* y descripción de una especie nueva del grupo *argillaceus* (Sauria: Iguanidae) para Cuba. *Solenodon* 1:66–75.

Near, T.J., P.A. Meylan, and H.B. Shaffer. 2005. Assessing concordance of fossil calibration points in molecular clock studies: An example using turtles. *American Naturalist* 165:137–146.

Nečas, P. 2004. *Chameleons: Nature's Hidden Jewels*, 2nd Ed. Chimaira Buchhandelsgesellschaft: Frankfurt, Germany.

Nee, S., A.Ø. Mooers, and P.H. Harvey. 1992. Tempo and mode of evolution revealed from molecular phylogenies. *Proceedings of the National Academy of Sciences of the United States of America* 89:8322–8326.

Nei, M., T. Maruyama, and R. Chakraborty, 1975. The bottleneck effect and genetic variability in populations. *Evolution* 29:1–10.

Nicholson, K.E. 2002. Phylogenetic analysis and a test of the current infrageneric classification of *Norops* (beta *Anolis*). *Herpetological Monographs* 16:93–120.

Nicholson, K.E., R.E. Glor, J. J. Kolbe, A. Larson, S.B. Hedges, and J.B. Losos. 2005. Mainland colonization by island lizards. *Journal of Biogeography* 32:929–938.

Nicholson, K.E., L.J. Harmon, and J.B. Losos. 2007. Evolution of *Anolis* lizard dewlap diversity. *PLoS One* 2(3):e274.

Nicholson, K.E., A. Mijares-Urrutia, and A. Larson. 2006. Molecular phylogenetics of the *Anolis onca* Series: A case history in retrograde evolution revisited. *Journal of Experimental Zoology (Molecular Development and Evolution)* 306B:1–10.

Niswander, L. 2002. Interplay between the molecular signals that control vertebrate limb development. *International Journal of Developmental Biology* 46:877–881.

Norval, G., J.-J. Mao, H.-P. Chu, and L.-C. Chen. 2002. A new record of an introduced species, the brown anole (*Anolis sagrei*) (Duméril & Bibron, 1837), in Taiwan. *Zoological Studies* 41:332–336.

Nosil, P., B.J. Crespi, and C.P. Sandoval. 2002. Host-plant adaptation drives the parallel evolution of reproductive isolation. *Nature* 417:440–443.

Novo Rodríguez, J. 1985. Nido communal de *Anolis angusticeps* (Sauria: Iguanidae) en Cayo Francés, Cuba. 26:3–4.

Nunez, S.C., T.A. Jenssen, and K. Ersland. 1997. Female activity profile of a polygynous lizard (*Anolis carolinensis*): Evidence of intrasexual asymmetry. *Behaviour* 134:205–223.

O'Hara. R.J. 1988. Homage to Clio, or, toward an historical philosophy for evolutionary biology. *Systematic Zoology* 37:142–155.

O'Steen, S., A.J. Cullum, and A.F. Bennett. 2002. Rapid evolution of escape ability in Trinidadian guppies (*Poecilia reticulata*). *Evolution* 56:776–784.

Oakley, T.H., and C.W. Cunningham. 2000. Independent contrasts succeed where ancestor reconstruction fails in a known bacteriophage phylogeny. *Evolution* 54:397–405.

Ogden, R., and R.S. Thorpe. 2002. Molecular evidence for ecological speciation in tropical habitats. *Proceedings of the National Academy of Sciences of the United States of America* 99:13612–13615.

Okochi, I., M. Yoshimura, T. Abe, and H. Suzuki. 2006. High population densities of an exotic lizard, *Anolis carolinensis* and its possible role as a pollinator in the Ogasawara Islands. *Bulletin of FFPRI* 5:265–269.

Olesen, J.M., and A. Valido. 2003. Lizards as pollinators and seed dispersers: An island phenomenon. *Trends in Ecology and Evolution* 18:177–181.

Oliver, J.A. 1948. The anoline lizards of Bimini, Bahamas. *American Museum Novitates* 1383:1–36.

Opler, P.A., Frankie, G.W., and H.G. Baker. 1980. Comparative phenological studies of treelet and shrub species in tropical wet and dry forests in the lowlands of Costa Rica. *Journal of Ecology* 68: 167–188.

Ord, T.J., D.T. Blumstein, and C.S. Evans. 2001. Intrasexual selection predicts the evolution of signal complexity in lizards. *Proceedings of the Royal Society B* 268:737–744.

Ord, T.J., and E.P. Martins. 2006. Tracing the origins of signal diversity in anole lizards: Phylogenetic approaches to inferring the evolution of complex behaviour. *Animal Behaviour* 71:1411–1429.

Ord, T.J., R.A. Peters, B. Clucas, and J.A. Stamps. 2007. Lizards speed up visual displays in noisy motion habitats. *Proceedings of the Royal Society of London B* 274:1057–1062.

Ord, T.J., and J.A. Stamps. 2008. Alert signals enhance animal communication in 'noisy' environments. *Proceedings of the National Academy of Sciences of the United States of America* 105:18830–18835.

Orians, G.H. 1969. The number of bird species in some tropical forests. Ecology 50: 783–801.

Orians, G.H., and R.T. Paine. 1983. Convergent evolution at the community level. Pp. 431–458 in D.J. Futuyma and M. Slatkin, Eds., *Coevolution*. Sinauer Associates: Sunderland, MA.

Orrell, K.S., and T.A. Jenssen. 1998. Display behavior of *Anolis bahorucoensis*: An anole with a diminutive dewlap. *Caribbean Journal of Science* 34:113–125.

Orrell, K.S., and T.A. Jenssen. 2002. Male mate choice by the lizard *Anolis carolinensis*: A preference for novel females. *Animal Behaviour* 63:1091–1102.

Orrell, K.S., and T.A. Jenssen. 2003. Heterosexual signalling by the lizard *Anolis carolinensis*, with intersexual comparisons across contexts. *Behaviour* 140:603–634.

Ortiz, P.R., and T.A. Jenssen. 1982. Interspecific aggression between lizard competitors, *Anolis cooki* and *Anolis cristatellus*. *Zeitschrift für Tierpsychologie* 60:227–238.

Ortlund, E.A., J.T. Bridgham, M.R. Redinbo, and J.W. Thornton. 2007. Crystal structure of an ancient protein: Evolution by conformational epistasis. *Science* 317:1544–1548.

Østbye, K., P.-A. Amundsen, L. Bernatchez, A. Klemetsen, R. Knudsen, R. Kristoffersen, T.F. Naesje and K. Hindar. 2006. Parallel evolution of ecomorphological traits in the European whitefish *Coregonus lavaretus* (L.) species complex during postglacial times. *Molecular Ecology* 15:3983–4001.

Owens, I.P.F., P.M. Bennett, and P.H. Harvey. 1999. Species richness among birds: Body size, life history, sexual selection or ecology? *Proceedings of the Royal Society of London B* 266:933–939.

Pacala, S.W., and J. Roughgarden. 1982. Resource partitioning and interspecific competition in two two-species insular *Anolis* lizard communities. *Science* 217:444–446.

Pacala, S.W., and J. Roughgarden. 1984. Control of arthropod abundance by *Anolis* lizards on St. Eustatius (Neth. Antilles). *Oecologia* 64:160–162.

Pacala, S.W., and J. Roughgarden. 1985. Population experiments with the *Anolis* lizards of St. Maarten and St. Eustatius. *Ecology* 66:129–141.

Pacheco, N.N., and O.H. Garrido. 2004. Especie nueva de *Anolis* (Sauria: Lacertilia: Iguanidae) de la región Suroriental de Cuba. *Solenodon* 4:85–90.

Pagel, M., A. Meade, and D. Barker. 2004. Bayesian estimation of ancestral character states on phylogenies. *Systematic Biology* 53:673–684.

Panhuis, T.M., R. Butlin, M. Zuk, and T. Tregenza. 2001. Sexual selection and speciation. *Trends in Ecology and Evolution* 16:364–371.

Parent, C.E., and B.J. Crespi. 2006. Sequential colonization and diversification of the Galápagos endemic land snail genus *Bulimulus* (Gastropoda, Stylommatophora). *Evolution* 60:2311–2328.

Parker, G.A. 2006. Sexual conflict over mating and fertilization: An overview. *Philosophical Transactions of the Royal Society of London B* 361:235–259.

Parmelee, J.R., and C. Guyer. 1995. Sexual differences in foraging behavior of an anoline lizard, *Norops humilis. Journal of Herpetology* 29:619–621.

Parmesan, C. 2006. Ecological and evolutionary responses to recent climate change. *Annual Review of Ecology and Systematics* 37:637–669.

Passek, K.M. 2002. *Extra-Pair Paternity within the Female-Defense Polygyny of the Lizard,* Anolis carolinensis: *Evidence of Alternative Mating Strategies.* Ph.D. Dissertation, Virginia Polytechnic Institute and State University: Blacksburg, VA.

Paterson, A.V. 2002. Effects of an individual's removal on space use and behavior in territorial neighborhoods of brown anoles (*Anolis sagrei*). *Herpetologica* 58:382–393.

Paterson, A.V., and S. McMann 2004. Differential headbob displays toward neighbors and nonneighbors in the territorial lizard *Anolis sagrei. Journal of Herpetology* 38:288–291.

Paulay, G. 1994. Biodiversity on oceanic islands: Its origin and extinction. *American Zoologist* 34:134–144.

Paulissen, M.A., and H.A. Meyer. 2000. The effect of toe-clipping on the gecko *Hemidactylus turcicus. Journal of Herpetology* 34:282–285.

Pearman, P.B., A. Guisan, O. Broennimann, and C.F. Randin. 2008. Niche dynamics in space and time. *Trends in Ecology and Evolution* 23:149–158.

Peattie, A.M., and R.J. Full. 2007. Phylogenetic analysis of the scaling of wet and dry biological fibrillar adhesives. *Proceedings of the National Academy of Sciences of the United States of America* 104:18595–18600.

Pérez-Higareda, G., H.M. Smith, and D. Chiszar. 1997. Natural history notes: *Anolis pentaprion* (lichen anole). Frugivory and cannibalism. *Herpetological Review* 28:201–202.

Perkins, S.L. 2001. Phylogeography of Caribbean lizard malaria: Tracing the history of vector-borne parasites. *Journal of Evolutionary Biology* 14:34–45.

Perkins, S.L., A. Rothschild, and E. Waltari. 2007. Infections of malaria parasite, *Plasmodium floridense*, in the invasive lizard, *Anolis sagrei*, in Floida. *Journal of Herpetology* 41:750–754.

Perry, G. 1996. The evolution of sexual dimorphism in the lizard *Anolis polylepis* (Iguania): Evidence from intraspecific variation in foraging behavior and diet. *Canadian Journal of Zoology* 74:1238–1245.

Perry, G. 1999. The evolution of search modes: Ecological versus phylogenetic perspectives. *American Naturalist* 153:98–109.

Perry, G., B.W. Buchanan, R.N. Fisher, M. Salmon, and S.E. Wise. 2008. Effects of artificial night lighting on reptiles and amphibians in urban environments. Pp. 239–265 in J. C. Mitchell, R.E. Jung Brown, and R. Bartholomew, eds., *Urban Herpetology*. Society for the Study of Amphibians and Reptiles: Salt Lake City, UT.

Perry, G., and J. Lazell. 1997. Natural history notes: *Anolis stratulus* (saddled anole). Nectivory. *Herpetological Review* 28:150–151.

Perry, G., and J. Lazell. 2006. *Anolis pulchellus* (Grass Anole). Nectivory. *Herpetological Review* 37:218–219.

Persons, M.H., L.J. Fleishman, M.A. Frye, and M.E. Stimphil. 1999. Sensory response patterns and the evolution of visual signal design in anoline lizards. *Journal of Comparative Physiology* 184:585–607.

Persson, L., P.-A. Amundsen, A.M. De Roos, A. Klemetsen, R. Knudsen, and R. Primicerio. 2007. Culling prey promotes predator recovery–alternative states in a whole-lake experiment. *Science* 316:1743–1746.

Peters, R.H. 1983. *The Ecological Implications of Body Size.* Cambridge University Press: Cambridge, UK.

Peterson, A.T. 2001. Predicting species' geographic distributions based on ecological niche modeling. *Condor* 103:599–605.

Peterson, A.T., M.A. Ortega-Huerta, J. Bartley, V. Sánchez-Cordero, J. Soberón, R.H. Buddemeier, and D.R.B. Stockwell. 2002. Future projections for Mexican faunas under global climate change scenarios. *Nature* 416:626–629.

Peterson, J.A. 1974. Untitled. Pp. 37–43 in E.E. Williams, Ed., *The Second* Anolis *Newsletter*. Museum of Comparative Zoology, Harvard University: Cambridge, MA.

Peterson, J.A. 1983. The evolution of the subdigital pad in *Anolis*. I. Comparisons among the anoline genera. Pp. 245–283 in A.G.J. Rhodin and K. Miyata, Eds., *Advances in Herpetology and Evolutionary Biology: Essays in Honor of Ernest E. Williams.* Museum of Comparative Zoology, Harvard University: Cambridge, MA.

Peterson, J.A., and E.E. Williams. 1981. A case history in retrograde evolution: The *onca* lineage in anoline lizards. II. Subdigital fine structure. *Bulletin of the Museum of Comparative Zoology* 149:215–268.

Phillips, B.L., and R. Shine. 2004. Adapting to an invasive species: Toxic cane toads induce morphological change in Australian snakes. *Proceedings of the National Academy of Sciences of the United States of America* 101:17150–17155.

Pianka, E.R. 1986. *Ecology and Natural History of Desert Lizards: Analyses of the Ecological Niche and Community Structure.* Princeton, NJ: Princeton University Press.

Pigliucci, M. 2006. Genetic variance-covariance matrices: A critique of the evolutionary quantitative genetics research program. *Biology and Philosophy* 21:1–23.

Pindell, J.L. 1994. Evolution of the Gulf of Mexico and the Caribbean. Pp. 13–39 in S.K. Donovan and T.A. Jackson, Eds., *Caribbean Geology: An Introduction.* University of West Indies Publishers Association: Kingston, Jamaica.

Pindell, J., L. Kennan, K.P. Stanek, W.V. Maresch, and G. Draper. 2006. Foundations of Gulf of Mexico and Caribbean evolution: Eight controversies resolved. *Geologica Acta* 4:303–341.

Pinto, G., D.L. Mahler, L.J. Harmon, and J.B. Losos. 2008. Testing the island effect in adaptive radiation: rates and patterns of morphological diversification in Caribbean and mainland *Anolis* lizards. *Proceedings of the Royal Society B* 275: 2749–2757.

Poche, A.J., Jr., R. Powell, and R.W. Henderson. 2005. Sleep-site selection and fidelity in Grenadian anoles (Reptilia: Squamata: Polychrotidae). *Herpetozoa* 18:3–10.

Podos, J. 2001. Correlated evolution of morphology and vocal signal structure in Darwin's finches. *Nature* 409:185–188.

Poe, S. 1998. Skull characters and the cladistic relationships of the Hispaniolan dwarf twig *Anolis*. *Herpetological Monographs* 12:192–236.

Poe, S. 2004. Phylogeny of anoles. *Herpetological Monographs* 18:37–89.

Poe, S. 2005. A study of the utility of convergent characters for phylogeny reconstruction: Do ecomorphological characters track evolutionary history in *Anolis* lizards? *Zoology* 108:337–343.

Poe, S., and A.L. Chubb. 2004. Birds in a bush: Five genes indicate explosive evolution of avian orders. *Evolution* 58:404–415.

Poe, S., J.R. Goheen, and E.P. Hulebak. 2007. Convergent exaptation and adaptation in solitary island lizards. *Proceedings of the Royal Society of London B* 274:2231–2237.

Poe, S., and R. Ibañez. 2007. A new species of *Anolis* lizard from the Cordillera de Talamanca of western Panama. *Journal of Herpetology* 41:263–270.

Poe, S., and Yañez-Miranda. 2007. A new species of phenacosaur *Anolis* from Peru. *Herpetologica* 63:219–223.

Poinar, G. Jr., and R. Poinar. 2001. *The Amber Forest: A Reconstruction of a Vanished World.* Princeton University Press: Princeton, NJ.

Polcyn, M.J., J.V. Rogers II, Y. Kobayashi, and L.L. Jacobs. 2002. Computed tomography of an *Anolis* lizard in Dominican amber: Systematic, taphonomic, biogeographic, and evolutionary implications. *Palaeontologia Electronica* 5(1):13 pp.

Polis, G.A., C.A. Myers, and R.D. Holt. 1989. The ecology and evolution of intraguild predation: Potential competitors that eat each other. *Annual Review of Ecology and Systematics* 20:297–330.

Polly, R.D. 2001. Paleontology and the comparative method: Ancestral node reconstructions versus observed node values. *American Naturalist* 157:596–609.

Porter, W.P., J.W. Mitchell, W.A. Beckman, and C.B. DeWitt. 1973. Behavioral implications of mechanistic ecology. *Oecologia* 13:1–54.

Pough, F.H., R.M. Andrews, J.E. Cadle, M.L. Crump, A.H. Savitzky, and K.D. Wells. 2004. *Herpetology*, 3rd Ed. Pearson Education, Inc.: Upper Saddle River, NJ.

Poulin, B., G. Lefebvre, R. Ibañez, C. Jaramillo, C. Hernández, and A.S. Rand. 2001. Avian predation upon lizards and frogs in a neotropical forest understorey. *Journal of Tropical Ecology* 17:21–40.

Pounds, J.A. 1988. Ecomorphology, locomotion, and microhabitat structure: Patterns in a tropical mainland *Anolis* community. *Ecological Monographs* 58:299–320.

Pounds, J.A., M.R. Bustamante, L.A. Coloma, J.A. Consuegra, M.P.L. Fogden, P.N. Foster, E. La Marca, K.L. Masters, A. Merino-Viteri, R. Puschendorf, S.R. Ron, G.A. Sánchez-Azofeifa, C.J. Still, and B.E. Young. 2006. Widespread amphibian extinctions from epidemic disease driven by global warming. *Nature* 439:161–167.

Pounds, J.A., M.P.L. Fogden, and J.H. Campbell. 1999. Biological response to climate change on a tropical mountain. *Nature* 398:611–615.

Powell, R. 1999. Herpetology of Navassa Island, West Indies. *Caribbean Journal of Science* 35:1–13.

Powell, R., and R.W. Henderson. 2005. Conservation status of Lesser Antillean reptiles. *Iguana* 12:3–17.

Powell, R., and R.W. Henderson. 2008a. Avian predators of West Indian reptiles. *Iguana* 15:9–11.

Powell, R., and R.W. Henderson. 2008b. Urban herpetology in the West Indies. Pp. 389–404 in J.C. Mitchell, R.E. Jung Brown, and B. Bartholomew, Eds., *Urban Herpetology.* Society for the Study of Amphibians and Reptiles: Salt Lake City, UT.

Powell, R., D.D. Smith, J.S. Parmerlee, C.V. Taylor, and M.L. Jolley. 1990. Range expansion by an introduced anole: *Anolis porcatus* in the Dominican Republic. *Amphibia-Reptilia* 11:421–425.

Pregill, G. 1986. Body size of insular lizards: A pattern of Holocene dwarfism. *Evolution* 40: 997–1008.

Pregill, G.K. 1999. Eocene lizard from Jamaica. *Herpetologica* 55:157–161.

Pregill, G.K., D.W. Steadman, S.L. Olson, and F.V. Grady. 1988. Late Holocene fossil vertebrates from Burma Quarry, Antigua, Lesser Antilles. *Smithsonian Contributions in Zoology* 463:1–27.

Price, T. 1998. Sexual selection and natural selection in bird speciation. *Philosophical Transactions of the Royal Society of London B* 353:251–260.

Price, T. 2007. *Speciation in Birds*. Roberts and Company: Greenwood Village, CO.

Price, T., I.J. Lovette, E. Bermingham, H.L. Gibbs, and A.D. Richman. 2000. The imprint of history on communities of North American and Asian warblers. *American Naturalist* 156:354–367.

Propper, C.R., R.E. Jones, M.S. Rand, and H. Austin. 1991. Nesting behavior of the lizard *Anolis carolinensis*. *Journal of Herpetology* 25:484–486.

Protas, M.E., C. Hersey, D. Kochanek, Y. Zhou, H. Wilkens, W.R. Jeffery, L.I. Zon, R. Borowsky, and C.J. Tabin. 2006. Genetic analysis of cavefish reveals molecular convergence in the evolution of albinism. *Nature Genetics* 38:107–111.

Pulliam, H.R., and G.S. Mills. 1977. The use of space by wintering sparrows. *Ecology* 58:1393–1399.

Purvis, A., J.L. Gittleman, and H.-K. Luh. 1994. Truth or consequences: Effects of phylogenetic accuracy on two comparative methods. *Journal of Theoretical Biology* 167: 293–300.

Qualls, C.P., and R.G. Jaeger. 1991. Dear enemy recognition in *Anolis carolinensis*. *Journal of Herpetology* 25:361–363.

Radtkey, R.R. 1996. Adaptive radiation of day-geckos (*Phelsuma*) in the Seychelles Archipelago: A phylogenetic analysis. *Evolution* 50:604–623.

Radtkey, R.R., S.M. Fallon, and T.J. Case. 1997. Character displacement in some *Cnemidophorus* lizards revisited: A phylogenetic analysis. *Proceedings of the National Academy of Sciences of the United States of America* 94:9740–9745.

Rainey, P.B., A. Buckling, R. Kassen, and M. Travisano. 2000. The emergence and maintenance of diversity: Insights from experimental bacterial populations. *Trends in Ecology and Evolution* 15:243–247.

Rainey, P.B., and M. Travisano. 1998. Adaptive radiation in a heterogeneous environment. *Nature* 394:69–72.

Ramírez-Bautista, A., and M. Benabib 2001. Perch height of the arboreal lizard *Anolis nebulosus* (Sauria: Polychrotidae) from a tropical dry forest of México: Effect of the reproductive season. *Copeia* 2001:187–193.

Rand, A.S. 1962. Notes on Hispaniolan herpetology 5. The natural history of three sympatric species of *Anolis*. *Breviora* 154:1–15.

Rand, A.S. 1964a. Ecological distribution in anoline lizards of Puerto Rico. *Ecology* 45:745–752.

Rand, A.S. 1964b. Inverse relationship between temperature and shyness in the lizard *Anolis lineatopus*. *Ecology* 45:863–864.

Rand, A.S. 1967a. Communal egg laying in anoline lizards. *Herpetologica* 23:227–230.

Rand, A.S. 1967b. Ecology and social organization in the iguanid lizard *Anolis lineatopus*. *Proceedings of the United States National Museum* 122:1–79.

Rand, A.S. 1967c. The ecological distribution of anoline lizards around Kingston, Jamaica. *Breviora* 272:1–18.

Rand, A.S. 1969. Competitive exclusion among anoles (Sauria: Iguanidae) on small islands in the West Indies. *Breviora* 319:1–16.

Rand, A.S. 1999. Of FAN, SAN, and TAN—the WAN (Williams *Anolis* Newsletters). Pp. 1–5 in J.B. Losos and M. Leal, Eds., Anolis *Newsletter V*. Washington University: Saint Louis, MO.

Rand, A.S., G.C. Gorman, and W.M. Rand. 1975. Natural history, behavior, and ecology of *Anolis agassizi*. *Smithsonian Contributions in Zoology* 174:27–38.

Rand, A.S., and S.S. Humphrey. 1968. Interspecific competition in the tropical rain forest: Ecological distribution among lizards at Belém, Pará. *Proceedings of the United States National Museum* 125:1–17.

Rand, A.S., and P.J. Rand. 1967. Field notes on *Anolis lineatus* in Curaçao. *Studies on the Fauna of Curaçao and Other Caribbean Islands* 24:112–117.

Rand, A.S., and E.E. Williams. 1969. The anoles of La Palma: Aspects of their ecological relationships. *Breviora* 327:1–19.

Rand, A.S., and E.E. Williams. 1970. An estimation of redundancy and information content of anole dewlaps. *American Naturalist* 104:99–103.

Rassmann, K. 1997. Evolutionary age of the Galápagos iguanas predates the age of the present Galápagos islands. *Molecular Phylogeny and Evolution* 7:158–172.

Reagan, D.P. 1986. Foraging behavior of *Anolis stratulus* in a Puerto Rican rain forest. *Biotropica* 18:157–160.

Reagan, D.P. 1992. Congeneric species distribution and abundance in a three-dimensional habitat: The rain forest anoles of Puerto Rico. *Copeia* 1992:392–403.

Reagan, D.P. 1996. Anoline lizards. Pp. 322–345 in D.P. Reagan and R.B. Waide, Eds., *The Food Web of a Tropical Rain Forest*. University of Chicago Press: Chicago, IL.

Reagan, D.P., Camilo, G.R., and R.B. Waide. 1996. The community food web: Major properties and patterns of organization. Pp 461–510 in D.P. Reagan and R.B. Waide, Eds., *The Food Web of a Tropical Rain Forest*. University of Chicago Press: Chicago, IL.

Regalado, R. 1998. Approach distance and escape behavior of three species of Cuban *Anolis* (Squamata, Polychrotidae). *Caribbean Journal of Science* 34:211–217.

Reilly, S.M., McBrayer, L.M. and Miles D.B. 2007. *Lizard Ecology*. Cambridge University Press: Cambridge, UK.

Revell, L.J., L.J. Harmon, R.B. Langerhans, and J.J. Kolbe. 2007a. A phylogenetic approach to determining the importance of constraint on phenotypic evolution in the neotropical lizard *Anolis cristatellus*. *Evolutionary Ecology Research* 9:261–282.

Revell, L.J., M.A. Johnson, J.A. Schulte, II, J.J. Kolbe, and J.B. Losos. 2007b. A phylogenetic test for adaptive convergence in rock-dwelling lizards. *Evolution* 61:2898–2912.

Reznick, D.N., and C.K. Ghalambor. 2005. Selection in nature: Experimental manipulations of natural populations. *Integrative and Comparative Biology* 45:456–462.

Rice, W.R., and E.E. Hostert. 1993. Perspective: Laboratory experiments on speciation: What have we learned in forty years? *Evolution* 47:1637–1653.

Richman, A.D., T.J. Case, and T.D. Schwaner. 1988. Natural and unnatural extinction rates of reptiles on islands. *American Naturalist* 131:611–630.

Richman, A.D., and T. Price. 1992. Evolution of ecological differences in the old world leaf warblers. *Nature* 355:817–821.

Ricklefs, R.E. 2002. Splendid isolation: Historical ecology of the South American passerine fauna. *Journal of Avian Biology* 33:207–211.

Ricklefs, R.E. 2003. Global diversification rates of passerine birds. *Proceedings of the Royal Society of London B* 270:2285–2291.

Ricklefs, R.E., and E. Bermingham. 1999. Taxon cycles in the Lesser Antillean avifauna. *Ostrich* 70:49–59.

Ricklefs, R.E., and E. Bermingham. 2004. History and the species-area relationship in Lesser Antillean birds. *American Naturalist* 163:227–239.

Ricklefs, R.E., J.B. Losos, and T.M. Townsend. 2007. Evolutionary diversification of clades of squamate reptiles. *Journal of Evolutionary Biology* 20:1751–1762.

Ricklefs, R.E., and I.J. Lovette. 1999. The roles of island area per se and habitat diversity in the species-area relationships of four Lesser Antillean faunal groups. *Journal of Animal Ecology* 68:1142–1160.

Ricklefs, R.E., and J. Travis. 1980. A morphological approach to the study of avian community organization. *Auk* 97:321–338.

Riddle, B.R., D.J. Hafner, L.F. Alexander, and J.R. Jaeger. 2000. Cryptic vicariance in the historical assembly of a Baja California Peninsula desert biota. *Proceedings of the National Academy of Sciences of the United States of America* 97:14438–14443.

Ridley, M. 1983. *The Explanation of Organic Diversity: The Comparative Method and Adaptations for Mating.* Oxford University Press: Oxford, UK.

Rieppel, O. 1980. Green anole in Dominican amber. *Nature* 286:486–487.

Rieseberg, L.H., T.E. Wood, and E.J. Baack. 2006. The nature of plant species. *Nature* 440:524–527.

Rios-López, N., and A.R. Puente-Colón. 2007. Natural history notes. *Anolis cuvieri* (Puerto Rican giant anole). Reproduction. *Herpetological Review* 38:73–75.

Ripple, W.J., and R.L. Beschta. 2004. Wolves and the ecology of fear: Can predation risk structure ecosystems? *BioScience* 54:755–766.

Roach, D.A., and R.D. Wulff. 1987. Maternal effects in plants. *Annual Review of Ecology and Systematics* 18:209–235.

Robinson, E. 1994. Jamaica. Pp. 111–127 in S.K. Donovan and T.A. Jackson, Eds., *Caribbean Geology: An Introduction.* University of West Indies Publishers Association: Kingston, Jamaica.

Rocha, S., D. Posada, M.A. Carretero, and D.J. Harris. 2007. Phylogenetic affinities of Comoroan and East African day geckos (genus *Phelsuma*): Multiple natural colonisations, introductions and island radiations. *Molecular Phylogeny and Evolution* 43:685–692.

Rodda, G.H., T.H. Fritts, and J.D. Reichel. 1991. The distributional patterns of reptiles and amphibians in the Mariana Islands. *Micronesica* 24:195–210.

Rodrigues, M.T., V. Xavier, G. Skuk and D. Pavan. 2002. New specimens of *Anolis phyllorhinus* (Squamata, Polychrotidae): The first female of the species and of proboscid anoles. *Papéis Avulsos de Zoologia* 42:363–380.

Rodríguez Schettino, L. 1999. *The Iguanid Lizards of Cuba.* University of Florida Press: Gainesville, FL.

Rodríguez Schettino, L., and M.M. Reyes. 1996. Algunos aspectos de la ecología trófica de *Anolis argenteolus* (Sauria: Iguanidae) en una población de la costa suroriental de Cuba. *Biotropica* 28:252–257.

Rodríguez-Robles, J.A., T. Jezkova, and M.A. García. 2007. Evolutionary relationships and historical biogeography of *Anolis desechensis* and *Anolis monensis*, two lizards endemic to small islands in the eastern Caribbean Sea. *Journal of Biogeography* 34:1546–1558.

Rodríguez-Robles, J.A., and M. Leal. 1993. Life history notes. *Alsophis portoricensis* (Puerto Rican racer). Diet. *Herpetological Review* 24:150–151.

Roff, D.A. 2007. A centennial celebration for quantitative genetics. *Evolution* 61:1017–1032.

Rohlf, F.J. and F.L. Bookstein. 1990. *Proceedings of the Michigan Morphometrics Workshop.* University of Michigan, Museum of Zoology: Ann Arbor, MI.

Ronquist, F. 2004. Bayesian inference of character evolution. *Trends in Ecology and Evolution* 19:475–481.

Root, R.B. 1967. The niche exploitation pattern of the blue-gray gnatcatcher. *Ecological Monographs* 37:317–350.

Rose, B. 1982. Food intake and reproduction in *Anolis acutus. Copeia* 1982:323–330.

Ross, C.F. 2004. The tarsier fovea: Functionless vestige or nocturnal adaptation? Pp. 477–537 in C.F. Ross and R.F. Kay, Eds. *Anthropoid Origins: New Visions.* Kluwer Academic/Plenum Publishers: New York, NY.

Rothblum, L.M., J.W. Watkins, and T.A. Jenssen. 1979. A learning paradigm and the behavioral demonstration of audition for the lizard *Anolis grahami. Copeia* 1979: 490–494.

Roughgarden, J. 1972. Evolution of niche width. *American Naturalist* 106:683–718.

Roughgarden, J. 1974. Niche width: Biogeographic patterns among *Anolis* lizard populations. *American Naturalist* 108:429–442.

Roughgarden, J. 1989. The structure and assembly of communities. Pp. 203–226 in J. Roughgarden, R.M. May, and S.A. Levin, Eds., *Perspectives in Ecological Theory.* Princeton University Press: Princeton, NJ.

Roughgarden, J. 1992. Comments on the paper by Losos: Character displacement versus taxon loop. *Copeia* 1992:288–95.

Roughgarden, J. 1995. Anolis *Lizards of the Caribbean: Ecology, Evolution, and Plate Tectonics.* Oxford University Press: Oxford, UK.

Roughgarden, J., and E. Fuentes. 1977. The environmental determinants of size in solitary populations of West Indian *Anolis* lizards. *Oikos* 29:44–51.

Roughgarden, J., D. Heckel, and E. Fuentes. 1983. Coevolutionary theory and the biogeography and community structure of *Anolis.* Pp. 371–410 in R. Huey, E. Pianka and T. Schoener, Eds., *Lizard Ecology: Studies of a Model Organism.* Harvard University Press: Cambridge, MA.

Roughgarden, J., and S. Pacala. 1989. Taxon cycle among *Anolis* lizard populations: Review of the evidence. Pp. 403–432 in D. Otte and J. Endler, Eds., *Speciation and its Consequences.* Sinauer Associates: Sunderland, MA.

Roughgarden, J., S. Pacala, and J. Rummel. 1984. Strong present-day competition between the *Anolis* lizard populations of St. Maarten (Neth. Antilles). Pp. 203–220 in B. Shorrocks, Ed., *Evolutionary Ecology.* Blackwell Scientific: Oxford, UK.

Roughgarden, J., W. Porter, and D. Heckel. 1981. Resource partitioning of space and its relationship to body temperature in *Anolis* lizard populations. *Oecologia* 50:256–264.

Rüber, L., and D.C. Adams. 2001. Evolutionary convergence of body shape and trophic morphology in cichlids from Lake Tanganyika. *Journal of Evolutionary Biology* 14:325–332.

Rüber, L., and R. Zardoya. 2005. Rapid cladogenesis in marine fishes revisited. *Evolution* 59:1119–1127.

Ruby, D. 1984. Male breeding success and differential access to females in *Anolis carolinensis*. *Herpetologica* 40:272–280.

Ruedi, M., and F. Mayer. 2001. Molecular systematics of bats of the genus *Myotis* (Vespertilionidae) suggests deterministic ecomorphological convergences. *Molecular Phylogeny and Evolution* 21:436–448.

Ruibal, R. 1961. Thermal relations of five species of tropical lizards. *Evolution* 15:98–111.

Ruibal, R., and V. Ernst. 1965. The structure of the digital setae of lizards. *Journal of Morphology* 117:271–294.

Ruibal, R., and R. Philibosian. 1970. Eurythermy and niche expansion in lizards. *Copeia* 1970:645–653.

Ruibal, R., and R. Philibosian. 1974a. Aggression in the lizard *Anolis acutus*. *Copeia* 1974:349–357.

Ruibal, R., and R. Philibosian. 1974b. The population ecology of the lizard *Anolis acutus*. *Ecology* 55:525–537.

Ruiz, C.C., and J. Wade. 2002. Sexual dimorphisms in a copulatory neuromuscular system in the green anole lizard. *Journal of Comparative Neurology* 443:289–297.

Rummel, J.D., and J. Roughgarden. 1985. Effects of reduced perch-height separation on competition between two *Anolis* lizards. *Ecology* 66:430–444.

Rundle, H.D., S.F. Chenoweth, and M.W. Blows. 2006. The roles of natural and sexual selection during adaptation to a novel environment. *Evolution* 60:2218–2225.

Rundle, H.D., L. Nagel, J.W. Boughman, and D. Schluter. 2000. Natural selection and parallel speciation in sympatric sticklebacks. *Science* 287:306–308.

Rundle, H.D., and P. Nosil. 2005. Ecological speciation. *Ecology Letters* 8:336–352.

Rundle, H.D., S.M. Vamosi, and D. Schluter. 2003. Experimental test of predation's effect on divergent selection during character displacement in sticklebacks. *Proceedings of the National Academy of Sciences of the United States of America* 100:14943–14948.

Russell, A.P., and V. Bels. 2001. Digital hyperextension in *Anolis sagrei*. *Herpetologica* 57: 58–65.

Russell, A.P., and M.K. Johnson. 2007. Real-world challenges to, and capabilities of, the gekkotan adhesive system: Contrasting the rough and the smooth. *Canadian Journal of Zoology* 85:1228–1238.

Rutherford, S.Z., and S. Lindquist. 1998. Hsp90 as a capacitor for morphological evolution. *Nature* 346:336–342.

Rutschmann, F. 2006. Molecular dating of phylogenetic trees: A brief review of current methods that estimate divergence times. *Diversity and Distributions* 12:35–48.

Ryan, M.J., D.K. Hews, and W.E. Wagner, Jr. 1990. Sexual selection on alleles that determine body size in the swordtail *Xiphophorus nigrensis*. *Behavioral Ecology and Sociobiology* 26: 231–237.

Salzburg, M.A. 1984. *Anolis sagrei* and *Anolis cristatellus* in southern Florida: A case study in interspecific competition. *Ecology* 65:14–19.

Salzburger, W., T. Mack, E. Verheyen, and A. Meyer. 2005. Out of Tanganyika: Genesis, explosive speciation, key-innovations and phylogeography of the haplochromine cichlid fishes. *BMC Evolutionary Biology* 5:17.

Salzburger, W., and A. Meyer. 2004. The species flocks of East African cichlid fishes: Recent advances in molecular phylogenetics and population genetics. *Naturwissenschaften* 91:277–290.

Sanderson, M.J. 2003. r8s: Inferring absolute rates of evolution and divergence times in the absence of a molecular clock. *Bioinformatics* 19:301–302.

Sanger, T.J., P.M. Hime, M.A. Johnson, J. Diani, and J.B. Losos. 2008a. Laboratory protocols for husbandry and embryo collection of *Anolis* lizards. *Herpetological Review* 39:58–63.

Sanger, T.J., J.B. Losos, and J.J. Gibson-Brown. 2008b. A developmental staging series for the lizard genus *Anolis*: A new system for the integration of evolution, development, and ecology. *Journal of Morphology* 269:129–137.

Savage, J.M. 2002. *The Amphibians and Reptiles of Costa Rica*. University of Chicago Press: Chicago, IL.

Savage, J.M., and C. Guyer. 1989. Infrageneric classification and species composition of the anole genera, *Anolis*, *Ctenonotus*, *Dactyloa*, *Norops*, and *Semiurus* (Sauria, Iguanidae). *Amphibia-Reptilia* 10:105–116.

Savage, J.M., and C. Guyer. 1991. Nomenclatural notes on anoles (Sauria: Polychrotidae): Stability over priority. *Journal of Herpetology* 25:365–366.

Sax, D.F., J.J. Stachowicz, J.H. Brown, J.F. Bruno, M.N. Dawson, S.D. Gaines, R.K. Grosberg, A. Hastings, R.D. Holt, M.M. Mayfield, M.I. O'Connor and W.R. Rice. 2007. Ecological and evolutionary insights from species invasions. *Trends in Ecology and Evolution* 22:465–471.

Scales, J., and M. Butler. 2007. Are powerful females powerful enough? Acceleration in gravid green iguanas (*Iguana iguana*). *Integrative and Comparative Biology* 47:285–294.

Schall, J.J. 1992. Parasite-mediated competition in *Anolis* lizards. *Oecologia* 92:58–64.

Schall, J.J., and A.R. Pearson. 2000. Body condition of a Puerto Rican anole, *Anolis gundlachi*: Effect of a malaria parasite and weather variation. *Journal of Herpetology* 34:489–491.

Schall, J.J., and C.M. Staats. 2002. Virulence of lizard malaria: Three species of *Plasmodium* infecting *Anolis sabanus*, the endemic anole of Saba, Netherlands Antilles. *Copeia* 2002:39–43.

Schall, J.J., and S.P. Vogt. 1993. Distribution of malaria in *Anolis* lizards of the Luquillo Forest, Puerto Rico: Implications for host community ecology. *Biotropica* 25:229–235.

Scheffer, M., S. Carpenter, J.A. Foley, C. Folke, and B. Walker. 2001. Catastrophic shifts in ecosystems. *Nature* 413:591–596.

Schlaepfer, M.A. 2003. Successful lizard eggs in a human-disturbed habitat. *Oecologia* 137:304–311.

Schlaepfer, M.A. 2006. Growth rates and body condition in *Norops polylepis* (Polychrotidae) vary with respect to sex but not mite load. *Biotropica* 38:414–418.

Schlaepfer, M.A., C. Hoover, and C.K. Dodd, Jr. 2005. Challenges in evaluating the impact of the trade in amphibians and reptiles on wild populations. *Bioscience* 55:256–264.

Schlichting, C.D., and H. Smith. 2002. Phenotypic plasticity: Linking molecular mechanisms with evolutionary outcomes. *Evolutionary Ecology* 16:189–211.

Schlichting, C.D., and M. Pigliucci. 1998. *Phenotypic Evolution: A Reaction Norm Perspective.* Sinauer Associates: Sunderland, MA.

Schluter, D. 1986. Tests for similarity and convergence of finch communities. *Ecology* 67: 1073–1085.

Schluter, D. 1988a. Character displacement and the adaptive divergence of finches on islands and continents. *American Naturalist* 131:799–824.

Schluter, D. 1988b. The evolution of finch communities on islands and continents: Kenya vs. Galápagos. *Ecological Monographs* 58:229–249.

Schluter, D. 1990. Species-for-species matching. *American Naturalist* 136:560–568.

Schluter, D. 1994. Experimental evidence that competition promotes divergence in adaptive radiation. *Science* 266:798–801.

Schluter, D. 1996. Adaptive radiation along the lines of least resistance. *Evolution* 50: 1766–1774.

Schluter, D. 2000. *The Ecology of Adaptive Radiation.* Oxford University Press: Oxford, UK.

Schluter, D. 2001. Ecology and the origin of species. *Trends in Ecology and Evolution* 16:372–380.

Schluter, D. and P. R. Grant. 1984. Determinants of morphological patterns in communities of Darwin's finches. *American Naturalist* 123:175–196.

Schluter, D., T. Price, A. Ø. Mooers, and D. Ludwig. 1997. Likelihood of ancestor states in adaptive radiation. *Evolution* 51:1699–1711.

Schmalhausen, I.I. 1949. *Factors of Evolution.* Blakiston: Philadelphia, PA.

Schmidt, B.R., and J. Van Buskirk. 2004. A comparative analysis of predator-induced plasticity in larval *Triturus* newts. *Journal of Evolutionary Biology* 18:415–425.

Schneider, C.J. 1996. Distinguishing between primary and secondary intergradation among morphologically differentiated populations of *Anolis marmoratus*. *Molecular Ecology* 5:239–249.

Schneider, C.J. 2008. Exploiting genomic resources in studies of speciation and adaptive radiation of lizards in the genus *Anolis*. *Integrative and Comparative Biology* 98:520–526.

Schneider, C.J., J.B. Losos, and K. de Queiroz. 2001. Evolutionary relationships of the *Anolis bimaculatus* group from the Northern Lesser Antilles. *Journal of Herpetology* 35:1–12.

Schneider, C.J., and C. Moritz. 1999. Rainforest refugia and evolution in Australia's wet tropics. *Proceedings of the Royal Society of London B* 266:191–196.

Schneider, C.J., T.B. Smith, B. Larison, and C. Moritz. 1999. A test of alternative models of diversification in tropical rainforests: Ecological gradients vs. rainforest refugia. *Proceedings of the National Academy of Sciences of the United States of America* 96: 13869–13873.

Schneider, C.J., and S.E. Williams. 2005. Effects of Quaternary climate change on rainforest diversity: Insights from spatial analyses of species and genes in Australia's wet tropics. Pp. 401–424 in E. Bermingham, C.W. Dick, and C. Moritz, Eds., *Tropical Rainforests: Past, Present, and Future.* University of Chicago Press: Chicago, IL.

Schneider, K.R., J.S. Parmerlee, Jr., and R. Powell. 2000. Escape behavior of *Anolis* lizards from the Sierra de Baoruco, Hispaniola. *Caribbean Journal of Science* 36:321–323.

Schoener, T.W. 1967. The ecological significance of sexual dimorphism in size of the lizard *Anolis conspersus*. *Science* 155:474–478.

Schoener, T.W. 1968. The *Anolis* lizards of Bimini: Resource partitioning in a complex fauna. *Ecology* 49:704–726.

Schoener, T.W. 1969. Size patterns in West Indian *Anolis* lizards. I. Size and species diversity. *Systematic Zoology* 18:386–401.

Schoener, T.W. 1970a. Nonsynchronous spatial overlap of lizards in patchy habitats. *Ecology* 51:408–418.

Schoener, T.W. 1970b. Size patterns in West Indian *Anolis* lizards. II. Correlations with the size of particular sympatric species—displacement and convergence. *American Naturalist* 104:155–174.

Schoener, T.W. 1974. Resource partitioning in ecological communities. *Science* 185:27–39.

Schoener, T.W. 1975. Presence and absence of habitat shift in some widespread lizard species. *Ecological Monographs* 45:233–258.

Schoener, T.W. 1976a. Habitat shift in widespread *Anolis* lizard species. Pp. 369–378 in *National Geographic Society Reports, 1968 Projects*. National Geographic Society: Washington, DC.

Schoener, T.W. 1976b. The species-area relation within archipelagos: Models and evidence from island land birds. Pp. 629–642 H.J. Frith and J.H. Calaby, Eds., *Proceedings of the 16th International Ornithological Congress*. Australian Academy of Science: Canberra, Australia.

Schoener, T.W. 1977. Competition and the niche. Pp. 35–136 in C. Gans and D. Tinkle, Eds., *Biology of the Reptilia, Volume 7: Ecology and Behaviour A*. Academic Press: London, UK.

Schoener, T.W. 1979. Feeding, spacing, and growth in four species of Bimini *Anolis* lizards. Pp. 479–485 in *National Geographic Society Reports, 1970 Projects*. National Geographic Society: Washington, DC.

Schoener, T.W. 1983. Field experiments on interspecific competition. *American Naturalist* 122:240–285.

Schoener, T.W. 1985. Are lizard population sizes unusually constant through time? *American Naturalist* 126:633–641.

Schoener, T.W. 1986a. Patterns in terrestrial vertebrate versus arthropod communities: Do systematic differences in regularity exist? Pp. 556–586 in J. Diamond and T.J. Case, Eds., *Community Ecology*. Harper & Row: New York, NY.

Schoener, T.W. 1986b. Resource partitioning. Pp. 91–126 in D.J. Anderson and J. Kikkawa, Eds., *Community Ecology: Pattern and Process*. Blackwell Scientific Publications: Melbourne, Australia.

Schoener, T.W. 1988. Testing for non-randomness in sizes and habitats of West Indian lizards: Choice of species pool affects conclusions from null models. *Evolutionary Ecology* 2:1–26.

Schoener, T.W. 1989. The ecological niche. Pp. 79–113 in J.M. Cherrett, Ed., *Ecological Concepts: The Contribution of Ecology to an Understanding of the Natural World*. Blackwell Scientific: Oxford, UK.

Schoener, T.W. 1996. Foreword. Pp. 9–10 in R. Powell and R.W. Henderson, Eds., *Contributions to West Indian Herpetology: A Tribute to Albert Schwartz*. Society for the Study of Amphibians and Reptiles: Ithaca, NY.

Schoener, T.W., and G.C. Gorman. 1968. Some niche differences in three Lesser Antillean lizards of the genus *Anolis*. *Ecology* 49:819–830.

Schoener, T.W., J.B. Losos, and D.A. Spiller. 2005. Island biogeography of populations: An introduced species transforms survival patterns. *Science* 310:1807–1809.

Schoener, T.W., and A. Schoener. 1971a. Structural habitats of West Indian *Anolis* lizards. I. Jamaican lowlands. *Breviora* 368:1–53.

Schoener, T.W., and A. Schoener. 1971b. Structural habitats of West Indian *Anolis* lizards. II. Puerto Rican uplands. *Breviora* 375:1–39.

Schoener, T.W., and A. Schoener. 1976. The ecological context of female pattern polymorphism in *Anolis sagrei*. *Evolution* 30:650–658.

Schoener, T.W., and A. Schoener. 1978. Estimating and interpreting body-size growth in some *Anolis* lizards. *Copeia* 1978:390–405.

Schoener, T.W., and A. Schoener. 1980a. Densities, sex ratios, and population structure in four species of Bahamian *Anolis* lizards. *Journal of Animal Ecology* 49:19–53.

Schoener, T.W., and A. Schoener. 1980b. Ecological and demographic correlates of injury rates in some Bahamian *Anolis* lizards. *Copeia* 1980:839–850.

Schoener, T.W., and A. Schoener. 1982a. Intraspecific variation in home-range size in some *Anolis* lizards. *Ecology* 63:809–823.

Schoener, T.W., and A. Schoener. 1982b. The ecological correlates of survival in some Bahamian *Anolis* lizards. *Oikos* 392:1–26.

Schoener, T.W., and A. Schoener. 1983a. Distribution of vertebrates on some very small islands. I. Occurrence sequences of individual species. *Journal of Animal Ecology* 52: 209–235.

Schoener, T.W., and A. Schoener. 1983b. Distribution of vertebrates on some very small islands. II. Patterns in species number. *Journal of Animal Ecology* 52:237–262.

Schoener, T.W., and A. Schoener. 1983c. The time to extinction of a colonizing propagule of lizards increases with island area. *Nature* 302:332–334.

Schoener, T.W., J.B. Slade, and C.H. Stinson. 1982. Diet and sexual dimorphism in the very catholic lizard genus, *Leiocephalus* of the Bahamas. *Oecologia* 53:160–169.

Schoener, T.W., and D.A. Spiller. 1996. Devastation of prey diversity by experimentally introduced predators in the field. *Nature* 381:691–694.

Schoener, T.W., and D.A. Spiller. 1999. Indirect effects in an experimentally staged invasion by a major predator. *American Naturalist* 153:347–358.

Schoener, T.W., D.A. Spiller, and J.B. Losos. 2001. Natural restoration of the species–area relation for a lizard after a hurricane. *Science* 294:1525–1528.

Schoener, T.W., D.A. Spiller, and J.B. Losos. 2002. Predation on a common *Anolis* lizard: Can the food-web effects of a devastating predator be reversed? *Ecological Monographs* 72:383–408.

Schoener, T.W., D.A. Spiller, and J.B. Losos. 2004. Variable ecological effects of hurricanes: The importance of timing for survival of lizards on Bahamian islands. *Proceedings of the National Academy of Sciences of the United States* 101:177–181.

Schulte, J.A. II, J.B. Losos, F.B. Cruz, and H. Núñez. 2004. The relationship between morphology, escape behaviour and microhabitat occupation in the lizard clade *Liolaemus* (Iguanidae: Tropidurinae*: Liolaemini). *Journal of Evolutionary Biology* 17:408–420.

Schulte, J.A. II, J.P. Valladares, and A. Larson. 2003. Phylogenetic relationships within Iguanidae inferred using molecular and morphological data and a phylogenetic taxonomy of iguanian lizards. *Herpetologica* 59:399–419.

Schulte, J.A. II, J.R. Macey, A. Larson, and T.J. Papenfuss. 1998. Molecular tests of phylogenetic taxonomies: A general procedure and example using four subfamilies of the lizard family Iguanidae. *Molecular Phylogeny and Evolution* 10:367–376.

Schulte, J.A. II, J.R. Macey, and T.J. Papenfuss. 2006. A genetic perspective on the geographic association of taxa among arid North American lizards of the *Sceloporus magister* complex (Squamata: Iguanidae: Phrynosomatinae). *Molecular Phylogenetics and Evolution* 39:873–880.

Schwartz, A. 1968. Geographic variation in *Anolis distichus* Cope (Lacertilia, Iguanidae) in the Bahama Islands and Hispaniola. *Bulletin of the Museum of Comparative Zoology* 137:255–310.

Schwartz, A. 1973. A new species of montane *Anolis* (Sauria, Iguanidae) from Hispaniola. *Annals of the Carnegie Museum* 44:183–195.

Schwartz, A. 1978. A new species of aquatic *Anolis* (Sauria, Iguanidae) from Hispaniola. *Annals of the Carnegie Museum* 47:261–279.

Schwartz, A. 1989. A review of the cybotoid anoles (Reptilia: Sauria: Iguanidae) from Hispaniola. *Contributions in Biology and Geology, Milwaukee Public Museum* 78:1–32.

Schwartz, A., and R.W. Henderson. 1991. *Amphibians and Reptiles of the West Indies: Descriptions, Distributions, and Natural History*. University of Florida Press: Gainesville, FL.

Schwenk, K. 2000. Feeding in lepidosaurs. Pp. 175–291 in K. Schwenk, Ed., *Feeding: Form, Function and Evolution in Tetrapod Vertebrates*. Academic Press: San Diego, CA.

Schwenk, K., and G.C. Mayer. 1991. Tongue display in anoles and its evolutionary basis. Pp. 131–140 in J.B. Losos and G.C. Mayer, Eds., Anolis *Newsletter IV*. Division of Amphibians and Reptiles, National Museum of Natural History, Smithsonian Institution: Washington, DC.

Schwenk, K., and G.P. Wagner. 2003. Constraint. Pp. 52–61 in B.K. Hall and W.M. Olson, Eds., *Keywords and Concepts in Evolutionary Developmental Biology*. Harvard University Press: Cambridge, MA.

Schwenk, K., and G.P. Wagner. 2004. The relativism of constraints on phenotypic evolution. Pp. 390–408 in M. Pigliucci and K. Preston, Eds., *Phenotypic Integration: Studying the Ecology and Evolution of Complex Phenotypes*. Oxford University Press: Oxford, UK.

Scott, M. 1984. Agonistic and courtship displays of male *Anolis sagrei*. *Breviora* 479:1–22.

Scott, N.J., D.E. Wilson, C. Jones, and R.M. Andrews. 1976. The choice of perch dimensions by lizards of the genus *Anolis* (Reptilia, Lacertilia, Iguanidae). *Journal of Herpetology* 10:75–84.

Seaman, G.A., and J.E. Randall. 1962. The mongoose as a predator in the Virgin Islands. *Journal of Mammalogy* 43:544–546.

Seebacher, F., and C.E. Franklin. 2005. Physiological mechanisms of thermoregulation in reptiles: A review. *Journal of Comparative Physiology B* 175:533–541.

Seehausen, O. 2006. African cichlid fish: A model system in adaptive radiation research. *Proceedings of the Royal Society of London B* 273:1987–1998.

Seehausen, O., and J.J.M. van Alphen. 1999. Can sympatric speciation by disruptive sexual selection explain rapid evolution of cichlid diversity in Lake Victoria? *Ecology Letters* 2:262–271.

Servedio, M.R., and M.A.F. Noor. 2003. The role of reinforcement in speciation: Theory and data. *Annual Review of Ecology and Systematics* 34:339–364.

Settle, W.H., and L.T. Wilson. 1990. Invasion by the variegated leafhopper and biotic interactions: Parasitism, competition, and apparent competition. *Ecology* 71:1461–1470.

Sever, D.M., and W.C. Hamlett. 2002. Female sperm storage in reptiles. *Journal of Experimental Zoology* 292:187–199.

Sexton, O.J., J. Bauman, and E. Ortleb. 1972. Seasonal food habits of *Anolis limifrons*. *Ecology* 53:182–186.

Sexton, O.J. and H. Heatwole. 1968. An experimental investigation of habitat selection and water loss in some anoline lizards. *Ecology* 49:762–767.

Sexton, O.J., E.P. Ortleb, L.M. Hathaway, R.E. Ballinger, and P.E. Licht. 1971. Reproductive cycles of three species of anoline lizards from the Isthmus of Panama. *Ecology* 52: 201–215.

Shafir, S., and J. Roughgarden. 1994. Instrumental discrimination conditioning of *Anolis cristatellus* in the field with food as a reward. *Caribbean Journal of Science* 30:228–233.

Shapiro, M.D., M.A. Bell, and D.M. Kingsley. 2006. Parallel genetic origins of pelvic reduction in vertebrates. *Proceedings of the National Academy of Sciences of the United States of America* 103:13753–13758.

Shapiro, M.D., M.E. Marks, C.L. Peichel, B.K. Blackman, K.S. Nereng, B. Jónsson, D. Schluter, and D.M. Kingsley. 2004. Genetic and developmental basis of evolutionary pelvic reduction in threespine sticklebacks. *Nature* 428:717–723.

Shaw, F.H., R.G. Shaw, G.S. Wilkinson, and M. Turelli. 1995. Changes in genetic variances and covariances: G Whiz! *Evolution* 49:1260–1267.

Shaw, P.W., G.F. Turner, M.R. Idid, R.L. Robinson and G.R. Carvalho. 2000. Genetic population structure indicates sympatric speciation of Lake Malawi pelagic cichlids. *Proceedings of the Royal Society of London B* 267:2273–2280.

Sheridan, R.E., H.T. Mullins, J.A. Austin, Jr., M.M. Ball, and J.W. Ladd. 1988. Geology and geophysics of the Bahamas. Pp. 329–364 in R.E. Sheridan and J.A. Grow, Eds., *The Atlantic Continental Margin: U.S.* Geological Society of America: Boulder, CO.

Shew, J.J., S.C. Larimer, R. Powell, and J.S. Parmerlee, Jr. 2002. Sleeping patterns and sleep-site fidelity of the lizard *Anolis gingivinus* on Anguilla. *Caribbean Journal of Science* 38:136–138.

Shochat, D., and H.C. Dessauer. 1981. Comparative immunological study of albumins of *Anolis* lizards of the Caribbean islands. *Comparative Biochemistry and Physiology* 68A: 67–73.

Shuster, S.M., and M.J. Wade. 1991. Equal mating success among male reproductive strategies in a marine isopod. *Nature* 350:608–610.

Sibley, C.G., and J.E. Ahlquist. 1990. *Phylogeny and Classification of Birds*. Yale University Press: New Haven, CT.

Sifers, S.M., M.L. Yeska, Y.M. Ramos, R. Powell, and J.S. Parmerlee, Jr. 2001. *Anolis* lizards restricted to altered edge habitats in a Hispaniolan cloud forest. *Caribbean Journal of Science* 37:55–62.

Simmonds, F.J. 1958. The effect of lizards on the biological control of scale insects in Bermuda. *Bulletin of Entomological Research* 49:601–612.

Simmons, P.M., B.T. Greene, K.E. Williamson, R. Powell, and J.S. Parmerlee, Jr. 2005. Ecological interactions within a lizard community on Grenada. *Herpetologica* 61:124–134.

Simpson, G.G. 1951. The species concept. *Evolution* 5:285–298.

Simpson, G.G. 1953. *The Major Features of Evolution.* Columbia University Press: New York, NY.

Sinclair, E.A., R.L. Bezy, K. Bolles, J.L. Camarillo R., K.A. Crandall, and J.W. Sites, Jr. 2004. Testing species boundaries in an ancient species complex with deep phylogeographic history: Genus *Xantusia* (Squamata: Xantusiidae). *American Naturalist* 164:396–414.

Sinervo, B., P. Doughty, R.B. Huey, and K. Zamudio. 1992. Allometric engineering: A causal analysis of natural selection on offspring size. *Science* 258:1927–1930.

Sinervo, B., and J.B. Losos. 1991. Walking the tight rope: Arboreal sprint performance among *Sceloporus occidentalis* lizard populations. *Ecology* 72:1225–1233.

Singhal, S., M.A. Johnson, and J.T. Ladner. 2007. The behavioral ecology of sleep: Natural sleeping site choice in three *Anolis* lizard species. *Behaviour* 144:1033–1052.

Slabbekoorn, H., and T.B. Smith. 2002. Bird song, ecology and speciation. *Philosophical Transactions of the Royal Society of London B* 357:493–503.

Slinker, B.K., and S.A. Glantz. 1985. Multiple regression for physiological data analysis: The problem of multicollinearity. *American Journal of Physiology* 249 (*Regulatory and Integrative Comparative Physiology* 18): R1–R12.

Slowinski, J.B., and C. Guyer. 1989. Testing the stochasticity of patterns of organismal diversity: An improved null model. *American Naturalist* 134:907–921.

Smith, H.M., G. Sinelnik, J.D. Fawcett, and R.E. Jones. 1972. (1973). A survey of the chronology of ovulation in anoline lizard genera. *Transactions of the Kansas Academy of Science* 75:107–120.

Smith, J.W., and Benkman, C.W. 2007. A coevolutionary arms race causes ecological speciation in crossbills. *American Naturalist* 169:455–465.

Smith, K.T. 2006. A diverse new assemblage of late Eocene squamates (Reptilia) from the Chadron Formation of North Dakota, U.S.A. *Palaeontologia Electronica* 9: 5A: 44pp.

Smith, T.B. 1993. Disruptive selection and the genetic basis of bill size polymorphism in the African finch *Pyronestes*. *Nature* 363:618–620.

Smith, T.B., C.J. Schneider, and K. Holder. 2001. Refugial isolation versus ecological gradients. *Genetica* 112–113:383–398.

Sneath, P.H.A., and R.R. Sokal. 1973. *Numerical Taxonomy.* W.H. Freeman: San Francisco, CA.

Snorrason, S.S., and S. Skúlason. 2004. Adaptive speciation in northern freshwater fishes. Pp. 210–229 in U. Dieckmann, M. Doebeli, J.A.J. Metz, and D. Tautz, Eds., *Adaptive Speciation.* Cambridge University Press: Cambridge, UK.

Socci, A.M., M.A. Schlaepfer, and T.A. Gavin. 2005. The importance of soil moisture and leaf cover in a female lizard's (*Norops polylepis*) evaluation of potential oviposition sites. *Herpetologica* 61:233–240.

Spezzano, L.C. Jr., and B.C. Jayne. 2004. The effects of surface diameter and incline on the hindlimb kinematics of an arboreal lizard (*Anolis sagrei*). *Journal of Experimental Biology* 207:2115–2131.

Spiller, D.A., and T.W. Schoener. 1988. An experimental study of the effect of lizards on web-spider communities. *Ecological Monographs* 58:57–77.

Spiller, D.A., and T.W. Schoener. 1990. A terrestrial field experiment showing the impact of eliminating top predators on foliage damage. *Nature* 347:469–472.

Spiller, D.A., and T.W. Schoener. 1996. Food-web dynamics on some small subtropical islands: effects of top and intermediate predators. Pp. 160–169 in G.A. Polis and K.O. Winemiller, Eds., *Food Webs: Integration of Patterns and Dynamics*. Chapman and Hall: New York, NY.

Spiller, D.A., and T.W. Schoener. 1998. Lizards reduce spider species richness by excluding rare species. *Ecology* 79:503–516.

Springer, M.S., J.A.W. Kirsch, and J.A. Chase. 1997. The chronicle of marsupial evolution. Pp. 126–161 in T.J. Givnish and K.J. Systma, Eds., *Molecular Evolution and Adaptive Radiation*. Cambridge University Press: Cambridge, UK.

Staats, C.M., and J.J. Schall. 1996. Malarial parasites (*Plasmodium*) of *Anolis* lizards: Biogeography in the Lesser Antilles. *Biotropica* 28:388–393.

Stadelmann, B., L.-K. Lin, T.H. Kunz, and M. Ruedi. 2007. Molecular phylogeny of New World *Myotis* (Chiroptera, Vespertilionidae) inferred from mitochondrial and nuclear DNA genes. *Molecular Phylogeny and Evolution* 43:32–48.

Stamps, J.A. 1975. Courtship patterns, estrus periods and reproductive conditions in a lizard, *Anolis aeneus*. *Physiology and Behavior* 14:531–535.

Stamps, J.A. 1976. Egg retention, rainfall and egg laying in a tropical lizard *Anolis aeneus*. *Copeia* 1976:759–764.

Stamps, J.A. 1977a. The function of the survey posture in *Anolis* lizards. *Copeia* 1977:756–758.

Stamps, J.A. 1977b. The relationship between resource competition, risk, and aggression in a tropical territorial lizard. *Ecology* 58:349–358.

Stamps, J.A. 1983a. Sexual selection, sexual dimorphism, and territoriality. Pp. 169–204 in R.B. Huey, E.R. Pianka, and T.W. Schoener, Eds., *Lizard Ecology: Studies of a Model Organism*. Harvard University Press: Cambridge, MA.

Stamps, J.A. 1983b. The relationship between ontogenetic habitat shifts, competition and predator avoidance in a juvenile lizard (*Anolis aeneus*). *Behavioral Ecology and Sociobiology* 12:19–33.

Stamps, J.A. 1987. The effect of familiarity with a neighborhood on territory acquisition. *Behavioral Ecology and Sociobiology* 21:273–277.

Stamps, J.A. 1988. Conspecific attraction and aggregation in territorial species. *American Naturalist* 131:329–347.

Stamps, J.A. 1990. Starter homes for young lizards. *Natural History* 100(10):40–44.

Stamps, J.A. 1994. Territorial behavior: Testing the assumptions. *Advances in the Study of Behavior* 23:173–231.

Stamps, J.A. 1995. Using growth-based models to study behavioral factors affecting sexual size dimorphism. *Herpetological Monographs* 8:75–87.

Stamps, J.A. 1999. Relationships between female density and sexual size dimorphism in samples of *Anolis sagrei*. *Copeia* 1999:760–765.

Stamps, J.A. 2001. Learning from lizards. Pp. 149–168 in L.A. Dugatkin, Ed., *Model Systems in Behavioral Ecology: Integrating Conceptual, Theoretical, and Empirical Approaches*. Princeton University Press: Princeton, NJ.

Stamps, J.A., and G.W. Barlow. 1973. Variation and stereotypy in the displays of *Anolis aeneus* (Sauria: Iguanidae). *Behaviour* 47:67–94.

Stamps, J.A., and D.P. Crews. 1976. Seasonal changes in reproduction and social behavior in the lizard *Anolis aeneus*. *Copeia* 1976:467–476.

Stamps, J.A., and S.M. Gon. 1983. Sex-biased pattern variation in the prey of birds. *Annual Review of Ecology and Systematics* 14:231–253.

Stamps, J.A., J.B. Losos, and R.M. Andrews. 1997. A comparative study of population density and sexual size dimorphism in lizards. *American Naturalist* 149:64–90.

Stamps, J.A., and S. Tanaka. 1981. The influence of food and water on growth rates in a tropical lizard (*Anolis aeneus*). *Ecology* 62:33–40.

Stamps, J.A., S. Tanaka, and V.V. Krishnan. 1981. The relationship between selectivity and food abundance in a juvenile lizard. *Ecology* 62:1079–1092.

Stayton, C.T. 2006. Testing hypotheses of convergence with multivariate data: Morphological and functional convergence among herbivorous lizards. *Evolution* 60:824–841.

Steffen, J.E., and K.J. McGraw. 2007. Contributions of pterin and carotenoid pigments to dewlap coloration in two anole species. *Comparative Biochemistry and Physiology B* 146: 42–46.

Stenson, A.G., A. Malhotra, and R.S. Thorpe. 2002. Population differentiation and nuclear gene flow in the Dominican anole (*Anolis oculatus*). *Molecular Ecology* 11: 1679–1688.

Stenson, A.G., R.S. Thorpe, and A. Malhotra. 2004. Evolutionary differentiation of *bimaculatus* group anoles based on analyses of mtDNA and microsatellite data. *Molecular Phylogeny and Evolution* 32:1–10.

Steppan, S.J., C. Zawadzki, and L.R. Heaney. 2003. Molecular phylogeny of the endemic Philippine rodent *Apomys* (Muridae) and the dynamics of diversification in an oceanic archipelago. *Biological Journal of the Linnean Society* 80:699–715.

Stiassny, M.J., and A. Meyer. 1999. Cichlids of the rift lakes. *Scientific American* 280(2): 64–69.

Stinchcombe, J.R., and H.E. Hoekstra. 2008. Combining population genomics and quantitative genetics: Finding the genes underlying ecologically important traits. *Heredity* 100: 158–170.

Stockwell, E.F. 2001. Morphology and flight manoeuverability in new world leaf-nosed bats (Chiroptera: Phyllostomidae). *Journal of Zoology* 254:505–514.

Stopper, G.F., and G.P. Wagner. 2005. Of chicken wings and frog legs: A smorgasbord of evolutionary variation in mechanisms of tetrapod limb development. *Developmental Biology* 288:21–39.

Storm, E.E., T.V. Huynh, N.G. Copeland, N.A. Jenkins, D.M. Kingsley, and S.-J. Lee. 1994. Limb alterations in *brachypodism* mice due to mutations in a new member of the TGF β-superfamily. *Nature* 368:639–643.

Streelman, J.T., M. Alfaro, M.W. Westneat, D.R. Bellwood, and S.A. Karl. 2002. Evolutionary history of the parrotfishes: Biogeography, ecomorphology, and comparative diversity. *Evolution* 56:961–971.

Streelman, J.T., and P.D. Danley. 2003. The stages of evolutionary radiation. *Trends in Ecology and Evolution* 18:126–131.

Strong, D.R., D. Simberloff, L.G. Abele, and A. Thistle. Eds. *Ecological Communities: Conceptual Issues and the Evidence.* Princeton University Press: Princeton, NJ.

Stuart, B.L., A.G.J. Rhodin, L.L. Grismer, and T. Hansel. 2006. Scientific description can imperil species. *Science* 312:1137.

Stuart-Fox, D., and I.P.F. Owens. 2003. Species richness in agamid lizards: Chance, body size, sexual selection or ecology? *Journal of Evolutionary Biology* 16:659–669.

Stuart-Fox, D.M., A. Moussalli, N.J. Marshall, and I.P.F. Owens. 2003. Conspicuous males suffer higher predation risk: Visual modelling and experimental evidence from lizards. *Animal Behaviour* 66:541–550.

Sucena, E., I. Sdelon, I. Jones, F. Payre, and D.L. Stern. 2003. Regulatory evolution of *shavenbaby/ovo* underlies multiple cases of morphological parallelism. *Nature* 424:935–938.

Sullivan, B.K., and M.A. Kwiatkowski. 2007. Courtship displays in anurans and lizards: Theoretical and empirical contributions to our understanding of costs and selection on males due to female choice. *Functional Ecology* 21:666–675.

Sultan, S.E. 1992. What has survived of Darwin's theory? *Evolutionary Trends in Plants* 6:61–71.

Summers, C.H., W.J. Korzan, J.L. Lukkes, M.J. Watt, G.L. Forster, O. Overli, E. Hoglund, E.T. Larson, P.J. Ronan, J.M. Matter, T.R. Summers, K.J. Renner, and N. Greenberg. 2005. Does serotonin influence aggression? Comparing regional activity before and during social interaction. *Physiological and Biochemical Zoology* 78:679–694.

Suzuki, A., and M. Nagoshi. 1999. Habitat utilization of the native lizard, *Cryptoblepharis boutonii nigropunctatus*, in areas with and without the introduced lizard, *Anolis carolinensis*, on Hahajima, the Ogasawara Islands, Japan. Pp. 155–168 in H. Ota, Ed., *Tropical Island Herpetofauna: Origin, Current Diversity, and Conservation*. Elsevier: Amsterdam, Netherlands.

Svensson, E.I., F. Eroukhmanoff, and M. Friberg. 2006. Effects of natural and sexual selection on adaptive population divergence and premating isolation in a damselfly. *Evolution* 60:1242–1253.

Swofford, D.L. 1991. When are phylogeny estimates from molecular and morphological data incongruent? Pp. 90–128 in M.M. Miyamoto and J. Cracraft, Eds., *Phylogenetic Analysis of DNA Sequences*. Oxford University Press: Oxford, UK.

Swofford, D.L. and W.P. Maddison. 1987. Reconstructing ancestral character states under Wagner parsimony. *Mathematical Biosciences* 87:199–229.

Talbot, J.J. 1977. Habitat selection in two tropical anoline lizards. *Herpetologica* 33:114–123.

Talbot, J.J. 1979. Time budget, niche overlap, inter- and intraspecific aggression in *Anolis humilis* and *A. limifrons* from Costa Rica. *Copeia* 1979:472–481.

Taylor, E.B., and J.D. McPhail. 2000. Historical contingency and ecological determinism interact to prime speciation in sticklebacks, *Gasterosteus*. *Proceedings of the Royal Society of London B* 267:2375–2384.

Telford, S.R. Jr. 1974. The malarial parasites of *Anolis* species (Sauria: Iguanidae) in Panama. *International Journal of Parasitology* 4:91–102.

Templeton, A.R. 1998. Species and speciation: Geography, population structure, ecology, and gene trees. Pp. 32–43 in D.J. Howard and S.H. Berlocher, Eds., *Endless Forms: Species and Speciation*. Oxford University Press: New York, NY.

Tewksbury, J.J., R.B. Huey, and C.A. Deutsch. 2008. Putting the heat on tropical animals. *Science* 320:1296–1297.

Thomas, R., and A. Schwartz. 1967. The *monticola* group of the lizard genus *Anolis* in Hispaniola. *Breviora* 261:1–27.

Thomas, R., and S.B. Hedges. 1991. Rediscovery and description of the Hispaniolan lizard *Anolis darlingtoni* (Sauria: Iguanidae). *Caribbean Journal of Science* 27:90–93.

Thomas, Y., M.-T. Bethenod, L. Pelozuelo, B. Frérot, and D. Bourguet. 2003. Genetic isolation between two sympatric host-plant races of the European corn borer, *Ostrinia nubilalis* Hubner. I. Sex pheromone, moth emergence timing and parasitism. *Evolution* 57: 261–273.

Thompson, J.N. 2006. *The Geographic Mosaic of Coevolution*. University of Chicago Press: Chicago, IL.

Thorpe, R.S. 1984. Primary and secondary transition zones in speciation and population differentiation: A phylogenetic analysis of range expansion. *Evolution* 38:233–243.

Thorpe, R.S. 2002. Analysis of color spectra in comparative evolutionary studies: Molecular phylogeny and habitat adaptation in the St. Vincent anole (*Anolis trinitatis*). *Systematic Biology* 51:554–569.

Thorpe, R.S., and C.M. Crawford. 1979. The comparative abundance and resource partitioning of two green-gecko species (*Phelsuma*) on Praslin, Seychelles. *British Journal of Herpetology* 6:19–24.

Thorpe, R.S., A.G. Jones, A. Malhotra, and Y. Surget-Groba. 2008. Adaptive radiation in Lesser Antillean lizards: Molecular phylogenetics and species recognition in the Lesser Antillean dwarf gecko complex, *Sphaerodactylus fantasticus*. *Molecular Ecology* 17:1489–1504.

Thorpe, R.S., D.L. Leadbeater, and C.E. Pook. 2005a. Molecular clocks and geological dates: Cytochrome *b* of *Anolis extremus* substantially contradicts dating of Barbados emergence. *Molecular Ecology* 14:2087–2096.

Thorpe, R.S., A. Malhotra, A.G. Stenson and J.T. Reardon. 2004. Adaptation and speciation in Lesser Antillean anoles. Pp. 322–344 in U. Dieckmann, M. Doebeli, J.A.J. Metz and D. Tautz. Eds., *Adaptive Speciation*. Cambridge University Press: Cambridge, UK.

Thorpe, R.S., J.T. Reardon, and A. Malhotra. 2005b. Common garden and natural selection experiments support ecotypic differentiation in the Dominican anole (*Anolis oculatus*). *American Naturalist* 165:495–504.

Thorpe, R.S., and A.G. Stenson. 2003. Phylogeny, paraphyly and ecological adaptation of the colour and pattern in the *Anolis roquet* complex on Martinique. *Molecular Ecology* 12:117–132.

Thorpe, R.S., Y. Surget-Groba, and H. Johansson. 2008. The relative importance of ecology and geographic isolation for speciation in anoles. *Philosophical Transactions of the Royal Society B* 363:3071–3081.

Tickle, C. 2002. Vertebrate limb development and possible clues to diversity in limb form. *Journal of Morphology* 252:29–37.

Tilman, D., P.B. Reich, J. Knops, D. Wedin, T. Mielke, and C. Lehman. 2001. Diversity and productivity in a long-term grassland experiment. *Science* 294:843–845.

Timmerman, A., B. Dalsgaard, J.M. Olesen, L.H. Andersen, and A.M. Martínez González. 2008. Natural history notes. *Anolis aeneus* (Grenadian bush anole), *Anolis richardii* (Grenadian tree anole). Nectarivory/pollination. *Herpetological Review* 39:84–85.

Toft, C.A., and T.W. Schoener. 1983. Abundance and diversity of orb spiders on 106 Bahamian islands: Biogeography at an intermediate trophic level. *Oikos* 41:411–426.

Tokarz, R.R. 1985. Body size as a factor determining dominance in staged agonistic encounters between male brown anoles (*Anolis sagrei*). *Animal Behaviour* 33:746–753.

Tokarz, R.R. 1988. Copulatory behaviour of the lizard *Anolis sagrei*: Alternation of hemipenis use. *Animal Behaviour* 36:1518–1524.

Tokarz, R.R. 1992. Male mating preference for unfamiliar females in the lizard, *Anolis sagrei*. *Animal Behaviour* 44: 843–849.

Tokarz, R.R. 1995. Mate choice in lizards: A review. *Herpetological Monographs* 8:17–40.

Tokarz, R.R. 1998. Mating pattern in the lizard *Anolis sagrei*: Implications for mate choice and sperm competition. *Herpetologica* 54:388–394.

Tokarz, R.R. 1999. Relationship between copulation duration and sperm transfer in the lizard *Anolis sagrei*. *Herpetologica* 55:234–241.

Tokarz, R.R. 2002. An experimental test of the importance of the dewlap in male mating success in the lizard *Anolis sagrei*. *Herpetologica* 58:87–94.

Tokarz, R.R. 2007. Changes in the intensity of male courtship behavior following physical exposure of males to previously unfamiliar females in brown anoles (*Anolis sagrei*). *Journal of Herpetology* 41:501–505.

Tokarz, R.R., and S.J. Kirkpatrick. 1991. Copulation frequency and pattern of hemipenis use in males of the lizard *Anolis sagrei* in a semi-natural enclosure. *Animal Behaviour* 41: 1039–1044.

Tokarz, R.R., S. McMann, L.C. Smith, and H. John-Alder. 2002. Effects of testosterone treatment and season on the frequency of dewlap extensions during male-male interactions in the lizard *Anolis sagrei*. *Hormones and Behavior* 41:70–79.

Tokarz, R.R., and J. Slowinski. 1990. Alternation of hemipenis use as a behavioural means of increasing sperm transfer in the lizard *Anolis sagrei*. *Animal Behaviour* 40:374–379.

Tolson, P.J., and R.W. Henderson. 2006. An overview of snake conservation in the West Indies. *Applied Herpetology* 6:345–356.

Toro, E., A. Herrel, and D. Irschick. 2004. The evolution of jumping performance in Caribbean *Anolis* lizards: Solutions to biomechanical trade-offs. *American Naturalist* 163: 844–856.

Toro, E., A. Herrel, and D.J. Irschick. 2006. Movement control strategies during jumping in a lizard (*Anolis valencienni*). *Journal of Biomechanics* 39:2014–2019.

Townsend, T.M., A. Larson, E. Louis, and J.R. Macey. 2004. Molecular phylogenetics of Squamata: The position of snakes, amphisbaenians, and dibamids, and the root of the squamate tree. *Systematic Biology* 53:735–757.

Travisano, M., J.A. Mongold, A.F. Bennett, and R.E. Lenski. 1995. Experimental tests of the roles of adaptation, chance, and history in evolution. *Science* 267:87–90.

Treglia, M.L., A.J. Muensch, R. Powell, and J. S. Parmerlee, Jr. 2008. Invasive *Anolis sagrei* on St. Vincent and its potential impact on perch heights of *Anolis trinitatis*. *Caribbean Journal of Science* 44:251–256.

Trivers, R. 1976. Sexual selection and resource-accruing abilities in *Anolis garmani*. *Evolution* 30:253–269.

Trivers, R. 1985. *Social Evolution*. Benjamin/Cummings Publishing Co.: Menlo Park, CA.

Turner, D. 2005. Local underdetermination in historical science. *Philosophy and Science* 72:209–230.

Ugueto, G.N., G.R. Fuenmayor, T. Barros, S.J. Sánchez-Pachecho, and J.E. García-Pérez. 2007. A revision of the Venezuelan anoles I: A new species from the Andes of Venezuela with the redescription of *Anolis jacare* Boulenger 1903 (Reptilia: Polychrotidae) and the clarification of the status of *Anolis nigropunctatus* Williams 1974. *Zootaxa* 1501:1–30.

Underwood, G. 1970. The eye. Pp. 1–97 in C. Gans, Ed., *Biology of the Reptilia, Vol. 2*. Academic Press: New York, NY.

Underwood, G., and E. Williams. 1959. The anoline lizards of Jamaica. *Bulletin of the Institute of Jamaica, Science Series* 9:1–48.

Valido, A. 2006. *Anolis allisoni* (Allison's anole/Camaleón azul). Nectar Feeding. *Herpetological Review* 37:461.

Vamosi, S.M. 2005. On the role of enemies in divergence and diversification of prey: A review and synthesis. *Canadian Journal of Zoology* 83:894–910.

van Berkum, F.H. 1986. Evolutionary patterns of the thermal sensitivity of sprint speed in *Anolis* lizards. *Evolution* 40:495–604.

Van Damme, R., and T.J.M. Van Dooren. 1999. Absolute versus per unit body length speed of prey as an estimator of vulnerability to predation. *Animal Behaviour* 57:347–352.

Van der Klaauw, C.J. 1948. Ecological morphology. *Bibliotheca Biotheoretica D* 4:27–111.

Vanderpoorten, A., and B. Goffinet. 2006. Mapping uncertainty and phylogenetic uncertainty in ancestral character state reconstruction: An example in the moss genus *Brachytheciastrum*. *Systematic Biology* 55:957–971.

Vanhooydonck, B., P. Aerts, D.J. Irschick, and A. Herrel. 2006a. Power generation during locomotion in *Anolis* lizards: An ecomorphological approach. Pp. 253–270 in A. Herrel, T. Speck, and N.P. Rowe, Eds., *Ecology and Biomechanics*. Taylor and Francis: Boca Raton, FL.

Vanhooydonck, B., A. Herrel, and D.J. Irschick. 2006b. Out on a limb: The differential effect of substrate diameter on acceleration capacity in *Anolis* lizards. *Journal of Experimental Biology* 209:4515–4523.

Vanhooydonck, B., A. Herrel, and D.J. Irschick. 2007. Determinants of sexual differences in escape behavior in lizards of the genus *Anolis*: A comparative approach. *Integrative and Comparative Biology* 47:200–210.

Vanhooydonck, B., A. Herrel, R. Van Damme, and D.J. Irschick. 2005. Does dewlap size predict male bite performance in Jamaican *Anolis* lizards? *Functional Ecology* 19:38–42.

Vanhooydonck, B., A. Herrel, R. Van Damme, and D.J. Irschick. 2006c. The quick and the fast: The evolution of acceleration capacity in *Anolis* lizards. *Evolution* 60:2137–2147.

Vanhooydonck, B., R. Van Damme, and P. Aerts. 2002. Variation in speed, gait characteristics and microhabitat use in lacertid lizards. *Journal of Experimental Biology* 205:1037–1046.

Vasemägi, A., and C.R. Primmer. 2005. Challenges for identifying functionally important genetic variation: The promise of combining complementary research strategies. *Molecular Ecology* 14:3623–3642.

Velasco, J.A., and A. Herrel. 2007. Ecomorphology of *Anolis* lizards of the Chocó region in Colombia and comparisons with Greater Antillean ecomorphs. *Biological Journal of the Linnean Society* 92:29–39.

Vellend, M., L.J. Harmon, J.L. Lockwood, M.M. Mayfield, A.R. Hughes, J.P. Wares, and D.F. Sax. 2007. Effects of exotic species on evolutionary diversification. *Trends in Ecology and Evolution* 22:481–488.

Vermeij, G.J. 1974. Adaptation, versatility, and evolution. *Systematic Zoology* 22:466–477.

Vermeij, G.J. 1987. *Evolution and Escalation: An Ecological History of Life.* Princeton University Press: Princeton, NJ.

Verwaijen, D., R. Van Damme, and A. Herrel. 2002. Relationships between head size, bite force, prey handling efficiency and diet in two sympatric lacertid lizards. *Functional Ecology* 16:842–850.

Vilella, F.J. 1998. Biology of the mongoose (*Herpestes javanicus*) in a rain forest of Puerto Rico. *Biotropica* 30:120–125.

Vinson, J., and J.-M. Vinson. 1969. The saurian fauna of the Mascarene Islands. *Mauritius Institute Bulletin* 6:203–320.

Vitt, L.J. 1995. The ecology of tropical lizards in the Caatinga of northeast Brazil. *Occasional Papers of the Oklahoma Museum of Natural History* 1:1–29.

Vitt, L.J., T.C.S. Avila-Pires, M.C. Espósito, S.S. Sartorius and P.A. Zani. 2003a. Sharing Amazonian rain-forest trees: Ecology of *Anolis punctatus* and *A. transversalis* (Squamata: Polychrotidae). *Journal of Herpetology* 37:276–285.

Vitt, L.J., T.C.S. Avila-Pires, P.A. Zani, and M.C. Espósito. 2002. Life in shade: The ecology of *Anolis trachyderma* (Squamata: Polychrotinae) in Amazonian Ecuador and Brazil, with comparisons to ecologically similar anoles. *Copeia* 2003:275–286.

Vitt, L.J., T.C.S. Avila-Pires, P.A. Zani, S.S. Sartorius, and M.C. Espósito. 2003b. Life above ground: Ecology of *Anolis fuscoauratus* in the Amazon rain forest, and comparisons with its nearest relatives. *Canadian Journal of Zoology* 81:142–156.

Vitt, L.J., and C. Morato de Carvalho. 1995. Niche partitioning in a tropical wet season: Lizards in the lavrado area of northern Brazil. *Copeia* 1995:305–329.

Vitt, L.J., and E.R. Pianka. 2003. *Lizards: Windows to the Evolution of Diversity.* University of California Press: Berkeley, CA.

Vitt, L.J., S.S. Sartorius, T.C.S. Avila-Pires, and M.C. Espósito. 2001. Life on the leaf litter: The ecology of *Anolis nitens tandai* in the Brazilian Amazon. *Copeia* 2001:401–412.

Vitt, L.J., D.B. Shepard, G.H.C. Vieira, J.P. Caldwell, G.R. Colli, and D.O. Mesquita. 2008. Ecology of *Anolis nitens brasiliensis* in Cerrado woodlands of Cantão. *Copeia* 2008:144–153.

Vitt, L.J., P.A. Zani, and R.D. Durtsche. 1995. Ecology of the lizard *Norops oxylophus* (Polychrotidae) in lowland forest of southeastern Nicaragua. *Canadian Journal of Zoology* 73:1918–1927.

Vitt, L.J., P.A. Zani, and M.C. Espósito. 1999. Historical ecology of Amazonian lizards: Implications for commuity ecology. *Oikos* 87:286–294.

Vitt, L.J., P.A. Zani, and A.A. Monteiro do Barros. 1997. Ecological variation among populations of the gekkonid lizard *Gonatodes humeralis* in the Amazon basin. *Copeia* 1997: 32–44.

Vitt, L.J., and P.A. Zani. 1996a. Ecology of the South American lizard *Norops chrysolepis* (Polychrotidae). *Copeia* 1996:56–68.

Vitt, L.J., and P.A. Zani. 1996b. Organization of a taxonomically diverse lizard assemblage in Amazonian Ecuador. *Canadian Journal of Zoology* 74:1313–1335.

Vitt, L.J., and P.A. Zani. 1998a. Ecological relationships among sympatric lizards in a transitional forest in the northern Amazon of Brazil. *Journal of Tropical Ecology* 14:63–86.

Vitt, L.J., and P.A. Zani. 1998b. Prey use among sympatric lizard species in lowland rain forest of Nicaragua. *Journal of Tropical Ecology* 14:537–559.

Vitt, L.J., and P.A. Zani. 2005. Ecology and reproduction of *Anolis capito* in rain forest of southeastern Nicaragua. *Journal of Herpetology* 39:36–42.

Vogel, P. 1984. Seasonal hatchling recruitment and juvenile growth of the lizard *Anolis lineatopus*. *Copeia* 1984:747–757.

Vogel, P., and D.A.P. Bundy. 1987. Helminth parasites of Jamaican anoles (Reptilia: Iguanidae): Variation in prevalence and intensity with host age and sex in a population of *Anolis lineatopus*. *Parasitology* 94:399–404.

Vogt, R.C., J.L. Villareal-Benítez, and G. Pérez-Higareda. 1997. Lista anotada de anfibios y reptiles. Pp. 507–522 in E. González-Soriano, R. Dirzo, and R.C. Vogt, Eds., *Historia Natural de Los Tuxtlas*. Instituto de Biología, UNAM-Conabio, Mexico.

Voight, B.F., S. Kudaravalli, X. Wen, and J.K. Pritchard. 2006. A map of recent positive selection in the human genome. *PLoS Biology* 4(e72):446–458.

Waddington, C.H. 1975. *The Evolution of an Evolutionist*. Cornell University Press: Ithaca, NY.

Wade, J. 2005. Current research on the behavioral neuroendocrinology of reptiles. *Hormones and Behavior* 48:451–460.

Wade, J.K., A.C. Echternacht, and G.F. McCracken. 1983. Genetic variation and similarity in *Anolis carolinensis* (Sauria: Iguanidae). *Copeia* 1983:523–529.

Wagner, G.P., and L. Altenberg. 1996. Complex adaptations and the evolution of evolvability. *Evolution* 50:967–976.

Wagner, P.J., and D.H. Erwin. 1995. Phylogenetic patterns as tests of speciation models. Pp. 87–122 in D.H. Erwin and R.L. Anstey, Eds., *New Approaches to Speciation in the Fossil Record*. Columbia University Press: New York, NY.

Waide, R.B., and D.P. Reagan. 1983. Competition between West Indian anoles and birds. *American Naturalist* 121:133–138.

Wainwright, P.C. 1994. Functional morphology as a tool in ecological research. Pp. 42–59 in P.C. Wainwright and S.M. Reilly, Eds., *Ecological Morphology: Integrative Organismal Biology*. University of Chicago Press: Chicago, IL.

Wainwright, P.C. 2007. Functional versus morphological diversity in macroevolution. *Annual Review of Ecology and Systematics* 38:381–401.

Wainwright, P.C., and S.M. Reilly. 1994. *Ecological Morphology: Integrative Organismal Biology*. University of Chicago Press: Chicago, IL.

Wainwright, P.C., M.E. Alfaro, D.I. Bolnick, and C.D. Hulsey. 2005. Many-to-one mapping of form to function: A general principle in organismal design? *Integrative and Comparative Biology* 45:256–262.

Wake, D.B. 2006. Problems with species: Patterns and processes of species formation in salamanders. *Annals of the Missouri Botanical Garden* 93:8–23.

Waldschmidt, S., and C.R. Tracy. 1983. Interactions between a lizard and its thermal environment: Implications for sprint performance and space utilization in the lizard *Uta stansburiana*. *Ecology* 64:476–484.

Warheit, K.I., J.D. Forman, J.B. Losos, and D.B. Miles. 1999. Morphological diversification and adaptive radiation: A comparison of two diverse lizard clades. *Evolution* 53: 1226–1234.

Webster, A.J., and A. Purvis. 2002. Testing the accuracy of methods for reconstructing ancestral states of continuous characters. *Proceedings of the Royal Society of London B* 269: 143–149.

Webster, T.P. 1969. Ecological observations on *Anolis occultus* Williams and Rivero (Sauria, Iguanidae). *Breviora* 312:1–5.

Webster, T.P., and J.M. Burns. 1973. Dewlap color variation and electrophoretically detected sibling species in a Haitian lizard, *Anolis brevirostris*. *Evolution* 27:368–377.

Webster, T.P., R.K. Selander, and S.Y. Yang. 1972. Genetic variability and similarity in the *Anolis* lizards of Bimini. *Evolution* 26:523–535.

Weinreich, D.M., N.F. Delaney, M.A. DePristo, and D.L. Hartl. 2006. Darwinian evolution can follow only very few mutational paths to fitter proteins. *Science* 312:111–114.

Weisrock, D.W., J.R. Macey, I.H. Ugurtas, A. Larson, and T.J. Papenfuss. 2001. Molecular phylogenetics and historical biogeography among salamandrids of the "true" salamander clade: Rapid branching of numerous highly divergent lineages in *Mertensiella huschani* associated with the rise of Anatolia. *Molecular Phylogeny and Evolution* 18:434–448.

Werner, Y.L. 1972. Temperature effects on inner-ear sensitivity in six species of iguanid lizards. *Journal of Herpetology* 6:147–177.

West-Eberhard, M.J. 1983. Sexual selection, social competition, and speciation. *Quarterly Review of Biology* 58:155–183.

West-Eberhard, M.J. 2003. *Developmental Plasticity and Evolution*. Oxford University Press: Oxford, UK.

Wetmore, A. 1916. The birds of Porto Rico. *U.S. Department of Agriculture Bulletin* 326:1–140.

White, G.L., and A. Hailey. 2006. The establishment of *Anolis wattsi* as a naturalized exotic lizard in Trinidad. *Applied Herpetology* 3:11–26.

Whitfield, S.M., and M.A. Donnelly. 2006. Ontogenetic and seasonal variation in the diets of a Costa Rican leaf-litter herpetofauna. *Journal of Tropical Ecology* 22:409–417.

Whitfield, S.M., K.E. Bell, T. Philippi, M. Sasa, F. Bolaños, G. Chaves, J.M. Savage, and M.A. Donnelly. 2007. Amphibian and reptile declines over 35 years at La Selva, Costa Rica. *Proceedings of the National Academy of Sciences of the United States of America* 104: 8352–8356.

Whittaker, R.J., and J.M. Fernández-Palacios. 2007. *Island Biogeography: Ecology, Evolution and Conservation*, 2nd Ed. Oxford University Press: Oxford, UK.

Whittall, J.B., C. Voelckel, D.J. Kliebenstein, and S.A. Hodges. 2006. Convergence, constraint and the role of gene expression during adaptive radiation: Floral anthocyanins in *Aquilegia*. *Molecular Ecology* 15:4645–4657.

Whittemore, A.T. 1993. Species concepts: A reply to Mayr. *Taxon* 42:573–583.

Wichman, H.A., M.R. Badgett, L.A. Scott, C.M. Boulianne, and J.J. Bull. 1999. Different trajectories of parallel evolution during viral adaptation. *Science* 285:422–424.

Wiens, J.A. 1989. *The Ecology of Bird Communities. Volume 1: Foundations and Patterns.* Cambridge University Press: Cambridge, UK.

Wiens, J.J., M.C. Brandley, and T.W. Reeder. 2006. Why does a trait evolve multiple times within a clade? Repeated evolution of snakelike body form in squamate reptiles. *Evolution* 60:123–141.

Wilcove, D.S., D. Rothstein, J. Dubow, A. Phillips, and E. Losos. 1998. Quantifying threats to imperiled species in the United States. *Bioscience* 48:607–615.

Wilcox, B.A. 1978. Supersaturated island faunas: A species-age relationship for lizards on post-Pleistocene land-bridge islands. *Science* 199:996–998.

Wiley, E.O. 1978. The evolutionary species concept reconsidered. *Systematic Zoology* 27:17–26.

Wiley, E.O. 1981. *Phylogenetics: The Theory and Practice of Phylogenetic Systematics.* John Wiley and Sons: New York.

Wiley, J.W. 2003. Habitat association, size, stomach contents, and reproductive condition of Puerto Rican boas (*Epicrates inornatus*). *Caribbean Journal of Science* 39:189–194.

Wilkins, J.F. 2004. A separation-of-timescales approach to the coalescent in a continuous population. *Genetics* 168:2227–2244.

Williams, E.E. 1965. The species of Hispaniolan green anoles (Sauria, Iguanidae). *Breviora* 227:1–16.

Williams, E.E. 1969. The ecology of colonization as seen in the zoogeography of anoline lizards on small islands. *Quarterly Review of Biology* 44:345–389.

Williams, E.E. 1972. The origin of faunas. Evolution of lizard congeners in a complex island fauna: A trial analysis. *Evolutionary Biology* 6:47–89.

Williams, E.E. 1975. *Anolis marcanoi* new species: Sibling to *Anolis cybotes*: Description and field evidence. *Breviora* 430:1–9.

Williams, E.E. 1976a. South American anoles: The species groups. *Papeís Avulsos de Zoologia* 29:259–268.

Williams, E.E. 1976b. West Indian anoles: A taxonomic and evolutionary summary 1. Introduction and a species list. *Breviora* 440:1–21.

Williams, E.E. 1977a. An anecdote. P. iv in E.E. Williams, Ed., *The Third* Anolis *Newsletter.* Museum of Comparative Zoology: Cambridge, MA.

Williams, E.E. 1977b. Species problems. Pp. 132–151 in E.E. Williams, Ed., *The Third* Anolis *Newsletter.* Museum of Comparative Zoology: Cambridge, MA.

Williams, E.E. 1983. Ecomorphs, faunas, island size, and diverse end points in island radiations of *Anolis*. Pp. 326–370 in R.B. Huey, E.R. Pianka, and T.W. Schoener, Eds., *Lizard Ecology: Studies of a Model Organism.* Harvard University Press: Cambridge, MA.

Williams, E.E. 1989. A critique of Guyer and Savage (1986): Cladistic relationships among anoles (Sauria: Iguanidae): Are the data available to reclassify the anoles? Pp. 433–477 in C.A. Woods, Ed. *Biogeography of the West Indies: Past, Present, & Future.* Sandhill Crane Press: Gainesville, FL.

Williams, E.E., and S.M. Case. 1986. Interactions among members of the *Anolis distichus* complex in and near the Sierra de Baoruco, Dominican Republic. *Journal of Herpetology* 20:535–546.

Williams, E.E., and M.K. Hecht. 1955. "Sunglasses" in two anoline lizards from Cuba. *Science* 122:691–692.

Williams, E.E., and J.A. Peterson. 1982. Convergent and alternative designs in the digital adhesive pads of scincid lizards. *Science* 215:1509–1511.

Williams, E.E., and A.S. Rand. 1977. Species recognition, dewlap function, and faunal size. *American Zoologist* 17:261–270.

Wilson, A.C., S.S. Carlson, and T.J. White. 1977. Biochemical evolution. *Annual Review of Biochemistry* 46:573–639.

Wilson, B., and P. Vogel. 1999. Exotic predator control in the Hellshire Hills, Jamaica. *West Indian Iguana Specialist Group Newsletter* 2(2):5–6.

Wilson, D., R. Heinsohn, and J.A. Endler. 2007. The adaptive significance of ontogenetic colour change in a tropical python. *Biology Letters* 3:40–43.

Wilson, L.D., and L. Porras. 1983. *The Ecological Impact of Man on the South Florida Herpetofauna*. University of Kansas Press: Lawrence, KS.

Wingate, D.B. 1965. Terrestrial herpetofauna of Bermuda. *Herpetologica* 21:199–219.

Wittkopp, P.J., B.L. Williams, J.E. Selegue, and S.B. Carroll. 2004. *Drosophila* pigmentation evolution: Divergent genotypes underlying convergent phenotypes. *Proceedings of the National Academy of Sciences of the United States of America* 100:1808–1813.

Wolcott, G.N. 1923. The food of Porto Rican lizards. *Journal of the Department of Agriculture of Porto Rico* 7:5–43.

Wright, S.J. 1981. Extinction-mediated competition: The *Anolis* lizards and insectivorous birds of the West Indies. *American Naturalist* 117:181–192.

Wright, S.J., R. Kimsey, and C.J. Campbell. 1984. Mortality rates of insular *Anolis* lizards: A systematic effect of island area? *American Naturalist* 123:134–142.

Wunderle, J.M. Jr. 1981. Avian predation upon *Anolis* lizards on Grenada, West Indies. *Herpetologica* 37:104–108.

Wyles, J.S., and G.C. Gorman. 1980. The classification of *Anolis*: Conflict between genetic and osteological interpretation as exemplified by *Anolis cybotes*. *Journal of Herpetology* 14:149–153.

Yamagishi, S., M. Honda, K. Eguchi, and R. Thorstrom. 2001. Extreme endemic radiation of the Malagasy vangas (Aves: Passeriformes). *Journal of Molecular Evolution* 53:39–46.

Yang, E.-J., S.M. Phelps, D. Crews, and W. Wilczynski. 2001. The effects of social experience on aggressive behavior in the green anole lizard (*Anolis carolinensis*). *Ethology* 107:777–793.

Yang, S.Y., M. Soulé, and G.C. Gorman. 1974. *Anolis* lizards of the eastern Caribbean: A case study in evolution. I. Genetic relationships, phylogeny, and colonization sequence of the *roquet* group. *Systematic Zoology* 23:387–399.

Yeska, M.L., R. Powell, and J.S. Parmerlee, Jr. 2000. The lizards of Cayo Pisaje, Dominican Republic, Hispaniola. *Herpetological Review* 31:18–20.

Yoder, A.D., L.E. Olson, C. Hanley, K.L. Heckman, R. Rasoloarison, A.L. Russell, J. Ranivo, V. Soarimalala, K.P. Karanth, A.P. Raselimanana, and S.M. Goodman. 2005. A multidimensional approach for detecting species patterns in Malagasy vertebrates. *Proceedings of the National Academy of Sciences of the United States of America* 102:6587–6594.

Yoder, A.D., R.M. Rasoloarison, S.M. Goodman, J.A. Irwin, S. Atsalis, M.J. Ravosa, and J.U. Ganzhorn. 2000. Remarkable species diversity in Malagasy mouse lemurs (Primates, *Microcebus*). *Proceedings of the National Academy of Sciences of the United States of America* 97:11325–11330.

Yorks, D.T., K.E. Williamson, R.W. Henderson, R. Powell, and J.S. Parmerlee, Jr. 2004. Foraging behavior in the arboreal boid *Corallus grenadensis*. *Studies on Neotropical Fauna and Environment* 38:167–172.

Young, R.L., T.S. Hasselkorn, and A.V. Badyaev. 2007. Functional equivalence of morphologies enables morphological and ecological diversity. *Evolution* 61:2480–2492.

Zani, P.A. 2000. The comparative evolution of lizard claw and toe morphology and clinging performance. *Journal of Evolutionary Biology* 13:316–325.

Zani, P.A. 2001. Clinging performance of the western fence lizard, *Sceloporus occidentalis*. *Herpetologica* 57:423–432.

Zimmerman, E.C. 1970. Adaptive radiation in Hawaii with special reference to insects. *Biotropica* 2:32–38.

Zippel, K.C., R. Powell, J.S. Parmerlee, Jr., S. Monks, A. Lathrop, and D.D. Smith. 1996. The distribution of larval *Eutrombicula alfreddugesi* (Acari: Trombiculidae) infesting *Anolis* lizards (Lacertilia: Polychrotidae) from different habitats on Hispaniola. *Caribbean Journal of Science* 32:43–49.

INDEX

Page numbers followed by *f* refer to figures, *t* refer to tables and *n* refer to notes.

INDEXER: Indexing Solutions
ILLUSTRATOR: Electronic Illustrators Group (eiG)
COMPOSITION: Michael Bass Associates
TEXT: 9.5/14 Scala